MEMBRANE TOXICITY

ADVANCES IN EXPERIMENTAL MEDICINE AND BIOLOGY

Editorial Board:

Nathan Back	***State University of New York at Buffalo***
N. R. Di Luzio	***Tulane University School of Medicine***
Bernard Halpern	***Collège de France and Institute of Immuno-Biology***
Ephraim Katchalski	***The Weizmann Institute of Science***
David Kritchevsky	***Wistar Institute***
Abel Lajtha	***New York State Research Institute for Neurochemistry and Drug Addiction***
Rodolfo Paoletti	***University of Milan***

Recent Volumes in this Series

MEMBRANE TOXICITY

Edited by

Morton W. Miller
and Adil E. Shamoo

The University of Rochester
Rochester, New York

PLENUM PRESS • NEW YORK AND LONDON

Library of Congress Cataloging in Publication Data

Rochester International Conference on Environmental Toxicity, 9th, 1976.
Membrane toxicity.
(Advances in experimental medicine and biology; 84)

Includes index.
1. Toxicology, Experimental – Congresses. 2. Membranes (Biology) – Congresses. 3. Pathology, Molecular – Congresses. I. Miller, Morton W., 1936- II. Shamoo, Adil E. III. Title.
RA199.R62 1976 615.9 77-1562
ISBN 0-306-39084-1

Proceedings of the Ninth Annual Rochester International Conference on Environmental Toxicity held in Rochester, New York, May 24–26, 1976

© 1977 Plenum Press, New York
A Division of Plenum Publishing Corporation
227 West 17th Street, New York, N.Y. 10011

All rights reserved

No part of this book may be reproduced, stored in a retrieval system, or transmitted, in any form or by any means, electronic, mechanical, photocopying, microfilming, recording, or otherwise, without written permission from the Publisher

Printed in the United States of America

PREFACE

Research on the study of membrane toxicity has advanced a great deal in a relatively short period of time, prompting scientists to re-examine the problems associated with carriers, receptors and reactors to toxic substances. This book presents current research on the responses of membranes to toxic substances both by direct observation of macromolecules reacting with these substances and by inference from data on the biochemical responses of cells and cell fractions.

Two basic areas of membrane toxicity are analyzed by a large number and variety of scientists in this field. The first area is the effect of various toxic substances on membrane structure and function. The second area is where the membrane serves as the site of rate limiting step of the transport of various toxic substances.

The aim of the book is to evaluate present concepts of membrane structure and function in relation to exposure to environmental toxicants. The book is divided into five sessions:

1. Xenobiotics and Membrane Transport
2. Cellular Responses to Toxins
3. Effects of Membranes and Receptors
4. Modification of Membrane Function by Toxicological Agents
5. Toxic Chemicals as Molecular Probes of Membrane Structure and Function

Each paper is generally concluded by an edited discussion which contains many useful and interesting additional insights in each subject area.

ACKNOWLEDGEMENT

We gratefully acknowledge the support provided for this conference by the U. S. Energy Research and Development Administration, the Environmental Protection Agency and the University of Rochester.

The Conference Committee wishes to express its appreciation to Judy Havalack, the conference secretary, for her efficient secretarial service and coordination of arrangements for the conference and to Florence Marsden for assistance with travel arrangements.

The Conference Committee
Adil Shamoo, Chairman
John Brand
Morton W. Miller
Thomas Clarkson

CONTENTS

WELCOMING REMARKS

J. Lowell Orbison, M.D., Dean

School of Medicine and Dentistry

It is a pleasure this morning on behalf of the University and the Medical Center to extend to you a warm welcome to this Ninth Conference on Environmental Toxicity. It is a special pleasure to welcome back old friends who have been with us in the past as students or fellows or faculty, and I would like to add to that a special welcome to those of you who came from other countries to lend your expertise and your special knowledge and viewpoints to our discussions here. The toxicity conferences are now in their ninth year and they are an internationally recognized activity of this school and of the Department of Radiation Biology and Biophysics. The school has reason to be proud of these conferences and we are happy to be a host once again to this, the ninth one. These conferences originated in the toxicology program which was sponsored by the Department of Radiation Biology and Biophysics, and by the Department of Pharmacology and Toxicology. Recently the activities of one part of this program have been awarded the status of an environmental health sciences center by the National Institutes of Environmental Health Sciences and we anticipate that the other part of the program will be funded soon as a special program to study the toxicological effects of therapeutic agents. So, from having a single program we have moved to two; one focusing on environmental toxicology and the other focusing on the toxicology of therapeutic agents. In both of these, education is a very important activity and in the conference which you are now beginning there will be graduate students in the audience, and I hope you will take time to visit them in their laboratories and let them have a chance to meet you and talk with you about their interests and yours.

The conference which is beginning today under the topic of membrane toxicity will be a second look at some of the problems discussed at the second conference seven years ago. The earlier conference discussed experiments in cell physiology in which information about the targets of toxic substances were inferred from data on biochemical responses of cells and cell fractions. Research in the field of membranes has advanced a great deal even in the short time since that conference, and it is therefore fitting to re-examine that problem and in this ninth conference to emphasize molecular biophysics and biochemistry in which the macromolecules attacked by the toxic substances are identified and studied directly. The research program now emphasizes the direct characterization of the effect of toxic agents on a defined membrane target, either as a part of an intact system or as a model system. Membrane components that are responsible for membrane function are now isolated in their molecular interactions with foreign substances and directly studied. This change of outlook and technology reminds me, a pathologists, of an advance in my own field that was a long time ago. A young hematologist at that time, Rudolf Virchow first gained scientific emminence by his studies on blood clotting and they were based on physiologic experiments and chemical analyses. But in 1845 he announced a change of outlook, a new experimental approach to pathology, and instead of inferring mechanisms of damage from physiologic responses he proposed to identify the change by observing it directly on the target substance or part of the body and in that particular instance he was talking about the cell as the target. That was a new view and a new approach which has made it possible for us to make many more advances since that time, and that incidentally was 130 years ago. That new viewpoint and new approach was really the beginning of cellular pathology. Virchow's work advanced the study of pathology and of all biologic science at that time to a new and more fundamental level of investigation. I think your conference here today promises to mark still another advance from the response of specialized cells to the response of specialized molecules. These as the carriers, receptors and reactors to the toxic substances. And so it is my pleasure to welcome you to this ninth conference. I hope that you will find that your stay here in Rochester is not only pleasant, but is also stimulating and professionally profitable.

Xenobiotics and Membrane Transport

EFFECT OF P-CHLOROMERCURIBENZOATE (pCMB), OUABAIN AND 4-ACETAMIDO-4'ISO-THIOCYAMATOSTILBENE-2,2'-DISULFONIC ACID (SITS) ON PROXIMAL TUBULAR TRANSPORT PROCESSES

K. J. Ullrich, G. Capasso, G. Rumrich and K. Sato

Max-Planck-Institut für Biophysik

Kennedyallee 70, 6000 Frankfurt/Main, Germany

ABSTRACT

Using microperfusion techniques the following transport parameters of the proximal tubule were measured: 1. Isotonic fluid (Na^+) absorption (J_{Na}). 2. Zero net flux concentration (electrochemical potential) differences which are proportional to the respective active transport rates of H^+(glycodiazine), D-glucose (α-methyl-D-glycoside), L-histidine, inorganic phosphate and calcium ions. 3. Transtubular and transcellular electrical potential differences and transcellular resistances.

The following was found: 1. Ouabain (1mM) applied peritubularly in golden hamsters inhibited J_{Na} incompletely and the sodium-coupled (secondary active) transport processes of glucose, histidine, phosphate and Ca^{++} by more than 80%. The H^+(glycodiazine) transport was not affected. Ouabain (1mM) plus acetazolamide (0.2 mM) inhibited J_{Na} completely.
2. In the rat, pCMB (0.2 mM) when applied long enough, inhibits J_{Na} completely. At a time of pCMB application when J_{Na} is reduced to 1/3, this substance inhibits the active H^+(glycodiazine) transport which must be considered to be a direct action of pCMB. Furthermore it inhibits the secondary active phosphate transport either directly or via the inhibition of Na^+ and/or H^+ transport. The secondary active glucose, histidine and Ca^{++} transport are little affected by pCMB. pCMB reduces the cell potential, reversibly, yet leaves unchanged the resistance ratio of the luminal to peritubular cell membrane.
3. In the rat SITS inhibits J_{Na} moderately but the active H^+(glycodiazine) transport strongly. It does not affect the glucose transport.

On the basis of these and other results a hypothesis of the interaction of Na^+ and $H^+(HCO_3^-)$ transport is given.

Introduction

Recently it has been found that in the proximal tubule the transport mechanisms of glucose, amino acids, phosphate and calcium are sodium-dependent (18, 19, 1, 21). Subsequently, it was shown with isolated plasma membrane vesicles that in the brush border a cotransport of Na^+ together with glucose, amino acids and phosphate takes place and in the contraluminal cell membrane, countertransport of Na^+ against Ca^{++} (10, 2, 8, 7) (Fig. 1). The Na^+-glucose and Na^+-amino acid cotransport could also be seen by changes of the transcellular and transtubular electrical potential difference (5, 13). Furthermore, it was observed that the sodium and bicarbonate reabsorption depend strongly on each other (15, 20). In a bicarbonate-free system sodium reabsorption is largely reduced, and in a sodium-free system, bicarbonate reabsorption. By inhibiting the contraluminal Na^+ transport with ouabain and the buffer transport with acetazolamide Frömter and Geßner (6) have revealed that both transport processes have electrogenic components, which are independent of each other, the bicarbonate transport making the lumen

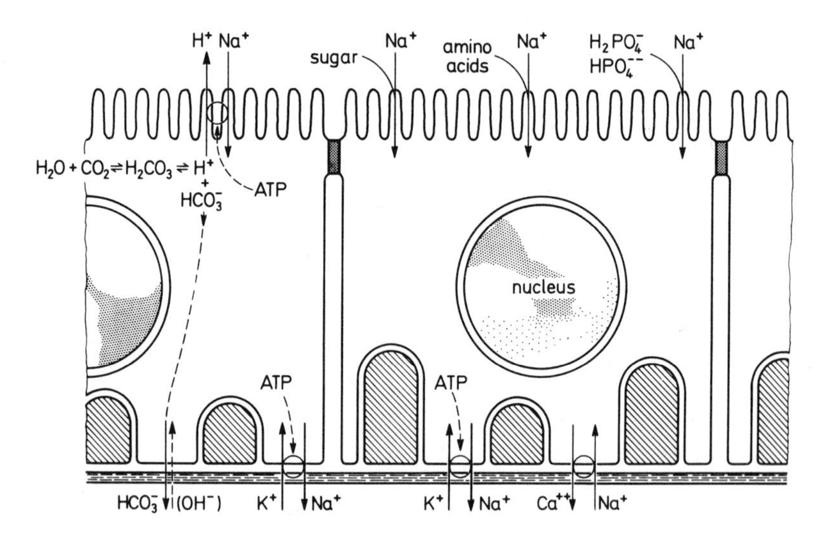

Figure 1. Scheme of proximal tubular cells and location of the transport processes.

and the sodium transport making the interstitium more positive. With brush border vesicles again Murer et al. have observed a Na^+-H^+-countertransport (12). To elucidate further the interdependence of the proximal tubular transport processes we applied ouabain. Because of the low sensitivity of the rat for this substance we used golden hamsters, assuming that between both species only quantitative and no qualitative differences exist in their respective transport mechanisms. On this basis we then tried to evaluate the mode of action of pCMB and SITS on the proximal tubular transport processes.

Methods

In all experiments the method of the double perfused proximal tubule was used (14, 4, 18) (Fig. 2). The measured parameters were: the isotonic volume Na^+ reabsorption (J_v or J_{Na}) (14), the transtubular concentration (electrochemical potential) difference at zero net flux (4) and the cell electrical potential difference (3). The transtubular concentration (electrochemical potential) difference is under the chosen conditions and at constant permeability a measure of the active transport rate of the respective substance: $J_{act} = P(\Delta c + \bar{c}\Delta\psi\ zF/RT)$. Except for the bivalent Ca^{++} the electrical term $\bar{c}\Delta\psi\ zF/RT$, $\bar{c}$ being the mean concentration and $\Delta\psi$ the electrical potential difference, was neglected because of the small transtubular $\Delta\psi$s. The transtubular concentration difference was measured with isotopes assuring that the specific activity was the same all over. For this condition in the case of Ca^{++} and P_i, the tubular lumen and the peritubular capillaries were perfused 3 - 4 minutes before the actual measuring period was started. The inhibitors were added to the capillary and/or the luminal perfusate.

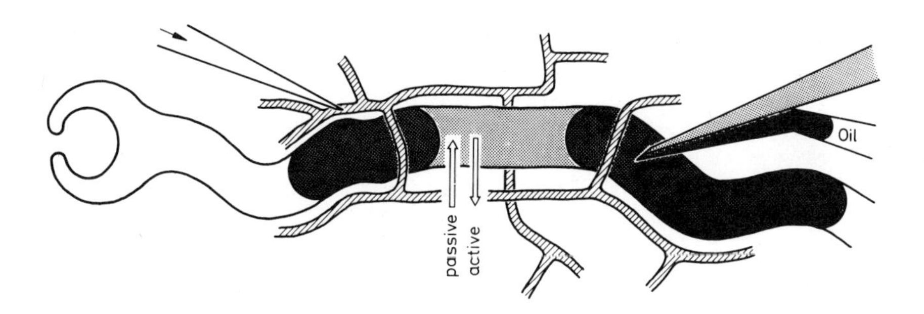

Figure 2. Scheme of the stop flow microperfusion of a proximal tubule with zero net flux.

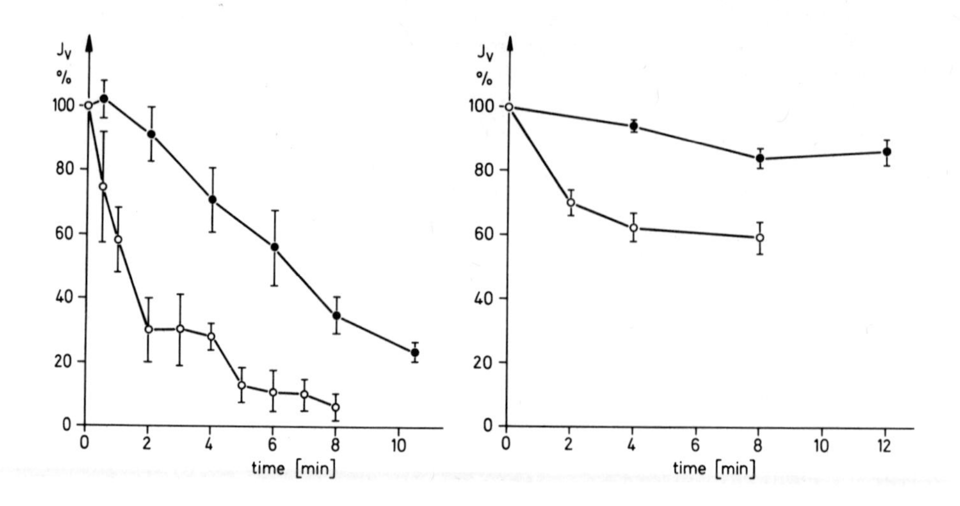

Figure 3. Effect of pCMB or pCMPS (0.2 mM left side) or SITS (1 mM right side) on proximal isotonic fluid (Na^+) absorption. Open circles = the tested compounds added to the capillary perfusate, closed circles = added to the luminal perfusate only.

Results and Discussion

As shown in Fig. 3 left side, isotonic Na^+ absorption is rapidly inhibited if 10^{-4}M pCMB or pCMPS (p-chloromercuriphenylsulfonate) is added to the capillary perfusate, half inhibition being reached after approximately one minute (16). If the mercuricompounds were added to the luminal perfusate only, the onset of inhibition was much slower, half inhibition being reached after 5 1/2 minutes. In both cases, however, an almost complete inhibition of the Na^+ transport was reached if the reagent acted long enough. When 1 mM SITS was applied from the peritubular side, half maximal inhibition was similarly achieved after 1.3 minutes, but the maximal inhibition was not very much more than 40% (Fig. 3 right side). When added to the luminal perfusate SITS did not inhibit. The behaviour of ouabain was between that of pCMB (pCMPS) and SITS. The maximal inhibition of isotonic Na^+ reabsorption seen in the rat was around 58% with 25 mM ouabain (17) and in the more sensitive golden hamster 85% with 1 mM ouabain. If in the latter experiments 0.2 mM acetazolamide was added in addition, a 96% inhibition was seen. Tentatively we explain these experiments in the following way: In the golden hamster ouabain inhibits the active Na^+ absorption completely. The Na^+ reabsorption which remains after ouabain is due to the lumen positive electrical potential difference (4, 6) caused by H^+ secretion. It is abolished by acetazolamide. Because of the almost complete inhibition of the net Na^+ transport, seen in the rat, pCMB must affect more than the contraluminal active Na^+ transport mechanism (Na^+K^+-ATPase) - as documented below it inhibits the buffer and phosphate reabsorption, too -. SITS inhibits either the active Na^+ reabsorption or the buffer absorption more or less completely or both partially - as we will show below it inhibits the H^+ transport strongly.

Here the "leaky membrane hypothesis" of mercurial action should be discussed. In 1957 Kleinzeller and Cort (11) observed that kidney slices release protein if mercurials were added to the incubation medium. Thus, they hypothesized that mercurials may inhibit the sodium transport by augmenting the leak permeability of the kidney cell membranes rather than by inhibition of the Na^+K^+-ATPase. To test this hypothesis we measured the resistance ratio of the luminal to peritubular cell membrane. A microelectrode was inserted into a cell and a current of 10^{-7}Amp. was pulsed with another electrode into the tubular lumen. The deflection of the intracellular potential caused by the current pulses is a relative measure of the permeability of the two cell membranes for ions. As can be seen in Fig. 4, a cell potential between 70 and 80 mV was recorded. During perfusion with $2 \cdot 10^{-4}$ pCMB the potential difference declined linearly at a rate of 5.5 mV/min. - In five other experiments, the rate of decline was between 4.7 and 8.0 mV/min and in one experiment declined very rapidly at 20.1 mV/min -. As soon as the pCMB perfusion stopped, the potential difference became stable and then repolarized again. The electrical potential deflections, however, were constant before (2.83 ± 0.16 mV) and after (2.66 ± 0.24 mV) adding $2 \cdot 10^{-4}$

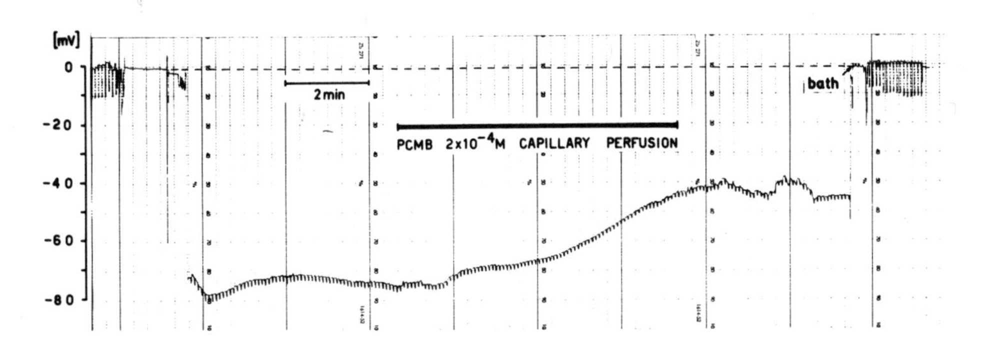

Figure 4. Intracellular electrical potential difference in the proximal tubular cell against interstitium. Current is pulsed into the lumen. The bar indicates when the capillaries were perfused with pCMB (0.2 mM).

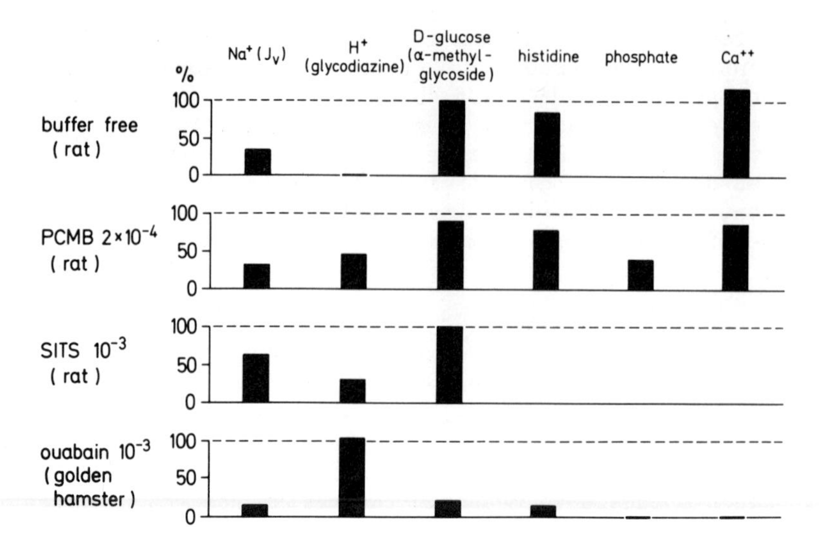

Figure 5. Effect of buffer omission, pCMB, SITS and ouabain on proximal isotonic fluid (Na^+) absorption and the zero net flux concentration differences of H^+(glycodiazine), D-glucose (α-methyl glycoside),histidine, inorganic phosphate and the electrochemical potential difference of Ca^{++}. 100 = control, the bars indicate the relative transport rates.

pCMB. From this one may conclude that pCMB stopped the pumps but left the cell permeability unchanged.

In a last set of experiments the effect of HCO_3^- buffer and the inhibitors ouabain, pCMB and SITS on up to six proximal transport processes was tested (Fig. 5). Omission of HCO_3^- inhibited the isotonic Na^+ transport by 66% but did not alter the Na^+ coupled glucose, histidine and Ca^{++} transport. Thus the remaining active Na^+ transport suffices to drive these coupled transport processes. The pCMB results taken at a time when the net sodium transport was inhibited to the same extent, fit with the results gained with HCO_3^--free solution. After 2 - 4 min capillary perfusion with pCMB the Na^+ coupled transport processes were not inhibited by more than 20% except the phosphate transport. With 1 mM SITS in the capillary perfusate the net Na^+ transport is inhibited by 40% and the glucose transport is unaltered. In all cases, except in the pCMB effect on phosphate transport, the remaining active Na^+ transport is - as already said - satisfactory to drive the Na^+ coupled transport processes. But when the active Na^+ transport is completely blocked by ouabain - the remaining net Na^+ transport is likely to be passive, driven by the unchanged H^+ secretion - all Na^+ coupled transport processes vanish. We face, therefore, a clear situation as concerns the coupled transport of glucose, amino acids, phosphate and Ca ions. But what can be said about the interaction of Na^+ and buffer (H^+) transport? First that ouabain, which blocked all other transport processes, does not affect the H^+ transport at all. Thus the H^+ secretion cannot proceed as secondary active Na^+-H^+ countertransport which was previously supposed by the H^+-Na^+ countertransport in brush border vesicles (12) and by the Na^+ dependence of the glycodiazine (H^+) transport. It must rather be driven by ATP, supposedly via a HCO_3^--sensitive ATPase which was found in brush border membranes (9). But how then can the mutual dependence of H^+ and Na^+ transport (15, 20) be interpreted which was mentioned in the introduction and also the Na^+-H^+ countertransport observed in brush border membrane vesicles. Our speculative working hypothesis is at the moment the following: under normal conditions, more H^+ ions than Na^+ ions are transported by the coupled luminal Na^+-H^+ transport mechanisms. This is the reason for the active transport potential for H^+ ion secretion (5, 6). The energy for H^+ secretion may come partially from the Na^+ gradient, otherwise from ATP. With ouabain, the Na^+ gradient is abolished as can be judged by the absence of the Na^+ coupled sugar-, amino acid-, phosphate- and Ca^{++}-transport. Therefore it is likely that with ouabain the Na^+ ions sitting on the H^+ carrier are not released within the cell and cycle back and forth. In this situation, ATP splitting must provide completely the energy for the net active H^+ secretion. There seems to be the possibility that this hypothesis can be tested with brush border membrane vesicles. Unfortunately, experiments with the HCO_3^--ATPase alone are not promising because the activity of this enzyme cannot be stimulated by Na^+.

REFERENCES

1 BAUMANN, K., DeROUFFIGNAC, C., ROINEL, N., RUMRICH, G., ULLRICH, K.J.: Pflügers Arch. 356 (1975) 287.

2 EVERS, J., MURER, H., KINNE, R.: Biochim. Biophys. Acta 426 (1976) 598.

3 FRÖMTER, E., MÜLLER, C.W. and WICK, T.: In Electrophysiology of Epithelial Cells, p. 119 (Giebisch, G., Ed.). F.K. Schattauer. Stuttgart-New York (1971).

4 FRÖMTER, E., RUMRICH, G., ULLRICH, K.J.: Pflügers Arch. 343 (1973) 189.

5 FRÖMTER, E., GEßNER, K.: Pflügers Arch. 351 (1974) 85.

6 FRÖMTER, E. and GEßNER, K.: Pflügers Arch. 357 (1975) 209.

7 GMAJ, P. and MURER, H. Personal communication.

8 HOFFMANN, N., THEES, M., KINNE, R.: Pflügers Arch. 262 (1976) 147.

9 KINNE-SAFFRAN, E., KINNE, R.: Proc. Soc. exp. Biol.Med. 146 (1974) 751.

10 KINNE, R., MURER, H., KINNE-SAFFRAN, E., THEES, M., SACHS, G.: J. Membrane Biol. 21 (1975) 375.

11 KLEINZELLER, A., CORT, J.H.: Biochem. J. 67 (1957) 15.

12 MURER, H., HOPFER, U., KINNE, R.: Biochem. J. 154 (1976) 597.

13 SAMARZIJA, I., FRÖMTER, E.: Pflügers Arch. 359 (1975) R 119.

14 ULLRICH, K.J., FRÖMTER, E., BAUMANN, K.: In Laboratory Technique in Membrane Biophysics, p.106 (Passow, H., Stämpfli, R., Eds.). Springer. Heidelberg (1969).

15 ULLRICH, K.J., RADTKE, H.W., RUMRICH, G.: Pflügers Arch. 330 (1971) 149.

16 ULLRICH, K.J., FASOLD, H., KLÖSS, S., RUMRICH, G., SALZER, M., SATO, K., SIMON, B., deVRIES, J. X.: Pflügers Arch. 344 (1973) 51.

17 ULLRICH, K.J., SATO, K., RUMRICH, G.: Transport Mechanisms in Epithelia, p. 560, (Usssing, H.H., Thorn, N.A., Eds.). Munksgard. Copenhagen (1973).

18 ULLRICH, K.J., RUMRICH, G., KLÖSS, S.: Pflügers Arch. 351 (1974) 35.

19 ULLRICH, K.J., RUMRICH, G., KLÖSS, S.: Pflügers Arch. 351 (1974) 49.

20 ULLRICH, K.J., RUMRICH, G., BAUMANN, K.: Pflügers Arch. 357 (1975) 149.

21 ULLRICH, K.J., RUMRICH, G., KLÖSS, S.: Pflügers Arch. in press.

DISCUSSION

ROTHSTEIN: These experiments of Dr. Ullrich, so elegantly done, show that modifying agents like the disulfonic stilbenes can be used with epithelia as well as with red blood cells. I thought it was hard to work with simple cells such as the erythrocyte. Dr. Ullrich has shown that with good technology you can do similar kinds of experiments even with kidney cells. The paper is now open for discussion.

MILLER: What was the current density used to achieve your results?

ULLRICH: Approximately 200μA/cm^2.

MILLER: Was a step impulse field used?

ULLRICH: Yes.

MILLER: What was the frequency on your step?

ULLRICH: 0.2 to 0.3 Hertz.

LAKOWICZ: Dr. Ullrich, when you produce a system like that with an inhibitor, is it known how much of the inhibitor remains in the membrane or within the cell?

ULLRICH: Well, from ouabain we know that it acts specifically on a membrane bound transport protein, the Na^+K^+ - ATPase. SITS on the other hand does not penetrate the cell; so we can say, it acts on the membrane. From pCMB it is difficult to differentiate direct membrane from intracellular actions.

BRODSKY: Have you ever found an inhibitor that can eliminate the proton secretion?

ULLRICH: Unfortunately not. Our idea is if an inhibitor on proximal H^+ secretion would be available somebody would have applied it on the whole kidney and would have seen a strong effect on H^+ secretion. But this was, to our knowledge, not the case yet.

SZABO: Is it possible to demonstrate the hydrogen sodium exchange?

ULLRICH: Yes, a direct H^+ -Na^+-exchange was seen by Murer *et al.*(12) in isolated bruch border vesicles.

SZABO: Is there any exchange before that?

ULLRICH: No.

SZABO: Could you speculate on the mechanism. Would it be like a neutral exchange?

ULLRICH: Well, on one hand it was observed by Murer *et al.* (12) that the exchange of Na^+ against H^+ is electroneutral. On the other hand Fromter and Geßner (5) observed that the H^+ ion secretion is electrogenic in a sense that more H^+ must be secreted into the lumen than Na^+ ion transported into the cell. On the molecular mechanism I think it is premature to speculate.

SZABO: To what extent would this exchange affect previous measurements in which one has shown the presence of a gradient of sodium? Would that alter at all the measurement?

ULLRICH: No. We can estimate on what way the sodium comes from the lumen into the cell. We can make a balance sheet of how Na^+ crosses the luminal cell membrane. Sixty percent seems to move in exchange for H^+ and 10% in cotransport with glucose, amino acids and phosphate. The route of the remaining 30% we do not know. Certainly this part moves only to a certain extent through a amiloride sensitive channel.

FISCHBARG: If we can go back to the subject of the site of action of the modifying reagent, I wonder which range of concentrations have you tested, and whether you have seen actual stimulation at low concentrations. I am asking this because disulfide compounds in our system stimulate transport at low concentration.

ULLRICH: Unfortunately, we have not tested very small concentrations of mercurials and of SITS. But I think that biphasic effects, as you mentioned, could possibly be seen on kidney transport processes, too.

ROTHSTEIN: Have you tried to rationalize your experiments with pCMB and pCMBS with their effects on the isolated ATPase systems?

ULLRICH: We have not tested explicitly the effect of pCMB and pCMBS on the Na^+K^+-ATPase in isolated contraluminal plasma membranes. This was done by others and is explicitly discussed in one of our previous publications on that subject (16). In our experiments we were rather interested in the sidedness of attack on the tubule *in situ* and of the size of the inhibitory molecule that can reach the site of attack. Thus, we tested two iodoaceto compounds which were supposed to act on the same SH-groups as pCMB to inhibit the proximal Na^+ absorption. One compound (4-benzamido-4'iodoacetamido-stilbene-2,2'disulfonate) had a molecular weight of 690 and inhibited the Na^+ reabsorption when applied from the contraluminal cell side. The other compound (1.9-benzoxanthene-3,4-dicarboxylic N-iodoacetooligoprolyl amino-ethylimide) had a molecular weight of 1660 and did not inhibit. The same was the case when dextran with a molecular weight of 10000 was coupled to pCMB. Therefore, we conclude that the larger compounds cannot reach the site of attack _in vivo_, although they inhibit the Na^+K^+-ATPase _in vitro_.

HEAVY METALS AND MEMBRANE FUNCTIONS OF AN ALVEOLAR EPITHELIUM

John T. Gatzy

Department of Pharmacology
School of Medicine
University of North Carolina
Chapel Hill, North Carolina 27514

ABSTRACT

The anuran lung can be mounted as a planar sheet for assessment of bioelectric properties and solute and water permeation. The lung of the bullfrog exhibited a transmural electrical p.d. of 19 mV (pleura positive), a resistance of 700 ohm cm^2 and a short-circuit current (I_{sc}) of 27 $\mu a/cm^2$. I_{sc} reflected active Cl^- secretion into the lumen by a transport mechanism located at the luminal border of the alveolar epithelial cell layer. Other halides and SCN^- were secreted; other bathing solution ions and H_2O followed the predictions for passive diffusion. Exposure of the luminal surface of the lung or disaggregated epithelial cells to heavy metals revealed two patterns of response. $HgCl_2$ decreased I_{sc}, cell ciliary motility and QO_2; $CdCl_2$ and $NiCl_2$ inhibited only QO_2 and cilia. Effects of $CdCl_2$ were reversed by dimercaprol but not by albumin. SH agents limited but did not reverse $HgCl_2$ toxicity. None of the metals altered osmotic flow across the lung even though most increased flow across the urinary bladder from the same species. The effects of $HgCl_2$ on bioelectric properties reflected changes in ion permeation. $HgCl_2$ followed by a SH compound stimulated I_{sc} and the pleural to lumen Cl^- flux selectively. Without complexer, transient stimulation was followed by a decrease in I_{sc} and an indiscriminate increase in ion permeability. We conclude that responses to heavy metals separate functions of ciliated cells from Cl^- transport by the non-ciliated epithelium. Furthermore, non-specific cell damage by $HgCl_2$ can be limited by SH agents to a selective action on the Cl^- pump.

REPORT

Introduction

For years the lung has been the nearly exclusive province of the respiratory physiologist. However, the continuing search for natural hemodialysis interfaces (10) led to renewed interest in the barriers that separate blood from air. Most of the studies which resulted from this interest have fostered the notion that the alveolar epithelium, like many other epithelia, imposes the major resistance to solute and water movement between the capillary and alveolar lumen (e.g. 3, 16, 17, 18, 20). In spite of the strategic location of the alveolar epithelium and its importance as a barrier, many air-borne toxic agents induce morphologic changes in the capillary endothelium and swelling of the interstitium before visible damage to the epithelium and flooding of the alveolar lumen occur (9, 21). Early derangement of the epithelium and subsequent edema induced by ingestion or parenteral administration of the herbicide, paraquat, is a notable exception (19).

Exploration of the limiting permeability barrier between blood and air and of the modes of damage to this barrier induced by chemical agents has been hampered by the complex morphology of the mammalian lung. For example, some of the most convincing arguments for the relatively low permeability of the alveolar epithelium to solute and water are indirect and stem from a comparison of the osmotic effectiveness of blood borne solutes in the air-filled and fluid-filled lung (16). Even though indirect approaches have yielded an apparent functional separation of the series barriers that are interposed between blood and lumen, it is difficult to distinguish flow through an alveolar path from parallel translocation across the small bronchioles.

Morphology Of The Amphibian Lung

The lungs of amphibians offer an alternative investigative avenue. Each lung lobe is a single large alveolus lined with a continuous epithelial monolayer. Like the mammalian alveolar epithelium, the luminal lining is somprised mainly of squamous (Type I) and cuboidal (Type II) cells (12). Unlike the alveolus of the mammal, trabeculae underlay the epithelium so that the luminal surface resembles a shallow, open honeycomb. The portion of the continuous epithelium that covers the tips of these septa is made up of ciliated cells (7) which occupy approximately 10 to 20% of the total surface area. An interstitium with blood vessels, axons, smooth muscle bundles and fibers separates the epithelium from the continuous pleural mesothelium.

Bioelectric Properties

Each lobe of the lungs of the bullfrog or marine toad can be opened and mounted as a planar sheet between lucite, Ussing-type chambers. When each surface of the bullfrog lung is exposed to identical Ringer solutions a spontaneous transmural bioelectric p.d. of nearly 20 mV (pleura or serosa positive) is observed. (Table I). The transmural d.c. resistance is somewhat lower than that reported for "tight" epithelia such as frog skin and toad urinary bladder but is much higher than values listed for "leaky" epithelia such as the intestine (14). Surgical removal of one-tenth to one-sixth of the pleural covering does not affect the transmural p.d. and disaggregating agents such as $HgCl_2$ or EDTA decrease the p.d. and d.c. resistance much more rapidly and effectively after addition to the luminal (mucosal) bathing solution than after injection into the serosal bath (4). These results suggest that the pleural covering is neither the site of the biopotential nor the major resistance to ion flow and that these properies reside in the only other continuous cell layer - the epithelium.

TABLE I. Summary of Bioelectric Properties and Permeability of the Bullfrog Lung

BIOELECTRIC	PERMEABILITY (PASSIVE)	ACTIVE TRANSPORT
Spontaneous ΔE	$k = 1$ to 10×10^{-7} cm/s	Net Flow
19 mV, pleura +	Na^+, K^+, Ca^{2+} SO_4^{2-}, PAH^-, TcO_4^- gluconate	pleura to lumen Cl^-, Br^-, I^-, SCN^-
D.C.Resistance	$k = 80$ to 800×10^{-7} cm/s	Cl Flow
700 ohm cm^2	HCO_3^-, DNP^-, H_2O	pleura to lumen $k = 28 \times 10^{-7}$ cm/s
Short-Circuit Current		lumen to pleura $k = 17 \times 10^{-7}$ cm/s
27 $\mu a/cm^2$		net flow = $I_{short\text{-}circuit}$
Site of ΔE and R		
epithelium		a saturable process inhibited by DNP, N_2 reduced by Br^-

Passive Permeability And Active Transport

Passive permeability coefficients (k) for the unidirectional flux of water and most ions across the excised bullfrog lung are similar to values reported for other "tight" epithelia such as frog skin and toad bladder. In addition, where comparisons are possible, the coefficients for bullfrog and fluid-filled mammalian lung agree within an order of magnitude. However, in contrast to the frog skin and toad urinary bladder which reabsorb Na^+ actively, the bullfrog lung secretes halide anions into the lumen (6). Cl^- secretion satisfies many of the criteria for active transport, including net movement in the absence of an energy gradient, inhibition by metabolic inhibitors or by another halide, and net movement that is not linked to the flow of any other bathing solution constituent. Furthermore, after an hour with the transmural p.d. clamped at zero, Cl^- secretion equals the current required to maintain the clamp (short-circuit current). The physiological significance of halide secretion by the bullfrog lung is unclear, but fluid in the mammalian fetal lung is maintained by the active secretion of Cl^- (13). Thus, anion transport by the bullfrog lung may represent a functional analog of a step in the development of the mammalian organ.

Since the Cl^- concentration of bullfrog lung tissue is less than that of the bathing media (6), the Cl^- concentration of the average cell is lower. On the assumption that the interior of the epithelial cells is less electronegative than the voltage required for Cl^- equilibrium across the mucosal border, net secretion of Cl^- and a low intracellular Cl^- concentration would require the anion pump to be located in the apical membrane. We reasoned that selective alterations in Cl^- transport might serve as a sensitive assay for airborne agents which act at this site.

Effects of Inorganic Heavy Metal Salts

Inorganic Cd and Hg salts were initially selected for comparison because both metals react with tissue ligands, such as sulfhydryl groups, but only airborne Cd salts routinely induce lung damage. Furthermore, sulfhydryl groups are targets of chemical attack by oxidant air pollutants (11).

Bioelectric properties, O_2 consumption and ciliary motility. Exposure of the mucosal surface of the excised lung to solutions with Hg or Cd salts for one hour led to the dose-effect relationships depicted in Figures 1 and 2. Whereas $HgCl_2$ progressively decreased the short-circuit current as the concentration increased, $CdCl_2$ failed to alter bioeletric properties at a concentration close to the limit of solubility. When alveolar epithelial cells

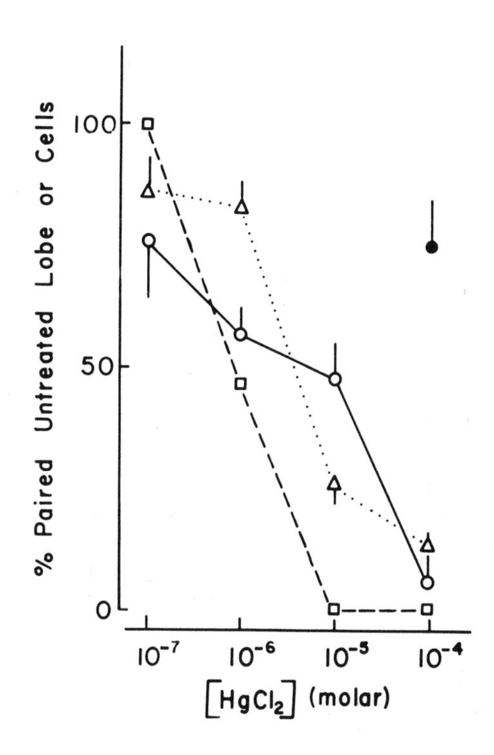

Figure 1. The relationship between the mercuric chloride concentration and functions of bullfrog lung. -O- = Q_{O_2} at 25°C, -□-= ciliary motility, ••Δ•• = short-circuit current and ● = osmotic flow. Each point represents the mean of results from three or more lungs. Vertical lines denote S.E. larger than the symbols. The metal salt was added to only the mucosal solution in the short-circuit current and osmotic flow studies and to solution that bathed separated epithelial cells in the Q_{O_2} and ciliary motility experiments. The gravimetric method of Bentley (1) was used to quantify volume loss from a lung sac induced by the addition of 100 mOsm/L of raffinose to the serosal bathing solution. (Na_2SO_4 Ringer). The same Na_2SO_4 Ringer (100 mOsm/L) bathed the mucosal surface of the lung. Ciliary activity was measured by the method of Scudi, et al (15). Ten fields in a wet mount of separated cells were examined under 500 x magnification for stopped cilia. Changes in O_2 tension were measured with a polarographic electrode.

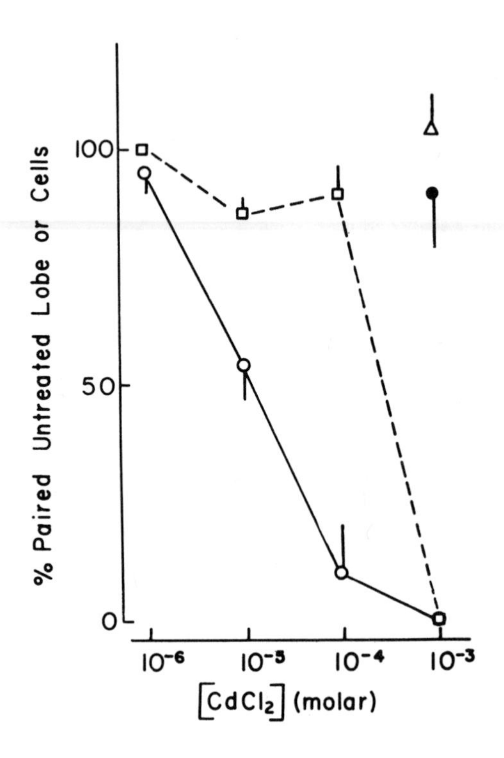

Figure 2. The relationship between cadmium chloride concentration and functions of bullfrog lung. —O— = Q_{O_2} at 25°C, - □ - = ciliary motility, Δ = short-circuit current, ● = osmotic flow. Other notations, conditions and methods are the same as those in Figure 1.

that were separated from the connective tissue after disaggregation with collagenase were exposed to the metals, both $CdCl_2$ and $HgCl_2$ inhibited ciliary motility and the rate of oxygen consumption. Similar studies with $PbCl_2$ and $NiCl_2$ demonstrated that these metals induced a pattern of response similar to that observed with $CdCl_2$ but were less than one-tenth as potent. $ZnCl_2$ (10^{-3}M) did not affect any of the measurements.

One objection to the apparent selective action of $CdCl_2$ on ciliary motility and Qo_2 centers on the fact that the entire plasma membrane of separated cells was exposed to the metal whereas during the measurement of bioelectric properties exposure was limited to the mucosal surface. However, when treatment in the Ussing chamber was followed by a Ringer wash, measurement of Qo_2 of the intact tissue and an assessment of ciliary motility of separated cells, inhibition of the cilia and Qo_2 persisted. Furthermore, this inhibition was at least partially reversed by washing epithelial cells with an excess of dimercaprol but not with Ringer solution or bovine serum albumin (Table II). Thus, it appears that the targets linked with ciliary motion are not accessible to an extracellular sulfhydryl agent. Moreover, the functional dissection of ciliary activity from the bioelectric properties of the lung by $CdCl_2$ suggests that most of the Cl^- transport is carried out by the non-ciliated epithelium.

TABLE II. Effect of Sulfhydryl Agents on Changes in Disaggregated Alveolar Cell Function Induced by $CdCl_2$

Paired Experiment	n	Qo_2 (µl/mg dry wt hr)		Ciliary Motility (% fields without stopped cilia)
		(mean ± S. E.)		
Untreated	4	4.1	(±0.8)	100
$CdCl_2$ +Ringer wash		1.6	(±0.3)†	0†
$CdCl_2$ +Ringer wash	3	1.2	(±0.6)	0
$CdCl_2$ + albumin wash		1.1	(±0.5)	0
$CdCl_2$ + Ringer wash	3	1.3	(±0.5)	0
$CdCl_2$ + BAL wash		3.4	(±0.6)†	33 (±3)†

* 10^{-3}M/L
† significantly different from the paired control

Osmotic flow. Surprisingly, none of the heavy metals altered volume flow appreciably in response to the addition of 100 mOsm/L of raffinose to the serosal solution (Figures 1 and 2) even though all inhibited transmural p.d. and all but $PbCl_2$ stimulated osmotic flow across urinary bladders excised from the same species (Table III). Since resting volume flow out of the lung sac averaged 200 mg/hr, about five times the flow across the bladder, the possibility that the pleural covering contributed significantly to the transmural barrier to net water movement was explored. If the hydraulic permeability coefficients of the epithelial (L_e) and pleural (L_p) barriers are in series, then the transmural coefficient L_t is given by:

$$\frac{1}{L_t} = \frac{1}{L_e} + \frac{1}{L_p} .$$

TABLE III. Effects of Heavy Metals on Transmural P.D. and Volume Flow Across Excised Bullfrog Urinary Bladder

Mucosal Agent	n†	Transmural P.D. $(\Delta E'_A/\Delta E_A - \Delta E'/\Delta E) \times 100$*	Volume Flow $(J_V^{A'}/J_V^A - J'_V/J_V) \times 100$*
		(mean $\pm$ S.E.)	
$HgCl_2$ (10^{-4}M)	5	-75 (±21)§	928 (± 318)§
$ZnCl_2$ (10^{-4}M)	4	-61 (± 9)§	185 (± 49)§
$CdCl_2$ (10^{-4}M)	7	-35 (± 7)§	611 (± 214)§
$PbCl_2$ (10^{-4}M)	6	-31 (±12)§	302 (± 182)

Tissues were exposed to a Na_2SO_4 Ringer (70 mEq Na^+/L ; 100 mOsm/L). Osmotic flow was established by the addition 100 mOsm/L of raffinose to the serosal solution.

* $\Delta E'_A$ and $J_V^{A'}$ represent the p.d. and volume flow after addition of metal to the test lobe, $\Delta E'$ and J'_V are values obtained from the control lobe at the same time. ΔE, ΔE_A, J_V and J_V^A refer to control measurements in the same lobes before the period of metal exposure.

† Paired lobes

§ Significantly different from zero (P <.05)

Flow through the pleural barrier is assumed to take place through channels with an aggregate area of one percent of the total. Surgical removal of the pleura from a region over the pulmonary vein increased volume flow about 40% (Table IV). The area removed by dissection was conservatively estimated at 10% of the total. When the increase in area available for flow and the change in volume flow are put into an expression for the dissected state, both equations can be solved for the relative value of L_e. Hydraulic permeability of the epithelial layer was calculated to be less than one-half of the value for the undissected pleura. Thus, metal-induced changes in L_e of the magnitude observed for the urinary bladder wall should have been detected.

Another objection to the comparison between metal actions on lung and bladder focuses on the gravimetric measurement of volume flow. Since the average lung weighed eight times more than the bladder, inhibition of metabolism and subsequent cell swelling could have resulted in a gain in tissue weight that masked volume flow out of the sac. In the absence of an osmotic gradient exposure of the lung to a battery of metabolic inhibitors increased sac weight but this change was only 20% of the resting osmotic flow. Furthermore, mucosal $HgCl_2$ (10^{-4} M) failed to affect the weight of lung sacs that were not exposed to an osmotic gradient. These observations and the estimates of hydraulic permeability suggest that the lung is much more resistant to the actions of metals on osmotic flow than the urinary bladder, an organ which, like the lung, develops as an evagination of the primitive gut.

TABLE IV. The Effect of the Removal of a Section of Pleura on Osmotic Flow Across the Excised Bullfrog Lung

6 Paired Lobes

Conditions	Volume Flow $(J_V'/J_V) \times 100$*
Intact	127
Pleura Dissected	169
Δ	42 (±13)†

See footnote to Table III for solution composition

*J_V =initial resting flow; J_V' represents flow from both lobes after removal of a portion of the pleura from the experimental lobe.
†significantly different from zero ($P < .05$).

Comparison Between The Actions of $HgCl_2$ And of Amphotericin B

Bioelectric Properties. It was noted during the study of the dose-effect relationships for $HgCl_2$ that low concentrations (10^{-7} to 10^{-6}M) induced a transient increase in transmural P.D. whereas higher concentrations(e.g., 10^{-5}M) resulted in an initial increase followed by a gradual fall to zero. The rate at which this pattern developed was related to the concentration as illustrated in Figure 3. Since conductance rose slightly (resistance fell) or did not change during the voltage increase, short-circuit current also rose. The fall in p.d. was paralleled by a progressive increase in conductance and decrease in bioelectric current (7). However, the initial increase in p.d. and short-circuit current after 10^{-5}M $HgCl_2$ were even greater and more prolonged if a ten-fold molar excess of a sulfhydryl agent such as cysteine or dimercaprol was added to the mucosal solution approximately one minute after the addition of the metal. Moreover, the expected late decrease in p.d. and short-circuit current was prevented completely. Addition of sulfhydryl compounds during the decline in p.d. tended to limit the fall but did not reverse it. Voltage responses of the lung to $HgCl_2$ alone and to $HgCl_2$ followed by a sulfhydryl complexer are reminiscent of those reported for the intestine (2).

To learn more about the mode of $HgCl_2$ action, the effects of the metal were compared with those of amphotericin B, a polyene antibiotic which increases the ion permeability of natural and artificial membranes that contain cholesterol. The time courses of biopotential for lungs that were exposed to mucosal solutions with different concentrations of amphotericin B are shown in Figure 4. Like $HgCl_2$ the voltage response to low doses of the antibiotic was characterized by an increase but the increase persisted for at least 45 min. Higher concentrations induced a transient rise followed by a fall in p.d. and, usually, in short-circuit current. Conductance was increased significantly during all phases of the bioelectric response regardless of the concentration of the drug.

Effects of metabolic inhibitors. When lungs were pretreated with 100% N_2 or 10^{-4}M dinitrophenol and sodium cyanide, the 25% increase in p.d. that was expected after the addition of $HgCl_2$ and cysteine to the mucosal solution was blocked completely (Table V). In contrast, inhibition of oxidative metabolism enhanced the voltage increase induced by amphotericin B in the mucosal solution. Since the conductance of the lung decreased in the presence of N_2 or metabolic inhibitors the increase in p.d. can be explained if exposure to amphotericin B induced a

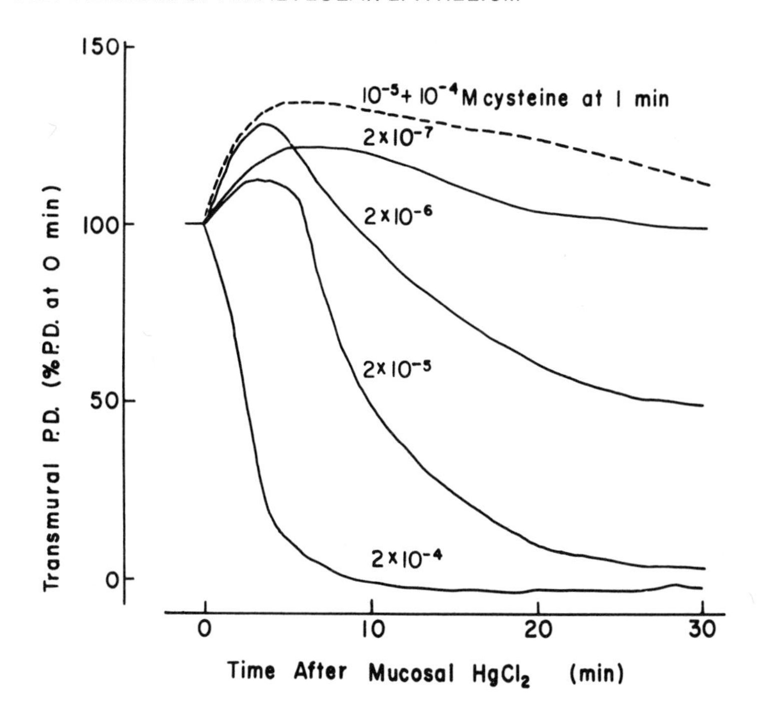

Figure 3. The time course of the mercuric chloride effect on transmural bioelectric p.d. of excised bullfrog lung after addition of the metal to the mucosal bathing solution. Each line represents the average of responses from two lungs. Numbers above the line denote the initial concentration of $HgCl_2$ in the mucosal solution.

similar flow of ions across a raised resistance. Furthermore, oxygen consumption of the lung was increased about 25% by concentrations of $HgCl_2$ that were associated with an increase in transmural p.d. (8). An equi-effective concentration of amphotericin B failed to affect Q_{O_2}. These results suggest that the initial bioelectric response to $HgCl_2$ requires metabolic energy; the p.d. increase after amphotericin B does not.

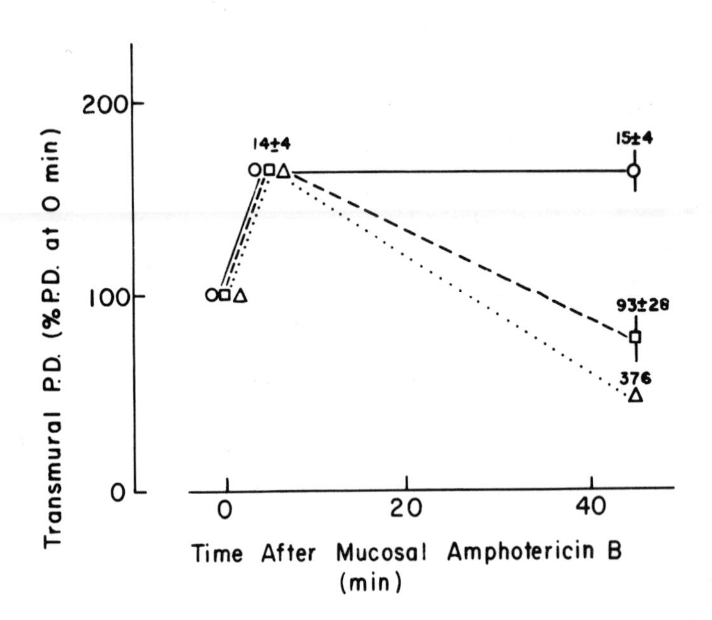

Figure 4. The time course of the amphotericin B effect on transmural bioelectric p.d. of the excised bullfrog lung after addition of the drug to the mucosal solution. —O— = 2.5×10^{-6} M, -□- = 10^{-5} and ··△·· = 10^{-4} M amphotericin B. Numbers indicate the percentage change in d.c. conductance that accompanied each voltage. Each time course is the mean of measurements from three lungs except for 10^{-4} M amphotericin where n=2. Vertical lines represent S.E. larger than the symbols.

TABLE V. Effect of Metabolic Inhibitors on Changes in the Spontaneous Bioelectric P.D. of Bullfrog Lung Induced by $HgCl_2$ - Cysteine or Amphotericin B.

Pretreatment	Mucosal Agent(s)	n	Spontaneous Transmural P.D. (% untreated) mean (±SE): pretreatment*	Δagent(s)†
None	$HgCl_2$ (10^{-5}M)	7	78(±9)	25 (±3)†
100% N_2	followed by	4	43(±3)	3(±1)
DNP + NaCN	cysteine (10^{-4}M)	3	20(±4)	-7 (±1)
None	amphotericin	5	104(±5)	58(±11)§
100% N_2	B	2	20	71
DNP + NaCN	($2.5x10^{-6}$M)	3	29(±7)	113(±19)§

* Steady-state p.d.

†peak p.d. recorded 5 min. after addition of the agent(s) to the mucosal solution.

§significantly increased from pretreatment alone ($p<.05$)

Effects of bathing solution composition. Changes in the composition of the mucosal bathing solution reversed the pattern that was delineated by the inhibition of oxidative metabolism (Table VI). Replacement of Na^+ in the bath by choline or Mg^{2+} did not affect resting transmural p.d. or its stimulation by $HgCl_2$ and cysteine. Substitution of SO_4^{2-} for Cl^- doubled the resting voltage, as had been noted before (7). The voltage change after $HgCl_2$ and cysteine was of the same magnitude as the control response but the results with mucosal sulfate Ringer were more variable. On the other hand, replacement of bathing solution Na^+ by choline or Mg^{2+} blocked the effect of amphotericin B on transmural p.d. completely. From these results we conclude that the bioelectric response to amphotericin B requires mucosal Na^+ whereas the actions of $HgCl_2$ are affected minimally by the composition of the luminal solution.

Unidirectional ion fluxes. An examination of simultaneous unidirectional fluxes of radiosodium and-chloride across the short-circuited lung revealed the changes in ion flow that were responsible for the above observations. Like the open circuit experiments the addition of $HgCl_2$ (10^{-5}M) to the mucosal solution

TABLE VI. Effect of Mucosal Bathing Solution Composition On Changes in Spontaneous Bioelectric P.D. of Bullfrog Lung Induced by $HgCl_2$ - Cysteine or Amphotericin B.

Mucosal Ringer Solution major salt	Mucosal Agent(s)	n	Spontaneous Trans-mural P.D. (% NaCl Ringer) Ringer* mean (SE)	Δ agent(s)† mean (SE)
NaCl	$HgCl_2$ (10^{-5}M)	9	90 (± 4)	31 (± 4)§
$MgCl_2$	followed	3	86 (± 8)	25 (± 5)§
Choline Cl	by	3	123 (±15)	33 (± 7)§
Na_2SO_4	cysteine (10^{-4}M)	3	273 (±41)	28 (± 9)
NaCl	amphotericin	5	92 (±10)	64 (±13)§
$MgCl_2$	B	2	98	-3
Choline Cl	(2.5×10^{-6}M)	3	77 (±17)	-4 (±10)

* steady-state p.d. after the change to a modified Ringer solution.

† peak p.d. recorded 5 min after addition of the agent(s) to the mucosal solution.

§ significantly increased from the value for Ringer alone ($p < .05$)

that bathed lungs clamped at zero p.d. induced a transient increase in short-circuit current followed by a gradual decrease to near zero. The initial stimulation of short-circuit current was accompanied by a significant increase of the serosal to mucosal Cl^- flux. Na^+ flow in the same direction and Na^+ and Cl^- movement in the mucosal to serosal direction were unaffected during the first 20 min. Later, as the short-circuit current fell and conductance increased, the fluxes of both ions in either direction increased several fold (7).

In light of the very large nonspecific increases in ion permeability that eventually resulted from $HgCl_2$ treatment, the early selective increase in serosal to mucosal Cl^- transport seemed tenuous. To study only the relationship between the ion fluxes

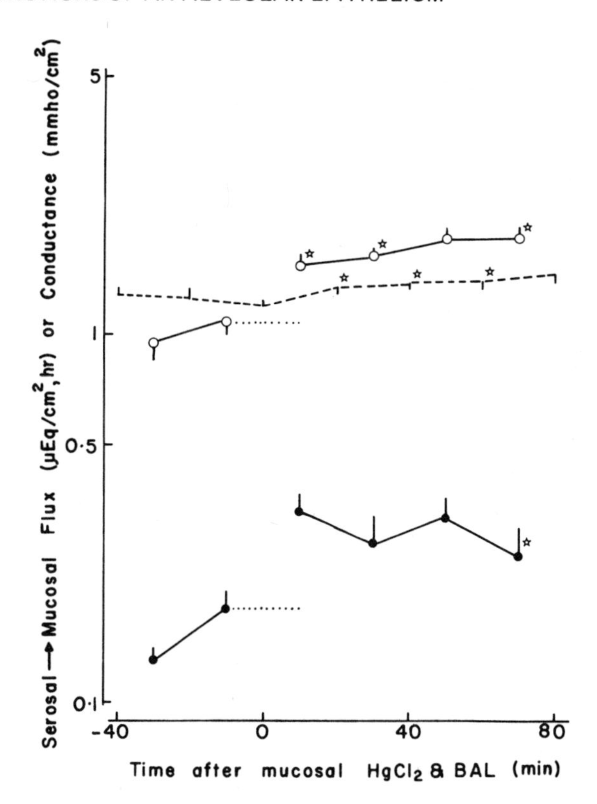

Figure 5. The effect of mucosal $HgCl_2$ (10^{-5}M) followed by dimercaprol (10^{-4}M) on simultaneous serosal to mucosal fluxes of ^{36}Cl (-O-) and ^{22}Na (-●-) and on d.c. transmural conductance (dashed lines). Conductances were measured at the end of each collection from the sink solution. Fluxes are plotted at the mid-point between collection periods. Each point is the mean of four experiments. Vertical lines represent the S.E. Stars indicate significant changes when calculated as a percentage of the value before the addition of $HgCl_2$.

and the metal's stimulation of short-circuit current, we prolonged and increased by more than 60% the early rise by $HgCl_2$ treatment followed by a ten-fold molar excess of dimercaprol. The time course of conductance and unidirectional flux are plotted in Figures 5 and 6. The units of conductance were chosen so that the overall ion flow can be compared directly with individual (partial) ion flows. Short-circuit current stimulation was paralleled by a doubling of the serosal to mucosal Cl^- flux. Conductance tended to increase but the change for any group of the tissues was not always significant. Both fluxes of Na^+ tended to increase but significant

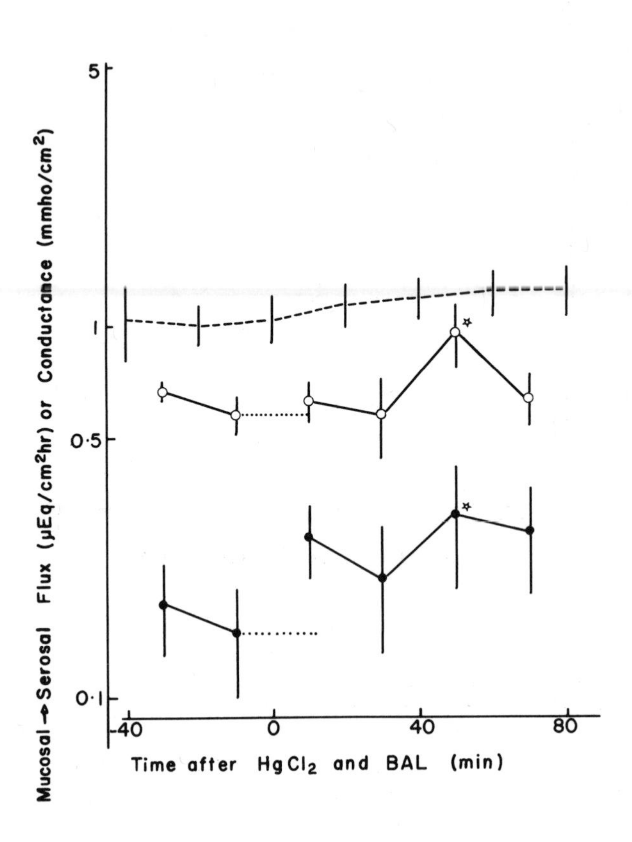

Figure 6. The effect of mucosal $HgCl_2$ (10^{-5}M) followed by dimercaprol (10^{-4}M) on simultaneous mucosal to serosal fluxes of ^{36}Cl (-0-) and ^{22}Na (-●-) and on d.c. teansmural conductance (dashed lines.) Conditions and notations are the same as in Figure 5.

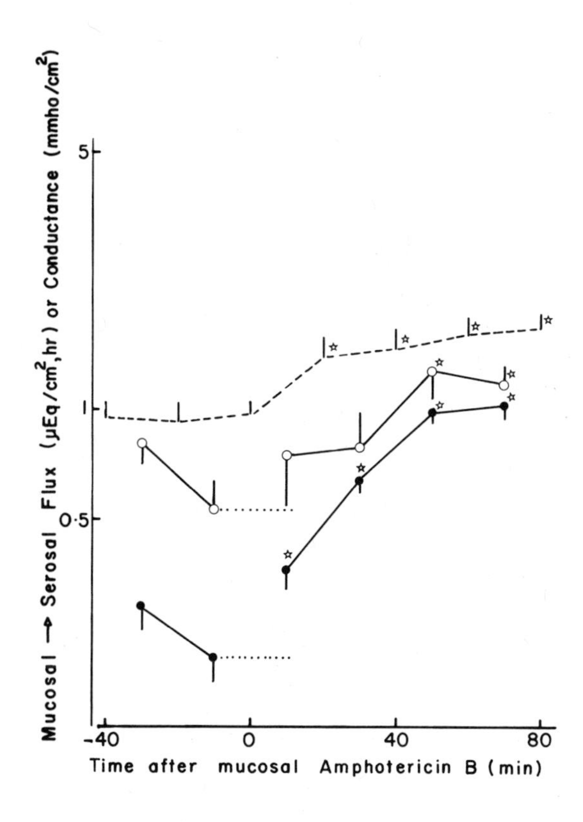

Figure 7. The effect of mucosal amphotericin B ($5x10^{-6}M$) on the simultaneous mucosal to serosal fluxes of ^{36}Cl (-0-) and ^{22}Na (-●-) and on d.c. transmural conductance (dashed lines.) Conditions and notations are the same as in Figure 5.

changes were not measured for 50 min. Compared to the change in serosal to mucosal Cl^- flux, the changes in Na^+ flow were quantitatively small. Except for a single measurement at 50 min. the mucosal to serosal flux of Cl^- was unaffected. These results indicate that the stimulation of short-circuit current induced by $HgCl_2$ is the result of a relatively selective increase in active Cl^- transport into the lumen. The rapid onset of stimulation of Cl^- secretion by $HgCl_2$ and its enhancement by sulfhydryl complexers added within one minute are compatible with an action of the metal at the mucosal surface of the alveolar epithelium. Later effects of the metal (in the absence of complexer) suggest a general breakdown in the resistance of the barrier to ion flow.

Repeating these measurements with amphotericin B led to the results illustrated in Figure 7. Addition of the antibiotic to the mucosal solution was followed by an increase in conductance and in the mucosal to serosal Na^+ flux. Similar changes were observed when Na^+ flow in the opposite direction was determined (7). Alterations in Cl^- flux lagged at least 30 min. behind the initial increase in Na^+ permeability. Again, the most plausible site of drug action is the apical surface of the epithelium but, in this case, amphotericin B appears to selectively increase the passive permeability of the membrane to Na^+, thereby increasing the rate of luminal Na^+ entry into the alveolar cell, the conductance, the bioelectric p.d. and the short-circuit current.

Conclusions

We conclude that cellular and subcellular sites for functions of the amphibian alveolar epithelium can be separated by the action of heavy metals. In this study selective alterations in functions of the ciliated and non-ciliated epithelium were inferred from a comparison of the effects of $HgCl_2$ with those of $CdCl_2$. These results suggested that the ciliated epithelium does not contribute significantly to Cl^- transport across the bullfrog lung. In addition, the spectrum of effects on transport and ion permeability usually induced by $HgCl_2$ can be limited by sulfhydryl agents to a selective stimulation of the Cl^- secretory pump located in the luminal membrane of the alveolar epithelial cells.

Acknowledgements

The author thanks Dr. R.E. Gosselin for suggesting the method for measurement of ciliary motility.

These studies were supported, in part, by U.S.Public Health Service Grants HL-12246 and HL-16674.

REFERENCES

1 BENTLEY, P.J.: The effects of neurohypophysial extracts on water transfer across the wall of the isolated urinary bladder of the toad Bufo marinus. J. Endocrinol. 17 (1958) 201.

2 CLARKSON, T.W.: Action of heavy metals on transport across the intestine, Ch. 6, Intestinal Transport of Electrolytes, Amino, Acids, and Sugars (Armstrong, W.McD., Nunn,A.S. Eds.). Thomas. Springfield (1971).

3 CROSS, C.E., RIEBEN, P.A. and SALISBURY, P.F.: Urea permeability of the alveolar membrane; hemodynamic effects of liquid in the alveolar spaces. Am. J. Physiol. 198, (1960) 1029.

4 GATZY, J.T.: Bioelectric properties of the isolated amphibian lung. Am. J. Physiol. 213 (1967) 425.

5 GATZY, J.T.: Ion transport across the excised bullfrog lung. Am. J. Physiol. 228 (1975) 1162.

6 GATZY, J.T.: Ion transport across amphibian lung, p.179, Lung Liquids (Ciba Found. Symp. 38). Elsevier. New York (1976).

7 GATZY, J.T. and STUTTS, M.J.: Ion transport and O_2 consumption of excised bullfrog lung after treatment with $HgCl_2$ or amphotericin B. Fed. Proc. 34 (1975) 753.

8 KILBURN, K.H.: Alveolar microenvironment. Arch. Intern. Med. 126 (1970) 435.

9 KYLSTRA, J.A.: Lavage of the lung. Acta Physiol. Pharmacol. Neerl. 7 (1957) 163.

10 MENZEL, D.B.: Toxicity of ozone, oxygen and radiation. Ann. Rev. Pharmacol. 10 (1970) 379.

11 NAGAISHI, C.: Functional Anatomy and Histology of the Lung. p. 46. University Park. Baltimore (1972).

12 OLVER, R.E., REYNOLDS, O.R . and STRANG, L.B.: Foetal lung liquid, p. 186, Foetal and Neonatal Physiology (Cross, K.W., Ed.) Cambridge University Press. London (1973).

13 QUAY, J.F. and ARMSTRONG, W. McD.: Sodium and chloride transport by isolated bullfrog small intestine. Am. J. Physiol. 217 (1969) 694.

14 SCUDI, J.V., KIMURA, E.T. and REINHARD, J.F.: Study of drug action on mammalian ciliated epithelium. J.Pharmacol. Exper. Ther. 102 (1951) 132.

15 TAYLOR, A.E. and GAAR, K.A.,JR.: Estimation of equivalent pore radii of pulmonary capillary and alveolar membranes. Am. J. Physiol. 218 (1970) 1133.

16 TAYLOR, A.E., GUYTON, A.C. and BISHOP, V.S.: Permeability of the alveolar membrane to solutes. Circ. Res. 16 (1965) 353.

17 THEODORE, J., ROBIN, E.D., GUIDIO, R. and ACEVEDO, J.: Transalveolar transport of large polar solutes (sucrose, inulin, and dextran). Am. J. Physiol. 229 (1975) 989.

18 VIJEYARATNAM, G. S. and CORRIN, B.: Experimental paraquat poisoning: a histological and electron-optical study of changes in the lung. J. Path. 103 (1971) 123.

19 WANGENSTEEN, O.D., WITTMERS, L.E., JR. and JOHNSON, J.A.: Permeability of the mammalian blood-gas barrier and its components. Am. J. Physiol. 216 (1969) 719.

20 WEIBEL, E.R. : Oxygen effect on lung cells. Arch. Intern. Med. 128 (1971) 54.

DISCUSSION

GENNARO: I would like to comment though that although the amphilian alveolar epithelium is much less complicated than that of the mammal, it still contains what could be called an alveolar type I and also an alveolar type II cell. The ciliated cell is a minor component of the epithelium.

GATZY: I didn't mean to imply that the cell morphology was simpler. I meant that the lung wall could be set up under conditions where one could stringently evaluate transport functions of the epithelium.

GENNARO: Then this begins to be complex. We have here a cell which is heavily endowed with lipid inclusions which it seems to be producing and secreting at the luminal surface (the alveolar type II cell), and then a different cell which is a very thin one, presumably very much like the cells of squamous endothelium. Which of these is responsible for the effects you see?

GATZY: I wish I could answer that question. We might be able to get some insight into the problem if we could disaggregate cells and separate them into different populations. But, in contrast to mammalian epithelial cells, these cells seem to clump together so that procedures that are commonly used to disperse mammalian cells don't work in this preparation.

GENNARO: However, they are very, very different in size. It might be that just a simple Ficoll step density gradient might separate them.

GATZY: That might work, if you could get them unclumped.

GENNARO: Another problem may arise from the use of these cells in a chamber. We have noticed that, when lung tissue is fixed for microscopy by perfusion through the vascular endothelium, cell shapes are greatly changed over what is obtained when the tissue is fixed by perfusion through the pulmonary lumen. That is, even slight pressure in introducing the fixative through the vascular pathway will stretch the cells greatly. This may produce the same kind of technical problem seen by people who work with bladder epithelia. These, when placed in chambers and stretched, show markedly different transport properties than when unstretched. Do you find great variation in the transport rates between the chamber and sac preparations?

GATZY: No, we don't. Bioelectric properties are similar. We have used the sack preparation to measure volume flow which, incidentally, is not affected by any heavy metal that we tested. These results are in contrast to results from urinary bladders, from the same animal. A number of agents including cadmium affect volume flow across this epithelium.

NARAHASHI: Did you use crystallized samples or clinical samples of amphotericin B?

GATZY: We used both, depending on the available supply.

NARAHASHI: But, did you get the same results?

GATZY: Yes.

NARAHASHI: That is good, because we tried the clinical sample on our nerve preparations and found very intriguing effects for production of repetitive discharges, but it turned out that the effect was produced by deoxycholate contained in the sample and amphotericin B had no effect whatsoever(Wu, Sides and Narahashi, Biophys. J. 15: 263a, 1975).

GATZY: In most of the experiments we used crystalline amphotericin B powder dissolved in dimethylsulfoxide and used dimethylsufoxide blanks as the control. DMSO had no effect.

SZABO: Just to add to the confusion with amphotericin B. On both sides it increases chloride permeability. I guess it is a surprise that on biological membranes it seems to increase cationic permeability, and I don't understand why. The second thing is that we have done some work with amphotericin B. The effect on monoolein bilayers seems to depend on the type of solvent that you use to dissolve the amphotericin B. DMSO is all right, and gives rather interesting results, but if you use dimethylsufoxide or methanol or ethanol, you get quantitatively different results.

GATZY: You mean deoxycholate rather than dimethylsufoxide?

SZABO: No, this is without deoxycholate. This is just dissolving amphotericin B in some kind of vehicle.

GATZY: You said DMSO was okay, and then you said dimethylsufoxide and ethanol produced different results, is that what you mean?

SZABO: No. Well, a number of other solvents - methanol, ethanol. Those give different results. So, it seems to depend on the degree of aggregation of the amphotericin B that you have in the solution. That is the only solution I can give. The second question I have is quite different. Do you have any idea of the edge damage?

GATZY: For years we have been publishing data which indicate that the bioelectric properties of the toad urinary bladder, particularly the transmural potentials, are as high as those measured by anyone. Now, if you consult Arthur Finn's paper (Finn, A. L. and Hutton, S. A., Am. J. Physiol. 227: 950-953, 1974) on edge damage, a potential of 80 millivolts represents minimal edge damage. These lungs were mounted in the same chambers, so by inference we assume that the lungs are not damaged, but, of course, they are different tissues.

SZABO: What kind of mounting technique did you use?

GATZY: This is the same technique that I have used for years for the toad bladder. It involves pinning the bladder or the lung to a cork platform that is positioned around one half-chamber. A large screw between a holding frame and the back of the other half-chamber moves the chambers together, compressing the periphery of the tissue in a V-groove.

SZABO: The reason why I am asking this question is that Simon Lewis (Lewis, S., and Diamond, J.M., J. Membrane Biol. 28: 1-40, 1976) has found that if you use silicone grease you can decrease the edge damage tremendously, under this condition I think he has been able to show that the urinary bladder is extremely resistant.

GATZY: Well, there are a number of problems with grease, at least one study (Higgins, Jr., J. T., Cesaro, L., Gebler, B. and Fromter, E., Pfluger Arch. 351: 41-56, 1975) indicates that resistance progressively rises. There is a suggestion that this is due to the spread of grease over the surface of the bladder. Attempts to minimize edge damage are often difficult to assess in tissues where the transporting epithelium comprises the luminal surface. An exception is the frog's skin where the apical surface of the transporting cells is protected by the cornified layer.

LAKOWICZ: I have two questions. The first one is that oxygen is probably the most important passive permeating molecule in the lung - is the barrier imposed by the cell membrane ever significant and if so does the heavy metal have any effect upon that barrier?

GATZY: You mean in terms of oxygen transport?

LAKOWICZ: Yes.

GATZY: I don't know.

LAKOWICZ: Contamination by mercury in the environment is, typically methylmercury, a more toxic species, although it is thought to be a neurotoxin it does have to get to the site of action. Have you done any studies on the effects of methylmercury?

GATZY: No, we haven't looked at methylmercury. Initially, we examined other organomercurials such as PCMB and PCMBS. PCMB's effects are characterized by a slow decline in p.d. and short-circuit current which begin forty minutes after the addition of the agent. In contrast to mercuric chloride, there is no initial stimulation. PCMBS doesn't have any effect that we can measure. Perhaps, this means that it doesn't penetrate into the cell.

SHAMOO: Have you changed the amount of calcium in your medium?

GATZY: Yes.

SHAMOO: Does it have any effect?

GATZY: Omission of calcium salts from the solution has no effect. Addition of EDTA to a calcium-free solution induces changes in bioelectric properties similar to those observed with other disaggregating agents.

SIEGEL: Did you look at the effects of other ions and interaction with the chloride transport and the methylmercury which apparently is not so?

GATZY: We have checked the action of other halides with the chloride transport system. For example, bromide and chloride may compete for the same mechanism. However, the bioelectric properties and the flux measurements do not give a clear-cut answer. From Cl^-flux measurements the K_i for bromide was estimated at about 60 milliequivalents per liter, which is approximately three times larger than the K_m for chloride. Yet, if you replace chloride with bromide there are no changes in the bioelectric properties of the tissue. These results predict that Br^- and Cl^- should have similar K_m's if the anions are transported by the same mechanism.

SIEGEL: What about cations?

GATZY: Cations don't seem to have much effect. You can replace any cation except, perhaps, serosal K^+ with any other cation which you consider to be inert and there is no change in bioelectric properties.

SIEGEL: Why have you chosen cadmium and mercury?

GATZY: We have also looked at a number of other metals including lead, nickel, zinc and cobalt. Nickel behaves very much like cadmium. Lead may also share some of cadmium's actions but it is not very efficacious. I chose cadmium and mercury originally because both interact with sulfhydryl groups but only cadmium had been reported to routinely induce acute lung damage (Browning, E., "Toxicity of Industrial Metals," Appleton-Century-Crofts, New York, 1969, pp. 102-104), although there are a few isolated cases of people who have inhaled mercury vapor and developed lung edema (Natelson, E.A., Blumenthal, B. J., and Fred, H. L., Chest 59: 667-678, 1971; Milne, S., Christophers, A. and DeSilva, P., Brit. J. Industr. Med. 27: 334-338, 1970). We reasoned that if there was some difference in the effects of the metals on the alveolar epithelium we might get some insight into the modes or action.

SIEGEL: Did you say that the short circuit current was fully accounted for by the metachloride transportation?

GATZY: After one hour.

SIEGEL: How much before that?

GATZY: Fifty percent.

SIEGEL: And what carried the current there?

GATZY: I don't know. It may be due to an asymmetric diffusion of an ion out of the tissue. We do know that you can see this kind of response if you replace chloride with sulfate. The short circuit current and voltage double and then fall. If you measure the chloride content in the cell, say after 90 minutes, you find that there is a direct relationship between chloride in the tissue and the short-circuit current that was measured just before the tissue was removed from the chamber. We conclude from these experiments that this response is due to the diffusion of chloride across the luminal membrane into the lumen. Since the passive flow of chloride out of the epithelium would give rise to a diffusion p.d. of the appropriate polarity, it seems unnecessary to postulate active transport of another ion or increased active transport of chloride in the presence of sulfate. In fact, the unidirectional flux of chloride from serosa to lumen does not change when luminal chloride is replaced by sulfate.

THE LOCALIZATION OF ION-SELECTIVE PUMPS AND PATHS IN THE PLASMA MEMBRANES OF TURTLE BLADDERS

William A. Brodsky and Gerhard Ehrenspeck

Mount Sinai School of Medicine/CUNY, Dept. Physiol. & Biophys., 100th Street and 5th Ave, N.Y., N.Y. 10029

ABSTRACT

Turtle bladders actively transport Na, Cl, and HCO_3 to the serosal fluid; and each ionic flux is independent of the others under short-circuiting conditions. This behavior mimics that of a parallel network of ion-selective, electrically-conductive paths and pumps in each membrane, - a picture consistent with recent evidence along three independent lines. (1) The potential response to increases in mucosal Na concentration indicates that the Na conductance of the apical membrane is 70% of the transepithelial conductance and that the Na transfer across this membrane occurs via an electrically-charged carrier operation. (2) The sidedness and selectivity of transport changes induced by certain agents are the following. Acting from the mucosal side only, amiloride blocks passive Na transfer; and catecholamines (or imidazoles or theophylline) accelerate active anion transport. Acting from the serosal side only, ouabain blocks active Na transport; and disulfonic stilbenes or acetazolamide block passive anion transfers. (3) The surface charge density of the apical membrane differs from that of basal-lateral during free-flow electrophoresis (FFE) of a mixed membrane fraction of epithelial cells. Basal-lateral membrane fragments (containing ouabain-sensitive ATPase and a stilbene-binding protein) migrate toward the positive electrode while apical membrane fragments (contain nor-epinephrine-sensitive adenylate cyclase and cAMP-activated protein kinase) migrate toward the negative electrode. (4) Thus, ouabain, nor-epinephrine, and a disulfonic stilbene are shown to be useful membrane probes for the Na pump, the anion pumps, and the passive anion transfer paths, respectively.

INTRODUCTION

Background. The epithelial cell layer of the turtle bladder is a specialized, nearly-homogeneous cell system (Figure 1) containing active transport mechanisms (ion pumps) for the translocation (reabsorption) of ion-pairs (NaCl and $NaHCO_3$) from the mucosal fluid to the serosal fluid (Brodsky and Schilb, 1965,1966). The effect of these translocations is to dilute and acidify the mucosal fluid (urine) and to contribute toward the maintenance of a constant state in the ionic composition and acid-base balance of the body fluids. At the cellular level, the effect of these ion pumps, particularly the Na pump, is to maintain transmembrane gradients of ion concentration, whereby the intracellular fluid is higher in [K] and lower in [Na] than the extracellular fluid (Schilb and Brodsky, 1970) and transmembrane gradients of electrical potentials whereby the serosal fluid is 60-80 mvolts positive to the mucosal fluid (Brodsky and Schilb, 1966; Gonzalez et al, 1967a). The cell fluid is 10-20 mvolts negative to mucosal fluid (Hirshhorn and Frazier, 1971), and hence the serosal fluid, 70-100 mvolts positive to the cell fluid. The magnitude and orentation of these electrochemical potential gradients suggest that anion pumps (for Cl and HCO_3) should be located in the apical membrane in order to drive NaCl and $NaHCO_3$ from the mucosal to the cellular fluid while a Na pump should be located in the basal-lateral membrane in order to drive the same ion-pairs from the cellular to the serosal fluid.

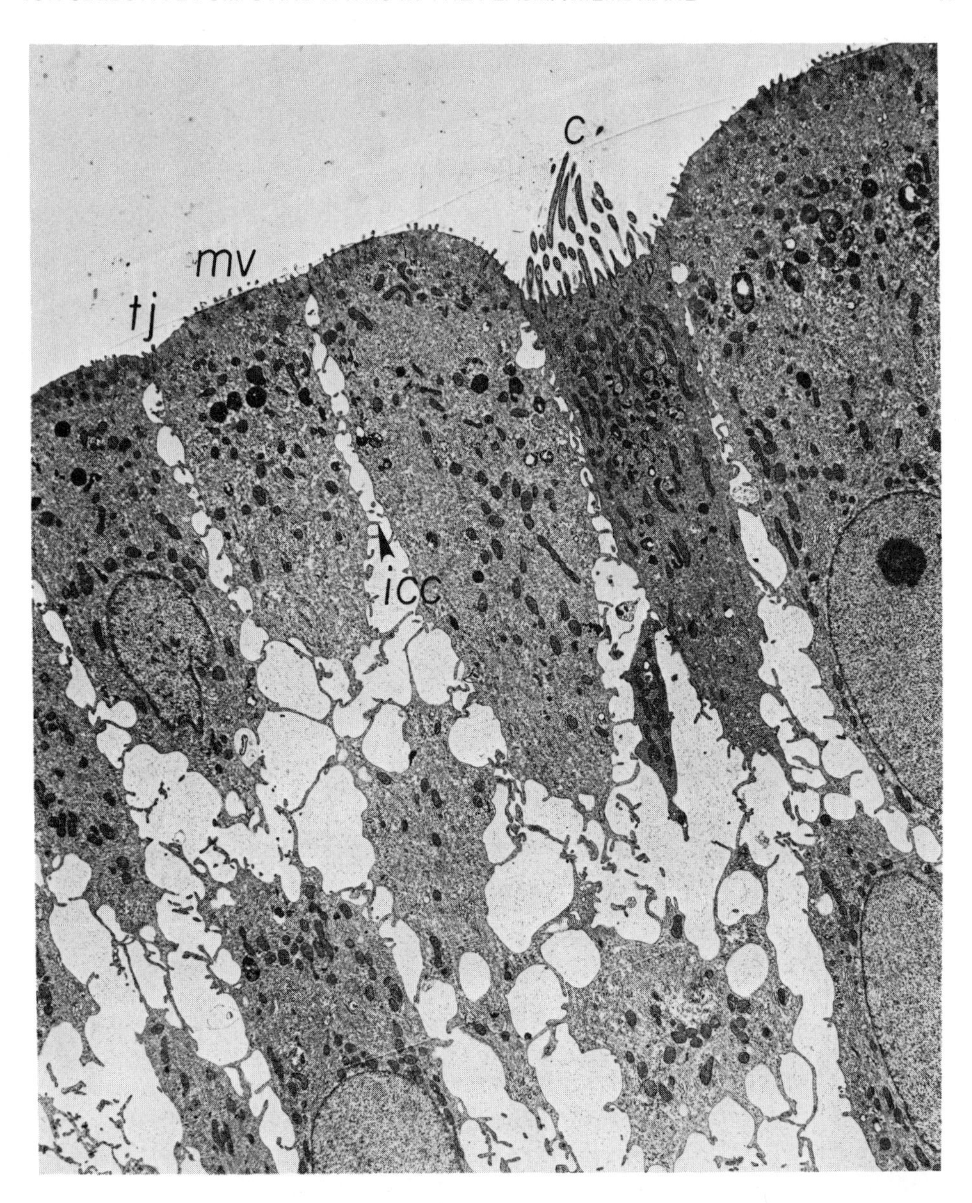

Figure 1 - Electron micrograph (x5000) of turtle bladder epithelium after two hours of incubation in a Na-rich Ringer solution containing HCO_3 and Cl and gassed continuously with 98% O_2, 2% CO_2. The electron-dense cell with the ciliated apical membrane (c) comprises no more than 5% of the epithelial cells. The remainder of cells (ca. 95%) contains: a microvillous type of apical membrane (mv); tight junctions (tj); and long narrow intercellular channels interrupted by projections of the basal membrane (icc). (Courtesy of Prof. J. G. Gennaro, Jr., Dept. Biol., NYU).

A. LUMPED MEMBRANE PARAMETERS

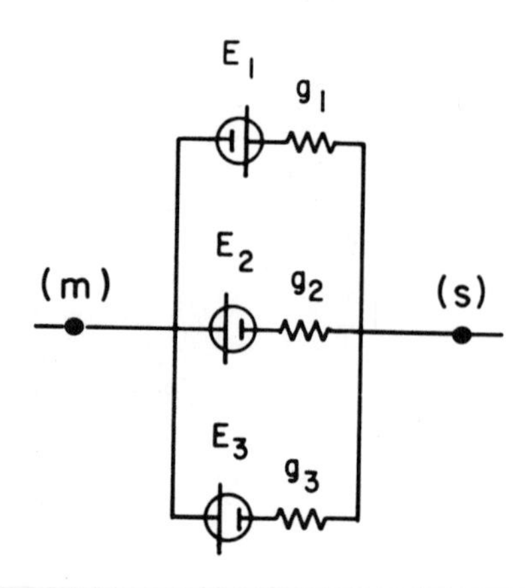

B. DISTRIBUTED MEMBRANE PARAMETERS

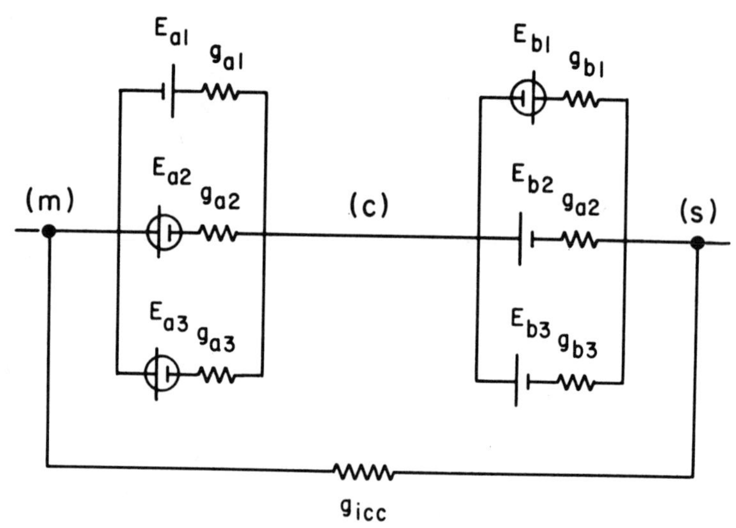

Figure 2. A). Oversimplified electrical analogue of bladder. E designates ion-specific electromotive forces and g, the corresponding ion-specific conductances. The subscripts 1, 2, and 3 denote Na, Cl, and HCO_3 respectively. The letters m and s denote mucosal and serosal fluid.

B). More realistic electrical analogue of bladder. The letters (E's and g's) and the numerical subscripts (1, 2, and 3) have the same meaning as those in panel A.

Circles about E denote pump EMF's and no circles denote diffusional EMF's (i.e. due to transmembrane ion gradients). The letters, m, c, and s denote mucosal cell, and serosal fluids; and icc denotes the tight junction - intercellular shunt pathways.

Further evidence for the presence of three distinct ion pumps (Na, Cl, and HCO_3) in this epithelium comes from data on the net mucosal-to-serosal fluxes of Na and Cl across short-circuited bladders bathed by Na-rich and Na-free Ringer solutions as well as by Cl-free and/or HCO_3-free Ringer solutions. The net Na current remains unchanged after Cl and HCO_3 are added to or removed from the mucosal fluid (Gonzalez et al, 1967b). The net Cl current remains unchanged after Na is added to or removed from the mucosal (and/or serosal) fluid (Gonzalez et al, 1967a). The short-circuiting current (I_{sc}) remains equal to the algebraic sum of the individual ionic currents under any or all of the aforementioned bathing conditions. Such data require that the ionic flows are in parallel and independent of each other.

Therefore, the epithelial cell layer can be represented as an electrical network of three ion-selective limbs in parallel (Figure 2A). Each limb contains an ion-specific driving element (EMF) in series with an ion-specific conductance element. This model accounts for the data on ion fluxes under open-circuit conditions and under short-circuiting conditions but not for those on the trans-apical or trans-basal-lateral membrane gradients of electrical potential or ion concentration.

A more realistic model, that of Figure 2B, represents the turtle bladder as two electric networks in series--one containing all of the ion-specific pumps and conductances of the apical membrane and the other, all of the ion-specific pumps and conductances of the basal-lateral membrane. The high-resistance shunt between the two networks denotes the path through the tight junction and intercellular channels. This model can account for: (i) all of the ionic fluxes under open-circuiting and short-circuiting conditions; (ii) the low [Na] and high [K] of the cell fluid; (iii) the electronegativity of the cell fluid relative to mucosal fluid; and (iv) the electronegativity of the serosal fluid with respect to the mucosal fluid under Na-free bathing conditions (Brodsky & Schilb, 1966, Gonzalez et al, 1967a). Furthermore, the model has provided a conceptual starting-off point in the design of our subsequent experiments.

EXPERIMENTAL APPROACHES

In order to evaluate some of the parameters that are distributed between the apical and basal-lateral membrane, three independent approaches were used.

First, an electrophysiological approach was used on intact bladders under open-circuit conditions. This approach depends on the transient electrical response to step-increases of the ionic concentrations in each of the bathing fluids. From such data, it is possible to evaluate the partial ionic conductance(s) or electrogenic pump activity of each membrane.

Second, a combined pharmacologic and electrophysiologic approach was used on intact bladders under short-circuiting conditions. This approach depends on the exclusive nature of a change in the transport of a single ion induced by the presence of an "effector" substance (inhibitor or accelerator) in one of the two bathing fluids. Such transport effectors (drugs, neurotransmitters, amino acids, etc) can serve as useful markers of ion-selective paths, carriers, or binding sites in isolated membrane preparations.

Third, a biochemical approach was used on isolated membrane fragments derived from turtle bladder cells. This approach, combining the principles and techniques of Skou (1957), Hannig (1969, 1972), and Heidrich et al (1972), requires a physical separation of the apical from the basal-lateral membranes. The data indicate the apical or basal-lateral location of certain ion-selective transport-related enzymes and their unique reactions with specific transport effector substances.

RESULTS

A. Partial Na-conductance (Open-circuit conditions)

The total transepithelial electrical conductance of the turtle bladder decreases by 50% to 70% after choline is substituted for Na in the bathing fluids (Gonzalez et al, 1967a) or after the Na flux has been suppressed by ouabain (Solinger et al, 1968). This implies that the Na-specific conductance accounts for over half of the total electrical conductance of the bladder; and was confirmed by independent experiments on the transient electrical potential response (ΔPD_{sm}) to step-increases in the Na concentration of the mucosal fluid ($[Na]_m$) and/or of the serosal fluid ($[Na]_s$). In these experiments, increments of Na were substituted for equimolar decrements of choline in the mucosal fluid and then in the serosal fluid. By this method, the Na-selectivity of the bladder was elicited. (Wilczewski & Brodsky, 1975).

Figure 3 is a plot of values of the electrical potential (PD_{sm}) versus the logarithm of the mucosal Na concentration ($\log[Na]_m$). Data are from control and ouabain-treated bladders that had been pre-incubated in a choline Ringer solution for 2 hours prior to the addition of Na to (and simultaneous removal of choline from) the mucosal fluid.

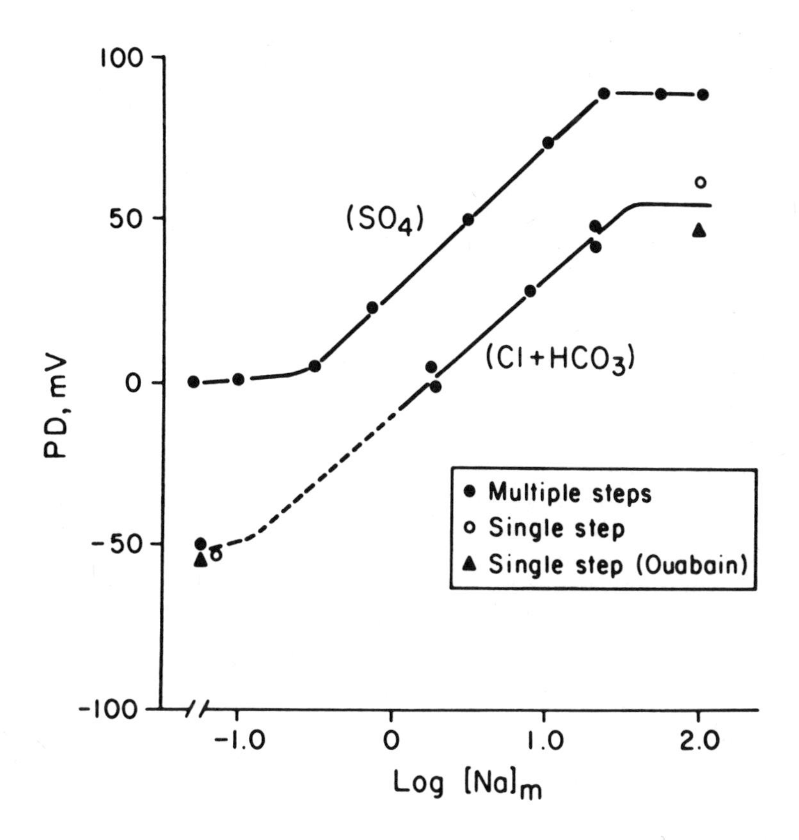

Figure 3. PD versus logarithm of mucosal Na concentration in bladders bathed by Na-free media prior to step increases in mucosal [Na]. PD values from 3 bladders subjected to multiple step increases of mucosal [Na] denoted by solid circles. Mean values of PD from 8 paired experiments before and after a single step increase of mucosal [Na] in untreated hemibladders denoted by open circles and in ouabain-treated hemibladders denoted by solid triangles.

When the mucosal Na concentration ($[Na]_m$) was increased from 0.5 to 20mM, the evoked transepithelial potential was a straight-line function of the $\log[Na]_m$. Evaluation of the slope of this line indicates that the partial Na conductance ($\bar{g}_{Na}$) of the epithelium accounts for 70% of the total transepithelial electrical conductance (G_t). However, when the serosal Na concentration ($[Na]_s$) was similarly increased, the magnitude of the electrical potential response was small; and evaluation of the slope of ΔPD on $\log[Na]_s$ indicated that the apparent g_{Na} accounts for ca. 5-7% of the total electrical conductance of the bladder. The mucosal fluid sidedness of this electrical potential response suggests that most of the partial Na conductance of the bladder resides in the apical membrane (see the element designated $g_{a,1}$ in figure 2B).

On the other hand, when $[Na]_m$ was increased from 20 to 100mM, no further increases in PD_{sm} were evoked, i.e., PD_{sm} became independent of $[Na]_m$. This type of zeroth-order function is electrically-"silent" and consistent with the flow of a fully-occupied (saturated) Na carrier-complex within the boundaries of the apical membrane. Since the Na conductance of the apical membrane is <u>not</u> apt to vanish with these (or any other) increases in $[Na]_m$, the electrically-"silent" function cannot be construed as a non-conductive (electroneutral) function. The electrically-conductive nature of this function is made apparent from the net Na transport that occurs when the bladder is short-circuited in the same bathing fluids (Wilczewski & Brodsky, 1975). Therefore, we are forced to infer that one intramembrane component of a limited (hence saturable) quantity of a Na-carrier system is electrically-charged and mobile. For example, the flow of an electropositive, Na-occupied-carrier, say a $(Na \cdot X)^+$ complex, from the mucosal to the cellular interface of the membrane, would be an electrically-conductive process even if all the Na sites on X were occupied (or saturated).[1]

When the bladder is bathed by Na-Ringer solutions, the sum of the ion-specific conductances other than Na (e.g., Cl, HCO_3, H, and K) must account for 30% to 40% of the total electrical conductance

---

[1] <u>Footnote.</u> This is apparently indistinguishable from the flow of an electronegative unoccupied carrier, say X^-, from the cellular to the mucosal interface of the bladder. However, Heinz and Geck (1976) have recently developed a kinetic basis for making such a distinction in the case of the Na-driven transport of amino acids into Ehrlich cells.

in order to satisfy Kirchoff's law of conservation of electric charge. But when the bladder is bathed by Na-free (choline) Ringer solutions, the sum of the partial ionic conductances due to Cl, HCO_3, K and H is significantly less than the total electrical conductance of the bladder (Ehrenspeck, Brodsky and Durham - unpublished data, 1976). This means that the partial conductance due to choline ion is finite and that that due to any other single ion (other than Na) is apparently no more than 20% of the total conductance. Thus, the electrical potential response (ΔPD) to a step-increase in the concentration of any one of these ions (from $C_{i,1}$ to $C_{i,2}$) is small and difficult to evaluate precisely - as can be seen by inspection of the conventionally-defined expression:

$$\Delta PD = -\bar{g}_i \frac{RT}{Z_iF} \ln \frac{C_{i,1}}{C_{i,2}} ,$$

where - $(RT/ZF)\cdot \ln (C_{i,1}/C_{i,2})$ is equal to E_i°, the Nernst equilibrium potential of the _ith_ ion (or the state of zero flow of the _ith_ ion); and where the magnitude of the concentration-dependent potential becomes small when $\bar{g}_i$ approaches near-zero levels.

B. Membrane-selective, ion-selective "effectors" (short-circuiting conditions)

The aforementioned considerations led to the use of a second approach for evaluating the ion-selective properties of these epithelial cell membranes. This approach depends on the selectivity of ion transport changes induced by certain pharmacological "effector" substances (see below). The experimental conditions imposed include those of short-circuiting, which is well-suited for monitoring steady-state changes in ion transport--thus eliminating the need of evaluating small, rapid transient responses.

Operationally defined, a _"transport-effector"_ substance is one that increases or decreases the transepithelial ionic flow (or flows) following its addition to the mucosal and/or serosal fluid that bathes the membranes of an epithelial cell layer. Such an effector is said to be "membrane-selective" when a change in transport is evoked by the presence of the effector in one bathing fluid--and explicitly _not_ in the opposite bathing fluid. An effector is said to be cation-selective when it induces a change in cationic transport, but not in anionic transport (and vice-versa for an anion-selective effector).

Table 1. Sidedness and selectivity of ion transport effectors in turtle bladders.

Effector	Transport Process (ms)	Induced Change
In Mucosal Fluid Only		
Amiloride	Na	↓
Catecholamine	Cl, HCO_3	↑
Imidazole	Cl, HCO_3	↑
Theophylline	Cl, HCO_3	↑
In Serosal Fluid Only		
Ouabain	Na	↓
DS-Stilbene	Cl, HCO_3	↓
Acetazolamide	Cl, HCO_3	↓

Table 1 presents a categorization of such membrane-selective, ion-selective transport-effectors that have been found in the case of the turtle bladder. This listing includes the following: (i) neurotransmitters such as nor-epinephrine and isoproterenol (Brodsky, et al, 1976); (ii) diuretic drugs such as ouabain (Solinger et al, 1968), ethacrynic acid (Wilczewski & Brodsky - unpublished data, 1976), theophylline (Brodsky et al, 1976), and acetazolamide (Gonzalez & Schilb, 1969; Gonzalez, 1969; Steinmetz, 1975); (iii) imidazole compounds such as histidine, histamine, and imidazole (Brodsky, et al, 1976); and (iv) organic negative ions such as the 4,4'-substituted disulfonic stilbenes, SITS or DIDS (Ehrenspeck & Brodsky, 1976).

Figure 4, showing plots of short-circuiting current (I_{sc}) versus time in bladders bathed by Na Ringer solutions, illustrates the mucosal and serosal sidedness of the effects of two substances (amiloride and ouabain) that are known to retard Na transport but not anion transport. The presence of amiloride in the mucosal fluid (but not in the serosal fluid) is required to produce its inhibitory effect on I_{sc} and Na transport (figure 4A). This required location of amiloride contrasts with that of ouabain--the

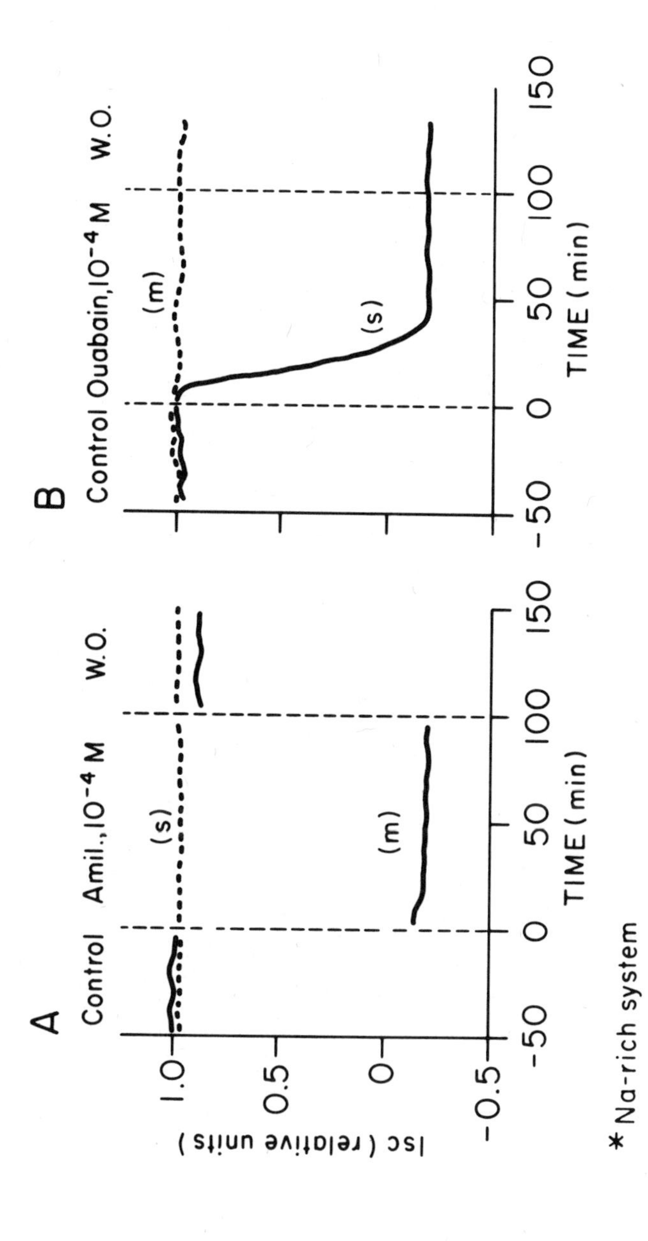

Figure 4. Sidedness of effect of Na-selective inhibitors. (A) Amiloride; (B) Ouabain.

Relative short-circuiting current (I_{sc}) versus time in minutes. A relative unit is the value of I_{sc} at any time, t, divided by that at zero time when the inhibitor was added to mucosal fluid (m) or to the serosal fluid (s).

presence of which in the serosal fluid (but not in the mucosal fluid) is required for eliciting its effect on I_{SC} and Na transport (Figure 4B).

Figure 5, showing analogous data plots in bladders bathed by Na-free (choline) Ringer solutions, illustrates the mucosal or serosal sidedness of the effects of two substances that change anion transport but not cation (Na) transport. The presence of nor-epinephrine in the mucosal fluid (not in the serosal fluid) is required for eliciting its stimulatory[2] action on the anion-dependent (Cl and HCO_3) moiety of the I_{SC} (Figure 5A).

In contrast to the mucosal-sidedness of the nor-epinephrine effect, figure 5B shows that the presence of the disulfonic stilbene, SITS, in the serosal fluid (not in the mucosal fluid) is required for eliciting its inhibitory effect on I_{SC} and on the related transport of Cl and HCO_3 (Ehrenspeck et al, - unpublished data, 1976).

These data suggest that there are discrete sets of "effector-specific" binding sites at or close to each of the ion-selective paths in the apical as well as in the basal-lateral membranes, and that certain transport-related enzymes and/or proteins should be present in the membrane fraction(s) of the epithelial cell homogenates.

C. Transport-related enzymes and proteins (isolated membranes)

Previous work has shown that there is a ouabain-sensitive, (Na + K)·ATPase in the isolated membrane fraction of turtle bladder epithelial cells (Solinger et al, 1968; Shamoo & Brodsky, 1970). This in-vitro enzymatic activity can be correlated by analogy to the in-vivo action of ouabain. But the (Na + K)·ATPase resides in a sub-cellular fraction that contains both apical and basal-lateral membrane fragments, which means that the in-vitro effect of ouabain is insufficient to account for the serosal-sidedness of its action (see table 1 and figure 4B) or for the assumed basal-lateral location of its membrane-binding site(s). Thus, it was necessary to separate the apical from the basal-lateral membranes; and this was achieved by using the Hannig (1969) technique of free-flow electrophoresis in the manner designed for various epithelial cell system by Heidrich et al, (1972).

---

[2] Footnote. We have not yet found an inhibitor of anion transport that works uniquely from the mucosal fluid side.

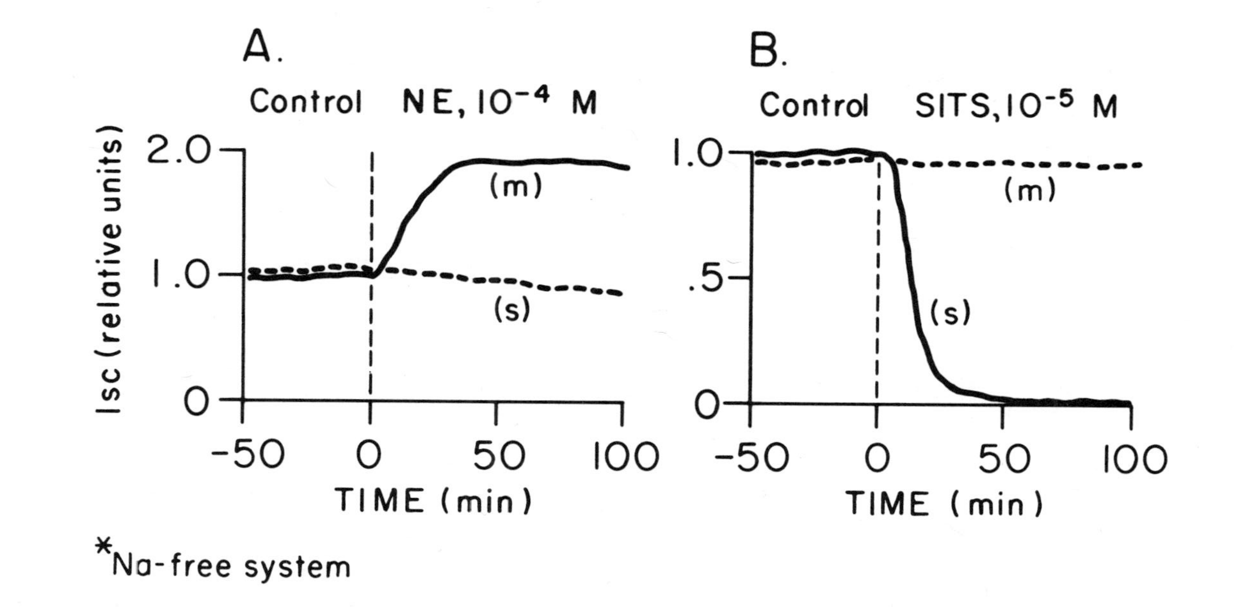

Figure 5. Sidedness of effect of anion-selective effectors. (A) Norepinephrine; (B) SITS.

Relative short-circuiting current (I_{sc}) versus time in minutes. Relative units of I_{sc} and the zero time of addition of effectors to m or s are as described in previous figure.

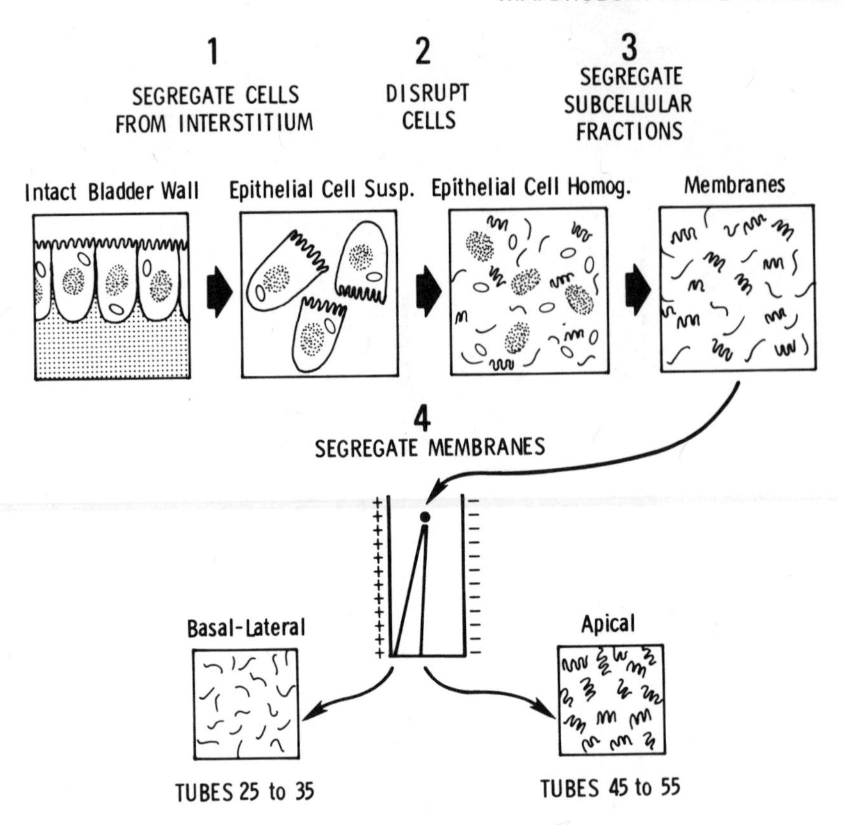

Figure 6. Consecutive steps (1 to 4) in separating the apical from the basal-lateral membranes--starting with the intact bladder and terminating in the effluent of the free-flow electrophoresis.

Figure 6 is a schematic view of the procedures used to obtain a mixture of apical and basal-lateral membranes (steps 1 to 3 inclusive); and those used to separate this mixture (by free-flow electrophoresis) into its apical and basal-lateral components (step 4). The whole bladder, in the form of a closed sac (serosal surface out), is first filled with and immersed in EDTA-containing Ringer solutions for one hour--a procedure that releases intact epithelial cells (free of the underlying sub-mucosal connective tissue) into the mucosal fluid (step 1). The cells are then homogenized in a Dounce homogenizer at 0°C (step 2) and carried through the previously described (Solinger et al, 1968) ultracentrifugal steps to obtain a mixture of microsomal membranes (step 3). This membrane suspension (7.5 to 15mg of membrane proteins in 5ml) was then injected into the entry porthole (step 4) of the free-flow electrophoresis apparatus (FF-V), the effluent volume from which drained into 100 tubes (3ml each). Most if not all of the membrane-protein was found in tubes 20 to 60 inclusive.

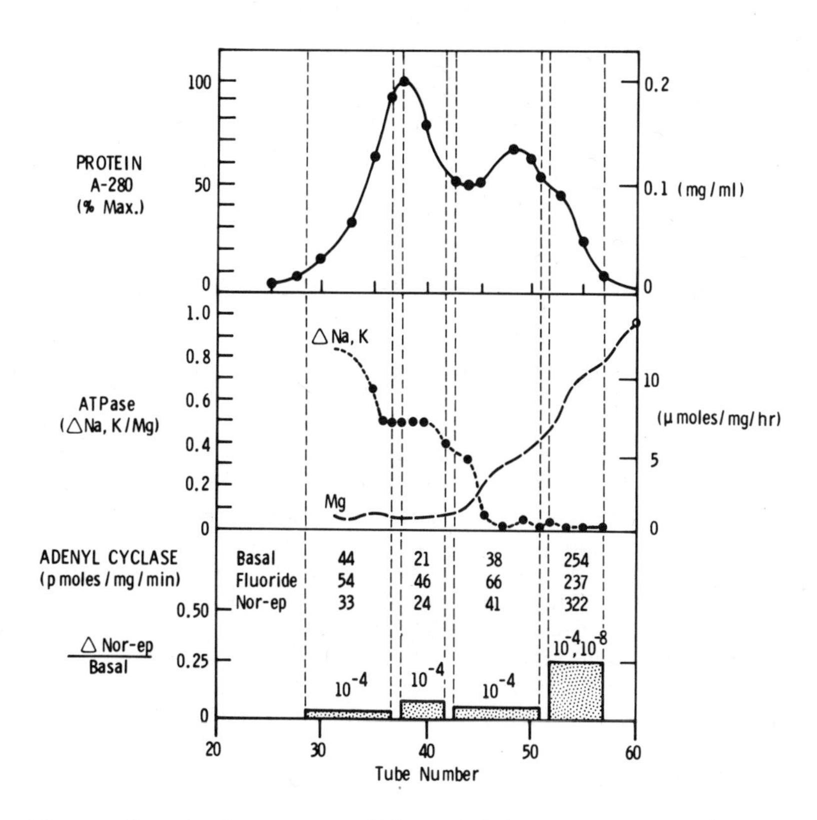

Figure 7. Analyses on effluent (the tube numbers) from free-flow electrophoresis. Concentration of protein (upper panel); ATPase activity per mg protein (Mg) and fractional increase in Mg·ATPase due to addition of Na + K (middle panel). Adenylate cyclase activity (basal; with F^-; and with nor-epinephrine) per mg protein--indicated by numbers. Fractional increase in adenylate cyclase due to nor-epinephrine--indicated by columns (lower panel).

Figure 7 shows the electrophoretic distribution of the membrane proteins, the Mg·ATPase and the (Na + K)-increment of Mg·ATPase, and the nor-epinephrine-sensitive adenylate cyclase.

These patterns indicate that membrane fragments containing ouabain-sensitive (Na + K)·ATPase migrated more rapidly toward the electropositive pole (appearing mainly in tubes 30 to 42), than did those containing a nor-epinephrine-sensitive adenylate cyclase (which appeared mainly in a pooled fraction of tubes 52 to 58). It should be noted that 10^{-8}M nor-epinephrine induces as much stimulation of this adenylate cyclase as did 10^{-4}M nor-epinephrine. Thus the nor-epinephrine sensitivity of the adenylate cyclase was 100 to 1,000 times greater than that of the anion transport (see figure 5A). It is also of interest to note that histidine, 10^{-4}M, failed to induce any detectable change in the adenylate cyclase activity of these fractions--which did not correspond to what might have been expected from the transport changes noted in table 1.

Since the nor-epinephrine-induced stimulation of anion transport (shown in figure 5A) could have been mediated by an adenylate cyclase mechanism in the apical membrane (as suggested by the membrane distribution shown in figure 7), a cAMP-sensitive protein kinase activity was looked for and found (figure 8).

Figure 8 shows the electrophoretic distribution of four pooled membrane fractions containing (Na + K)·ATPase, cAMP-sensitive protein kinase, and cGMP-sensitive protein kinase (the latter assay having been conducted as part of a determination on the cyclic-nucleotide preference of the protein kinase). These patterns indicate that the pool of membranes containing (Na + K)·ATPase migrated more rapidly toward the electropositive pole (tubes 25 to 31 inclusive) than did those containing the cyclic-nucleotide-sensitive protein kinase activity (tubes 45 to 55 inclusive). It should be noted that the phosphoryl receptor of this kinase resides on the membranes per se, i.e., the kinase system is endogenous--as has been recently shown for other protein kinase systems (Schwartz et al, 1974; Dowd & Schwartz, 1975).

The next set of experiments was part of an effort to localize the membrane binding sites of the disulfonic stilbenes (SITS or DIDS), which retard anion transport inhibitors only when added to the serosal fluid (see figure 5B and reference by Ehrenspeck & Brodsky, 1976). The procedure involved the use of the tritiated stilbene, ^{3}H·DIDS, of high specific activity. In two experiments, the serosal surfaces of twenty bladders were each exposed to 2ml. of Na Ringer solution containing ^{3}H·DIDS (5×10^{7} cpm in 10^{-5}M DIDS) for 10 minutes after which the radioactive stilbene was washed away with an albumin-containing Ringer solution. The twenty bladders were then processed for the free-flow electrophoresis of the membrane fraction in the manner depicted in figure 6.

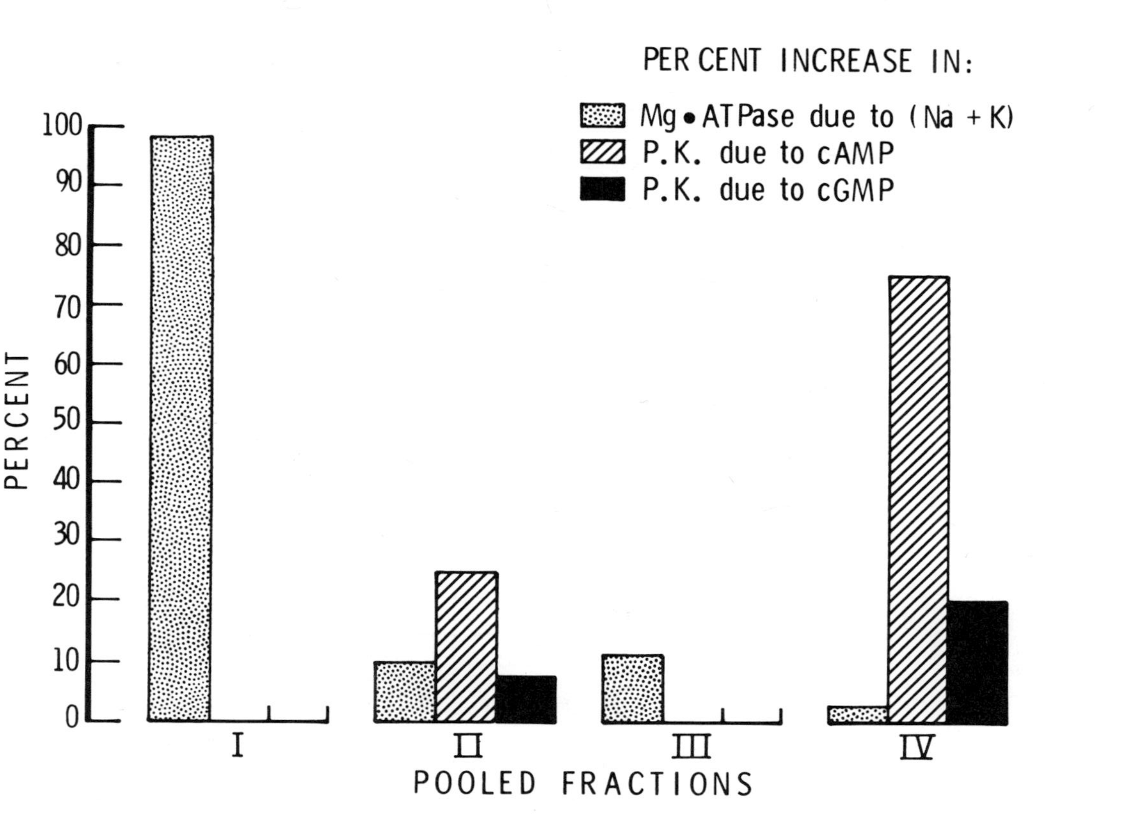

Figure 8. The (Na + K)-induced increment of Mg·ATPase and the cAMP and cGMP-induced increments of protein kinase activity (PK) in pooled effluents (I to IV) from the free-flow electrophoresis procedure. Fraction I, tubes 25-35; Fraction II, tubes 36-43; Fraction III, tubes 44-52; Fraction IV, tubes 53-58.

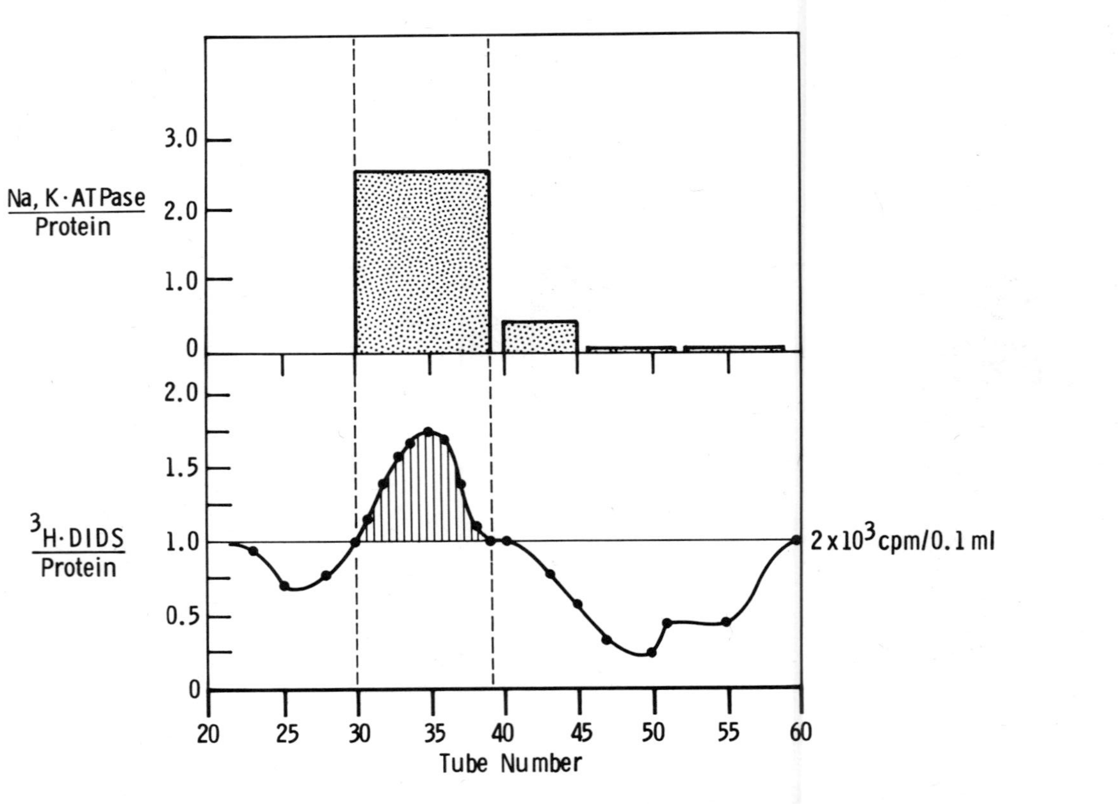

Figure 9. Co-migration of membrane fragments containing (Na + K)·ATPase and ^{3}H·DIDS-binding protein(s) in tubes 30 to 40 of the free-flow electrophoretic effluent system. Each value is a ratio of percentage of maximal ATPase or ^{3}H binding to the percentage of maximal protein concentration. When this ratio exceeds unity, there is "enrichment" of the membrane with enzyme or stilbene binding protein.

Figure 9 shows that the electrophoretic migration of membranes that are preferentially labelled by ^{3}H·DIDS was the same as those containing the ouabain-sensitive, (Na + K)·ATPase. This suggests that the high-affinity binding sites for DIDS are located on the same membrane (presumably the basal-lateral) as is (Na + K)·ATPase. However, the high-affinity stilbene-binding sites on the basal-lateral membrane are discrete and spatially separate from the (Na + K)·ATPase sites on the same membrane - as shown by the mutually exclusive nature of the transport changes evoked by the serosal application of the stilbene and ouabain. For example, the stilbene inhibits anion transport without changing Na transport while ouabain inhibits Na transport without changing anion transport (see table 1 and figures 4 and 5).

Not shown are similar data obtained in two experiments after labelling of the mucosal surface with ^{3}H·DIDS. In these cases, we failed to find any preferential (high-affinity) ^{3}H-labelling in any of the membrane fractions.

Figure 10, based on the data shown here, is a schematic diagram, depicting our current views on the ion-selective pathways in the apical and basal-lateral membranes of the turtle bladder epithelial cell.

DISCUSSION

Figure 10, based on the data shown here, is a schematic diagram, depicting our current views on the ion-selective pathways in the apical and basal-lateral membranes of the turtle bladder epithelial cell.

Passive elements. With respect to the paths through which the ions flow passively (downward arrows), there are no aqueous-filled conductance channels other than the long, narrow intercellular spaces. The nature of the path through the tight-junction is not yet known. The basis for invoking an electrically-conductive, passive carrier-mediated, amiloride-sensitive transport of Na across the apical membrane has been reported here and elsewhere (Wilczewski & Brodsky, 1975). But we have not yet localized any amiloride-sensitive membrane bound enzyme in the effluents from the free-flow electrophoretic of turtle bladder membranes and microsomes. Since the transport of Cl and HCO_3 are selectively blocked by the presence of acetazolamide (Gonzalez & Schilb, 1969; Gonzalez, 1969; Steinmetz, 1975) or disulfonic stilbenes (Ehrenspeck & Brodsky, 1976) in the serosal fluid, the presumed basal-lateral site of the preferential stilbene binding (shown in figure 9) suggests that the anion-selective paths in this membrane are discrete and probably involve a carrier operation. In this connection, Rothstein et al (1975) have found in erythrocyte membranes a stilbene-binding protein that increases the anionic (SO_4) permeability of synthetically-made vesicular membranes.

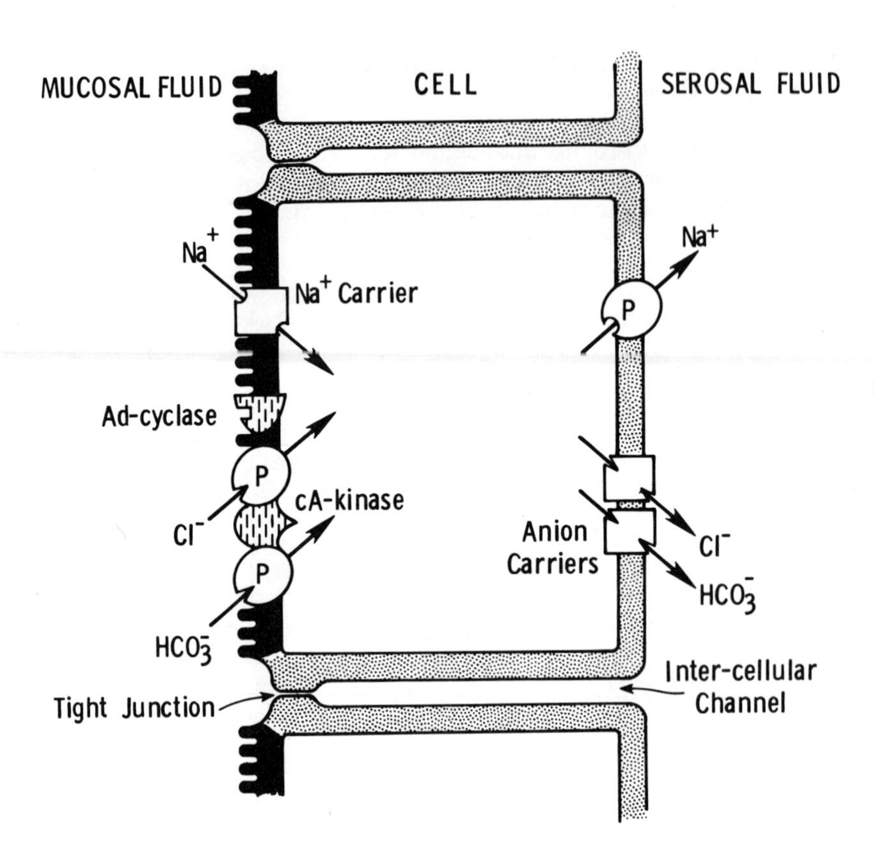

Figure 10. Ion-specific pumps and paths in membranes of turtle bladder epithelial cell. Upward-pointing arrows denote ion pumps and downward-pointing arrows, passive ion flows. The notch on the adenylate cyclase (facing the mucosal fluid) denotes its nor-epinephrine receptor site and the concavities on the cAMP-protein kinase (facing the cytoplasm) denotes the cAMP receptor sites. The ATP receptor sites of the kinase, not shown, also face toward the cytoplasm.

Active elements. Although the physiological and electrophoretic data shown and cited in this report suggest that adenylate cyclase and cAMP-activated protein kinase are regulators of an anion pump mechanism, the chemical or enzymatic nature of the anion pump itself remains unknown. The HCO_3-stimulated and SO_3-stimulated ATPases in electrophoretically separated fractions from turtle bladder membranes are not localized in any particular fraction. Evidently these enzymes may be present in both the apical and basal-lateral membranes and hence cannot as yet be correlated with a putative anion pump in the apical membrane.

Problems remaining include: (a) the isolation and/or identification of the anion pumps or the proton pump (Steinmetz, 1975) in the apical membrane; (b) the isolation and/or identification of the Na-selective carriers in the apical membrane; (c) the characterization of the preferential disulfonic stilbene-binding protein, presumably the anion carrier of the basal-lateral membrane (see figure 9); and (d) a completed analysis of the conductive and/or non-conductive nature of the transmembrane fluxes of the penetrant anions and cations.

ACKNOWLEDGMENT

This authors hereby express their gratitude to Ms. Cristina Matons and Susan Ehrenspeck for their technical work; to John Durham, Barry Cohen and Prof. Jos. G. Gennaro, Jr. of New York University for the histochemical and electron microscopic work; to Dr. Rolf Kinne of the Max Planck Institute, Frankfurt a Main for his guidance and help in the use of free-flow electrophoresis; and to Dr. Ioav Cabantchik, Life Sciences Foundation, Jersualem Israel, for the synthesis, preparation, and use of the tritiated disulfonic stilbenes.

This project was supported in part by the NIH AM-16928-03 and NSF PCM76-02344.

REFERENCES

Brodsky, W.A. and T.P. Schilb
Osmotic properties of the isolated turtle bladder
Am. J. Physiol., 208:46-57, 1965

Brodsky, W.A. and T.P. Schilb
Ionic mechanisms for sodium and chloride transport across turtle bladders
Am. J. Physiol., 210:997-1008, 1966

Brodsky, W.A., Schilb, T.P. and J.L. Parkes
Moderators of anion transport in the isolated turtle bladder
Symposium on Acidification; International Congress of Physiology (In Press), 1976

Dowd, F. and A. Schwartz
The presence of cyclic AMP-stimulated protein kinase substrates and evidence for endogenous protein kinase activity in various Na, K-ATPase preparations from brain, heart and kidney
J. Molec. & Cell. Cardiol., 7:483-497, 1975

Ehrenspeck, G. and W.A. Brodsky
Effects of 4-acetamido-4'-isothiocyano-2,2'-disulfonic stilbene on ion transport in turtle bladders
Biochim. Biophys. Acta, 419:555-558, 1976

Gonzalez, C.F.
Inhibitory effect of acetazolamide on the active chloride and bicarbonate transport mechanisms across short-circuited turtle bladders
Biochim. Biophys. Acta, 193:146-158, 1969

Gonzalez, C.F., and T.P. Schilb
Acetazolamide sensitive short-circuiting current versus mucosal bicarbonate concentration in turtle bladder
Biochim. Biophys. Acta. 193:419-429, 1969

Gonzalez, C.F., Shamoo, Y.E. and W.A. Brodsky
Electrical nature of active chloride transport across short-circuited turtle bladders
Am. J. Physiol., 212:641-650, 1967a

Gonzalez, C.F., Shamoo, Y.E., Wyssbrod, H.R., Solinger, R.E., and W.A. Brodsky
Electrical nature of sodium transport across the isolated turtle bladder
Am. J. Physiol., 213:333-340, 1967b

Hannig, K.
The application of free-flow electrophoresis to the separation of macromolecules and particles of biological importance.
Handbuch: Modern Separation Methods of Macromolecules and Particles,
Ed., Th. Gerritsen, John Wiley & Sons, Inc., Vol. 2, S. 45, 1969

Hannig, K.
Electrophoretic separation of Cells and particles by continuous free-flow electrophoresis
Handbuch: Techniques of Biochemical and Biophysical Morphology,
Ed., D. Glick and R. Rosenbaum, John Wiley & Sons, Inc. N.Y., 1972

Heidrich, H.G., Kinne, R., Kinne-Saffran, E. and R. Hannig
The polarity of the proximal tubule cell in rat kidney
J. Cell Biol., 54:232-245, 1972

Heinz, E. and P. Geck
The electrical potential difference as a driving force in Na-linked cotransport of organic solutes. Symposium: "Coupled Transport Phenomena in Cells and Tissues", Raven Press, Inc., N.Y., N.Y. (In Press), 1976

Hirschhorn, N., and H.S. Frazier
Intracellular electrical potential of the epithelium of turtle bladder
Am. J. Physiol., 220:1158-1161, 1971

Rothstein, A. Cabantchik, Z.I., Balshin, M. and Juliano, R.
Enhancement of anion permeability in lecithin vesicles by hydrophobic proteins extracted from red blood cell membranes.
Biochim. Biophys. Res. Commum. 64:144-150, 1975

Schilb, T.P. and W.A. Brodsky
Transient acceleration of transmural water flow by inhibition of sodium transport in turtle bladders
Am. J. Physiol., 219:590-596, 1970

Schwartz, I.L., Shlatz, L.J., Kinne-Saffran, E., and R. Kinne
Target cell polarity and membrane phosphorylation in relation to the mechanism of action of antidiuretic hormone
Proc. Nat. Acad. Sci., 71:2595-2599, 1974

Shamoo, Y.E., and W.A. Brodsky
The (Na + K)-dependent adenosine triphosphatase in the isolated mucosal cells of turtle bladder.
Biochim. Biophys. Acta., 203:111-123, 1970

Skou, J.C.
The influence of some cations on an adenosine triphosphatase from peripheral nerve
Biochim. Biophys. Acta, 23:394-401, 1957

Solinger, R.E., Gonzalez, C.F., Shamoo, Y.E., Wyssbrod, H.R. and W.A. Brodsky
Effect of ouabain on ion transport mechanisms in the isolated turtle bladder
Am. J. Physiol., 215:249-261, 1968

Steinmetz, P.R.
Cellular mechanisms of urinary acidification
Physiol, Rev., 54:890-956, 1975

Wilczewski, T. and W.A. Brodsky
Effect of ouabain and amiloride on Na pathways in turtle bladders
Am. J. Physiol., 228:781-790, 1975

DISCUSSION

ELDEFRAWI: Your finding on the presence of a general adenylate cyclase activity (stimulated by catecholamines) was an open invitation to look for the effect of neuro-transmitter such as acetylcholine. Have you looked for or considered an effect of acetylcholine?

BRODSKY: Schilb has found that 10^{-4}M acetyl-betamethylcholine added to the serosal fluid produces a 30 to 50 percent inhibition of the net sodium transport across the turtle bladder. This effect is reversible or blockable by adding equivalent quantities of atropine to the serosal fluid.

ELDEFRAWI: So, in effect then you do have the muscarinic kind of antagonism.

BRODSKY: Yes, but, I cannot yet say whether acetylcholine compounds react with adenylate cyclase or guanylate cyclase in the turtle bladder. However, other compounds known to activate adenylate cyclase (e.g., the histamine compounds and nor-epinephrine) stimulate anion transport in the intact bladder. Of these compounds, only nor-epinephrine (at concentration of 10^{-8}M) stimulates adenylate cyclase in the *in vitro* membranes derived from these bladders. The histamine group produces no effect on the *in vitro* adenylate cyclase. It remains possible that the histamines would stimulate the guanylate cyclase activity - which may be independent system.

SIEGEL: Is the transport from the mucosal surface in any way sensitive to cyclic nucleotides?

BRODSKY: We have not yet found an effect with c-AMP up to 1.0 mM. We have not tried the dibutyryl form, nor have we tried the cGMP[1]. Therefore, we can only infer that the nor-ep action on transport occurs *via* the cyclase kinase mechanism.

SIEGEL: Is the electrophoretic distribution of adenylate cyclase the same when you use other stimulating agents like fluoride? Is that the only adenylate cyclase in the preparation?

BRODSKY: No, the distribution of the adenylate cyclases differ from each other. "Basal" adenylate cyclase (that with no fluoride or nor-epinephrine) is in most of the electrophoretically separated membrane fragments. The membrane fragments containing a fluoride-stimulated cyclase activity move more rapidly toward the electropositive pole than do those containing a nor-epinephrine-stimulated cyclase activity.

In the slide shown, a fluoride-stimulated activity was not found in those membrane fragments containing maximal degrees of nor-epinephrine stimulation specifically in those fragments that migrate toward the electronegative pole.

SIEGEL: You mentioned that you have not found yet an inhibitor for anion transport that acts from the mucosal side. Would not ethacrynic be such an inhibitor?

BRODSKY: One of our group, Thaddeus Wilczewski, has recently shown that ethacrynic acid inhibits only sodium transport and only when it is added to the serosal side of the turtle bladder. Electrophysiologically, its effect looks identical to that of ouabain. It produces no detectable effect on the anion transport related portion of the short-circuiting current in the turtle bladder.

---

(1) cGMP - Since the data of this symposium, Ehrenspeck and Durham of our group have found that c-CMP does stimulate anion transport *in vivo* after it has been added to the mucosal fluid.

MEMBRANE WATER CHANNELS AND SH-GROUPS

R.I. Sha'afi and M.B. Feinstein

University of Connecticut Health Center

Farmington, Connecticut USA 06032

ABSTRACT

The transport of water across human red cell membranes is commonly interpreted in terms of small aqueous channels. This interpretation is based largely on indirect evidence. In this report, two sets of experiments providing more direct evidence for this idea is presented.

(1) The effect of various SH-reactive reagents on the movement of water was studied. Using these compounds we attempted to localize and characterize those membrane SH-groups which are important for water transport.

(2) Experimental evidence which suggests that these channels are assembled from aggregates of specific membrane protein(s) is presented.

REPORT

Classical interpretations of the mechanism of water transport across mammalian red cell membranes assume the existence of aqueous membrane channels (33,31). As the permeability coefficient to water measured under an osmotic pressure gradient is usually significantly higher than the corresponding value measured under diffusional flow ($g>1$), the human red cell membrane is thought to act both as a selective solvent and a molecular sieve. Its ability to function as a molecular sieve depends on the existence of the channels which could be assembled from aggregates of integral membrane proteins which span the membrane thickness. In this respect, a great deal of interest has been focused on the nature of these and other membrane proteins, their interactions with lipids and their functional importance in the regulation of transport across biolog-

TABLE I. Effect of PCMBS on the permeability coefficients of human red cell membrane to water

Condition	Permeability Coefficients (cm/sec) x 10^3		Ratio
	Under Osmotic Flow, P_f	Under diffusional, P_d	P_f/P_d
Red Cells	17.30 (27)	5.30 (26)	3.3
+1 mM PCMBS	1.44 (21)	1.35 (21)	1.1
Ghosts	17.9 (8)	----	---
+1 mM PCMBS	<<17.9 (8)	----	---

ical membranes. Recently, Macey and Farmer (20) and Naccache and Sha'afi (25) have shown that the organic mercurial p-chloromercuriphenyl sulfonic acid, PCMBS, and certain other sulfhydryl-reactive reagents inhibit significantly the transport of water across human red cell as well as other biological membranes. The results dealing with human red cell membranes are summarized in Table I. The inhibitory effect of PCMBS develops slowly reflecting the time which is needed for the compound to diffuse to and react with membrane sulfhydryl groups that are important for water movement. This inhibition can be reversed completely by the addition of an excess amount of cysteine or glutathione (Figure 1). If one accepts the hypothesis that water molecules cross the red blood cell membrane via aqueous channels, then the effect of these sulfhydryl-reactive reagents, including PCMBS, is to shut off these channels.

Effect of various sulfhydryl-reactive reagents on water flux across human red cell membranes

In order to get a better understanding of the molecular mechanism of water movement in these cells and the inhibitory action of PCMBS, we have studied the effect of various sulfhydryl-reactive compounds on the transport of water in these cells. The results are summarized in Table II. As shown in the table, these mercury containing compounds are very potent inhibitors of water movement, and the inhibitory potency of the compound can be enhanced by the addition of an electron withdrawing group on the molecule. This is best illustrated when comparing the compounds p-aminophenylmercuric acetate and phenylmercuric acetate. The addition of the amino group, an electron withdrawing group, to the latter compound increases significantly its inhibitory potency. Masking the SH-sensitive part

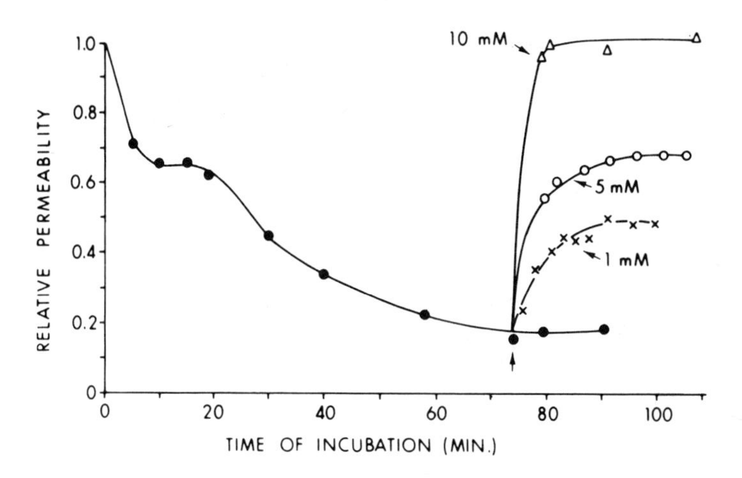

Figure 1. The time course of PCMBS, 1 mM, effect on water transport across human red cell membrane and the reversal of this inhibition by cysteine. The arrow indicates the time at which cysteine was added (the results are taken from Naccache and Sha'afi (25)).

of the compound by the addition of an ethyl group as in ethylmercurithiosalicylic acid abolishes the effectiveness of the compound. In spite of the effectiveness of all other mercury compounds tested, mersalyl is inactive as an inhibitor or water movement. The striking chemical difference between the active mercurials and mersalyl is the much greater distance of the Hg atom from the aromatic ring in the latter case. This suggests that the active SH-groups lie in very close proximity to a hydrophobic protein region, possibly containing an aromatic amino acid.

It is also apparent from the results given in the table that membrane sulfhydryl groups which are both accessible to and reactive with NEM and iodoacetamide, IAM, are not directly involved in the control of water transfer. This is based on their inability to inhibit water transport or to interfere with the inhibitory action of the active compounds. It is known that a large component (>70%) of membrane SH-groups are accessible to and reactive with these two compounds suggesting that only a small fraction of membrane-bound SH-groups are involved in the control of water transfer. One possible explanation for this which is consistent with the data derived from the studies with mercurials, is that those sulfhydryl groups which are involved in water transport are located in a hydrophobic environment. Consistent with this it is frequently found that polar organomercurials or more hydrophobic maleimides are much more effective than hydrophilic reagents such as NEM and IAM. In a study of yeast alcohol dehydrogenase, sulfhydryl reactivity was shown to

TABLE II Effect of various sulfhydryl-reactive reagents on the permeability of human red cell membrane to water

Compound	Structure	Relative permeability
Control	—	1.00
p-chloromercuriphenyl sulfonate	$ClHg-C_6H_4-SO_3^-$	0.20
p-chloromercuribenzoate	$ClHg-C_6H_4-COO^-$	0.23
Phenylmercuric acetate	$C_6H_5-Hg-OOC-CH_3$	0.43
Phenylmercuric chloride	C_6H_5-HgCl	0.39
p-Aminophenylmercuric acetate	$H_2N-C_6H_4-Hg-OOC-CH_3$	0.27
Fluoresceinmercuric acetate	HO, =O, $(CH_3COO)Hg$, $Hg(OOC-CH_3)$, COOH	0.20
Ethylmercurithiosalicyclic acid	OOC, $SHgC_2H_5$	1.01
Mersalyl	OCH_2COOH; $C(=O)-NH-CH_2-CH(OCH_3)-CH_2-HgOH$	0.85
N-ethylmaleimide	=O, NC_2H_5, =O	1.15
Iodoacetamide	$NH_2-C(=O)-CH_2-I$	1.10

parallel increasing chain lengths of a series of N-alkylmaleimides (13). N-butylmaleimide, for example, was more reactive than N-ethylmaleimide. Similar results were obtained with porcine heart fumarase (28). Likewise, Fernandez-Diez et al. (11) have shown that the sulfhydryl-group reactivity of β-lactoglobulin in its native state is 1.9 sulfhydryl per mole protein to PCMB, 1.0 to DTNB and 0 to NEM. With this in mind, we have studied the effect of increasing chain lengths of a series of N-alkylmaleimides (N-methylmaleimide, N-ethylmaleimide, N-butylmaleimide, N-cyclohexylmaleimide, N-phenylmaleimide, and N-benzylmaleimide) on the transport of water across human red cell membranes. None of these compounds had any significant effect on the movement of water in these cells. These results suggest that a more likely explanation for the inability of NEM and the more hydrophobic maleimides to inhibit water movement is that the SH-groups which are involved in the control of water movement are accessible, but not reactive with these agents.

Reactivity of membrane SH-groups important for the control of water flux

Based on the preceding discussion, it appears that the membrane SH-groups which control the water movement are not very reactive. In order to further test the reactivity of these groups, we have studied the effect of various disulfide compounds on the transport of water in human red cell membranes. The results are summarized in Table III. All these compounds are known to react with SH-groups in simple compounds such as cysteine or in proteins and to form mixed disulfides with the SH compound. Each disulfide was tested over a wide range of pH values (from 5.8 - 8.6 in steps of 0.3 pH units). The time of incubation with the test compound was also varied. As is evident from the table, none of the aliphatic disulfides and only two aromatic disulfide compounds, 2,2'-Dithiobis-(5-nitropyridine) and 3-nitrophenyldisulfide, have any inhibitory effect on the movement of water. This suggests that membrane sulfhydryl groups which are involved in the control of water transport in red cell membranes are much less reactive than those of small SH-containing molecules, such as cysteine, and various water soluble proteins (4). In addition, the presence of a NO_2 group in the phenyl ring increases the potency of the disulfide reagent as a water transport inhibitor. This may be due to its electron withdrawing power or its ability to engage in the formation of hydrogen bonds.

Membrane proteins related to water transport in human erythrocytes

It is generally agreed that among the various membrane proteins present in red cell membranes, there are two major species which seem to span the entire thickness of the membrane (3,32). These

TABLE III Effect of various disulfide compounds on the transport of water in human red cell membrane

Compound	Structure	% inhibition
2,2' Dithiodiglycolic acid	$S-CH_2-COOH$ / $S-CH_2-COOH$	<10
3,3' Dithiodipropionic acid	$S-CH_2-CH_2-COOH$ / $S-CH_2-CH_2-COOH$	<10
3,3' Carboxypropyl-disulfide	$S-CH_2-CH_2-CH_2-COOH$ / $S-CH_2-CH_2-CH_2-COOH$	<10
2,2' Dithiobis-(5-nitropyridine)	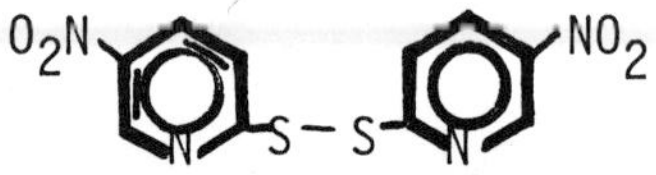 O_2N N S – S N NO_2	60
4,4' Dithiodipyridine	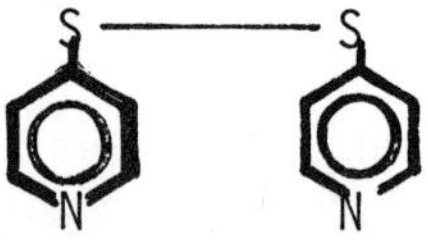 S — S N N	0
2,2' Dipyridyl-disulfide	 S — S N N	0
6,6' Dithiodinicotinic acid	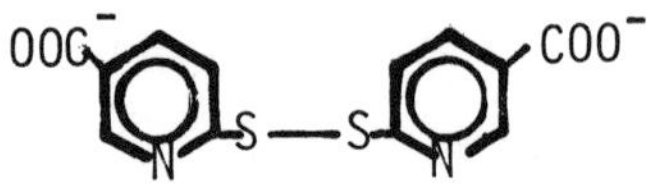 OOC^- N S — S N COO^-	0
3,3' nitrophenyl-disulfide	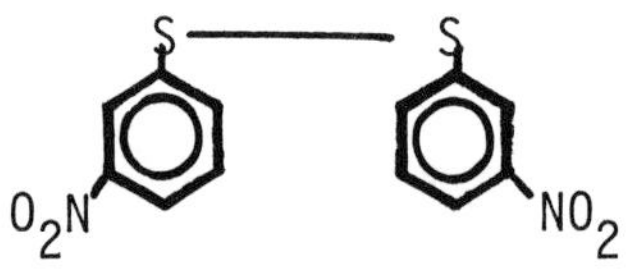 S — S O_2N NO_2	30

two integral proteins are known as band 3 (molecular weight ~95,000 daltons) and PAS-1 (glycophorin) (10,23). It appears that the integral protein(s) identified by gel electrophoresis as band 3 are involved in the control of the movements into and out of the cell of many substances including Na^+, K^+, glucose, Cl^-, water and maybe many others (17,6,29,30,5). In the remaining portion of this article, we will discuss some experimental data which suggest that one of these two major proteins or both may be involved in the formation of hydrophilic pathways for water transport. In addition, we will discuss whether band 3 protein is heterogenous with several species of polypeptide chains of closely related molecular weight involved in separate specific functions, or whether band 3 polypeptide chains are homogenous, but arranged within the membrane in a configuration that would allow for the assembly of more than one type of transmembrane channels.

There are four pieces of experimental data which are consistent with the idea that one or both of these integral proteins are related to water transport across human red cell membranes. First, it has been suggested that these proteins correspond to the intramembranous particles seen in freeze-etch images and freeze-fracture experiments (2). DeSilva has indicated that these membrane-interrelated particles could provide a structural basis for the postulated hydrophilic pathways in human red cell membranes (9). Second, incorporation of glycophorin prepared by trypsin hydrolysis of human erythrocytes into black lipid membranes increases significantly the permeability of these membranes to water (19). However, this effect could be nonspecific. It must be pointed out from the start that if the "aqueous channels" for water transport are made exclusively from glycophorin then it would be difficult to understand why certain sulfhydryl-reactive reagents are potent inhibitors of water movement. Purified glycophorin from human red cell membranes has been prepared and sequenced (24). It has no cysteine and, therefore, it is hard to imagine how sulfhydryl-reactive reagents will interact directly with this protein to inhibit water transport. Third, using polyacrylamide gel electrophoresis, we found that band 3 can be selectively labelled by water transport inhibitors. In these experiments, red blood cells were washed twice with isotonic sodium phosphate buffered solution (pH = 7.0). The cell suspension was then divided into two equal portions. To one portion, 1 mM each of IAM, NEM and mersalyl were added, and the two suspensions were incubated for 15 minutes at room temperature. At the end of this period, a known amount of cold and C^{14}-PCMB or C^{14}-DTNB was added to each of the two suspensions (final concentration of PCMB or DTNB = 1 mM). After 30 minutes of incubation at room temperature, IAM, NEM and mersalyl were added at 1 mM each to the suspension which did not contain these compounds. The cells were then all brought to 0°C in an ice bath and lysed by addition of 1.0 ml portions of cells into 40 ml of ice-cold 5 mM sodium phosphate pH 8.0 containing 1 mM each of IAM, NEM and mersalyl. The suspensions were centrifuged at 20,000 g for 15 min., and the supernatants were removed by aspiration. The translucent ghost pellets were removed from the underlying opaque "buttons" to minimize contamination with proteases (10). The ghosts were washed two more additional times. It was essential to add IAM, NEM and mersalyl prior to, and during lysis to block as many SH groups not related to water transport as possible and thereby minimize non-specific labelling, and to prevent removal of protein-bound radioactivity by thiol containing compounds released during cell lysis.

For ordinary SDS-gel electrophoresis of red cell membranes, the membrane ghost suspensions were dissolved in 1% SDS, 40 mM DTT, and heated for several minutes in a boiling water bath. The solutions were made 5% with glucose and either pyronin Y or bromphenol blue were added as tracking dyes. However, in those experiments in which membranes were labelled with either C^{14}-PCMB or C^{14}-DTNB, thiol reducing agents could not be employed since such treatment cleaves the mixed disulfides produced by reaction of proteins with DTNB and also removes PCMB from proteins. However, we found that membrane proteins dissolved by heating in 1% SDS gave essentially the same pattern of bands as in the reducing system when freshly run on SDS-polyacrylamide gels.

Polyacrylamide gel electrophoresis of erythrocyte membrane proteins was carried out in (a) 5.6% acrylamide, 0.21% bis acrylamide and 1% SDS disc gels according to the method of Fairbanks et al. (10); (b) slab gels with a polyacrylamide linear gradient from 4% to 10% and 1% SDS; or (c) the SDS-disc gel system of Maizel (22). The latter is a discontinuous gel buffer system containing Tris-HCl pH 8.9 in the resolving gel, Tris glycine as the electrode buffer, and Tris-HCl pH 6.7 in the 3% acrylamide spacer gel. Density gradient gel slabs or disc gels containing red cell membrane proteins labelled with C^{14}-DTNB or C^{14}-PCMB were cut into 1-3 mm wide slices, which were digested in scintillation vials containing 1.0 ml NCS-H_2O (9:1) at 50°C overnight. After cooling and addition of 10 ml Aquasol per vial the radioactivity of each gel slice was measured in a Packard Tri-Carb liquid scintillation counter. Adjacent sections of the gel slabs or separate disc gels run at the same time were stained with Coomassie blue R-250 (10). In another series of experiments, the exact location of bands to be cut out for measurement of radioactivity was accomplished by prior dansylation of the dissolved proteins and visualization under U.V. light (36).

The extent and specificity of labelling of red cell membrane proteins by C^{14}-PCMB was strongly dependent upon whether intact cells or unsealed ghosts were used, and whether IAM, NEM and mersalyl were present. In the absence of other SH-blocking agents, essentially all the principle protein bands were rendered radioactive by incubation of unsealed ghosts with C^{14}-PCMB or DTNB. However, if the ghosts were pretreated with NEM, IAM, and mersalyl, band 3 remained strongly labelled, but that of the other protein bands was greatly diminished. On the other hand, in intact red cells reacted with PCMB alone heavy labelling of bands 3 and significant amounts of C^{14}-PCMB labelling were found in the other bands (Figure 2). However, by pre-incubation of intact washed erythrocytes with IAM, NEM, and mersalyl, subsequent labelling by C^{14}-PCMB was restricted

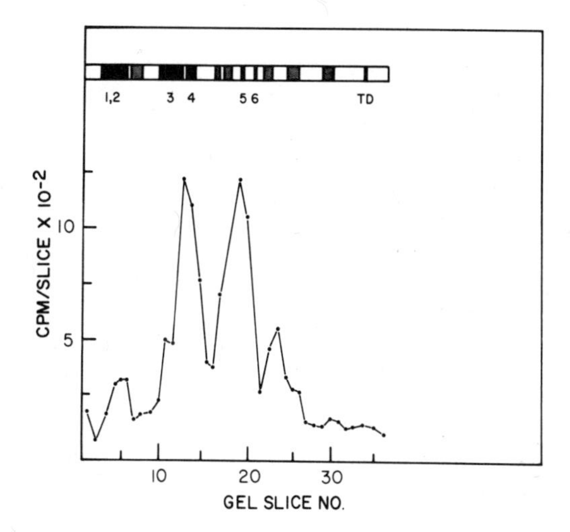

Figure 2. Labelling profiles of erythrocyte membrane proteins isolated from C^{14}-PCMB-labelled intact red cells. Ghosts isolated from red cells were analyzed for protein and radioactivity distribution by polyacrylamide SDS gel electrophoresis. The gels contained 5.6% acrylamide, 0.21% bis acrylamide and 1% SDS disc gel according to the method of Fairbanks et al. (10).

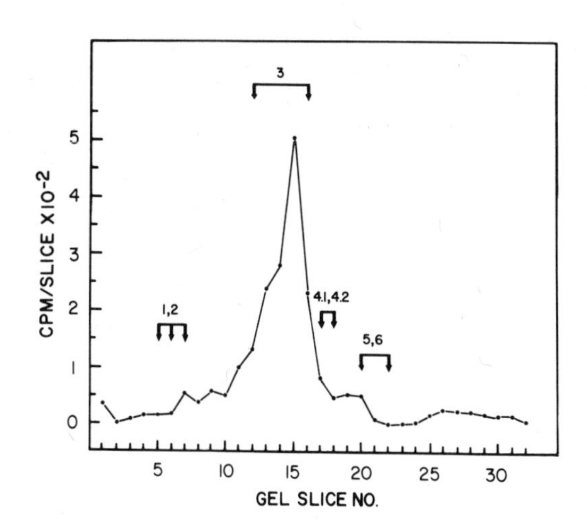

Figure 3. Labelling profiles of erythrocyte membrane proteins isolated from C^{14}-PCMB-labelled intact red cells. The same procedure as in Figure 2 was followed except that intact red cells were preincubated with 1 mM each of IAM, NEM and mersalyl for 15 minutes before the addition of C^{14}-PCMB.

to band 3 (Figure 3). As mentioned above it is important to recall that these three compounds IAM, NEM and mersalyl, did not inhibit water transport across human red cell membrane nor did they interfere with the inhibitory action of PCMB. Thus, under conditions in which water transport is inhibited, PCMB was found bound only to band 3 protein. The addition of thiol reagents, which reverses the inhibition of water transport, entirely removed labelled PCMB from band 3. Similar results were previously obtained using DTNB (5) and were reconfirmed under the conditions of our latest experiments.

Fourth, in a preliminary study we have found that water transport in liposomes prepared from egg lecithin and containing incorporated band 3 can be inhibited by PCMBS. For these experiments, solutions enriched in band 3 protein were obtained by sequential extraction at 0°C of ghosts with 5 mM sodium phosphate, pH 8.0, 51 mM sodium phosphate, pH 8.0, 36 mM sodium phosphate, pH 7.5, and finally with 0.5% Triton X-100 in 36 mM sodium phosphate, pH 7.5 (39). The original procedure of Yu and Steck (39) was modified to carry out the extraction with 51 mM phosphate, pH 8.0 overnight to enhance removal of proteins other than band 3. Variable, but usually very small, amounts of bands 1,2,4.2 and 6 were found in the enriched band 3 preparations. The best extractions yielded Triton X-100 solutions in which band 3 accounted for more than 90% of the Coomassie blue-stainable protein (Figures 4 and 5). It should be borne in mind however that this procedure also preferentially extracts the other membrane glycoproteins which stain with PAS (29). Before incorporation into liposomes, Triton X-100 was substantially removed from the enriched band 3 protein solutions by adsorption to Bio-Beads SM-2 (Bio-Rad Laboratory, 32nd and Griffin Avenue, Richmond, California 94804), a neutral porous styrene-divinylbenzene copolymer, according to the batch procedure of Holloway (15).

In conclusion, although other interpretations can be found for these results, the simplest and most straightforward explanation is to postulate that water molecules cross human red cell membrane via aqueous channels which are assembled from aggregates of the integral membrane protein(s) known as band 3.

Is band 3 protein heterogenous?

If one assumes that band 3 protein is involved in the regulation of both monovalent anion and water movement across the human red cell membrane and that based on inhibitor studies these movements appear to be independent of each other (5,29) then this would lead to the following intriquing question. Is band 3 protein *heterogenous* with several species of closely related molecular weight polypeptide chains involved in separate specific functions, or are the band 3 polypeptide chains *homogenous*, but arranged within the

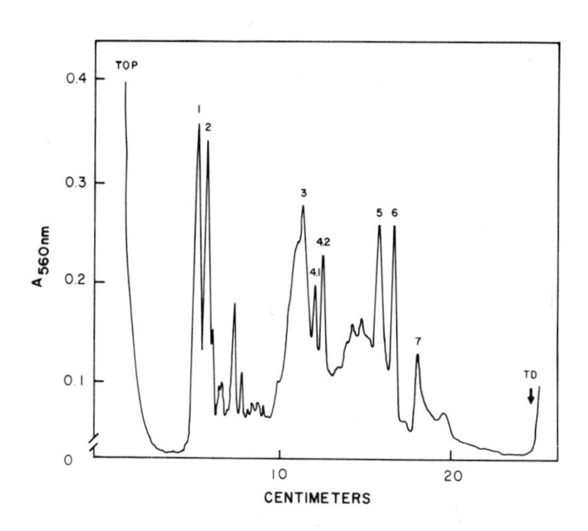

Figure 4. Erythrocyte membrane polypeptides. Densitometry of Coomassie blue stained gel. Gels were scanned in a Gilford spectrophotometer with a model 2410 linear transport accessory at 560 nm.

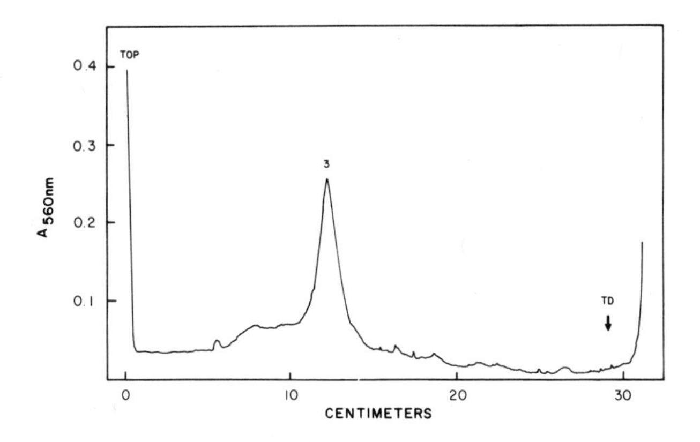

Figure 5. Densitometry of Coomassie blue stained gel of Triton X-100 extracts of human red cell ghost.

membrane in a configuration that would allow for the assembly of more than one type of transmembrane channel? Heterogeneity has been suggested by the unusual width and skewness of the band 3 staining pattern in SDS-polyacrylamide gels. Furthermore, as many as three separate bands have been observed in the residual band 3 region after partial pronase digestion (1,7). On the other hand, the reported existence of several NH_2-terminal amino acids (18) has been strongly disputed (14,37), and it is believed that the amino termini are blocked. It has been suggested that the apparent heterogeneity of band 3 may result from differences in the carbohydrate moiety of this glycoprotein (12).

Recent experiments concerning the configuration of band 3 protein *in situ* based on identification of the polypeptide fragments arising from the action of several proteases gave no evidence which would conclusively prove heterogeneity of the polypeptide chains (16,35). Furthermore, Jenkins and Tanner have made an interesting reconstruction of the protein configuration *in situ* as an S-shaped structure which transverses the membrane twice (16). This complex polypeptide structure, especially if it exists additionally in the form of a dimer (34) or tetramer (38) as has been suggested as the native state of the protein within the membrane, would allow for the possibility of formation of several transmembrane channels of differing function.

Acknowledgement

We would like to express our thanks to Mr. M. Volpi for his excellent assistance. This work was supported by NIH grant "GM-20268-02.

REFERENCES

1 BENDER, W.W., GARAN, H., and BERG, H.C.: J. Mol. Biol. 58 (1971) 783.

2 BRANTON, D.: Phil. Trans. R. Soc. London, B 261 (1971) 133.

3 BRETSCHER, M.S.: Science 181 (1973) 622.

4 BROCKLEHURST, K. and LITTLE, G.: Biochem. J. 133 (1973) 67.

5 BROWN, A.P., FEINSTEIN, M.B. and SHA'AFI, R.I.: Nature 254 (1975) 523.

6 CABANTCHIK, Z.I. and ROTHSTEIN, A.: J. Membr. Biol. 15 (1974) 207.

7 CABANTCHIK, Z.I. and ROTHSTEIN, A.: J. Membr. Biol. 15 (1974) 227.

8 COLOMBE, B., WINOCUR and MACEY, R.I.: Biochim. Biophys. Acta 263 (1974) 226.

9 DeSILVA, PEDRO, PINTO: Proc. Nat. Acad. Sci. U.S.A. 70 (1973) 1339.

10 FAIRBANKS, G., STECK, T.L. and WALLACH, D.F.H.: Biochemistry 10 (1971) 2617.

11 FERNANDEZ-DIEZ, M.J., OSUGA, D.T. and FEENEY, R.E.: Arch. Biochem. Biophys. 107 (1964) 449.

12 FINDLAY, J.B.C.: J. Biol. Chem. 249 (1974) 4398.

13 HEITZ, J.R., ANDERSON, C.D. and ANDERSON, B.M.: Arch. Biochem. Biophys. 127 (1968) 627.

14 HO, M.K. and GUIDOTTI, G.: J. Biol. Chem. 250 (1975) 675.

15 HOLLOWAY, P.J.: Anal. Bioch. 53 (1973) 304.

16 JENKINS, R.E. and TANNER, M.J.A.: Biochem. J. 147 (1975) 393.

17 KNAUF, P.A., PROVERBIO, F. and HOFFMAN, J.F.: J. Gen. Physiol. 63 (1974) 305.

18 KNUFERMANN, H., BHAKDI, R., SCHMIDT-ULLRICH and WALLACH, D.F.H.: Biochim. Biophys. Acta 330 (1973) 356.

19 LEA, E.J.A., RICH, G.T. and SEGREST, J.P.: Biochim. Biophys. Acta 382 (1975) 41.

20 MACEY, R.I. and FARMER, R.E.L.: Biochim. Biophys. Acta 211 (1970) 104.

21 MACEY, R.I., KARAN, D.M. and FARMER, R.E.L.: In Passive Permeability of Cell Membranes (Kreuzer, F. and Slegers, T.F.G., Eds.). Plenum Press, New York, London (1972).

22 MAIZEL, J.V.: In Methods in Virology (Maramorosch, K. and Koprowski, H. Eds.). Volume V, p. 180. Academic Press (1971).

23 MARCHESI, V.T.: In Cell Membranes (Weissmann, G. Ed.) Chap. 5, HP Publishing Co., Inc. (1975).

24 MARCHESI, V.T.: In Cell Membrane, Hospital Practice (Weissmann, G. Ed.) Chap. 5, HP Publishing Co., Inc. (1975).

25 NACCACHE, P. and SHA'AFI, R.I.: J. Cell Physiol. 83 (1974) 449.

26 PAGANELLI, C.V. and SOLOMON, A.K.: J. Gen. Physiol. 41 (1957) 259.

27 RICH, G.T., SHA'AFI, R.I., ROMUALDEZ, A. and SOLOMON, A.K.: J. Gen. Physiol. 52 (1968) 941.

28 ROBINSON, G.W., BRADSHAW, R.A., KANAREK, L. and HILL, R.L.: J. Biol. Chem. 242 (1967) 2709.

29 ROTHSTEIN, A., CABANTCHIK, Z.I., BALSHIN, M. and JULIANO, R.: Biochem. Biophys. Res. Commun. 64 (1975) 144.

30 ROTHSTEIN, A., CABANTCHIK, Z.I. and KNAUF, P.: Fed. Proc. 35 (1976) 3.

31 SHA'AFI, R.I. and GARY-BOBO, C.M.: Progr. in Biophys. and Mol. Biol. 26 (1973) 103.

32 SINGER, S.J.: Ann. Rev. Bioch. 43 (1974) 805.

33 SOLOMON, A.K.: J. Gen. Physiol. 51 (1968) 335S.

34 STECK, T.L.: J. Mol. Biol. 66 (1972) 295.

35 STECK, T.L., RAMOS, B. and STRAPAZON, E.: Biochemistry 15 (1976) 1154.

36 TALBOT, D.M. and YPHANTIS, D.A.: Analy. Biochem. 44 (1971) 246.

37 TANNER, M.J. and BOXER, D.H.: Biochem. J. 129 (1972) 333.

38 WANG, K. and RICHARDS, F.M.: J. Biol. Chem. 249 (1974) 8005.

39 YU, J. and STECK, T.L.: J. Biol. Chem. 250 (1975) 9170.

DISCUSSION

SZABO: I noticed that you made some prior experiments with glycophorin, in order to increase water permeability. Was there also an increased ionic permeability?

SHA'AFI: I haven't done the experiment with a glycophorin; a friend of mine did it.

SZABO: Okay, what was his conclusion?

SHA'AFI: I believe there was an increase also in the potassium flux.

GENNARO: If you label the surface of the cell with PCMB's, do you end with a label in band III, 55 K or 38 K? Have you thought about the possibility of using an antibody to 38 K or 55 K to plug the hole?

SHA'AFI: This is intended in the second set of experiments.

LAKOWICZ: Professor Lovrien at the University of Minnesota has shown that the red blood cell will bind approximately 20 million phenylalcohol type molecules (Rex Lovrien, William Tisel and Paul Pesheck, Stoichiometry of Compounds Bound to Human Erythrocytes in Relation to Morphology. J. Biol. Chem. 250: 3136-3141, 1975), there is a phenyl attached to three carbon chain and these protect the red blood cell from hemolysis at low levels in this range up to around 20 million, at which point the cell morphology changes. Is there very much unreacted PCMB in the membrane that might be causing a nonspecific effect of the lipid bilayer which you later removed by reaction with a sulfhydral reagent which increased its water soluability?

SHA'AFI: I am sure there are some nonspecific bindings of PCMBS.

LAKOWICZ: Is it not important to separate the two effects, that is the reaction with proteins or perhaps simple partitioning or adsorptions to the membrane surface.

SHA'AFI: It is important, but I just don't know how.

BLUMENTHAL: Isn't the change of the ratio between osmotic water flow to tritiated labeled water flow upon addition of PCMB an indication that the PCMB acts directly on the channels rather than on the lipid structure as you suggested.

SHA'AFI: This is the simplest and most straighforward explanation. There are always other explanations. I find it very difficult to imagine that PCMB is interacting non-specifically and affecting water transport through osmosis much more than through diffusion if it is non-specific interaction. It is possible to imagine a mechanism like that. I am not saying it is impossible, but it is not very likely, I think.

KOZARICH: In the models that you have shown on the structure of the B and III protein to form your channel, it seems to suggest that there is a multimeric structure. Normally you show two "protein threes" to form a channel. Do you have any evidence that there are aggregations of protein three to form channels?

SHA'AFI: Well, the general understanding is that band III exists as a dimer. The latest paper shows that band III appears as an S-shaped structure (Jenkin, R. E. and Janner, M.J.A., Bioch. J. 147: 393, 1975).

KOZARICH: I wonder if the variability in the inhibition that you find with the lyposomes can be due to only partial aggregation?

SHA'AFI: That is one possibility. The other possibility is that the protein is not incorporated properly.

FEINSTEIN: A previous question asked whether the 15 K and the 38 K were accessible from the outside. According to Steck et al. (Steck T.L., B. Ramos and E. Strapason, Biochem. 15: 1154, 1976) both fragments are accessible from the outside of the membrane to agents that are restricted to the extra-cellular space.

WEINER: Have you tried liposome experiments with different lipid fractions?

SHA'AFI: No. We were very happy to find it in one case. You have to experiment with whether protein increases water transport and whether PCMBS inhibits this increase.

KATZ: How many other proteins have you looked at in the lipid cells to make sure that your effect was specific?

SHA'AFI: The only specificity as far as we are concerned with is that the increase is inhibited by PCMBS. The question you are raising about whether other proteins do the same things, we cannot answer that at the present time.

SHAMOO: I believe the effect of proteins on lipid bilayer in just increasing in conductance or water permeability are of a highly non-selective nature because practically any protein including alpha-chymotrypsin, which has nothing to do with membrane function, could increase water permeability and lipid bilayer conductance. So, these two parameters are no indication of specificity.

SHA'AFI: This is why we were not interested in the increase, but the inhibition by PCMBS. That is why I only showed the slide on the inhibition not the increase. I think there is some specificity if you inhibit this increase by a known inhibitor of water transport.

SIEGEL: The question of specificity is really very intrigueing, and it can become a philosophical debate. I would like to ask whether or not the inhibition effect is specific for the PCMBS? That is, do agents which do not inhibit water transport in the intact membrane inhibit in the lysosomal model? I wonder if you would consider the possibility that all or many proteins actually modify water transport through the membrane in the intact state. Why does one have to impose the constraint here of specificity for a water channel or for a water transporting device?

SHA'AFI: I am not concerned with specificity. All that I am saying is that this protein which is found in red cell membranes is involved in water transport. If there are other proteins which do the same thing and are not found in red cell membrane, that is a different story. I did not say what can increase water transport, or what can influence the movement of water transport. I say, what membrane components are associated with water transport.

WEINER: Is it at all possible that PCMB could interact with the lipid itself in the lipsosome?

SHA'AFI: I don't know.

MEMBRANE TRANSPORT OF ANTIFOLATES AS A CRITICAL DETERMINANT OF DRUG CYTOTOXICITY

I. David Goldman

Department of Medicine, Medical College of Virginia, Richmond, Virginia 23298

Introduction

There are few areas in therapeutics in which membrane transport of pharmacologic agents has been subjected to as intense study as in the treatment of malignant diseases with cytotoxic agents. Unlike the antibacterial or antifungal agents, which may selectively inhibit biochemical processes unique to the bacterial or fungal organism, the agents which comprise the current armamentarium of the cancer chemotherapist produce toxic effects on both tumor and susceptible host tissues. The lack of understanding of the basic biochemical differences between normal and malignant cells require the utilization of subtle techniques, often empirical, to achieve "selective" effects of the agents in an attempt to minimize toxicity to susceptible host tissues while maximizing toxicity to the tumor. The better understood of these techniques usually exploit quantitative rather than qualitative differences in the tumor vs. host cell rates of (a) cellular proliferation, (b) drug activation or inactivation, (c) drug transport across the cell membrane, and (d) drug interaction with target sites within the cell.

The recognition of the critical role of membrane transport as a determinant of cytotoxicity of anticancer agents came with the antifolates, antimetabolites which have played an important role in cancer chemotherapy since their development in the 1940's. This article will (a) review the evolution of the understanding of the relationship between transport of antifolates and cytotoxicity, and (b) the unique properties of the membrane transport process for the folate and antifolate compounds.

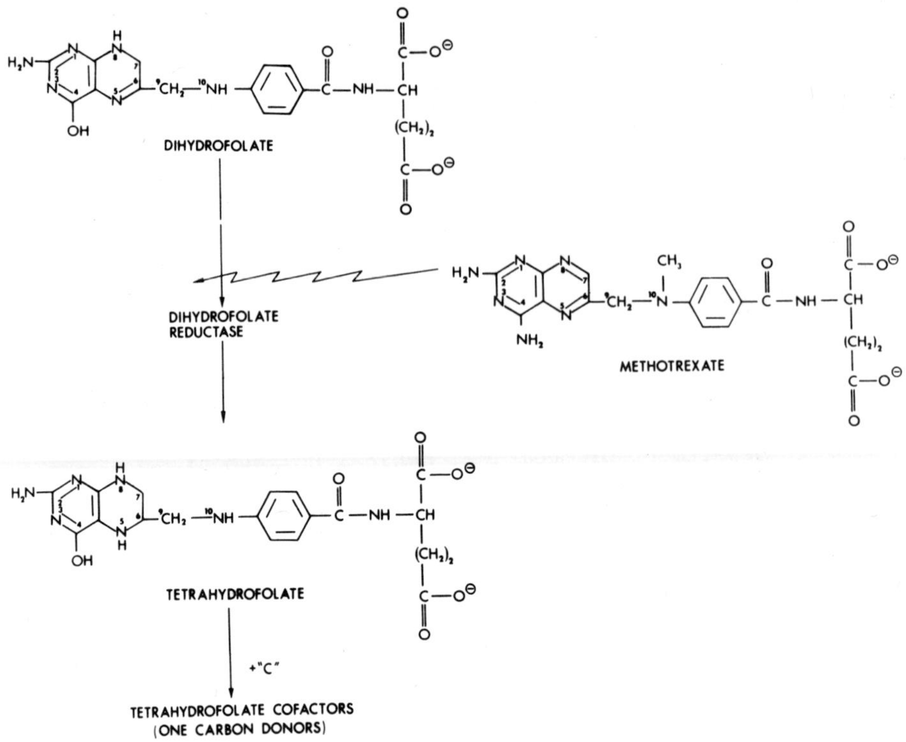

Figure 1. The structural and biochemical relationship between methotrexate (MTX) and the folate compounds.

1. The Metabolic Effects of MTX

Methotrexate (MTX) is an analog of folic acid which differs from folic acid by substitution of an amino for hydroxyl group at the pteridine ring four position and addition of a methyl group at the N^{10} position. It is the presence of this amino group which confers upon MTX an intense affinity for dihydrofolate reductase (DHFR), the enzyme which mediates the reduction of dihydrofolate (DHF) to tetrahydrofolate (THF). This is a key biosynthetic process since derivatives of THF which contain a carbon molecule at various oxidation states at the N^5, N^{10} or shared at both positions provide one-carbon moieties which play a crucial role in purine, pyrimidine and amino acid synthesis. Hence, in the presence of sufficient intracellular MTX, THF synthesis is blocked and THF-dependent DNA, RNA and protein synthesis is abolished.

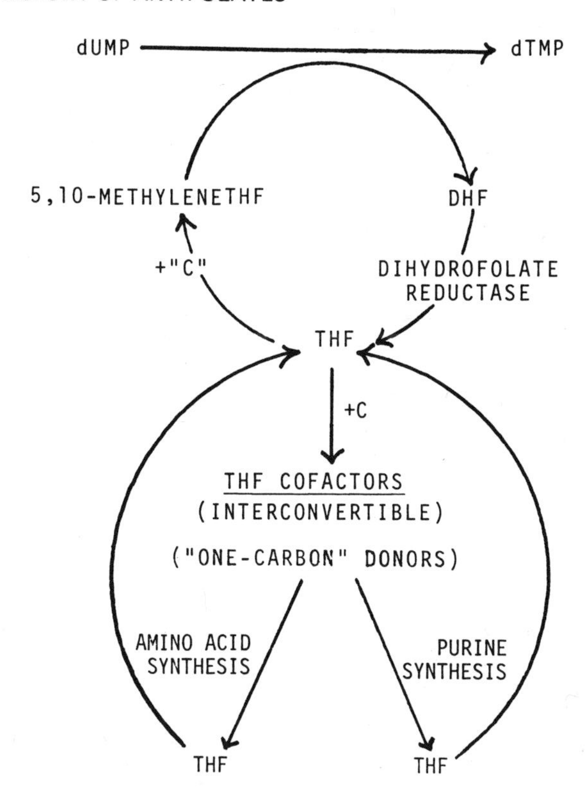

Figure 2. Utilization of THF cofactors in biosynthetic processes. With the synthesis of purines and amino acids, the THF moiety is unchanged and reassociation of THF with one-carbon molecules maintains the THF cofactor pool. However, with synthesis of deoxythymidine monophosphate (dTMP) from deoxyuridine monophosphate (dUMP), THF is oxidized to DHF and the THF cofactor pool can only be sustained by the regeneration of THF from DHF mediated by DHFR.

There are important differences, however, in the way THF cofactors are utilized in these biosynthetic processes. When one-carbon moieties associated with THF are utilized for amino acid and purine synthesis, the cofactor reverts to THF which is then capable of associating with another one-carbon moiety to regenerate the active THF cofactors (Fig. 2). Hence, purine and amino acid synthesis does not deplete the THF cofactor pool. However, the utilization of the carbon group in the synthesis of deoxythymidine monophosphate (dTMP) from deoxyuridine monophosphate (dUMP) results both in the consumption of the carbon moiety and the oxidation of THF to DHF. When dTMP is synthesized and DHF is generated, cellular THF cofactor pools can only be maintained by the DHFR-mediated intracellular regeneration of THF from DHF or the transport of THF cofactors into the cell from the extracellular compartment.

dUMP synthesis is the only known metabolic mode of consumption of THF and as such represents a potential for depletion of cellular THF under conditions in which regeneration of THF is inhibited. It can be seen from this then that the metabolic consequences of MTX and other inhibitors of DHFR is the impaired regeneration of THF from the DHF which is produced during dTMP synthesis. This results ultimately in the depletion of cellular THF cofactor stores. The consequent impairment of pyrimidine, purine and amino acid synthesis results in the suppression of THF-dependent DNA, RNA and protein synthesis (1).

The initial view of the interaction between MTX and its target enzyme, DHFR, was that this binding was so tight that it could be considered "stoichiometric." This implied that free MTX does not appear within the intracellular compartment until the drug is present within the cell at molar equivalence to the enzyme. Likewise, this would also be the level of intracellular drug which would abolish the reduction of DHF to THF. Indeed, studies from a number of laboratories indicated that transport of MTX into mammalian cells is rate limiting to binding to DHFR within the intracellular compartment and that free drug _cannot be detected_ until the intracellular MTX level is in excess of the enzyme binding capacity (2,3). From this it was assumed that the rate of penetration of MTX across the cell membrane would be a key element in drug cytotoxicity, a view which was strongly supported by early studies described in the next section.

2. The Relationship Between Membrane Transport of Antifolates and Cytotoxicity in Mammalian Cells

The most striking demonstration of the correlation between the membrane transport of MTX and cytotoxicity to tumor cells was the demonstration by Kessel _et al_. (4) that the percent increase in survival of mice bearing a variety of murine leukemias in the ascitic compartment was directly proportional to the uptake of MTX into the tumor cell _in vitro_ (Fig. 3). It was later pointed out (5) that "uptake" was measured under conditions which reflected the unidirectional flux of drug into the cells. Within the context of the concepts of the mechanism of action of MTX at that time, this data was interpreted to indicate that the critical element in the transport of MTX was its rate of translocation across the cell membrane, the rate limiting step in the interaction between MTX and its target enzyme within the cell. This parameter, as well as the capacity of DHFR binding sites within the cell would then determine the rate at which DHFR was inactivated by MTX and the interval required to achieve inactivation of total enzyme within the cell at any given extracellular drug concentration.

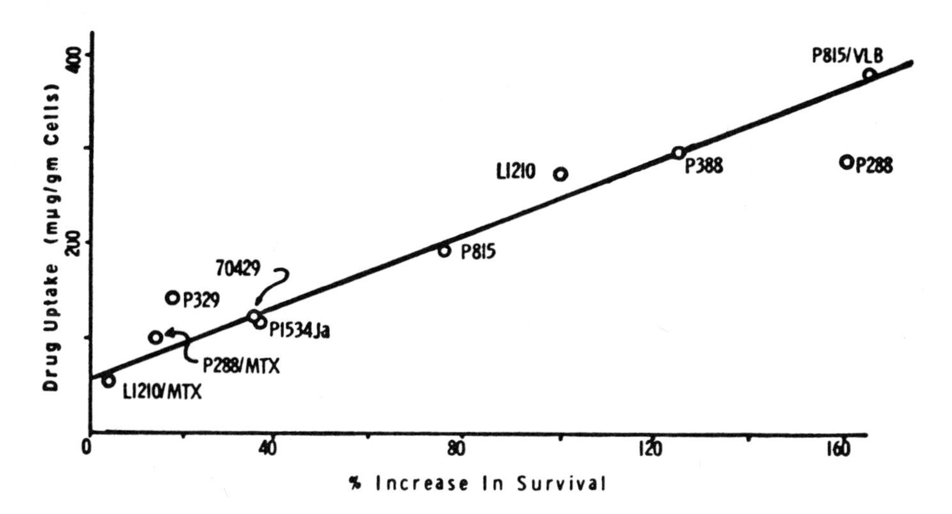

Figure 3. The relationship between uptake of MTX into murine leukemia cells _in vitro_ and the survival of tumor-bearing mice given MTX _in vivo_. The tumor was grown in the ascitic form within the intra-peritoneal compartment. The dose of MTX was 1.5 mg/kg administered intraperitoneally. Drug uptake under these conditions represents the unidirectional flux of MTX into the cell (from Kessel _et al._, ref. 4).

When this data was subjected to a quantitative analysis, however, it became apparent that this explanation was inadequate (5). Hence, under the conditions of the study by Kessel _et al._ (4), the interval of exposure of the tumor cells to drug was very long in comparison to the interval which would be required to achieve intracellular drug levels at molar equivalence to DHFR and generate and sustain appreciable levels of free drug within the cell. Hence, small differences in influx of MTX from one tumor cell line to another could only have a negligible effect on the total interval over which enzyme was inhibited and could not in itself account for differences in sensitivities among different tumors to this agent. However, as pointed out in this analysis (5), alterations in the unidirectional flux of MTX into the cell could not occur as an isolated event but would be accompanied by changes in the steady-state concentration and/or the unidirectional efflux of drug. Hence, assuming uphill transport of MTX (see below), if influx was reduced in a resistant tumor, the net transport of drug would be reduced or the unidirectional efflux reduced or there might be a lesser change in both parameters. Since a reduced rate of efflux would presumably increase cytotoxicity by slowing the rate of drug loss following drug administration, the possibility was raised that the critical element in the interaction between MTX and these tumor cells was not the rate of translocation of drug into the cell *per se*, but a parallel change in another parameter of the drug-cell interaction controlled by the membrane transport system - the level of free drug achieved and sustained with the intracellular compartment (5). However, the concept that appreciable levels of free drug within the cell could play an important role as a cytotoxic determinant of this agent seemed to be contrary to the view of a stoichiometric or, at the least, a very intense interaction between MTX and DHFR. Hence, studies were initiated to evaluate the possible role for free drug in achieving the biochemical effects of MTX within the cell. As will be described in Section 3, these studies clearly indicated that substantial levels of free MTX are required within the intracellular compartment to achieve cessation of THF synthesis, the basis for the observation that the generation and maintenance of free drug levels must be a critical determinant of drug cytotoxicity *in vivo* (see below).

As biochemical evidence was accumulating which supported a critical role for free intracellular drug in the suppression of THF synthesis and THF-dependent processes *in vitro* (6-10), other experimental approaches using *in vivo* systems supported the concept that the metabolic effects of MTX require the sustained exposure of tumor cells to free drug. When correlated with *in vitro* studies, these observations pointed towards the critical role of free intracellular drug. Margolis *et al*. in 1971 (11) demonstrated the rapid reversibility of the suppressive effect of MTX on ^{3}H-deoxyuridine (^{3}H-UdR) incorporation into DNA in mouse small intestine, an experimental finding which suggested that the metabolic consequences of the interaction between MTX and its target site(s) were rapidly reversible. Similarly, Chabner *et al*. (12) demonstrated rapid resumption of ^{3}H-UdR incorporation into DNA in murine tumor cells and bone marrow cells as the free extracellular MTX level fell to $<10^{-8}$ M. Sirotnak and Donsbach (13) correlated the relationship between the persistence of exposure of tumor and intestinal cells to extracellular MTX with the persistence of free intracellular drug and the suppression of ^{3}H-UdR incorporation into DNA following pulse administration of MTX to L1210 leukemia-bearing mice *in vivo*. As the free drug level fell within the intracellular compartment approaching the level bound to DHFR, the incorporation of ^{3}H-UdR into DNA resumed. Likewise, the greater toxicity of MTX to the L1210 leukemia cell relative to the intestinal cell appeared to be

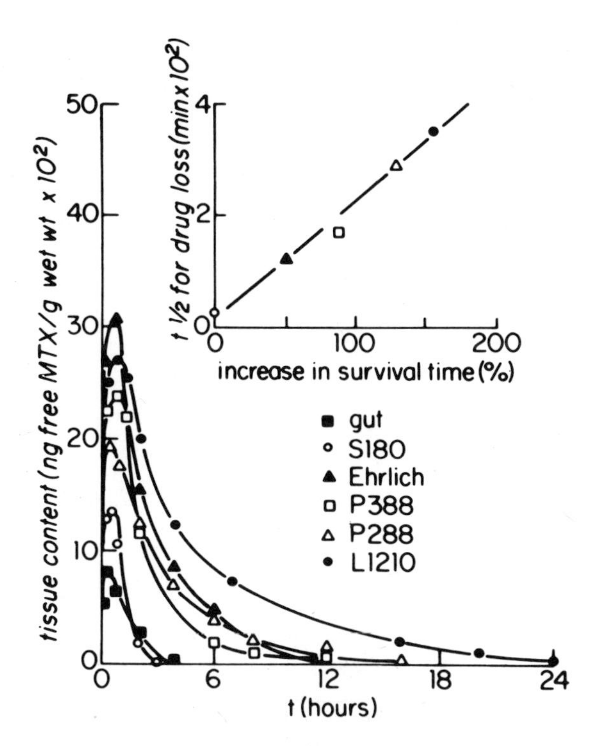

Figure 4. The Free MTX level within a variety of cells as a function of time following pulse administration of the drug *in vivo*. The inset describes the linear relationship between the half-time for free MTX lost from a variety of tumor cells *in vivo* as the extracellular MTX level falls as a function of the increase in survival of mice bearing these tumors *in vivo* (from Sirotnak and Donsbach, ref. 14).

related to the longer duration of maintenance of free intracellular drug levels in the former vs. latter cells (Fig. 4). The authors suggested that the critical role of the membrane transport system is to sustain free intracellular drug levels as the extracellular MTX level falls. They went on (14) to show that the half-time for the *net* loss of intracellular drug (Fig. 4) and the free intracellular drug level at the steady-state (Fig. 5) in a variety of mammalian cells were both directly proportional to the increase in survival time of tumor-bearing animals treated with MTX *in vivo* (14,15). This confirmed the earlier prediction (5) that it is

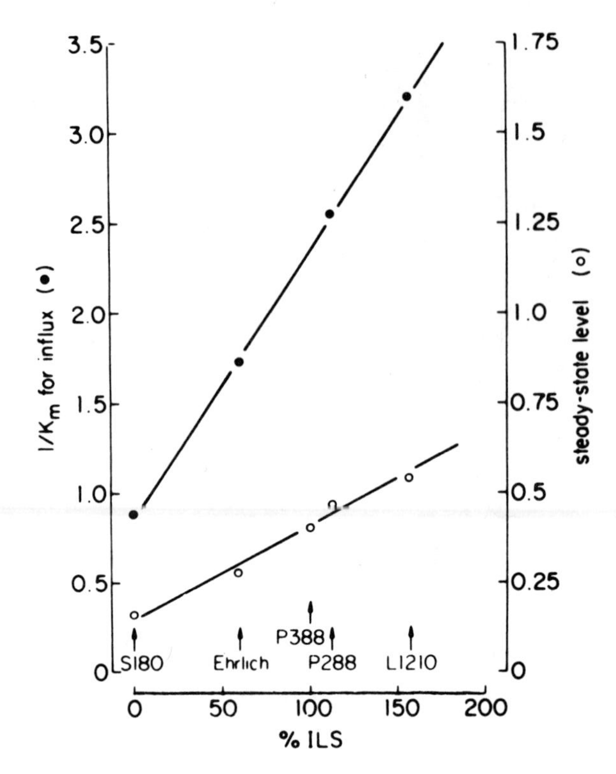

Figure 5. The relationship between the steady-state free intracellular MTX level (right ordinate) or the reciprocal of the influx K_m (left ordinate) in a variety of tumor cells as a function of the percent increase in survival of mice bearing these tumors *in vivo* (from Sirotnak and Donsbach, ref. 15).

the free intracellular drug, controlled by the membrane transport system, that is a critical factor in determining cytotoxicity. It was shown also that the relationship between the reciprocal of the K_m for influx was linearly related to survival of the tumor-bearing animal (15). The authors suggested on this basis that the major alteration in the transport system in tumor cell lines of varying sensitivities to MTX relates to differences in affinities of the carrier for MTX. However, this relationship does not always hold; there are large differences (∿50%) in the influx V_{max} among these tumor lines as well as considerable differences in the half-time for the unidirectional drug efflux (80%) (15). Hence, the molecular basis for the differences in accumulation of free drug among different tumors and different tissues is variable and may be caused by alterations in a number of the properties of the transport mechanism(s). This analysis becomes increasingly complex in view of substantial evidence which indicates that transport of folate compounds across the cell membrane is not mediated by a single transport carrier but is likely related to the translocation of folates across the cell membrane via multiple transport pathways (see below, Section 4.B).

3. The Biochemical Basis for the Requirement for Free Intracellular MTX in the Suppression of THF Synthesis

The interaction between MTX and DHFR represents one of the most intense interactions between an inhibitor and its target enzyme. At low pH the interaction is "stoichiometric" (16), i.e., the K_i is too low to measure; even at physiological pH, the K_i is $\sim 10^{-11}$ M (17). In spite of this, however, the effects of the drug are rapidly reversible and sustained inhibition of the target enzyme requires sustained exposure of the cell to drug with the sustained accumulation of free drug within the intracellular compartment. Initial studies from this laboratory were designed to evaluate the role of free intracellular MTX in the suppression of THF-dependent processes. This required an experimental design which would exclude the possibility that the effects of MTX were related to extracellular drug and the interaction between extracellular MTX and THF cofactors at the level of their common membrane carrier. This required *in vitro* studies in which THF cofactors could be eliminated from the extracellular compartment. The critical role for free intracellular MTX in *in vivo* studies in the inhibition of THF-dependent processes must be extrapolated from these *in vitro* experiments which ensured that extracellular MTX neither alters transport of THF cofactors into or out of the cell, leaving intracellular drug as the critical factor (6-8). Finally, the demonstration that free drug is required to inhibit THF synthesis, an intracellular event, proved the critical role of intracellular drug (9,10).

Initial studies which suggested a critical role for free intracellular drug were based upon the observation that the THF cofactor-dependent incorporation of ^{3}H-UdR or ^{14}C-formate into DNA along with the incorporation of ^{14}C-formate into RNA and protein required substantial levels of free intracellular MTX (6-9). Hence, when cells were loaded to levels of free intracellular MTX orders of magnitude above its K_i for DHFR and to a level in excess of the DHFR binding capacity, following which free drug was eliminated from the intracellular compartment, these THF-dependent processes were negligibly effected (6-8). As indicated in Fig. 6, suppression of these processes necessi-

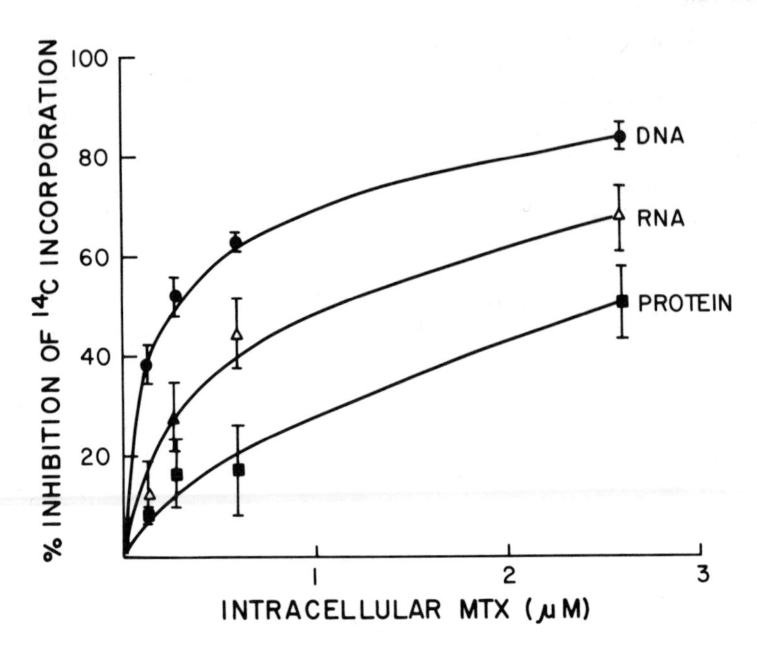

Figure 6. The relationship between the percent suppression of ^{14}C-formate incorporation into DNA, RNA and protein as a function of the free intracellular MTX level. Free drug is estimated as that intracellular drug component which leaves the cell upon resuspension of cells loaded with MTX into MTX-free medium. Treatment of cells to achieve intracellular drug levels in excess of the capacity of high-affinity binding sites following which free drug is removed has little effect on these processes (from White *et al.*, ref. 8; data was obtained with L-cell mouse fibroblasts).

tated the continuous exposure of these cells to extracellular and intracellular drug with 50% suppression requiring a free intracellular drug level of at least 0.3×10^{-7} M - a value four orders of magnitude above the MTX K_i (6-8). In further studies, it was demonstrated that this requirement for free drug in the inhibition of THF cofactor-dependent processes is related to the necessity for free intracellular drug in the suppression of ^3H-THF synthesis (9,10). As seen in Fig. 7, 50% suppression of THF synthesis in the Ehrlich ascites tumor *in vitro* requires a free intracellular MTX level of $\sim 10^{-7}$ M, while complete suppression of of this process requires a free drug concentration in excess of 10^{-6} M.

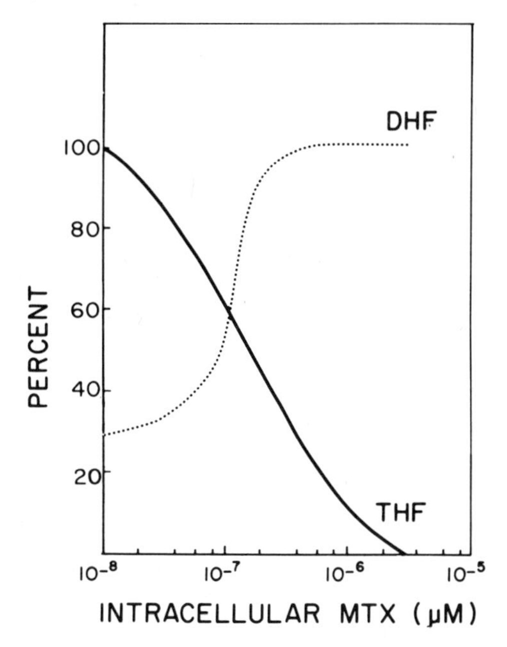

Figure 7. The relationship between the rising DHF and falling THF levels within the cell as the free intracellular MTX concentration is increased (from White & Goldman, ref. 9).

The basis for this requirement for free intracellular MTX in the suppression of THF synthesis is illustrated in Fig. 8, a computer simulation developed by Jackson and Harrap (18). DHF is an excellent substrate for DHFR and is so rapidly reduced to THF as it is generated during the synthesis of dTMP from dUMP that the usual intracellular level must be $\sim 10^{-8}$ M. This was confirmed in studies from this laboratory (9). Since the K_m for DHF is $\sim 10^{-6}$ M DHFR operates at a very low state of saturation with respect to this substrate. Hence, only a small percentage of total DHFR activity is required to meet usual cellular demands for THF. When cells are incubated with MTX and the drug enters the intracellular compartment, the following sequence of events would be expected to occur. Initially as MTX binds to DHFR, free drug does not accumulate within the cell since influx is rate limiting to binding and the ratio of free enzyme to intracellular drug is very great. Association of MTX with DHFR sites results in a small transient reduction in the rate of THF synthesis with a consequent increase in DHF substrate level. Since the intracellular DHF level is initially so much lower than its K_m, the interaction between DHF and DHFR is essentially first order and the increase in intracellular DHF results in a proportional

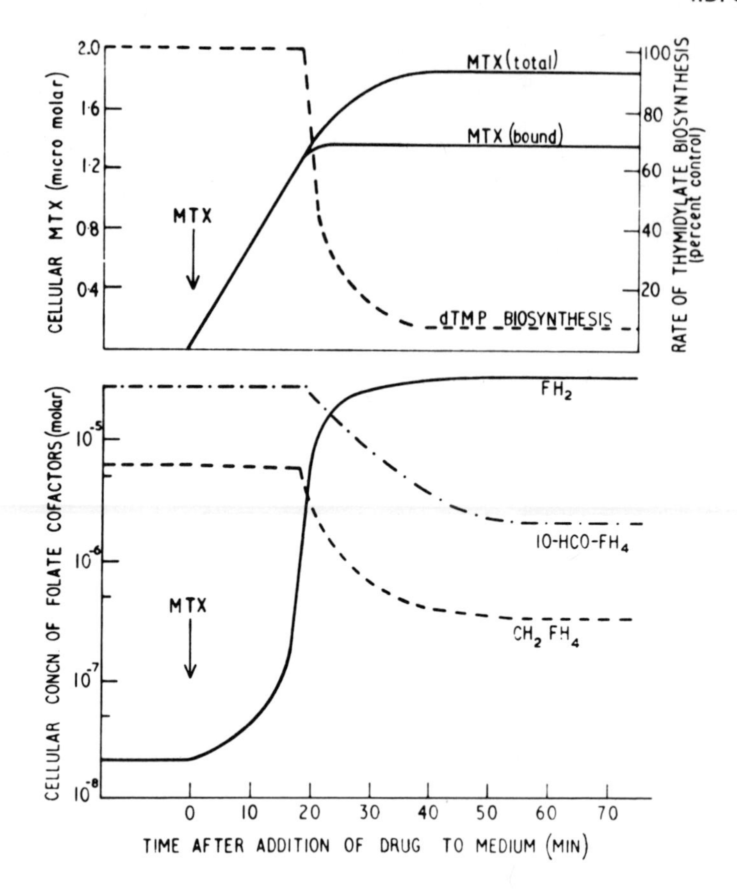

Figure 8. The changes in intracellular folate levels and dTMP biosynthesis as the level of intracellular MTX is increased. FH_2 is DHF; 10-HCO-FH_4 is 10-formylTHF; CH_2FH_2 is 5,10-methylene-THF (from Jackson & Harrap, ref. 18). These data were predicted by Jackson and Harrap based upon a computer simulation. Studies from this laboratory support this model (see refs. 9, 20 and the text).

increase in the rate of DHF reduction as the net rate of interaction between the higher level of DHF and those enzyme sites unassociated with MTX increases. Because of this, THF synthesis (and dTMP synthesis) is initially maintained at an essentially normal rate. However, when the major portion of DHFR sites become associated with MTX, the intracellular DHF concentration approaches its K_m and the rate of THF synthesis can no longer increase in proportion to the increase in the intracellular DHF level. At this point, the rate of THF synthesis falls, the level of cellular THF cofactors falls, and the rate of dTMP synthesis declines. Also, because the interaction between MTX and DHFR is not "stoichiometric" at physiological pH and because DHF has risen to high levels at a time when the free intracellular MTX level is still trivial, DHF competes with MTX for binding to the few remaining DHFR sites. Hence, free MTX accumulates within the cell at intracellular drug levels below molar equivalence to the DHFR binding capacity and to fully saturate DHFR sites with MTX and completely suppress THF synthesis requires levels of free intracellular MTX orders of magnitude above its K_i. Conversely, when cells which contain sufficient free MTX to abolish THF synthesis are suspended into drug-free media, as free MTX leaves the cells under conditions in which the cellular DHF level is high, DHF will displace MTX from a few percent of the total DHFR sites before resumption of DHF reduction decreases cell DHF to trivial levels once again. Cell MTX finally falls to a constant level which appears tightly bound within the cell but represents a value which is less than the total DHFR binding capacity.

Jackson and Harrap (18) predicted that 95% of total DHFR sites within exponentially growing L1210 leukemia cells could be inhibited before the cellular growth rate is appreciably reduced. This prediction was made utilizing a K_i for MTX of 1.9×10^{-11} M. In studies from this laboratory, the requirement for free intracellular MTX was determined *in vitro* in cells which are not capable of replication and would, therefore, have a lower basal rate of dTMP synthesis and DHF generation. Hence, under these conditions, even a smaller percentage of DHFR activity would be sufficient to meet cellular demands for THF and a higher concentration of free MTX would be required to suppress this basal rate of THF synthesis. This is analyzed further in ref. 20.

Rothenberg (19) has suggested that multiple forms of DHFR with a lower affinity for MTX but capable of reducing DHF to THF may be present in murine leukemia cells at physiological pH. Hence, the requirement for free intracellular MTX might be related, in addition, to an interaction between free drug and these other enzyme forms.

4. The Mechanism of the Membrane Transport of Folate and Antifolate Compounds in Mammalian Cells

A. Characteristics of a High-Affinity Transport Carrier for the Antifolates - MTX, Aminopterin, and the THF Cofactor Derivatives of Folic Acid

MTX shares with the THF cofactors a high-affinity carrier transport system in many mammalian cells (21,22). This laboratory has utilized MTX as a model substrate for evaluating the characteristics of this transport route because MTX is not metabolized, it is available commercially with a ^3H label of high specific activity which is easily purified and has satisfactory stability. In addition, because the translocation of MTX across the cell membrane is rate limiting to its binding to DHFR within the intracellular compartment, very accurate unidirectional influx measurements can be made when net uptake of the drug into the cell is monitored prior to saturation of the high-affinity binding sites (3,21). A number of aspects of the MTX carrier system have been well characterized. The unidirectional influx for MTX is largely (but not completely - see below) saturable at low concentrations with a K_t (concentration at which influx is one-half V_{max}) of ∿5 μM and a maximum influx velocity (V_{max}) of ∿1 - 2 μmoles per minute per liter cell water. The unidirectional influx of MTX is competitively inhibited by 5-methylTHF (the major circulating folate of man and rodents), 5-formylTHF, aminopterin, and folic acid (3,21). Based on the K_i's and K_t's for these agents, the affinity of this carrier for 5-methylTHF and aminopterin is greater than that for MTX, the affinity for 5-formylTHF is comparable to that of MTX, the affinity for folic acid is considerably lower. Hence, the K_i for folic acid is 200 μM and 400 μM in the L1210 leukemia and Ehrlich ascites tumor, respectively. Compatible also with a transport carrier is a high Q_{10} ($V_{37} - V_{27} = 5\text{-}7$) and the observation that the major portion of MTX influx is sulfhydryl dependent and inhibited by organic mercurials (3,21,23). The transport characteristics for 5-methyTHF and aminopterin have also been characterized (21,24,25). As indicated above, the K_t of the carrier transport system for these folates is lower than for MTX. The unidirectional influx of these folates is competitively inhibited by MTX or 5-formylTHF.

Consistent with a carrier model to account, in part, for the transport of MTX and the THF cofactors is the observation that the unidirectional influx of MTX is stimulated when either 5-methylTHF or 5-formylTHF is present within the trans-compartment (26). Similarly, the unidirectional influx of 5-methylTHF or aminopterin is stimulated when 5-formylTHF or 5-methylTHF is present within the trans-compartment (27). On the other hand, it has not been possible to demonstrate stimulation of the unidirectional influx of 5-methylTHF, MTX, or aminopterin by the presence of MTX or aminopterin in the trans-compartment (26,27).

Because MTX influx is rate limiting to its binding within the intracellular compartment, conditions can be achieved in which uptake of MTX is measured when free intracellular MTX is negligible. Further, since the affinity of DHFR for 5-formylTHF is >5 orders of magnitude below that of MTX, it was possible to measure the influx of MTX under conditions in which 5-formylTHF was present within the intracellular compartment. This facilitated a quantitative evaluation of the kinetics of trans-stimulation of MTX influx during heteroexchange diffusion with 5-formylTHF under conditions in which a possible inhibitory effect by 5-formylTHF on a unidirectional efflux component to the net uptake of MTX could be excluded. Fig. 9 illustrates the time-course of the trans-stimulation of MTX influx by 5-formylTHF. Fig. 10 indicates that trans-stimulation of MTX influx results in an augmentation of the influx V_{max} without a significant change in the influx K_t. This might be expected if the rate limiting step of the MTX carrier cycle is the rate of re-orientation of the unloaded carrier from the inner to outer boundary of the cell membrane.

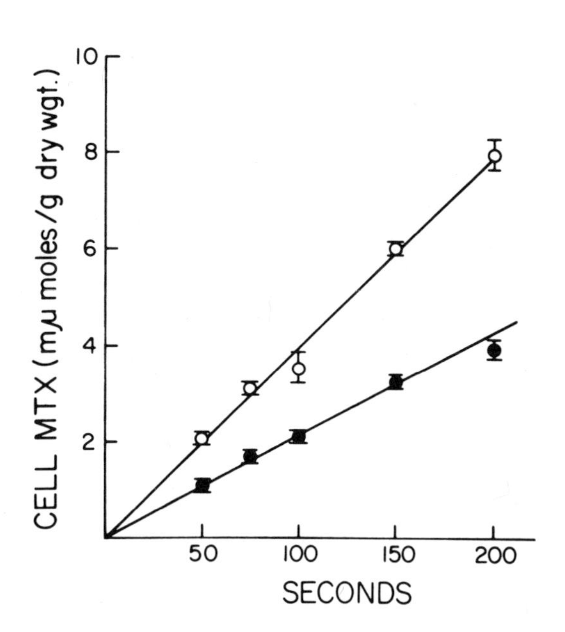

Figure 9. Trans-stimulation of the unidirectional influx of ^3H-MTX by 5-formylTHF. Cells were loaded with 5-formylTHF, washed with 0° buffer, resuspended into medium containing ^3H-MTX and the time-course of uptake of ^3H over 200 seconds was monitored (open circles). Control cells were treated similarly except for the omission of 5-formylTHF during the initial incubation (closed circles) (from Goldman, ref. 26).

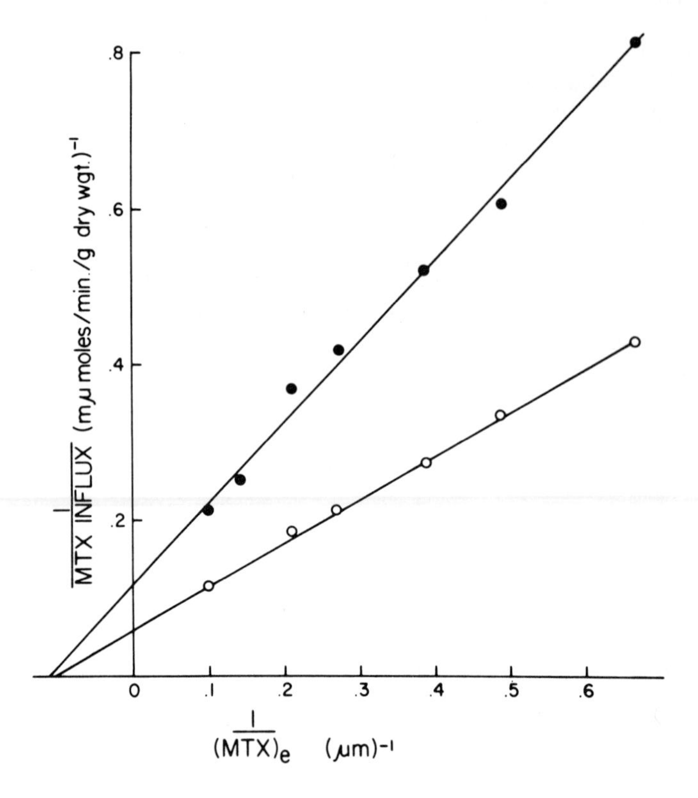

Figure 10. The kinetics of trans-stimulation of the unidirectional influx of ^{3}H-MTX by 5-formylTHF. (The flux interval was 30 or 50 sec. For a detailed description of the experimental procedures, see Goldman, ref. 26.) Closed circles represent control cells, open circles represent cells which contain 5-formylTHF within the trans-intracellular compartment.

B. Evidence for Multiple Transport Routes for Folate Compounds Across Mammalian Cell Membranes

Studies from this laboratory have shown that while the major portion of the MTX influx at low extracellular drug concentrations is saturable, there is a second component to the transport system (28). Recent studies from this laboratory confirm the presence of a second component to the MTX transport process which differs considerably from the high-affinity carrier-mediated component (27). Further, it is clear that the bulk of folic acid transport into the Ehrlich ascites tumor and the L1210 leukemia cell occurs by a mechanism distinct from the high-affinity MTX-THF cofactor carrier. Hence, while the major portion of the influx of MTX is inhibited by sulfhydryl reagents, the initial uptake velocity for folic acid is negligibly effected by these agents (23). Further, the measured initial uptake rate for folic acid far exceeds the predicted influx velocity if it is assumed that the folic acid K_i reflects its K_t for the high-affinity MTX-THF cofactor carrier mechanism and that this carrier is its sole route of translocation across the cell membrane (22,27,29). Finally, the initial uptake of folic acid is negligibly affected by levels of MTX or 5-formylTHF that should saturate the high-affinity MTX-THF cofactor carrier (22,27,29). Quantitatively, this would indicate that the major portion of folic acid transport across the cell membrane even at low extracellular drug concentrations is mediated by an alternate route. The significance of this route, and its properties are currently being characterized in this laboratory.

C. The Thermodynamics of MTX Transport in Mammalian Cells

The critical role of free intracellular MTX as a cytotoxic determinant of this agent was stressed in the first section of this paper. In view of this, it is of particular interest that mammalian cells markedly restrict the net accumulation of free MTX and other folate compounds within the intracellular compartment. Not only is the unidirectional influx of MTX to a large extent saturable as the extracellular MTX level is increased, but the steady-state accumulation of free intracellular drug follows an absorption isotherm (3). Hence, as the extracellular MTX concentration rises, the ratio of the concentration of free drug in the intracellular to extracellular compartment (distribution ratio) falls. Fig. 11 illustrates that the steady-state free intracellular MTX level is 6 μM when the extracellular MTX concentration is 85 μM. Indeed, at higher extracellular MTX levels the distribution ratio may fall below 0.05 μM under steady-state conditions. The low distribution ratio for MTX may be based, in part, on its negative charge (MTX is a bivalent anion - see Fig. 1). However, the extent to which this can be based upon passive electrical considerations depends upon the electrical-potential difference across the cell membrane; the latter is uncertain. Even the higher membrane potentials estimated for the Ehrlich ascites tumor, ∿24 mv (31), would not appear to account for the very low distribution ratios for MTX that have been observed in this laboratory in this cell. This suggests that there may be an exit pump for this agent. There is considerable evidence to support this. As illustrated in Fig. 12, when cells are brought to a steady-state with MTX then exposed to a variety of metabolic inhibitors (in this case azide), net accumulation of intracellular

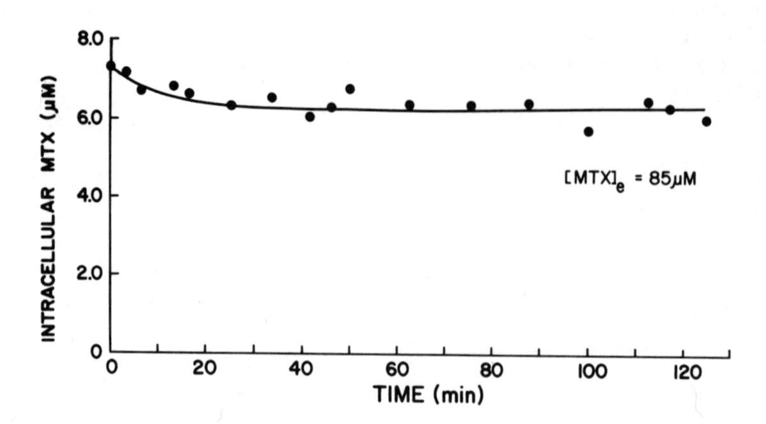

Figure 11. The steady-state level of free intracellular MTX at a high extracellular drug level. Ehrlich ascites tumor cells were incubated for 30 minutes with 175 μM MTX following which sufficient fresh buffer was added to the extracellular compartment at time zero to reduce the extracellular drug level to 85 μM. By this procedure, intracellular MTX falls to the steady state assuring that true steady-state conditions are achieved. Only the free intracellular drug content (total minus bound) is indicated on the graph. (From Gupta et al., ref. 30.)

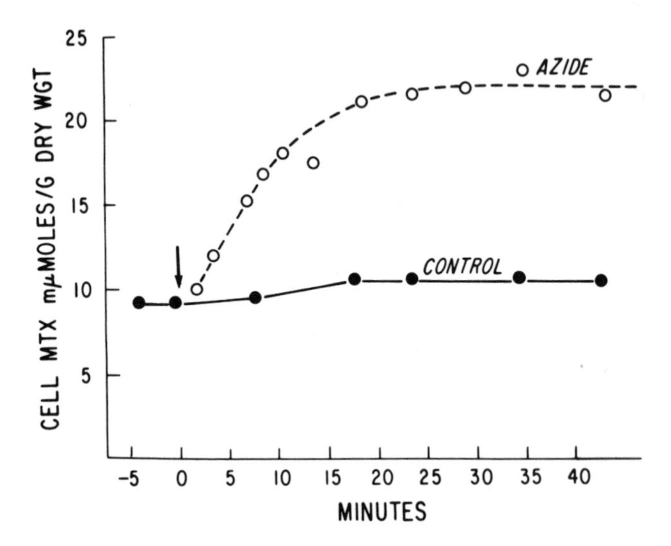

Figure 12. The effect of 10mM sodium azide added at the arrow on net uptake of MTX in L1210 leukemia cells (from Goldman, ref. 28).

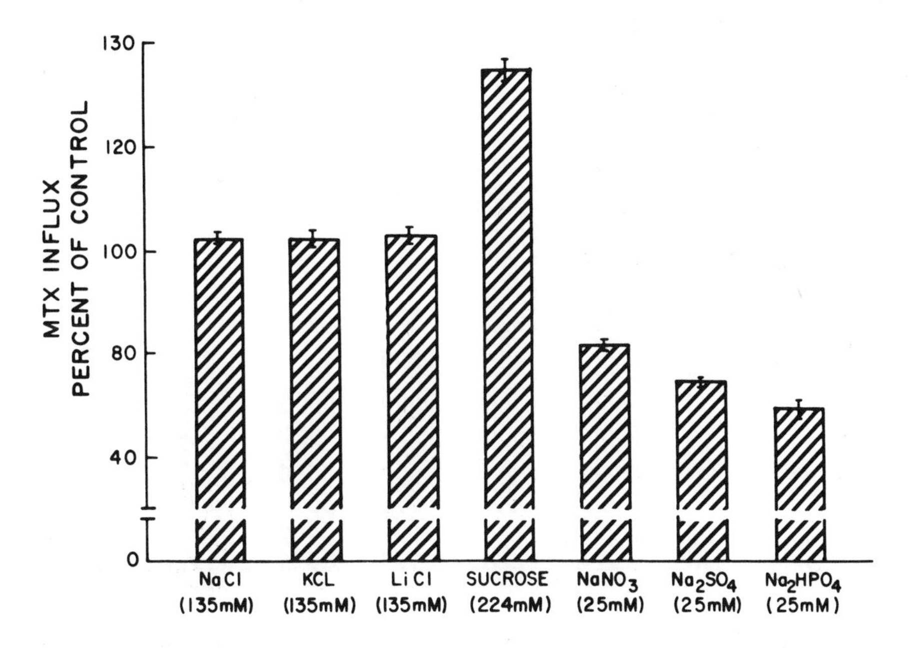

Figure 13. The effect of alterations in the ionic composition of the extracellular compartment on the 100 second unidirectional influx of MTX into the Ehrlich ascites tumor cell. All solute replacements were made isosmotically. Results are expressed as ± standard error of the mean (from Goldman, ref. 21).

drug is enhanced. This occurs with inhibitors of anaerobic or aerobic metabolism, singly or in combination (28). The major portion of this augmented drug uptake appears to be free within the intracellular water. This suggested that metabolic poisons may block an exit pump for MTX (28). Since the distribution ratio for MTX may exceed 3 in the presence of metabolic inhibitors, the data suggested that uphill transport of MTX into the cell can persist even when energy metabolism is blocked. The energy source which supports this presumed uphill transport of MTX into the cell under these conditions is unclear; however, a tentative model is suggested below.

D. A Possible Cotransport of MTX and Organic Phosphates to Account for Uphill Transport of MTX into the Cell

While many uphill carrier transport systems derive their energy from the electrochemical-potential differences for sodium, potassium, or hydrogen ions, across the cell membrane linked to the substrate carrier by a cotransport with these cations.

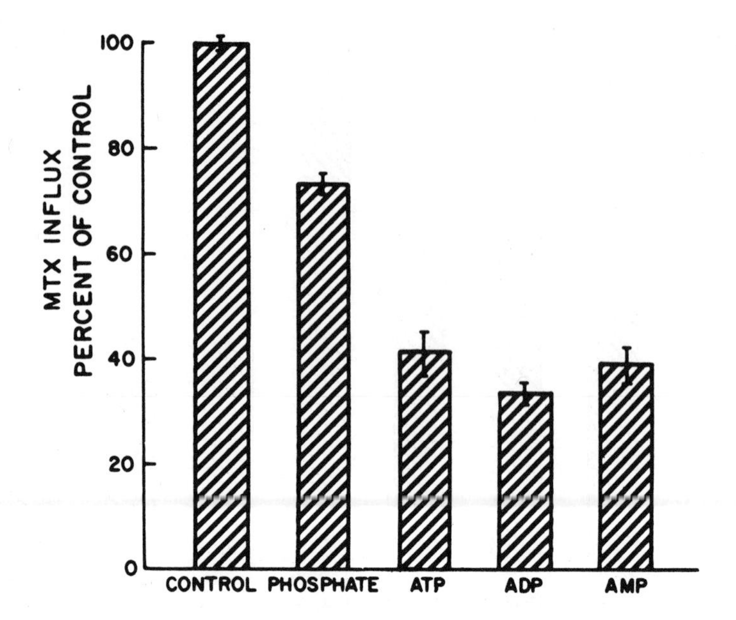

Figure 14. The effect of 5 mM phosphate or adenine nucleotides on the 100 second unidirectional influx of 0.5 μM ^{3}H-MTX in Ehrlich ascites tumor cells. Data are expressed as $\pm$ standard error of the mean, AMP - adenosinemonophosphate; ADP - adenosine diphosphate; ATP - adenosinetriphosphate.

(32), the transport of MTX appears to be sodium independent. As indicated in Fig. 13, when extracellular sodium is completely replaced by either potassium or lithium, the unidirection influx of MTX is unchanged. On the other hand, when both sodium and chloride are replaced with the nonionic sucrose, the unidirectional influx of MTX is prominently enhanced. This raised the possibility that the removal of chloride is stimulatory and that the anionic composition of the extracellular compartment might be critical. Indeed, when chloride is replaced by a variety of inorganic anions - nitrate, sulfate, or phosphate, the unidirectional influx of MTX is inhibited (21,33). Likewise, addition of phosphate to the extracellular compartment inhibits the influx of 5-methylTHF into the Ehrlich ascites tumor cell (33). This is observed not only for inorganic anions, but for organic anions as well. Of particular interest is the observation that organic phosphates are potent inhibitors of the unidirectional influx of MTX and that this inhibitory effect is independent of the energy potential for compounds. That is, AMP is at least as inhibitory to the unidirectional influx of MTX as is ADP and ATP (Fig. 14,

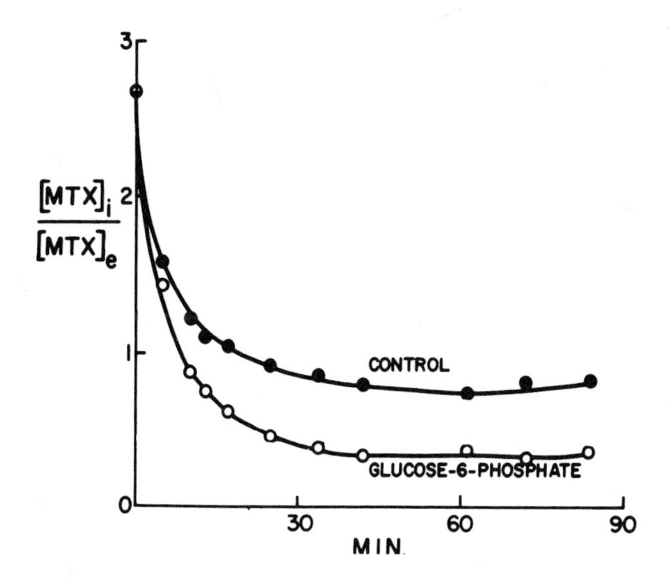

Figure 15. The effect of 50 mM extracellular glucose-6-phosphate on the steady-state distribution ratio for MTX in Ehrlich ascites tumor cells. The experimental design is described in the text. Cells were initially loaded to a high level of intracellular MTX, following which the extracellular MTX level was reduced and the cells suspended into control media or media containing 50 mM glucose-6-phosphate to permit the intracellular MTX to fall to the steady state. The final extracellular MTX level was 1 μM, the 50 mM glucose-6-phosphate replaced an isosmotic amount of sodium chloride.

ref. 21). Further, the effect of alterations in the anionic composision of the extracellular compartment is not restricted to changes in the unidirectional influx of MTX. As seen in Fig. 15, when cells are loaded to a high level of intracellular MTX then resuspended into a lower concentration of MTX in the presence or absence of 50 mM glucose-6-phosphate, the steady-state distribution ratio for MTX is markedly depressed. This occurs under conditions in which the Donnan potential should be reduced and changes in passive electrical forces affecting net transport of MTX should result in a rise rather than a fall in the intracellular MTX level. These observations suggested that structurally unrelated anionic compounds can alter the thermodynamics of MTX transport. These effects of organic phosphates do not appear to be related to the ability of these compounds to chelate bivalent anions. Hence, an 80% reduction of extracellular calcium results in a small increase rather than decrease in MTX influx, while a similar reduction in extracellular magnesium does not alter MTX influx at all (33).

It is unclear as to the role which this interaction between organic phosphates and the high-affinity MTX-THF cofactor transport system plays in regulating the net transport of these compounds under physiological conditions. However, given two assumptions, the high level of organic phosphates within the cell could represent an important source of potential energy for the generation of uphill transport of MTX and the THF cofactors into the cell (Fig. 16). These assumptions are as follows: (a) Organic phosphates normally present within the intracellular compartment inhibit the interaction between MTX or THF cofactors and their carrier at the inner boundary of the cell membrane as they appear to inhibit the interaction between these compounds and their carrier at the outer boundary of the cell membrane; (b) Organic phosphates can utilize this carrier themselves to move across the cell membrane either complexed to the carrier alone or in a ternary complex with carrier and MTX or carrier and THF cofactors. The usual asymmetrical distribution of organic phosphates across the cell membrane could then serve as a battery to drive MTX uphill into the cell as the downhill flow of organic phosphates out

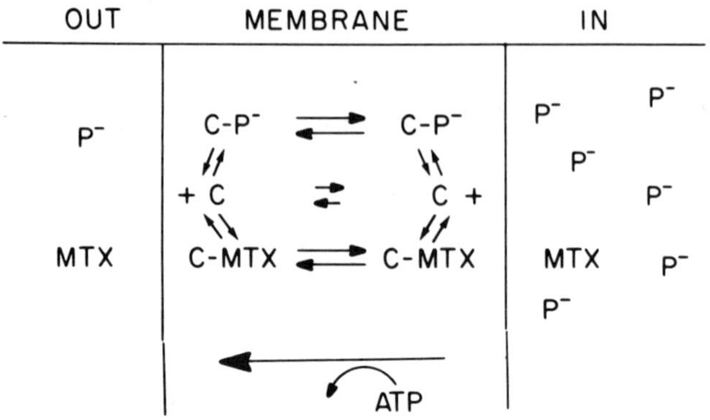

OUT MEMBRANE IN

P^- C-P^- ⇄ C-P^- P^- P^- P^-

+ C ⇄ C + P^-

MTX C-MTX ⇄ C-MTX MTX P^- P^-

ATP

Figure 16. A possible model to account for uphill transport of MTX and THF cofactors into the Ehrlich ascites tumor cell based upon a cotransport with intracellular anionic organic phosphates (P^-) via the high-affinity MTX-THF cofactor carrier. To account for the observation that metabolic poisons augment net uptake of MTX, an exit pump is also proposed as indicated by ATP. The energetics of this pump is based upon the release of free energy in the hydrolysis of ATP. This pump might be coupled in a variety of ways to the high-affinity carrier or might be coupled to another transport mechanism.

of the cell via the high-affinity MTX-THF cofactor carrier is coupled to the simultaneous uphill flow of MTX into the cell. It is of interest that this interaction would not be reduced by the immediate effects of metabolic poisons - the hydrolysis of ATP to ADP and AMP since this would result in the production of compounds at least as inhibitory to the influx of MTX as ATP (Fig. 14). Likewise, the reduction of NAD would produce the more inhibitory NADH (21). This might account for the observation that uphill transport of MTX into the cell can be sustained in the presence of metabolic poisons. The observation that uphill transport is augmented under these conditions would be related to the inhibition of an exit pump (a pump which is driven by the free energy released by the hydrolysis of ATP). Hence, the net level of MTX within the cell at any time would depend upon the relative potency of these processes. Although this model might appear contradictory to the observation that the cell membrane is poorly permeable to organic phosphates, it must be considered that the MTX carrier system is a very low capacity process (see above 1 - 2 μmoles per liter cell water per minute) while the total cell organic phosphate pool is in millimolar quantities. Hence, the loss of organic phosphates via the high-affinity MTX-THF cofactor carrier, while critical to MTX or THF cofactor fluxes, might negligibly affect the total intracellular organic phosphate pool.

E. Clinical Applications of the Thermodynamics of MTX Transport

Since the free intracellular MTX concentration is a critical determinant of cytotoxicity and accumulation of this intracellular component is markedly restricted in mammalian cells, techniques which lead to accumulation of higher levels of free drug in tumor cells should enhance cytotoxicity. One approach is the identification of MTX analogs which might result in compounds which are concentrated within the cell water to a greater extent than MTX. Hence, a compound with a higher affinity for the mechanism producing uphill transport into the cell or a lower affinity for the mechanism which may pump folates out of the cell might be clinically useful. This would be of particular value if the characteristics of the tumor cell transport mechanism were different from that of the susceptible host tissues so that the compound were concentrated to a greater extent in the former than the latter cell. Aminopterin, for instance, is a compound that is more cytotoxic than MTX. It would appear that this is related to the capacity of cells to generate higher net levels of free intracellular drug with a lower net rate of loss of aminopterin from the cell as the extracellular drug concentration falls in comparison to MTX (24,34,35). It has been suggested that the selective toxicity of aminopterin may be related to the greater rate of loss of free drug from small intestinal cells than L1210 leukemia cells after pulse administration of aminopterin to leukemia-bearing mice *in vivo* (34).

Another approach to the development of antifolate compounds might be to identify agents which utilize the transport process for folates which is distinct from the high-affinity MTX-THF cofactor carrier. It remains to be seen, however, how this route relates to the thermodynamics of transport. It will be particularly important to learn more about the apparent exit pump for folates, whether this is linked to the high-affinity transport carrier or whether this is associated with a second transport route.

One aspect of the thermodynamics of MTX transport which has received clinical attention relates to the observation that the periwinkle alkaloids, vincristine and vinblastine, augment the net accumulation of intracellular MTX in a variety of mammalian cells (36-38). This augmentation of the free intracellular MTX level appears to be related to the inhibition of the exit pump (36) and results in the augmentation of MTX inhibition of ^{3}H-UdR incorporation into DNA (7) and the cytotoxicity of MTX to the L1210 leukemia *in vivo* (37). In some clinical treatment programs vincristine is used concurrently with high doses of MTX based upon this interaction between these drugs (39).

The observation that vinca alkaloids augment net accumulation of MTX in much the same way as a variety of metabolic poisons suggested that these agents act in this context as inhibitors of energy metabolism. Further studies (4) showed that vinca alkaloids partially depress the uphill transport of α-aminoisobutyric acid in the Ehrlich ascites tumor cell. The latter observation is of particular interest since this inhibitory effect of vinca alkaloids occurs without a change in trans-membrane gradients of sodium, potassium, or hydrogen ion, and cannot be accounted for on the basis of changes in membrane potential as estimated from the chloride distribution ratio (41,42). These observations support the concept that there may be sources of energy which sustain uphill transport of amino acids in cells in addition to the known cationic cotransport phenomena.

REFERENCES

1. BLAKELY, R.L.: The Biochemistry of Folic Acid and Related Pteridines. North Holland. Amsterdam (1969).

2. HAKALA, M.T.: Biochim. Biophys. Acta 102 (1965) 198.

3. GOLDMAN, I.D., LICHTENSTEIN, N.S., AND OLIVERIO, V.T.: J. Biol. Chem. 243 (1968) 5007.

4. KESSEL, D., HALL, T., AND ROBERTS, D.: Science 150 (1965) 752.

5. GOLDMAN, I.D.: Uptake of Drugs and Resistance, Ch. 8, Drug Resistance and Selectivity: Biochemical and Cellular Basis (E. Mihich, ed.). Academic Press. New York (1973).

6. GOLDMAN, I.D.: Mol. Pharmacol. 10 (1974) 257.

7. GOLDMAN, I.D., and FYFE, M.J.: Mol. Pharmacol. 10 (1974) 275.

8. WHITE, J.C., LOFTFIELD, S., and GOLDMAN, I.D.: Mol. Pharmacol. 11 (1975) 287.

9. WHITE, J.C., and GOLDMAN, I.D.: Mol. Pharmacol. in press.

10. GOLDMAN, I.D.: Cancer Chemother. Rep. 6 (1975) 51.

11. MARGOLIS, S., PHILIPS, F.S., and STERNBERG, S.S.: Cancer Res. 31 (1971) 2037.

12. CHABNER, B.A., and YOUNG, R.C.: J. Clin. Invest. 52 (1973) 1804.

13. SIROTNAK, F.M. and DONSBACH, R.C.: Cancer Res. 33 (1973) 1290.

14. SIROTNAK, F.M., and DONSBACH, R.C.: Cancer Res. 35 (1975) 1737.

15. SIROTNAK, F.M. and DONSBACH, R.C.: Cancer Res. 36 (1976) 1151.

16. WERKHEISER, W.C.: J. Biol. Chem. 236 (1961) 888.

17. WHITE, J.C., POE, M., and GOLDMAN, I.D.: Unpublished data.

18. JACKSON, R.C., and HARRAP, K.R.: Arch. Biochem. Biophys. 158 (1973) 827.

19. ROTHENBERG, S.P., da COSTA, M., and IQBAL, M.P.: Proc. Am. Assoc. Cancer Res. 17 (1976) 106.

20. GOLDMAN, I.D.: Cancer Chemother. Rep., in press.

21. GOLDMAN, I.D.: Ann. N.Y. Acad. Sci. 186 (1971) 400.

22. GOLDMAN, I.D.: Cancer Chemother. Rep. 6 (1975) 63.

23. RADER, I., NIETHAMMER, C., and HUENNEKENS, F.M.: Biochem. Pharmacol. 23 (1974) 2057.

24. SIROTNAK, F.M., and DONSBACH, R.C.: Cancer Res. 32 (1972) 2120.

25. GOLDMAN, I.D.: Proc. Am. Assoc. Cancer Res. 17 (1976) 130.

26. GOLDMAN, I.D.: Biochim. Biophys. Acta 233 (1971) 624.

27. GOLDMAN, I.D., WHITE, J.C., and BAILEY, B.D.: Unpublished data.

28. GOLDMAN, I.D.: J. Biol. Chem. 244 (1969) 3779.

29. GOLDMAN, I.D., SNOW, R., and WHITE, J.C., Fed. Proc. 34 (1975) 807.

30. GOLDMAN, I.D., GUPTA, V., WHITE, J.C., and LOFTFIELD, S.: Cancer Res. 36 (1976) 276.

31. LASSEN, U.V., NIELSON, A.-M.T., and SIMONSEN, L.O.: J. Membrane Biol. 6 (1971) 269.

32. Na-Linked Transport of Organic Solutes (E. Heinz, ed.) Springer-Verlag. Heidelberg (1972).

33. JENNETTE, J.C., and GOLDMAN, I.D.: J. Lab. Clin. Med. 86 (1975) 834.

34. SIROTNAK, F.M., and DONSBACH, R.C.: Biochem. Pharmacol. 24 (1975) 156.

35. GOLDMAN, I.D.: Proc. Am. Assoc. Cancer Res. 17 (1976) 130.

36. FYFE, M.J., and GOLDMAN, I.D.: J. Biol. Chem. 248 (1973) 5067.

37. ZAGER, R.F., FRISBY, S.A., and OLIVERIO, V.T.: Cancer Res. 33 (1973) 1670.

38. BENDER, R.A., BLEYER, W.A., FRISBY, S.A.: Cancer Res. 35 (1975) 1305.

39. JAFFE, N., FREI, E. III, TRAGGIS, D., and BISHIP, Y.: New Engl. J. Med. 291 (1974) 994.

40. FYFE, M.J., LOFTFIELD, S., and GOLDMAN, I.D.: J. Cell. Physiol. 86 (1975) 201.

41. FYFE, M.J., GOLDMAN, I.D.: Fed. Proc. 34 (1975) 250.

42. SCHAFER, J.A., and GOLDMAN, I.D.: Fed. Proc. 35 (1976) 605.

DISCUSSION

GENNARO: Do you have any data on the effect of cooling on the loading and release of methotrexate?

GOLDMAN: The transport system is highly temperature sensitive. The $Q_{27°-37°}$ of the unidirectional fluxes is five to seven depending upon the specific cell. However, alterations in temperature have very little effect on the steady-state level of intracellular methotrexate.

GENNARO: How do you know that the membrane transport system is the major factor in determining the intracellular levels of methotrexate? Could binding play a role in the uptake of methotrexate?

GOLDMAN: In the last analysis, it is very difficult to prove that a compound is free within the intracellular water. We can easily measure and exclude tight binding of methotrexate to dihydrofolate reductase, but loose binding presents more difficult experimental problems. Since the compound is present within the intracellular water in micromolar levels, it is not possible to show cell swelling due to uptake of osmotically active solute - a definitive way of proving that a compound is free within the intracellular water. Hence, we always have to contend with the possibility of loose binding in interpretation of data which suggests uphill transport into cells. On the other hand, the data suggests, in addition, uphill transport out of cells. In this case, we are dealing with a very low intracellular electrochemical potential so that binding is not a factor. While we suggest an exit pump to explain these data, this may also be related to exclusion of drug from aqueous intracellular compartments, or a membrane potential that is much higher than our estimates.

PRESSMAN: Would you care to speculate on the normal function of this methotrexate carrier in view of the fact that methotrexate is a single xenobiotic substance?

GOLDMAN: We consider methotrexate a model for studying the transport characteristics of the tetrahydrofolate cofactors. It shares the same transport mechanism as the tetrahydrofolate cofactors with a comparable affinity for the carrier. The findings for methotrexate are, therefore, relevant to the tetrahydrofolate cofactors. Why, however, there should be an exit pump to drive substances which are so important to biosynthetic processes out of the cells eludes us.

SIEGEL: In kinetic studies using levels of substrates and inhibitors that are practically stoichiometric with the enzyme, an important consideration becomes the amount of enzyme that is present. What is the rate of biosynthesis of the enzyme in relation to the rate of methotrexate entry?

GOLDMAN: In the short experiments that I have shown you today, the rate of biosynthesis of dihydrofolate reductase is so slow as not to be measurable. Further, we have also done these studies with cycloheximide to be sure that there is no *de novo* enzyme synthesis and the results were similar.

LING: Are the concentrations that you use always well below the K_m of the transport carrier?

GOLDMAN: We studied transport of methotrexate at concentrations in the range of the carrier K_m, especially when evaluating exchange phenomena. Clinically, methotrexate may be used at levels orders of magnitude above its K_m. On the other hand, 5-CH_3-H_4-folate is present in the blood at concentrations an order of magnitude below its K_m.

LING: When you stimulate influx of methotrexate with metabolic poisons, is the carrier mediated part stimulated or is it perhaps stimulation of a diffusional part?

GOLDMAN: I did not have time to show you that there is transport heterogeneity for the folates; we are working on that now. However, the effect of metabolic poisons is on the saturable high-affinity carrier, not on a diffusional component. Passive diffusion is not important at these low drug levels.

BRODSKY: In the beginning, you had a slide which showed survival on the abscissa as a function of influx. What do you change that to now?

GOLDMAN: The relationship between half-time for net efflux of methotrexate *in vivo* or the steady-state free intracellular methotrexate concentration *in vitro* as well as methotrexate influx is also a linear function of survival (Sirotnak, F.M., Donsbach, R.C., Cancer Res. 35:1737, 1975 and Ibid. 36:1151 (1976).

Modification of Membrane Function

by Toxicological Agents

LIVER ENDOPLASMIC RETICULUM: TARGET SITE OF HALOCARBON METABOLITES

Edward S. Reynolds

Department of Pathology, Peter Bent Brigham Hospital and
Harvard Medical School
Boston, Massachusetts 02115

ABSTRACT

Initial injury produced by exposure of rats to carbon tetrachloride, halothane, vinyl chloride or trichloroethylene appears to involve the endoplasmic reticulum. First, there is dispersion of the ergastoplasm, then vacuolization and degranulation of the rough endoplasmic reticulum with concomitant retraction of the smooth endoplasmic reticulum into tightly clumped tubular aggregates. In addition, membranes in these tubular aggregates seem to undergo supra-molecular disassembly. Along with this structural disorganization, functional capacity of the organelle diminishes. Activation of these halocarbons to toxic species by functional elements of the endoplasmic reticulum is indicated by the enhancement of their toxicity by pretreatment with chemicals which induce components of the mixed function oxidase system and by the formation of certain metabolites and/or covalently bound products. Insight into the molecular mechanisms of membrane injury brought about by these halocarbon hepatotoxins has been provided by the characterization of chemical changes produced, such as increased lipid diene conjugate content, and the patterns of enzyme deactivation.

REPORT

Increasing numbers of small halogenated hydrocarbons including carbon tetrachloride, vinyl chloride, trichloroethylene and the anesthetic halothane have been demonstrated to cause acute liver injury. Carbon tetrachloride's hepatotoxicity is manifest in normal animals and exacerbated by pretreatments which induce the liver mixed function oxidase system (7,34,38). Halothane (14, 31,35,44) vinyl chloride (11,18,37) and trichloroethylene (6,22) cause liver injury in animals pretreated with phenobarbital or Aroclor 1254 (a polychlorinated biphenyl with 54% chlorine) potent inducers of cytochrome P-450 the terminal oxidase of this drug metabolizing enzyme system. Liver injury caused by these four halocarbons appears to involve the endoplasmic reticulum (ER) in a sequence of changes ranging from dispersion of the topographically distinct forms of this membraneous organelle, to loss of association between its supermolecular components, to collapse of its structure.

Before proceeding to describe the hepatotoxic effects of these halocarbons, certain parameters of the endoplasmic reticulum in normal cells will be reviewed.

Structure and Function of ER

Liver endoplasmic reticulum is a complicated network of membrane-lined cisternae permeating the cytoplasmic matrix. One form of ER, the rough (RER) forms broad sheets the outer surfaces of which are studded with myriads of ribosomes in rosettes and spirals (Figure 1A). The other form, the smooth (SER) is a branching, interconnecting, sparsely granulated network of tubules 50 to 80 nM in diameter (Figure 1B). Under normal conditions RER and SER are segregated; flat cisternae of the RER forming many layered stacks of "ergastoplasm" and the SER forming loosely woven webworks of tubules which permeate the cytoplasmic matrix. Protein synthesis takes place on the ribosomes of the RER, glycogen storage is associated with the SER and the entire membrane system of the ER appears to function in drug and steroid metabolism (43). Isolated liver endoplasmic reticulum, collectively known as microsomes, are rich in protein and phospholipids and relatively poor in cholesterol (protein : phospholipids : cholesterol = 30 : 15 : 1 by weight). Many drugs, such as phenobarbital, and synthetic chemicals, such as polychlorinated biphenyls, cause marked proliferation of the SER.

ER Denaturation by CCl_4

Electron microscopy has revealed the vulnerability of this organelle to injury by halogenated halocarbons. Within 2 hours

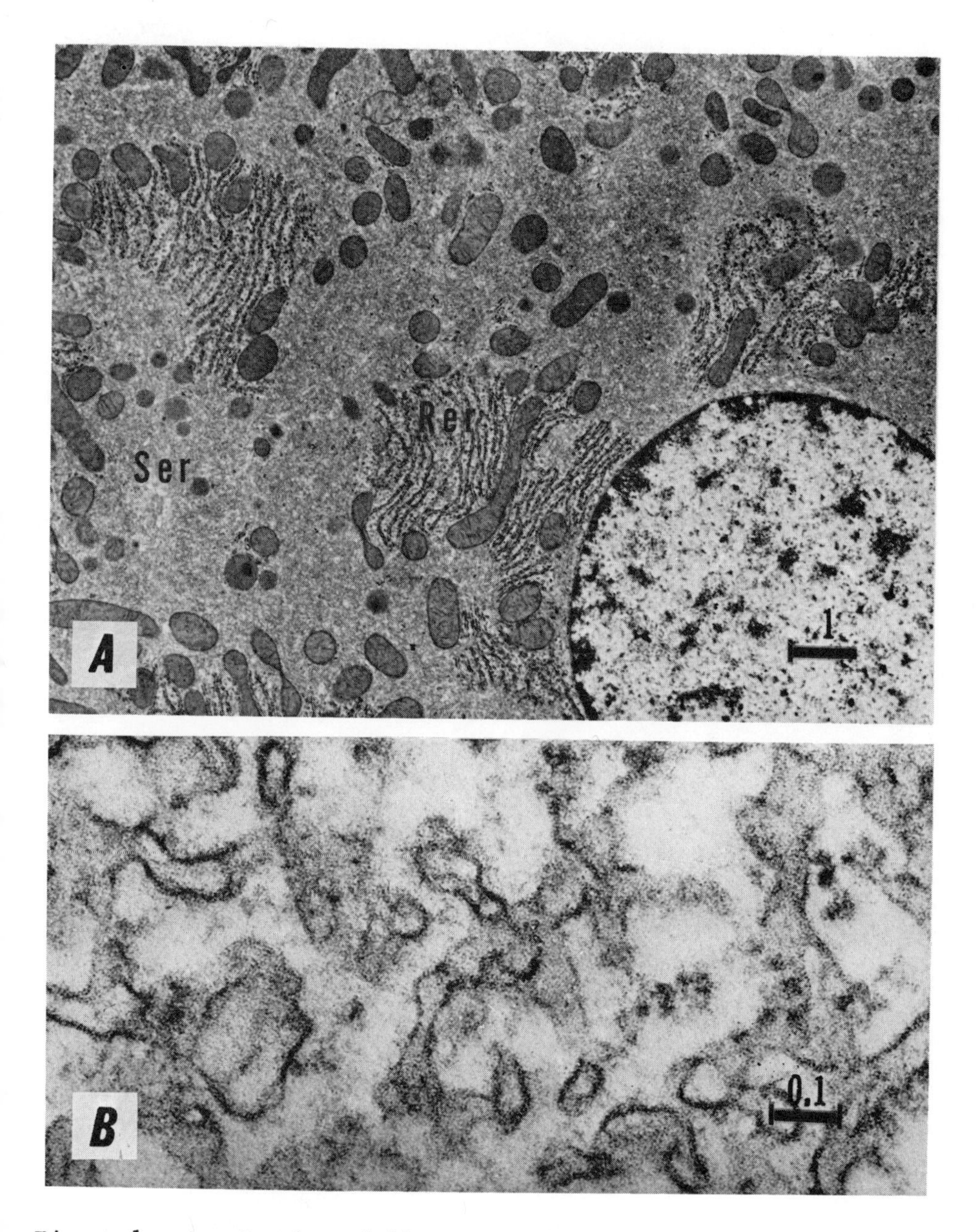

Figure 1. A. Portion of liver parenchymal cell of phenobarbital pretreated rat. Note segregation of rough endoplasmic reticulum (RER) from smooth endoplasmic reticulum (SER). B. Shows higher power of branching tubular profiles of SER. (scale in microns)

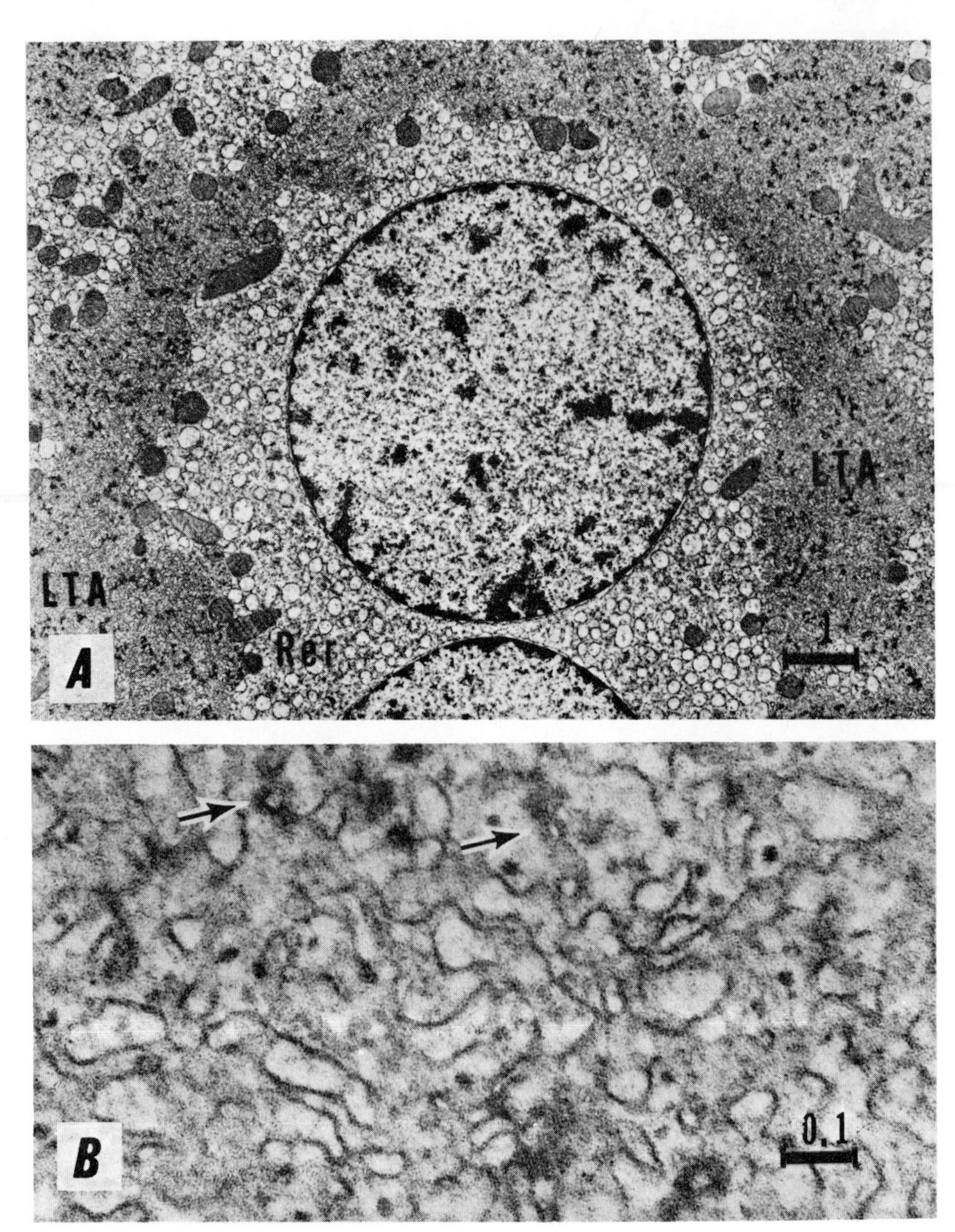

Figure 2. A. Portion of liver parenchymal cell of phenobarbital pretreated rat 2 h after CCl_4(26 mmoles/kg). Vacuolated RER forms "halo" about nucleus while SER forms prominent black-flecked labyrinthine tubular aggregates (LTA). B. Higher magnification of LTA with electron-opaque materials (arrows) apparent on the outer surfaces of tubular profiles. (scale in microns)

following the oral dosing of phenobarbital pretreated animals with CCl_4 (26 mmoles/kg), the endoplasmic reticulum of centrolobular parenchymal cells undergoes startling changes (Figure 2A). The ergastoplasm becomes indiscernable by light microscopy. Cisternae of the RER vacuolate and the outer surfaces of these membranes shed ribosomes. The SER coalesces into twisted masses of tubules liberally flecked with electron-opaque material of unknown composition. Changes in the appearance of the membranes of the SER are apparent as early as one hour. Tubular diameters in the compacted SER aggregates are markedly diminished. Electron-opaque areas considered to be regions of membrane collapse are readily apparent on the outer surfaces of the tubular profiles - particularly the areas of greatest tubular constriction (Figure 2B).

Enzyme Activation of CCl_4

The hepatotoxicity of CCl_4 and of numerous other chemicals can be attributed to the liver's ability to transform these agents into highly reactive molecular species. Carbon tetrachloride is postulated to be activated into a free radical within the endoplasmic reticulum. This may occur either through cleavage to a free radical by an electron transfer reaction (15) (Equation 1), or through interaction with pre-existent free radicals (R·) (Equation 2).

$$e^- + CCl_4 \longrightarrow \cdot CCl_3 + Cl^- \qquad (1)$$

$$R\cdot + CCl_4 \longrightarrow RCl + \cdot CCl_3 \qquad (2)$$

CCl_4 is known to be metabolized to $CHCl_3$, C_2Cl_6, CO_2 and ^{14}C or ^{36}Cl labeled CCl_4 covalently binds to liver constituents (4,12,24, 36). Fowler's (12) detection of C_2Cl_6, the dimerization product of $2\cdot CCl_3$ radicals, after CCl_4 administration is perhaps the strongest evidence for the activation of CCl_4 to a free radical. The activation of CCl_4 is considered to be an NADPH dependent mixed function oxidase process (42).

Figure 3 is a schematization of the multimolecular mixed function oxidase system of the liver considered primarily responsible for the biotransformation (oxidation) of steroids, fatty acids, and many drugs and zenobiotics, including halogenated alkanes and alkenes. Substrates including these halocarbon hepatotoxins bind to the central hemoprotein, and the complex is reduced by single electron transfers via specific flavoproteins from NADPH or NADH. Molecular oxygen reacts with the reduced complex in such a way that one of the oxygen atoms is reduced to water and the other introduced into the organic substrate. Fatty acid desaturation is associated with the cytochrome b_5 reductase - cytochrome b_5-arm. The

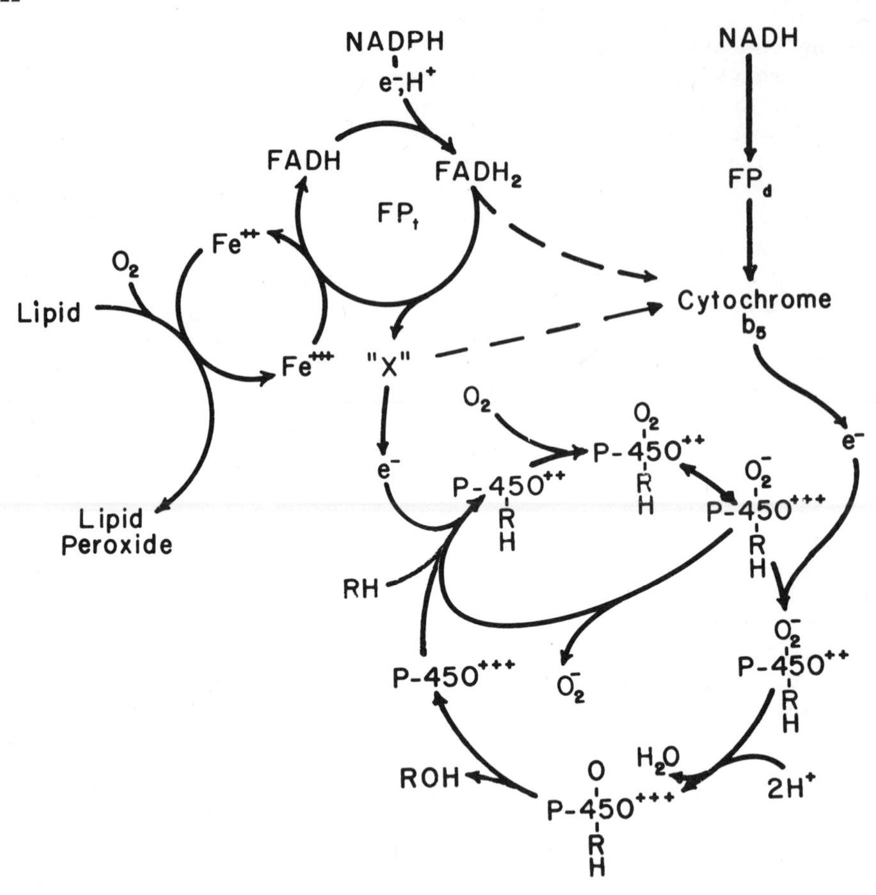

Figure 3. Schematization of liver mixed function oxidase system (37).

NADPH dependent flavoprotein, known as NADPH cytochrome c reductase because it is assayed by using cytochrome c as an electron acceptor, is involved in the peroxidation of unsaturated fatty acids (25). On the basis of this scheme (Figure 3) the R· of equation 2 could be a flavoprotein semiquinone free radical, a reduced P-450 superoxide, a P-450 singlet oxygen, ferrous iron or a pre-existent free radical such as the dissociation product of an organoperoxide.

Effects of CCl_4 Activation

Activation of CCl_4 and similar chemicals to free radicals should lead to specific and predictable alterations in hepatic function and chemical composition. Free radicals normally have short half-lives, are extremely reactive, and electrophilic.

Therefore in an organ attacked by free radicals, the interaction should be rapid, spatially confined near the point of origin and relatively specific for electrophilic submolecular regions - such as sulfhydryl groups on proteins (4) and hydrogens of methylene bridges of polyunsaturated fatty acids (2,17).

Within 30 min. after CCl_4 (26 mmoles/kg) functional capacity and associations between supramolecular components of endoplasmic reticulum are altered (34). *In vivo* protein synthesis decreases concomitant with the release of ribosomes from the membrane and dilitation of cisternae of RER. Certain components or activities of the mixed function oxidase system are also rapidly affected after carbon tetrachloride (28,34,45). Cytochrome P-450 and activities of oxidative-N-demethylase and arene hydrocarbon hydroxylase reactions diminish. Glucose-6-phosphatase activities, an ER enzyme not affiliated with the mixed function oxidase system, are also sharply decreased early in the course of injury. In contrast, other mixed function oxidase components are not diminished after CCl_4 (28,34,41). These include cytochrome b_5 content and the NADPH and β NADH cytochrome c reductase arms of the microsomal transport system. A similar selective pattern of deactivation and/or sparing of microsomal enzymes is found after *in vitro* peroxidation (16,28,51). These "deactivated" enzymes are considered particularly dependent on ER integrity. Similarities between the hepatotoxic effects of carbon tetrachloride and *in vitro* lipid peroxidation have led some investigators to attribute injury by carbon tetrachloride to CCl_4-initiated peroxidation (13,28,29). Increased lipid peroxidation following CCl_4 may, however, be indirect. Deactivation of cytochrome P-450 coupled with retention of the integrity of the NADPH cytochrome P-450 reductase arm could contribute to increased lipid peroxidation since electron transfer from NADPH would be shunted towards the endogenous NADPH-lipid peroxidation system.

Covalent Binding of CCl_4

Within one min. after intragastric feeding of $^{14}CCl_4$ (26 mmoles/kg rat), Rao and Recknagel (27) found non volatile ^{14}C became covalently bound to microsomal lipid and protein and to mitochondrial lipid in the ratio 11 : 3 : 1, with none recoverable in mitochondrial protein. Lipid labeling in microsomes was maximal by 5 min. and remained relatively constant for 90 min. With time, specific activities of mitochondrial protein approached that of microsomal protein.

We have found the pattern of $^{14}CCl_4$ labeling of subcellular constituents relatively stable at times up to 120 min. after $^{14}CCl_4$, with approximately one-quarter of the total bound to microsomal lipid and protein and relatively little to mitochondrial lipid (30,33). The total amount bound after a dose of 26 mmoles/kg $^{14}CCl_4$ is actually quite small (60 nanomole ^{14}C/g liver). Binding in microsomal phospholipid and protein components amounts to approximately 1 mole per 1000 moles phospholipid (estimated molecular weight 775g) and 1 mole per 200 moles protein (estimated molecular weight 50,000g). Virtually no bound CCl_4 metabolites are recoverable in nucleic acids (34).

What is the significance of this binding of labeled CCl_4 to liver constituents? Diaz-Gomez et al. (10) have found increased $^{14}CCl_4$ label in microsomal lipid after phenobarbital pretreatment. Even though phenobarbital pretreatment enhances the toxicity of CCl_4 and moderately increases the amount of ^{14}C- recoverable in liver lipids at 2 and 24 h (Table 1) it does not significantly increase the total liver incorporation of $^{14}CCl_4$. This increase in total ^{14}C- label in liver lipids reflects the pretreatment-induced increase of membrane phospholipid, for the specific activity of the ^{14}C- phospholipid label remains constant or diminishes. Although relatively small amounts of the total dose of $^{14}CCl_4$ are recovered covalently bound to lipids, we have found that the proportion of ^{14}C bound to lipids increases with successively higher doses of $^{14}CCl_4$ (Table 1).

TABLE I: Excretion and Binding to Liver of $^{14}CCl_4$

Pretreatment	Dose CCl_4	Time Sacrifice	24 h Urine	Liver	Protein	Lipid
	μmoles/100 g	h	mμmole ^{14}C/100 g animal			
none	0.83	2		29.0±1.6	7.2±0.6	7.0±0.6
PBT†	0.83	2		34.5±5.9	6.2±2.5	11.6±2.1
none	0.83	24	25.6±1.5	10.8±0.8	4.2±0.3	1.8±0.2
PBT	0.83	24	36.4±3.9	13.2±2.2	5.0±0.2	3.5±0.2
PBT	8.30	24	275±10	150±22	34±5	73±17
PBT	83.0	24	939±49	909±101	189±22	573±89

† phenobarbital 0.1% in drinking water for 30 days

Reactions of $\cdot CCl_3$ with Lipid

Possible reactions of the trichloromethyl radical with polyunsaturated fatty acids, are summarized in Figure 4. Two major metabolic pathways are shown. In the first, the hydrogen abstraction pathway, $\cdot CCl_3$ radicals abstract labile hydrogens from methylene bridges of polyunsaturated fatty acids with the formation of

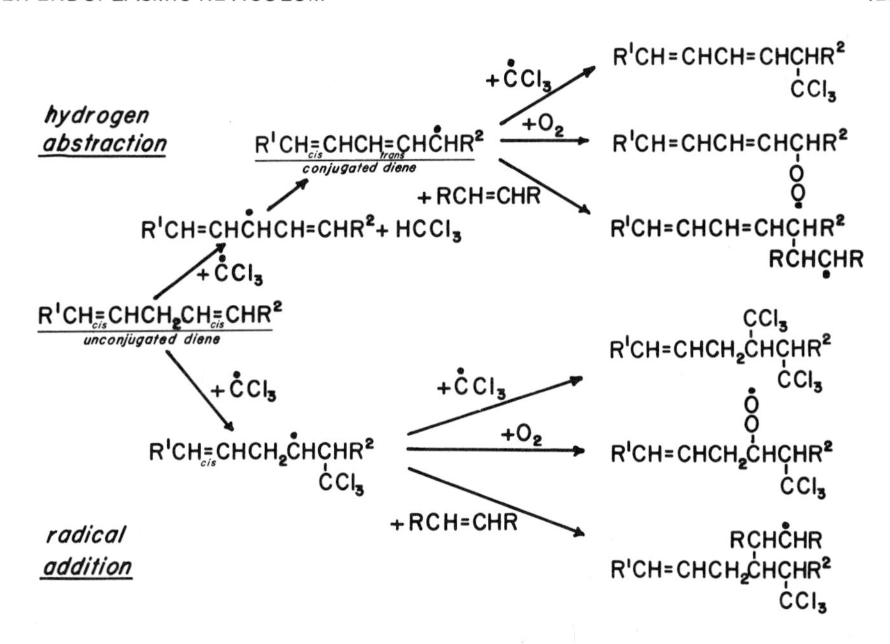

Figure 4. Hydrogen abstraction and chloromethyl radical addition pathways of CCl· reaction with polyunsaturated fatty acids, followed by subsequent radical annihilation, peroxidation and polymerization.

free radical lipid molecules and $CHCl_3$. One of the double bonds in the lipid free radical shifts to a conjugated diene configuration. Recknagel and Rao (26) detected an enhanced conjugated diene absorption spectra in lipids of rat liver microsomes 5 min. after CCl_4. This spectra rises to a peak value between 15 and 30 min., and progressively declines to normal levels by 12 h (19). Increases in diene conjugation occur following as little as 26 μmoles CCl_4/kg rat and the magnitude of increase (approximately 6 mole/1000 mole phospholipid) is relatively constant regardless of dose (33). Thus conjugated dienes must be considered as qualitative 'transient's stages in the alteration of ER membrane polyunsaturated fatty acids. As shown in Figure 4, the conjugated lipid diene could combine with a second $\cdot CCl_3$ radical in an annihilation reaction, with molecular oxygen in a peroxidation initiation reaction or with another unsaturated fatty acid in a polymerization reaction. $CHCl_3$ formed by the initial abstraction would either be expired, metabolized to form a mercapturic acid conjugate or to CO_2.

In the second pathway, the free radical addition pathway, the trichloromethyl radical would add across a double bond. Conjugated dienes would not result, subsequent reactions of the branch chain chloromethyl radical could be similar to the first.

According to this scheme (Figure 4), the initial reaction which results in diene conjugation is not the same as that responsible for covalent binding. Therefore, diene conjugation and $CHCl_3$ production would indicate the hydrogen abstraction pathway, while covalent binding to lipids would indicate an alternative pathway of radical attack.

Lipid Peroxidation and Injury

Polyunsaturated lipid radicals sequestering a molecule of diradical oxygen would be expected to undergo further peroxidative decomposition as indicated in the reaction scheme of Figure 5. Measureable products of peroxidative decomposition include alkyl aldehydes and alkyl fatty acids with terminal aldehydes (9,46) malonaldehyde and fluorescent conjugation products of malonaldehyde

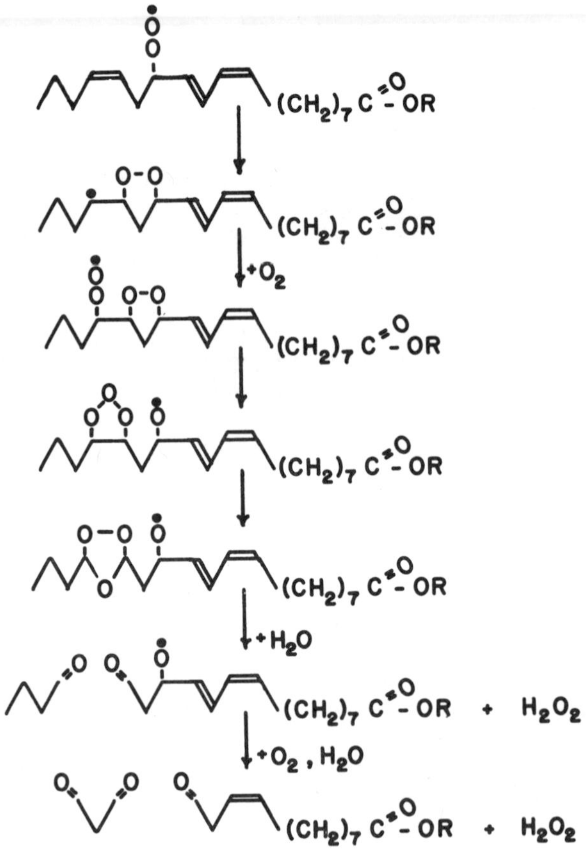

Figure 5. Reactions of polyunsaturated lipid radicals with oxygen (adapted from Dahle et al., (19)). Terminal products include alkyl aldehydes, alkyl fatty acid with terminal aldehydes and malonaldehyde.

with primary amines (1). Malonaldehyde is measurable only *in vitro* since it is metabolized rapidly by mitochondria (28). Peroxidative decomposition of the fatty acid "tails" of membrane fatty acids leads to loss of polyunsaturated fatty acids (Figure 5). Four h after CCl_4 (26mmoles/kg), Squotas (40) found 20% less arachidonic acid (20:4), a strikingly depletion of 20:5 and 22:5 polyunsaturated fatty acids, and a concomitant increase in shorter more saturated, linoleic (18:2) and stearic (18:0), fatty acids of rat liver lipids. Comporti et al. (8) demonstrated that this change in fatty acid composition occurred predominantly in microsomal phospholipids. Introduction of polar groups into the fatty acid tails could alter chemical properties of the membrane interior and result in the structural disorganization of the ER seen by electron microscopy.

In order to determine if lipid peroxidation is a major factor in CCl_4-induced liver injury, several investigators have looked at the effects of CCl_4 activation in microsomal fractions under conditions which minimize peroxidative reactions. When $^{14}CCl_4$ was added to NADPH reduced microsomes under anerobic conditions, $CHCl_3$ was formed and ^{14}C was covalently bound to microsomal proteins (47). When the reaction was run in air, $CHCl_3$ production and ^{14}C binding to proteins was markedly reduced (47). However when CCl_4 is activated in NADPH reduced microsomes in the presence of air, cytochrome P-450 is rapidly destroyed and lipids peroxidized (malonaldehyde produced) (13,29). In contrast, cytochrome P-450 contents are not depleted and lipids not peroxidized when the reaction is run under anerobic conditions in the presence of EDTA (13, 28). Thus *in vitro* lipid peroxidation but not $CHCl_3$ formation or covalent binding is associated with functional loss due to CCl_4 activation. These findings support "CCl_4-induced peroxidation" as a cause of cytochrome P-450 loss and membrane injury (13,28). However it should be pointed out that in these *in vitro* experiments, manipulations of oxygen and antioxidant concentrations have altered the normal "pathways" of CCl_4 activation, and not just minimized peroxidative reactions.

Significance of $CHCl_3$ Production In Vivo

In order to gain insight into the importance of the hydrogen abstraction pathway of CCl_4-induced liver injury under *in vivo* conditions, we are comparing the relationship between $CHCl_3$ exhalation and liver injury after CCl_4 administration in control and Aroclor 1254 induced rats. Carlson (7) has shown that pretreatment with this polychlorinated biphenyl potentiates CCl_4 hepatotoxicity. Male rats weighing 200 g were given 150 µmoles/kg Aroclor 1254 by gavage for 7 days while the controls were similarly pretreated with the administrative vehicle (water with traces of Tween 80). CCl_4

was given by gavage, animals placed in all glass metabolism cages and $CHCl_3$ expired into the chamber air monitored by gas chromatography at hourly intervals. Liver injury was estimated by measuring the activity of the liver enzyme glutamic oxalacetic transaminase in the serum at 24 h. Enhanced production of $CHCl_3$ by Aroclor 1254 animals in the first hour following CCl_4 seems to correlate with the exacerbation of CCl_4 induced liver injury (Table 2).

TABLE II: Enhancement of $CHCl_3$ Production and Liver Injury (SGOT)† Following 1.5 mmoles CCl_4/kg by Aroclor 1254 Pretreatment‡

Pretreatment	$CHCl_3$ in 1st h	SGOT at 24 h
	mμmoles/h	Karmen Units
Vehicle (6)*	18±8	1014±276
A-1254 (5)	235±109	5013±1052

† serum glutamic oxalacetic transaminase-control value 150-200

‡ 150 μmoles/kg p.o. for 7 days

* number of animals

Figure 6 illustrates ways the $\cdot CCl_3$ radical by hydrogen abstruction and chlorocarbon addition, could alter the integrity and components of a membrane. Chloromethylation, desaturation, polymerization, shortening, and oxidation of the fatty acid tails could radically alter the hydrophobic membrane interior, disrupt their surface properties, and impair the fluidity of lipid-protein associations necessary for enzyme function. Alterations in the physical-chemical properties of membrane surface could result in loss of ribosomes and thus impair protein synthesis. The destabilized endoplasmic reticulum membrane could then coalesce into non functional labyrinthine tubular aggregates.

ER Denaturation by Other Haloalkanes and Chloroethylenes

While it is not clear whether CCl_4 causes liver injury by covalent binding of free radical metabolites to tissue macromolecules or by initiating lipid peroxidation, the hepatocellular alterations produced are not unique for CCl_4. Two other halomethanes, CHI_3 (39) and $CBrCl_3$ (5,20) cause similar morphologic, chemical and functional changes in normal rats. The tendency of CHI_3 and $CBrCl_3$ to cleave homolytically - a measure of free radical reactivity - is similar to CCl_4 (20,39). Therefore these halomethanes may be activated by a similar mechanism.

The polyhalogenated anesthetic halothane ($CF_3CHBrCl$) may also be activated to a radical species following an electron capture

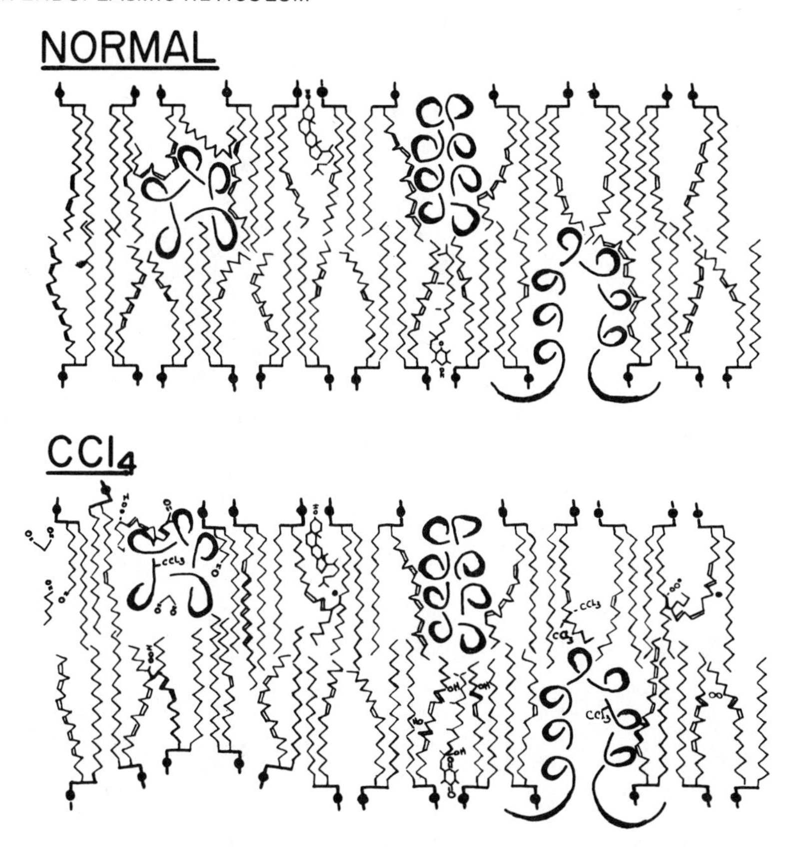

Figure 6. Conceptualization of alterations in intra- and intermolecular relationships in the membrane interior brought about by CCl_4.

---

reaction with loss of Br^- in liver ER. This proposed halothane radical could abstract labile hydrogens, initiate lipid peroxidation, or bind to cellular components in a manner analogous to that of CCl_4. Transient increases in lipid conjugated diene and decreased contents of cytochrome P-450 have been found in phenobarbital and Aroclor 1254 pretreated rats exposed to halothane (3,31, 35,44). Hepatic injury after halothane is focal in the phenobarbital animals (14,31,44) and widespread in Aroclor 1254 animals (35). Numerous *in vivo* and *in vitro* studies have shown that halothane labeled with ^{14}C or ^{36}Cl covalently binds to liver, particularly microsomal, proteins and lipids. Anerobic or hypoxic conditions enhance label incorporation (48,49,50). However, pretreatments

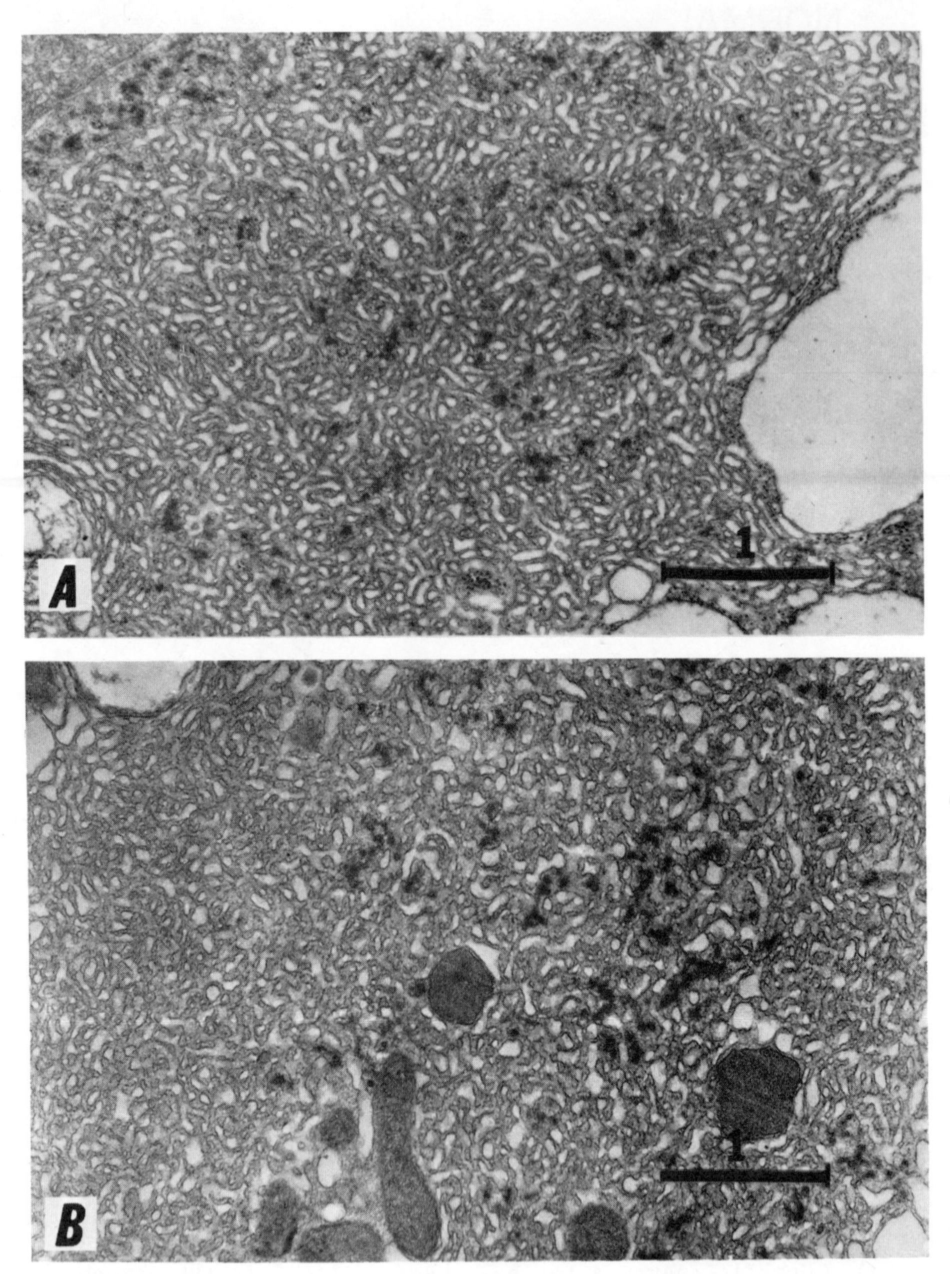

Figure 7. Tubular aggregates derived from endoplasmic reticulum of liver parenchymal cells of phenobarbital pretreated rats 24 h following exposure to A (halothane 0.85% x 5 h) and B (vinyl chloride 5% x 6 h). Note electron-opaque deposits. (scale in microns)

which potentiate halothane's hepatotoxicity have not been found to enhance the in vivo covalent binding of ^{14}C halothane metabolites to liver (32,35). Aroclor 1254 pretreated animals do not excrete more ^{14}C metabolite into their 24 h urine than non induced control animals fed a similar amount of ^{14}C- halothane (35). Morphologic changes of the ER in the phenobarbital or Aroclor 1254 pretreated animal after halothane appear similar to those caused by CCl_4 (Figure 7A).

Chloroethylenes are also thought to be metabolized by NADPH-dependent mixed function oxidase components but via reactive epoxides which can rearrange to aldehydes or be converted to alcohols or acids (37). We have found exposure of phenobarbital pretreated animals to vinyl chloride (5% x 6 h) (36) or trichloroethylene (1% x 2 h) (23) causes decreased P-450 contents and loss of oxidative N-demethylase activities - changes similar to those following CCl_4. Indeed the ultrastructural lesion produced by these chloroethylenes (Fig. 7B) is similar to that following CCl_4 (21,37). Enhancement of trichloroethylene's hepatotoxicity is associated with enhanced production of oxidized metabolites (22). Chemical changes produced by these two chloroethylenes have not been characterized. Although it is possible that similar structural-functional changes can be caused by different molecular mechanisms, it is not clear by what mechanisms chloroethylenes produce endoplasmic reticulum damage.

ACKNOWLEDGEMENTS

This work was supported by Grants AM-16183, HL-06370, GM-07309 from the National Institutes of Health. I would like to thank Mary Treinen Moslen, Hoe Jung Ree, Sandor Szabo, Paul Boor, Kathryn Bailey, Teri Paolini Kingsley and Mary Cook for their contributions to these studies.

REFERENCES

1 BIDLACK, W.R.: Ph.D. Dissertation (1972) p. 94.

2 BOLLAND, J.L. and KOCH, H.P.: J. Chem. Soc. (1945) p. 445.

3 BROWN, B.R.,Jr.: Anesthesiology 36 (1972) 458.

4 BUTLER, T.C.: J. Pharmacol. Exptl. Therap. 134 (1961) 311.

5 CALLIGARO, A., CONGIU, L., TOCCO, L., and VANNINI, V.: La Sperimentale 121 (1971) 121.

6 CARLSON, G.P.: Res. Comm. Chem. Pathol. Pharmacol. 7 (1974) 637.

7 CARLSON, G.P.: Toxicology 5 (1975) 69.

8 COMPORTI, M., LANDUCCI, G. and RAJA, F.: Separatum Experientia 27 (1971) 1155.

9 DAHLE, L.K., HILL, E.G., and HOLMAN, R.T.: Arch. Biochem. Biophys. 98 (1962) 253.

10 DIAZ-GOMEZ, M.I., CASTRO, J.A., deFERREYRA, E.C., D'ACOSTA, N., deCASTRO, C.R.: Toxicol. Appl. Pharmacol. 25 (1973) 534.

11 DREW, R.T., HARPER, C., GUPTA, B.N., and TALLEY, F.A.: Envirn. Health Persp. 11 (1975) 235.

12 FOWLER, J.S.L.: Brit. J. Pharmacol. 37 (1969) 733.

13 GLENDE, E.A.,Jr., and RECKNAGEL, R.O.: Fed. Proc. 33 (1974) 219.

14 GOPINATH, C., and FORD, E.J.H.: J. Pathol. 110 (1973) 333.

15 GREGORY, N.L.: Nature 212 (1966) 1460.

16 HÖGBERG, J., BERGSTRAND, A., and JAKOBSSON, S.V.: Europ. J. Biochem. 37 (1973) 51.

17 HOLMAN, R.T.: Progress in the Chemistry of Fats and Other Lipids, p. 51 (Holman, R.T., Landsberg, O., Malkin, T., Eds) Academic Press, New York (1954).

18 JAEGER, R.J., REYNOLDS, E.S., CONOLLY, R.B., MOSLEN, M.T., SZABO, S., and MURPHY, S.D.: Nature 252 (1974) 724.

19 KLAASSEN, C.D. and PLAA, G.L.: Biochem. Pharmacol. 18 (1969) 2019.

20 KOCH, R.R., GLENDE, E.A., Jr., RECKNAGEL, R.O.: Biochem. Pharmacol. 23 (1974) 2907.

21 MOSLEN, M.T., REYNOLDS, E.S., SZABO, S.: Fed. Proc. 35 (1976) 375.

22 MOSLEN, M.T., REYNOLDS, E.S., SZABO, S.: Biochem. Pharmacol. (in press).

23 MOSLEN, M.T., REYNOLDS, E.S., BOOR, P.J., SZABO, S. (in preparation).

24 PAUL, B.B., and RUBINSTEIN, D.: J. Pharmacol. Exptl. Therap. 141 (1963) 141.

25 PEDERSON, T.C. and AUST, S.D.,: Biochem. Biophys. Res. Comm. 48 (1972) 789.

26 RAO, K.S. and RECKNAGEL, R.O.: Exptl. Molec. Pathol. 9 (1968) 271.

27 RAO, K.S. and RECKNAGEL, R.O.: Exptl. Molec. Pathol. 10 (1969) 219.

28 RECKNAGEL, R.O., GLENDE, E.A., Jr.: Crit. Rev. Toxicol. 2 (1973) 263.

29 REINER, O., ATHANASSOPOULOUS, S., HELLMER, K.H., MURRAY, R.E., and UEHLEKE, H.: Arch. Toxicol. 29 (1972) 219.

30 REYNOLDS, E.S.: Pharmacol. Exptl. Therap. 155 (1967) 117.

31 REYNOLDS, E.S., and MOSLEN, M.T.: Biochem. Pharmacol. 23 (1974) 189.

32 REYNOLDS, E.S., and MOSLEN, M.T.: Biochem. Pharmacol. 24 (1975) 2075.

33 REYNOLDS, E.S. and REE, H.J.: Lab. Invest. 25 (1971) 269.

34 REYNOLDS, E.S., REE, H.J., MOSLEN, M.T.: Lab. Invest. 26 (1972) 290.

35 REYNOLDS, E.S., MOSLEN, M.T., SZABO, S.: Fed. Proc. 38 (1976) 376.

36 REYNOLDS, E.S., MOSLEN, M.T., SZABO, S., JAEGER, R.J.: Res. Comm. Chem. Pathol. Pharmacol 12 (1975) 685.

37 REYNOLDS, E.S., MOSLEN, M.T., SZABO, S., JAEGER, R.J., MURPHY, S.D.: Am. J. Pathol. 81 (1975) 219.

38 SEAWRIGHT, A.A. and McLEAN, A.E.M.: Biochem. J. 105 (1967) 1055.

39 SELL, D.A., and REYNOLDS, E.S.: J. Cell Biol. 41 (1969) 736.

40 SGOUTAS, D.S.: Metabolism 16 (1967) 382.

41 SLATER, T.F. and SAWYER, B.C.: Biochem. J. 111 (1969) 317.

42 SLATER, T.F. and SAWYER, B.C.: Biochem. J. 123 (1971) 805, 815 and 823.

43 SMUCKER, E.A. and ARCASOY, M.: Int. Rev. Exptl. Pathol. 7 (1969) 305.

44 STENGER, R.J. and JOHNSON, E.A.: Proc. Soc. Exptl. Biol. Med. 140 (1972) 1319.

45 STRIPP, B., HAMRICK, M.E. and GILLETTE, J.R.: Biochem. Pharmacol. 21 (1972) 745.

46 TAM, B.K. and McCAY, P.B.: J. Biol. Chem. 245 (1970) 2295.

47 UEHLEKE, H., HELLMER, K.H., TABARELLI, S.: Xenobiotica 3 (1973) 1.

48 UEHLEKE, H., HELLMER, K.H., and TABARELLI-POPLAWSKI, S.: Naunyn-Schmiedeberg's Arch. Pharmacol. 279 (1973) 39.

49 Van DYKE,R.A., and GANDOLFI, A.J.: Drug. Met. Disp. 4 (1976) 40.

50 WIDGER, L.A., GANDOLFI, A.J., and Van DYKE,R.A.: Anesthesiology 44 (1976) 197.

51 WILLS, E.D.: Biochem. J. 123 (1971) 983.

DISCUSSION

PRESSMAN: Isn't there a pathological condition of liver mitochondria, cloudy swelling, that is produced by chronic carbon tetrachloride exposure?

REYNOLDS: Cloudy swelling, a term coined by Virchow mentioned by Dr. Orbison this morning, is an increase in the opacity of thin slices of liver when they are held up to the light. This "archaic" pathological observation probably corresponds to swelling and vacuolization of the endoplasmic reticulum.

PRESSMAN: But, there is no doubt that carbon tetrachloride itself is a potent uncoupling agent of mitochondria and one of the questions I had is whether or not the effects that are seen at the very least are synergistically related to damage to the liver mitochondria which are then impaired in their ability to synthesize ATP, and some of the ATP requiring reactions for the maintenance of the ER.

REYNOLDS: At the time these early changes are seen there is no lesion in the mitochondria.

PRESSMAN: But all you have to do is to expose mitochondria to the carbon tetrachloride, you immediately uncouple them.

REYNOLDS: But, it requires a certain very low level of carbon tetrachloride. Uncoupling in mitochondria *in vitro* occurs at contents of about 15 micromoles/gm. mitochondria (Reynolds et al, J. Bidchem. 237, 3546,1972). Highest contents of carbon tetrachloride reached in the liver are around one micromole/gm. (Reynolds and Yee, Lab Invest 16, 591,1967). As a matter of fact, the mitochondrial lesion in carbon tetrachloride poisoned animals is essentially due to calcium uptake by the injured liver cells!

PRESSMAN: I just want to point out that your method of isolating mitochondria, which cannot metabolize carbon tetrachloride, would have essentially washed out all the carbon tetrachloride by the time you isolated them. The important thing is where the metabolites which are fixed end up, which we have to presume is the ER, but how much carbon tetrachloride was in the mitochrondria *in situ* which could have been considerably more than the labelled carbon tetrachloride that you found associated with the isolated mitochondria which were washed copiously with aqueous media during isolation.

REYNOLDS: I actually never measured the carbon tetrachloride content of mitochondria isolated from the animals, but the maximum content one obtains in liver following an oral dose of 26 millimoles/Kg animal is on the order of about 2 micromoles/gm. liver. One can drop the dose a hundred-fold and still obtain injury at dose levels where carbon tetrachloride is not detectable in the liver. As a matter of fact, in the case of Arochlor pretreated animals one can now produce injury with doses on the order of one microliter per kilogram or about 15 micromoles carbon tetrachloride.

PRESSMAN: Do nonhalogented hydrocarbons fail to produce any comparable lesions at all?

REYNOLDS: To the best of my knowledge they do not.

PRESSMAN: But you have used them as controls?

REYNOLDS: Yes.

LAKOWICZ: I would like to question you on the cause and effect of the damage to liver, the dosage of 26 millimoles per kilogram that you were using comes out to around 3 gms. per kilogram. That is probably roughly equivalent to the amount of phospholipid that is present in the liver, so that you probably have very significant mole fractions of carbon tetrachloride within the lipid bilayers themselves. Given the low solubility of carbon tetrachloride in water, I think the limit is about 2 millimolars, isn't there likely to be a combined effect in terms to toxicity? That is maybe there is some potentiation from phenobarbital treatment but there might be an equal amount of damage caused simply by the incorporation of a large amount of a hydrophobic molecules into the membrane.

REYNOLDS: Carbon tetrachloride content in the liver can rise to about 3 micromoles per gram at that dose level. I think that the amount of phospholipid in the liver cell at that level is about 40 micromoles per gram. This is a 10 to 1 ratio. In *in vitro* experiments with mitochondria, the ratio had to be almost equimolar before uncoupling occurred.

LAKOWICZ: How was the carbon tetrachloride added? I ask this because you seem to be exceeding the water solubility of the carbon tetrachloride.

REYNOLDS: These were added as saturated sucrose solutions.

SIEGEL: I don't understand exactly what you mean in using concentration units such as micromolar or millimolar since the material is probably not in a uniform homogeneous solution in the tissue, which is what the unit of moles per liter would imply.

REYNOLDS: Well, actually, the best way to probably compare these is by content in moles per gram.

SIEGEL: You really can't make a direct comparison anyway since in order to make any comparison you have to assume that the material was not sequestered substantially in any one particular compartment, and this assumption is really what you are questioning to begin with.

LAKOWICZ: It is certain that the carbon tetrachloride will be partitioned into the phospholipid bilayers because of its solubility properties; so there will be a several hundred-fold concentration difference between the aqueous phase carbon tetrachloride concentration and the lipid phase carbon tetrachloride concentration. Another point I think needs clarification is the loss of carbon tetrachloride because we find it is easily volutilized out of phospholipid bilayers. Membranes may pose no significant barrier to the loss of carbon tetrachloride.

REYNOLDS: Well, I haven't actually worked in *in vitro* systems for about 10 years. All this presented here is essentially *in vivo* animal experiments. Earlier *in vitro* experiments were all done in the cold and at that time I did not have the capabilities to measure carbon tetrachloride contents.

PASSOW: Is there any evidence that cell membranes are also affected?

REYNOLDS: You mean plasma membranes?

PASSOW: Yes, plasma membranes.

REYNOLDS: I think that is an open question. At the present time the answer requires their isolation and the separation of altered plasma membranes from normal plasma membranes. This is a difficult task!

THE ROLE OF MEMBRANE DAMAGE IN RADIATION-INDUCED CELL DEATH

Tikvah Alper, D.Sc.

Gray Laboratory of the Cancer Research Campaign

Mount Vernon Hospital, Northwood, HA6 2RN, U.K.

ABSTRACT

Radiation-induced cell death is probably mediated primarily through deposition of energy, in single events, in a few vital macromolecules, or targets, the integrity of which is indispensable for proliferation. The genome is customarily regarded as the main target, but several lines of evidence support the inference that there are important consequences of events in nuclear membranes in eukaryotes, and plasma membrane in bacteria.

The identification of a target depends to some extent on parallelism between modifications of biological damage to putative targets and to the cell as a whole. An important modifying procedure is removal of oxygen from the irradiated system. The presence of oxygen almost always sensitizes cells, but when model systems with biological function are irradiated extra-cellularly a high degree of sensitization by oxygen has been observed only with those in which membrane function is important. This makes sense because the lipid content of membranes renders them readily peroxidizable. When the quality of the radiation is changed, its effectiveness changes in opposite directions for subcellular model targets and for cells. This could be accounted for if interactions between lesions in membranes and in attached DNA play a substantial role in cellular radiation effects.

INTRODUCTION

"Cell Death"

Most radiobiological research is done in one of two contexts: radiotherapy (mainly of malignant disease), or protection against radiological hazards. It would seem reasonable, at least to scientists, that, as our understanding of basic mechanisms of radiation action increases, so will methods be evolved for the more effective use of radiation, or protection against it; and it is to be regretted that there is currently a tendency for the funding of research to favour 'practical' or 'clinical' radiobiology at the expense of more basic research. In this paper I shall be concerned with a very basic problem, namely the location of those primary events which, after a long series of biochemical steps, result in the death of cells.

In all that follows I attach a restricted meaning to the term 'viability' of a cell (whether lower or higher). This derives from well-established usage in microbiology: viable organisms are those capable of giving rise to colonies of similar ones. Thus an irradiated cell will be regarded as viable if it retains its capacity to proliferate and give rise to as many daughters as if it had not seen radiation; otherwise it will be regarded as 'dead'. According to this definition, a fully differentiated higher cell is, of course, no longer viable, and this sometimes seems an odd terminology to workers in other fields. However, the 'killing' of cells by radiation, in the proliferative sense, is the most important endpoint to be considered in radiotherapy-oriented radiobiology research, and it has its importance also in the protection field. In the latter, the danger that undesirable mutations may be induced is, of course, of prime importance when 'permissible levels' of radiation are being considered. But, with higher doses, an undesirable effect of radiation may be damage to the haemopoietic tissue, for example; and this damage can be ascribed to the killing of stem cells. Indeed, understanding the mechanism of mutation induction itself involves also an understanding of the mechanisms of cell killing since mutations can be observed only if the cells in which they are induced remain viable.

In the therapy of malignant disease the object of any treatment is, of course, precisely to stop the proliferation of the neoplastic cells. This is why ionizing radiation is used. But the magnitude of the dose given in radiotherapy is limited by the effects of radiation on normal tissue: the more we learn about these undesirable effects, the more clearly it emerges that these, too, can be ascribed to the proliferative death of 'clonogenic' cells. Thus radiobiologists understandably devote a great deal of attention to the mechanisms by which cells are killed: and to how radiation could at will be made both more and less effective in this respect.

The absorption of radiation by the matter through which it passes occurs in discrete quanta. With ionizing radiation, i.e. radiation of energy of more than one or two hundred electron volts, the biologically damaging events used to be referred to as ionizations, but this terminology is now out of favour, since it is not known whether the process by which energy is absorbed from radiation, in condensed media, is truly analogous with the ionizations that are observable in gases. In present day common usage, we tend rather to refer to 'energy-deposition events'. Conceptually, we identify the first event of importance as that which creates a free radical, i.e. a molecule with an unpaired electron. The question which we attempt to answer is this: does it matter in which cellular macromolecules free radicals are engendered; and, if it does matter, which are the target molecules or structures?

The Target Concept

Radiobiological experience supports the target concept, in the sense that the cell's ability to proliferate is lost as a consequence of radiation-induced lesions in one, or at most a few, vital structures. It is the essential function of these critical structures, rather than their unusual 'sensitivity' to radiation, which confers the status of 'target' upon them. Since the cells' hereditary continuity is a prerequisite of proliferation, it is self-evident that the genetic apparatus, as a whole, must constitute a target or a collection of targets, and it is not surprising that, since the 1940s, so much attention has been focussed on nuclear DNA as the target material for radiation damage, nor that attempts continue to be made to account for all cellular radiobiological phenomena in terms of the primary interaction of radiation with DNA. What tends to be forgotten is that the DNA does not float freely and independently within the cell, so that a primary event in another structure might well result in damage to the genome. Since DNA is attached to the nuclear membrane, in higher cells, and to the plasma membrane in bacteria, the integrity of those structures may also be critical for cell division, and there is inferential evidence - in my opinion good evidence - of the importance of energy deposition in cell membranes.

In this respect, it is useful to compare and contrast the effects of ionizing radiation with those of ultraviolet light in the 'germicidal' range, namely about 250 to 270 nm. Light of that wavelength is, of course, specifically absorbed by the nucleic acids, so it is entirely plausible that germicidal UV should exert its effects by virtue of the primary absorption of energy in DNA (or RNA). The killing of lower cells by UV is satisfactorily associated to a large extent with the formation of pyrimidine dimers, though it is not quite clear whether those products are also to be regarded as responsible for the killing of higher cells. On the other hand, there is no reason to expect that cell membranes should absorb UV in that range of wavelengths. With ionizing radiation, on the other hand, there is no specific energy absorption: a given quantum has equal probabilities of being absorbed anywhere, at least in biological material. Differences between effects of germicidal UV and of ionizing radiation are therefore useful as possible pointers to results of energy deposition in structures other than the genome itself. Other means of testing hypotheses concerning the targets for the biological action of radiation are afforded by the existence of methods for modifying the damage.

If any physical or chemical agent reduces or enhances the biological damage to the cell as a whole, and if the damage is attributable to the absorption of energy in a particular macro-molecular structure, then it is to be expected that the agent should similarly modify (i.e. repair, or irreparably fix) the lesion which is thought to be responsible for the biological effect. Physico-chemical tests of damage to biological macro-molecules are inadequate, unless we know how to assess the biological significance of the damage under investigation.

Two important methods of modification will be considered here; one is by changing the quality of the radiation, the other is by changing the oxygen concentration within the cells undergoing irradiation.

THE OXYGEN EFFECT IN RADIOBIOLOGY

Oxygen is perhaps the most important modifying agent of ionizing radiation damage to cells. In general, it is a sensitizing agent for almost every kind of damage with almost every kind of cell. That is to say, a greater dose of radiation is required to yield comparable effects on anoxic than on well-oxygenated cells: in some cases three, four or even five times as great a dose is required to bring about the same effect, when the irradiated cells are made anoxic. There is ample evidence that, irrespective of the respiratory state of the cell, it is the presence or absence of oxygen during irradiation (or within milliseconds thereafter) that is critical (9, 12, 34).

The oxygen effect is of more than academic interest. It has been recognized for many years that cancerous tumours are likely to contain some cells which are hypoxic, and therefore comparatively radioresistant. Certain methods of dealing with this problem are, indeed, in the process of clinical trial.

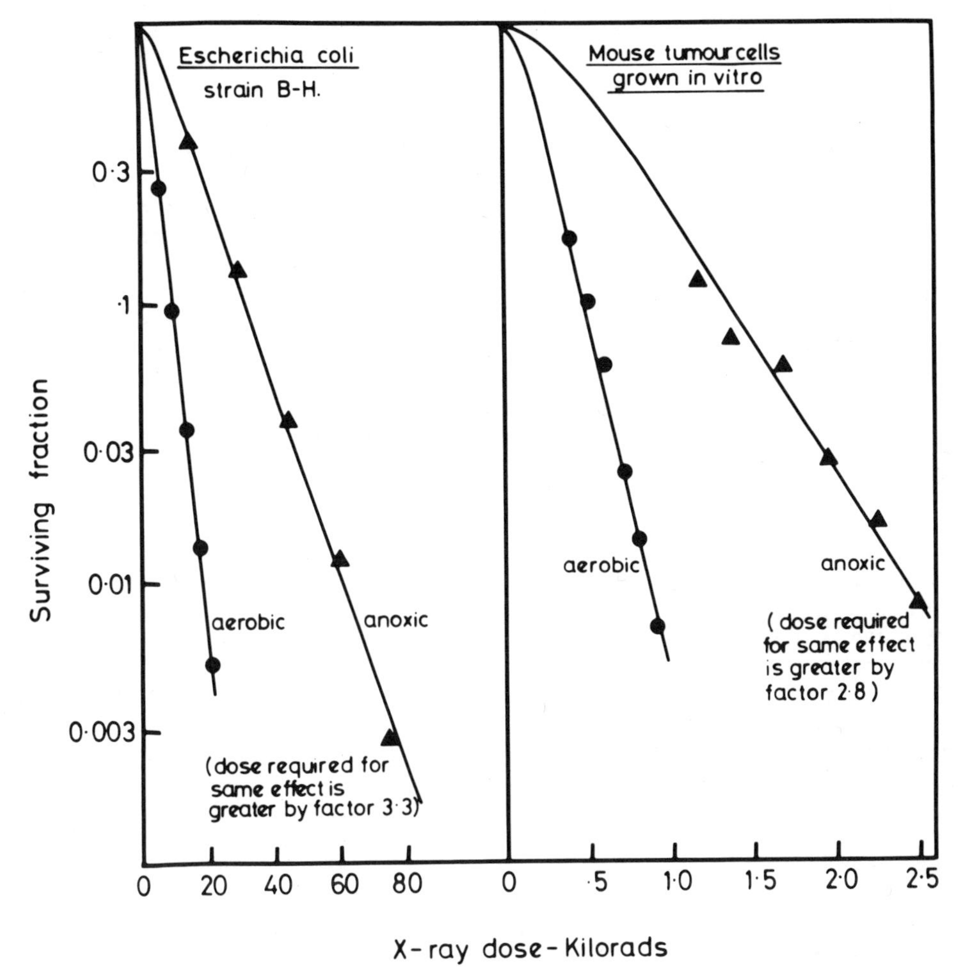

Figure 1. Typical cell survival curves to show radiosensitization by oxygen, with o.e.r's of 3.3 and 2.8 (acknowledgments to Dr. B.H.Cullen for mouse cell data).

It is customary and useful to quantify the radiosensitizing action of oxygen in terms of the dose-modification factor, i.e. the ratio of doses required to give the same effect in the absence and presence of oxygen. This is commonly known as the 'oxygen enhancement ratio' (o.e.r.) (Figure 1). When X- or γ-rays are used, the o.e.r. for the killing of mammalian cells is mostly in the range 2 to 3. A wider range of o.e.r's has been observed with vegetative bacteria, from near one, with radiosensitive mutant strains, to about 5.

Several hypotheses have been put forward to account for the oxygen effect, which has been recognized for nearly forty years as a dose-modifying agent. Some of these have been found inadequate, and most radiobiologists now envisage that, if oxygen is available, the most probable chemical reaction immediately consequent on the generation of a radical (the 'metionic reaction' (3)) will be the combination of oxygen with that radical, i.e. a peroxidation of the radical-molecule (3, 32). However, in the absence of oxygen (or another highly electron-affinic species), the metionic reaction would have a reasonably high probability of restoring to the target molecule its capacity for normal functioning. Evidence in support of that hypothesis has come from experiments involving the introduction of oxygen to an irradiated system very soon (i.e. within milliseconds) before or after a pulse of radiation (42, 51).

The Oxygen Effect and Nucleic Acid

With the widespread belief in DNA as the sole or main important target for cell-killing by radiation, the role of oxygen in modifying the effects of radiation on nucleic acids has naturally received a great deal of attention. Early work, in which suspensions of DNA were irradiated extracellularly, revealed that, of all the tests of damage that were then applied, only the destruction of bases was more effective when oxygen was present(49). This has also been found to be true of RNA, but it was found, also, that biological damage was not associated with that lesion (25). A technique currently in wide use is to examine irradiated DNA for single- and double-strand breaks. For extracellularly irradiated DNA, there is conflicting evidence as to whether or not oxygen is sensitizing for the latter (47, 56). If it is, the o.e.r. seems to be of the order of 2 (56), not enough to account for the considerably higher values found for cell killing. Sufficiently high o.e.r's for single strand breaks have been reported when cells have been irradiated, and the DNA extracted subsequently, but we have to remember that the DNA was attached to membranes at the time of irradiation. In any case, the role of DNA main chain breaks in inducing cell death is not at all clear.

It seems to me that, if the oxygen effect in cells is to be attributed to the interaction of oxygen with radicals formed in the nucleic acid, there should be a sensitizing action of oxygen for the loss of biological function, when suitable DNA or RNA preparations are irradiated extracellularly. There are two kinds of biologically active DNA with which this can be done, namely small DNA viruses and transforming DNA. Similarly, small RNA viruses can be irradiated, or RNA extracted from cells can be tested for radiation damage to various biological functions: for example that of messenger, or of ability to code for the synthesis of polypeptide chains. Although radiation experiments on <u>dried</u> DNA

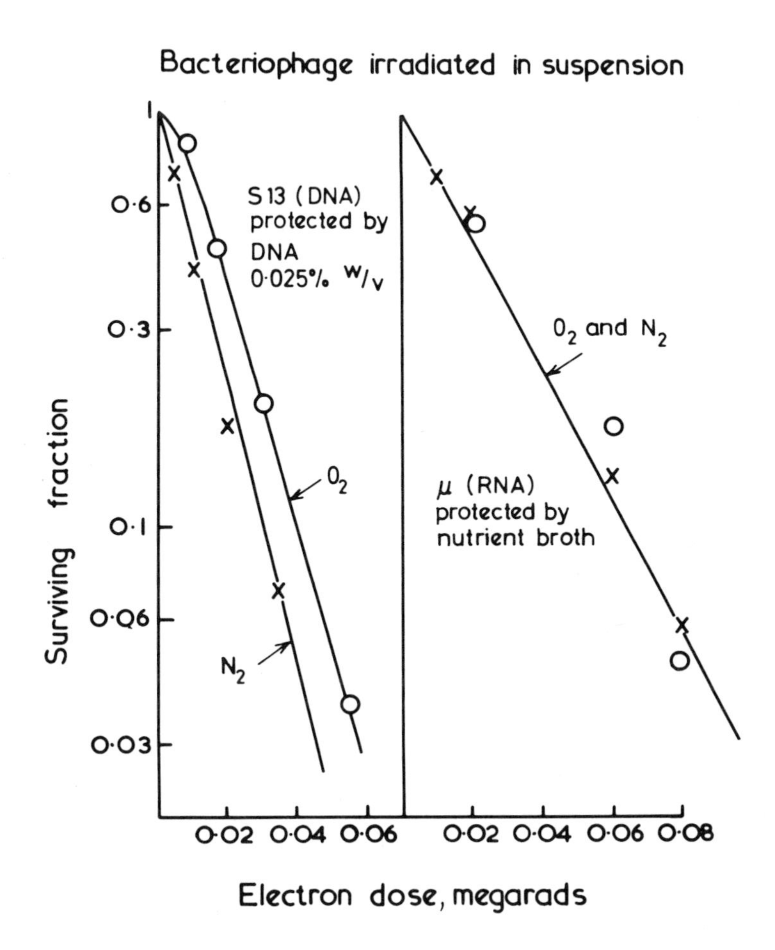

Figure 2. Failure of oxygen to act as radiosensitizer for bacteriophages in suspension containing organic matter.

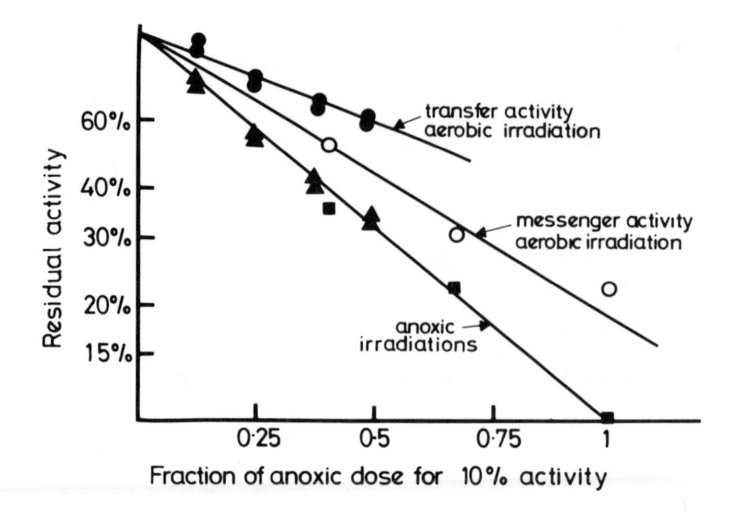

Figure 3. Redrawn from data of Ekert and his colleagues, to show that oxygen protects RNA against functional damage.

or RNA viruses have shown a sensitizing action of oxygen (8), no such action has been seen when biologically active DNA or RNA has been irradiated in suspension (23, 28). In certain circumstances, in fact, oxygen can be protective (2, 25, 26, 27) (Figures 2,3). Remembering that we are concerned with the radiosensitizing action of oxygen in vegetative, i.e. wet, cells, one may well doubt whether this is mainly attributable to the interaction of oxygen with radicals in the DNA.

This was only one of the reasons for my postulating, many years ago, that there must be at least one other important site of damage by ionizing radiation, and that this other site bears the main, if not the whole, responsibility for the sensitizing action of oxygen (4). The suggestion of at least two chemically different targets came from experiments showing that, with certain strains of bacteria, the post-irradiation conditions of growth affected the extent to which oxygen, present during irradiation, influenced the overall killing effect. The effects of germicidal UV, which are indeed mediated by energy absorbed in DNA, were somewhat similarly modified by the post-irradiation culture conditions, and those conditions which gave highest survival (i.e. in which repair of DNA was most effective) were also those in which the o.e.r. with ionizing radiation was maximal. Conversely, the overall sensitizing effect of oxygen was minimized when damage to DNA was permitted to contribute most to the overall killing.

The 'two types of target' hypothesis gained support from other experiments, in which the sensitizing action of oxygen was measured on radiosensitive bacterial strains (deficient in repair to DNA) and their wild-type parents. The o.e.r. was always greater with the latter, in which a great deal of damage to DNA was evidently repaired (5).

If my interpretation is correct, what is a likely candidate for the non-DNA target? Lipids are particularly subject to oxidative reactions, when exposed to radiation, so membranes are plausible structures to associate with radiosensitization by oxygen; and this association has been strengthened by the emergence of the importance of membranes as attachment sites for DNA, as well as of at least some of its replication (36, 38, 52, 53).

If the reasoning is correct, and o.e.r's for bacteria are correlated with effectiveness of repair to DNA, provided the contribution of membrane damage remains unchanged, the converse should hold: with the same effectiveness of repair to DNA, those strains with the highest capacity for repairing membrane damage should evince the lowest o.e.r's. In some exceptionally resistant bacterial strains, e.g. Micrococcus radiodurans, the o.e.r. is indeed rather low (about 2). Although no specific evidence for repair of membrane damage has been reported, it has been shown that the plasma membrane, after irradiation, is able to reattach the separated, broken DNA and this is thought to play an important part in the effective repair of DNA main chain breaks in that strain (16),

Unfortunately there has been very little experimental work on the biological effect of ionizing radiation on membranes - compared with the considerable effort that has gone into studying its effect on nucleic acids, irradiated both intra- and extra-cellularly. It appears difficult to devise meaningful methods for examining radiation damage to membranes: with these, as indeed with any biological molecules, the detection of changes by physico-chemical methods requires radiation doses that are very large, compared with those which produce biological effects. But it is likely that greater progress would have been made if not for the almost universal acceptance of what seems obvious, on the face of it: that DNA is the sole, or main target for radiation-induced cell killing.

Thus there is less to report on the involvement of membrane damage in that effect of ionizing radiation than I should have liked. The evidence in support is inferential only: but so, after all, is the evidence supporting the involvement of DNA!

Lysosomal Membranes

Radiation damage to lysosomal membranes may be assessed by measuring the release of bound enzymes (24). Wills and Wilkinson (59) tested the presence or absence of oxygen as a variable, but did not resolve the effect of oxygen present during irradiation from its role in the development of the lesion leading to enzyme

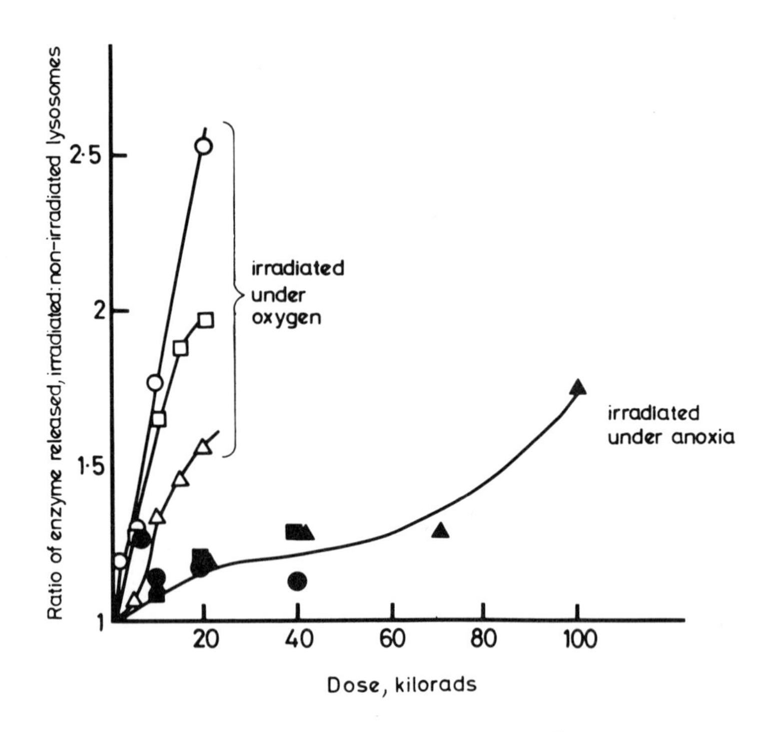

Figure 4. Release of lysosomal membranes (from Watkins, 56). Filled symbols refer to irradiation under anoxia. O,● β-glucuronidase; □,■ N-acetyl-β-glucosaminidase; △,▲ acid phosphatase.

release, during the incubation period required afterwards for observation of the effect. In order to examine the sensitizing action of oxygen as a metionic reactant, Watkins (57) irradiated lysosomes in the presence or absence of oxygen, then incubated the irradiated and unirradiated lysosomes aerobically, for the period required for maximum enzyme release. He used preparations of lysomes from rat spleens, having found these particularly suitable for examining radiation damage. After storage for 21 hours at 2°, activities of several solubilized enzymes were at their maximum as measured after 30 minutes' further incubation at 37°. The composite diagram of Figure 4 shows that anoxia was highly protective against radiation damage to the lysosomal membranes. The results suggest that all three enzymes were released to much the same extent, when no oxygen was present during irradiation. However, the degree of sensitization by oxygen depended somewhat

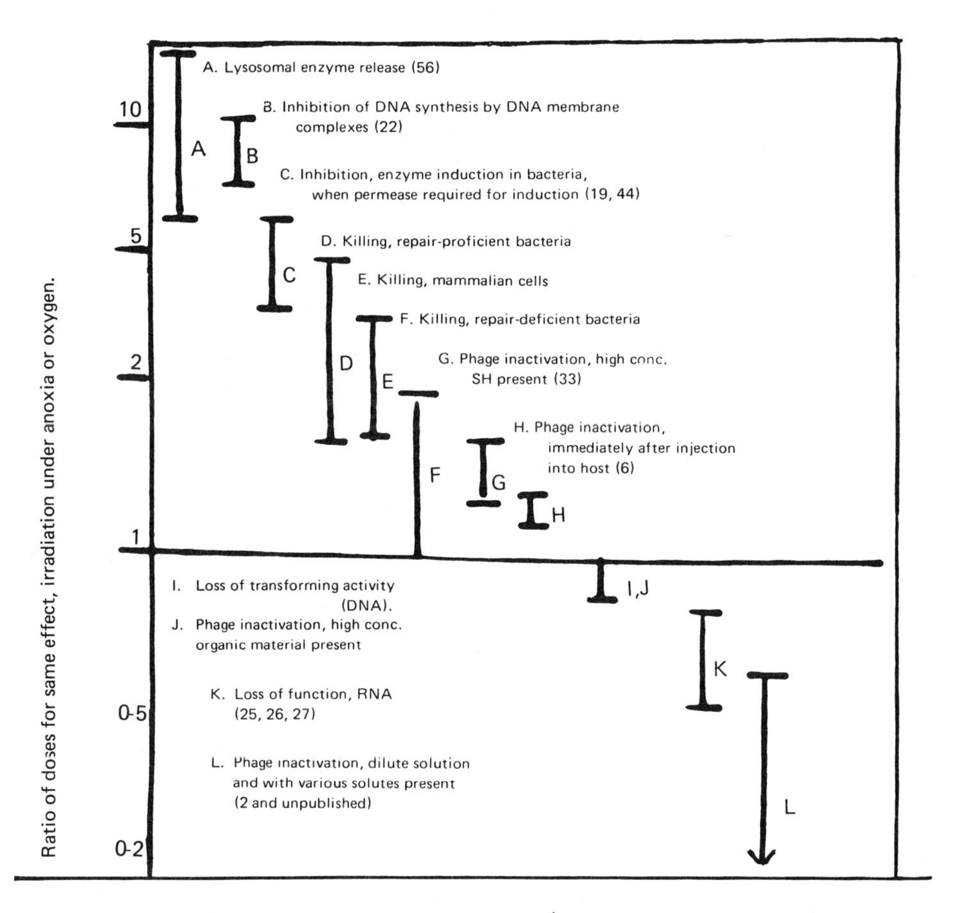

Figure 5. Sensitization by oxygen (dose ratio above 1) or protection (ratio below 1), various biological or biochemical effects.

on the enzyme being assayed. Even with acid phosphatase, for which the sensitization by oxygen was least, the o.e.r. was about 5, i.e. larger than values normally observed in the killing of mammalian cells (Figure 5).

Bacterial DNA-membrane Complexes

It was demonstrated that preparations of bacterial membranes to which DNA remained attached would, for a short time, retain the ability to synthesize new DNA at a rate comparable with that of intact bacteria (38, 51). The effect of radiation on the integrity of that system has been used by Cramp and his colleagues as a method for investigating biological damage to membranes (20, 21, 22).

If overall killing effects are attributable to unrepaired lethal events, both in membranes and in nucleic acid, it is plausible that a radiosensitive mutant of a given strain may differ from the wild type in respect of its ability to repair only one of the two types of lesion. The strain of bacteria _Escherichia Coli B_ has given rise to many mutants which differ from each other in respect of their capacity to repair or bypass damage to DNA - a phenomenon best recognized from the response to germicidal UV. It was found that, in general, UV-sensitive mutants displayed a lower o.e.r. than the wild type when exposed to ionizing radiation (5). This accorded with the prediction that the o.e.r. would be nearer to the intrinsic o.e.r. for membrane damage, the smaller the contribution to killing from damage to DNA. A further prediction would be that ability to repair membrane damage could be the same in the radiosensitive and in the resistant variants; and that the o.e.r. specifically relevant to membrane damage would be the same in both. The first radiosensitive mutant of _E.coli B_ to be isolated, known now as Bs-1 (31), was found to be deficient in its capacity to excise pyrimidine dimers (50). Cramp, Watkins and Collins (22) examined the effect of radiation on the DNA-synthetic ability of DNA-membrane complexes prepared from cultures of _E.coli_ B_{s-1} and the resistant variant _E.coli B/r_. If the bacteria were lysed before irradiation, the incorporation of DNA precursors was increased, if very large doses were given. But if intact bacteria were irradiated, and maintained for a period in buffer at room temperature, before they were lysed, the synthetic ability of the complexes was significantly _depressed_ by much lower doses, within the range that was used to construct a dose-effect curve for the biological end-point, cell death. The striking features of the results were, firstly, that the effect was precisely the same on the sensitive as the resistant variant, and secondly, that the o.e.r. was about 10: whereas, for killing, the o.e.r's were respectively 4.2 and 2.0 for the resistant and sensitive variants. Thus the predictions outlined above were fulfilled.

It may be inferred that the lesion resulting in depression of DNA-synthetic capacity results from energy deposition in a structure such that there will be very effective reaction with oxygen; and it is plausible that lipids should be implicated. Further experiments of Cramp and his colleagues support the inference that the crucial events take place within the membrane, rather than the attached DNA. Cramp and Walker (21) found no difference in the distribution of sizes of the pieces of DNA synthesized by irradiated and unirradiated complexes, so the radiation effect could not be attributed to breakage in the DNA main chain, with a resulting shorter template. Furthermore, irradiation of the bacteria by germicidal UV (absorbed strongly in the nucleic acid) gave quite different results: depression of DNA synthesis by the complexes was independent of the holding time after irradiation; and even enormous doses of UV failed to depress synthesis to less than 50% of normal (20).

Thus certain lesions formed by radiation both in lysosomal membranes and (putatively) in bacterial membranes are associated with o.e.r's which are high, compared with the ones measured for an overall effect like the killing of cells. It is noteworthy that, in both cases, time is needed for the development of the biological effect, which suggests that the initial event requires some biochemical attack before it is expressed as damage. Too little is known about the effects of radiation on biological membranes for interpretation of that observation.

QUALITY OF RADIATION

There is ample evidence that the biological effects of ionizing radiation result to a major extent from the absorption of energy in quanta at least large enough to generate active radicals: and the distance apart of these events varies with the type of radiation - for which characteristic the term quality is used. One measure is the energy lost per unit of path length - in kilo-electron volts per micron, for example - and this 'Linear Energy Transfer' (LET) is in common use as a description of quality. As a very rough guide, we may think of hard X-rays, or fast electrons, as losing about 1 to 2 KeV, and engendering perhaps 10 to 20 radicals, in one micron of tissue, whereas Polonium α-particles would give rise to about 100 times as many energy-deposition events.

It is convenient to compare the effects of radiations of different qualities in terms of their 'Relative Biological Effectiveness' or RBE, that parameter being defined as the ratio of doses of two kinds of radiation to cause the same effect. RBE values are usually based on X- or γ-rays. If, for example, the same numbers of cells of a given line are killed by one dose-unit of α-particles as by 3 dose-units of X-rays, the α-particles are 3 times as effective as the X-rays, so the RBE of the α-particles is 3: the dose of 'base-line' radiation required for the given effect becomes the numerator, when the ratio is calculated.

With almost every type of cell, the RBE initially increases with increasing LET, not only for killing, but also for other forms of damage like mutation induction. Such observations are in sharp contrast with those made on the inactivation of sub-cellular entities like enzyme molecules or viruses under extracellular irradiation. With these, effectiveness has been found to decrease with increasing LET (14, 18, 48), at least when biological or biochemical tests of damage have been used (Figure 6). The latter

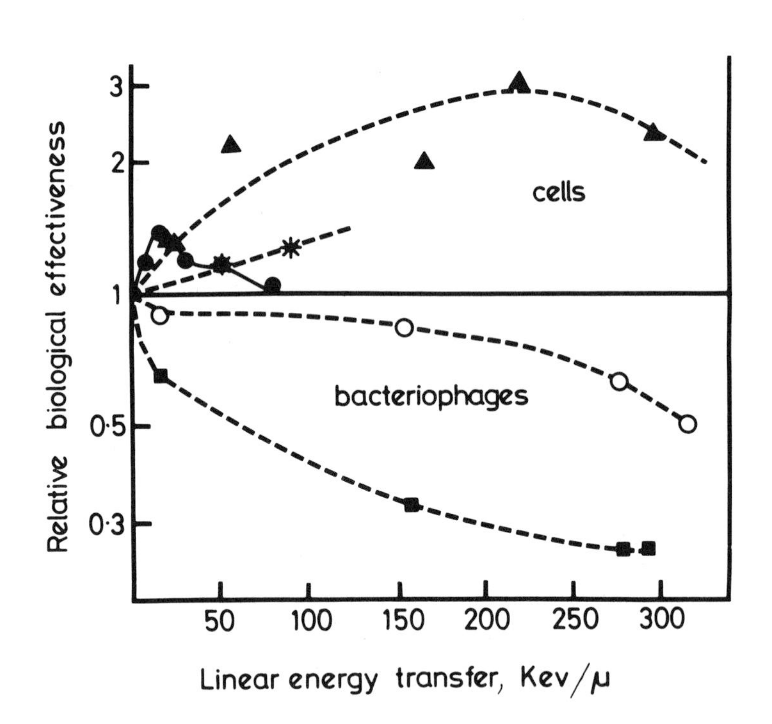

Figure 6. As LET increases, effectiveness increases initially for all cells, decreases continuously for sub-cellular units. ▲ Human kidney cells (54); ● Bacteria (<u>Shigella</u>) (11); * Alga (<u>Chlamydomonas</u>) (15);O ■ Double- and single-stranded DNA phages, T_1 and ØX174 (47). When RBE varies with dose, (mammalian cells and algae), minimum values are shown.

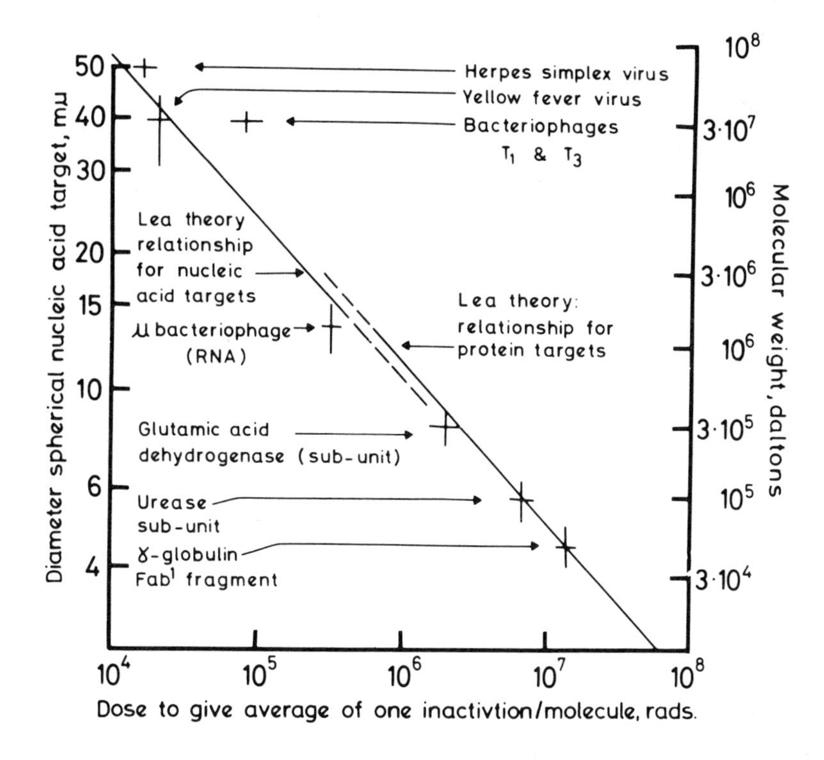

Figure 7. Agreement between molecular weights, conventionally measured, and radiation target sizes calculated from theory of Lea (40).

result is to be predicted in terms of the classical target theory of radiation action, according to which a single energy deposition within the target volume will destroy its biological function; the target volume being <u>defined</u> in that way. Provided appropriate conditions are used in the extracellular irradiation of macromolecules with biological function, there is excellent agreement between radiation target volumes, estimated from the doses required to deposit on average one inactivating event per unit, and molecular weights as determined by other methods (Figure 7).

If, then, a single event within the target is sufficient to inactivate it, we should expect that sparsely ionizing radiations should be the most effective: the closer togethei the sites of energy deposition, the greater the probability that, with a given dose, two or more events will occur within the same target volume and so be 'wasted'. Clearly some explanation is needed for the increase in RBE with LET, for cells, and the decrease, predicted by target theory, for all extracellularly irradiated macromolecules which might be targets for cell killing.

Many years ago it was postulated that in some targets within cells, like chromosomes, a single ionization would not suffice to inflict observable damage: for example, it was calculated that several ionizations in close proximity were required to break a chromosome (41). Co-operation or interaction between the sites of energy-deposition was envisaged as occurring at the metionic or physico-chemical level (3, 32). Within the past decade this concept has been challenged on different grounds (37, 46) and it has been inferred rather that RBE increases with LET because there may be interaction between sites of single events, resulting in a greater probability of damage fixation than if there were no interaction. As the density of ionization increases, there is also an increase in the probability that two events will occur close enough together in space and time to interact.

However, single events have a high probability of inactivating targets such as RNA or double-stranded DNA viruses, and probably also membranes (58). Some mechanism must therefore operate to reduce the probability of effectiveness of single energy-deposition events in intracellular targets. It is reasonable to attribute this reduction to biochemical processes. Indeed, a great deal of attention has been given to the enzymic repair of DNA lesions, particularly single and double strand breaks. This repair may occur over periods measured in tens of minutes, but some so-called ultra-fast repair has been detected in bacteria (55). If enzymes are present that can effect repair of potentially lethal lesions at individual sites, whereas lesions in adjacent sites fix the damage in either one, it must follow that repair can be inhibited also by some biochemical process, or that enzymic action may 'fix' a lesion that would otherwise be reparable. This is a reasonable expectation since the cell requires enzymes that will catalyse degradation as well as synthesis.

It was suggested long ago that one effect of radiation would be to cause the release of membrane-bound degradative enzymes (13); and Allison and Paton (1) showed that chromosome breaks could result from the destruction of lysosomes by light of a specifically absorbed wavelength, presumably because of the release of DNA-ase II. Since RBE effects are attributed to (usually unspecified) interactions, we might well expect even more effective damage to result from the release of degradative enzymes, if this is coupled with the infliction of potentially lethal lesions in DNA. There is, as yet, only inferential evidence for this hypothesis. We have already seen that oxygen enhancement ratios (for radiation of low LET) tend to be high, in bacteria that are proficient at repairing damage to DNA, and low, in repair-deficient ones (Figure 5). With the latter, we should expect it to make very little difference whether there is interaction of some other damage with that done to DNA, since lesions in the latter will tend in any case to be lethal. Thus there is little reason why the effectiveness of the radiation on such strains should increase, as the LET increases. On the other hand, RBE values considerably greater than one are observed with repair proficient strains (10, 45). That the interaction concerned is indeed between damaged sites in the membrane and the DNA is, of course, speculative; but the correlation shown in Table I seems to me to be suggestive.

TABLE I

Bacterial strain	Proficiency in repair of damaged DNA	o.e.r. X-rays	Effectiveness of neutrons relative to X-rays
E.Coli B strains:			
B-H	Proficient	3.4	1.4
B/r	Proficient	3.2	1.4
B :			
Condition a.	Proficient	3.2	1.3
b.	Reduced	2.5	1.2
B_{s-1}	Much reduced	1.8	1.1
B_{s-12}	None	1.0	1.0
E.coli K_{12} strains:			
AB1157	Proficient	3.5	1.5
AB2463	Much reduced	2.1	1.1

SCRAPIE AND ALLIED DISEASES: MEMBRANE FRAGMENTS AS INFECTIVE AGENTS?

Although this topic is peripheral to my main theme, it may be of interest in the context of this meeting. Also, it provides an example of how radiation, and, in particular, the synergistic or protective action of oxygen, may be useful in supplying inferential evidence on the nature of a biological entity that eludes normal procedures of identification.

Scrapie, primarily a disease of sheep, is one of a small group of transmissible encephalopathies, two of which affect humans. The development of symptoms takes many months, or years, after infection, so they are often classified as 'slow virus' diseases. It has not been possible to determine the nature of the agents by conventional techniques, since no method has yet been found for separating the infective entity from the tissues of affected animals. Quantitative work can be done only by measuring the activity of a given preparation in terms of the concentration required to kill 50% of a group of animals, so of course mice are the most suitable test animals. The best data have therefore been obtained by using scrapie strains which have been successfully adapted to mice (17).

There is a long list of properties of the scrapie and similar agents that distinguish them sharply from such viruses as are known to proliferate by replication of a nucleic acid core (35). In particular, the comparative transparency of mouse-adapted scrapie to 'germicidal' UV led us to suggest that the proliferative capacity of the agent could not depend on the integrity of a nucleic acid moiety (7). This inference found support in the subsequent observation that scrapie was much more readily inactivated by UV at about 240 than 260-270 nm (39), just the opposite of what has almost invariably been observed for inactivation of viruses and transforming DNA or induction of gene mutations. Consequent upon the suggestion that scrapie replicated without the involvement of a nucleic acid moiety (7), Gibbons and Hunter (30) postulated that the agent might be in the form of a small piece of cell membrane which could be integrated into the membranes of normal host cells (probably in nervous tissue). There is some experimental support for this hypothesis: for example, fractions separated from preparations of brains and spleens of affected mice were tested for activity, and a large fraction was found to be associated with the plasma membrane and the endoplasmic reticulum (43). This identification was supported in experiments in which enzyme membrane markers were found to be distributed coincidentally with most of the scrapie infectivity of mouse brain preparations (19).

Radiobiological support for the membrane hypothesis has come also from our experiments on the inactivation of scrapie in suspension. Where the response of nucleic acid targets - like viruses - was unaffected, or even reduced, by the presence of oxygen (Figures 2, 3, 5), the inactivation of the scrapie agent was greatly enhanced (Figure 8). As we have seen, the only other

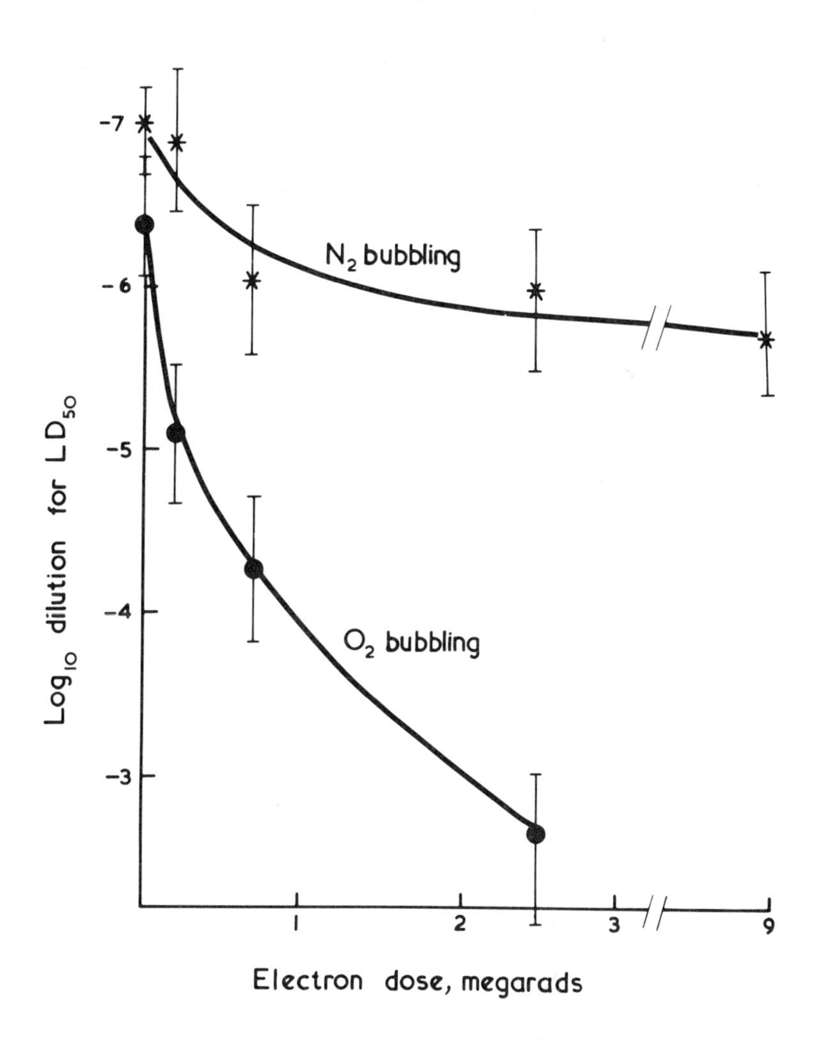

Figure 8. Inactivation of scrapie agent in suspension by radiation in presence and absence of oxygen.

extracellular systems which have so far demonstrated that characteristic are lysosomes and bacterial DNA-membrane complexes. We have, therefore, two hypotheses, relating to quite separate fields, each of which has its own experimental and/or inferential support: (1) that the radiosensitizing effect of oxygen, in vegetative cells, is primarily attributable to the deposition of energy in membranes; (2) that the agents of scrapie and other related diseases are membranous in character, and do not depend on a nucleic acid moiety for proliferative ability. These hypotheses find mutual support in the results of work on the irradiation of scrapie.

CONCLUSION

In investigations into the mechanisms of damage to cells by radiation, it has been an axiom that effects on nucleic acids have a predominant role. There has been mutual gain from those lines of work and that of molecular biologists not primarily interested in radiation: for example, evidence for the existence of specific enzymes involved in the 'repair' of DNA came from photo- and radiobiological experiments, and this introduced a new concept to biochemists studying the synthesis of nucleic acids. By contrast, there are very few radiation workers as yet interested in the role of cell membranes, and I doubt whether any expert on membrane biology has, as yet, taken account of radiation effects, or indeed used ionizing radiation as a means by which certain questions might be answered. Perhaps it would not be too optimistic to suggest that, before long, there will be mutually heuristic developments in both these interesting and important topics.

REFERENCES

1 ALLISON, A.C. and PATON, G.R.: Chromosome damage in human diploid cells following activation of lysosomal enzymes. Nature, Lond. 207 (1965) 1170.

2 ALPER, T.: Bacteriophage as indicator in radiation chemistry. Rad. Res. 2(1955)119

3 ALPER, T,: The modification of damage caused by primary ionization of biological targets. Rad. Res. 5 (1956) 573.

4 ALPER, T.: Lethal mutations and cell death. Physics in Med. and Biol.8(1963) 365.

5 ALPER, T.: A characteristic of the lethal effect of ionizing radiation on "Hcr⁻" bacterial strains. Mutat. Res. 4 (1967) 15.

6 ALPER, T.: Cell death and its modification: the roles of primary lesions in membranes and DNA. In Biophysical Aspects of Radiation Quality. STI/PUB/286 IAEA, Vienna (1971) 171.

7 ALPER, T., CRAMP, W.A., HAIG, D.A. and CLARKE, M.C.:Does the agent of scrapie replicate without nucleic acid? Nature, Lond. 214 (1967) 764.

8 ALPER, T. and HAIG, D.A.: Protection by anoxia of the scrapie agent and some DNA and RNA viruses irradiated as dry preparations. J. Gen. Virol. 3 (1968) 157.

9 ALPER, T. and HOWARD-FLANDERS, P.: The role of oxygen in modifying the radiosensitivity of E.coli B. Nature 178 (1956) 978.

10 ALPER, T. and MOORE, J.L.: The interdependence of oxygen enhancement ratios for 250 kVp X-rays and fast neutrons. Br. J. Radiol. 40 (1967) 843.

11 ALPER, T., MOORE, J.L. and BEWLEY, D.K. LET as a determinant of bacterial radiosensitivity, and its modification by anoxia and glycerol. Rad. Res. 32 (1967) 277.

12 ANDERSON, R.S. and TURKOWITZ, H.: The experimental modification of the sensitivity of yeast to roentgen rays. Amer. J. Roentgenol. and Radium Therapy 46 (1941) 537.

13 BACQ, Z.M. and ALEXANDER, P.: Fundamentals of Radiobiology. Butterworths, London (1955) 185ff.

14 BRUSTAD, T.: Heavy ions and some aspects of their use in molecular and cellular radiobiology. Advances Biol. Med. Phys. 8 (1962) 161.

15 BRYANT, P.E.: LET as a determinant of oxygen enhancement ratio and shape of survival curve for Chlamydomonas. Int. J. Radiat. Biol. 23 No. 3 (1973) 217.

16 BURRELL, A.D., FELDSCHREIBER and DEAN, C.J.: DNA-membrane association and the repair of double breaks in X-irradiated Micrococcus radiodurans. Biochim et Biophys. Acta 247 (1971) 38.

17 CHANDLER, R.L.: Experimental scrapie in the mouse. Res. Vet. Sci. 4 (1963) 276.

18 CHRISTENSEN, R.C., TOBIAS, C.A. and TAYLOR, W.D.: Heavy-ion-induced single- and double-strand breaks in ØX-174 replicative form DNA. Int. J. Radiat. Biol. 22 (1972) 457.

19 CLARKE, M.C. and MILLSON, G.C.: The membrane location of scrapie infectivity. J. Gen. Virol. (1976) in press.

20 CRAMP, W.A. and PETRUSEK, R.: The synthesis of DNA by membrane-DNA complexes from E.coli B/r and E.coli B_{s-1} after exposure to UV light: a comparison with the effects of ionizing radiation. Int. J. Radiat. Biol. 26 (1974) 277.

21 CRAMP, W.A. and WALKER, A.: The nature of the new DNA synthesized by DNA-membrane complexes isolated from irradiated E.coli. Int. J. Radiat. Biol. 25 (1974) 175.

22 CRAMP, W.A., WATKINS, D.K. and COLLINS, J.: Effects of ionizing radiation on bacterial DNA-membrane complexes. Nature New Biol. 235 No. 55 (1972) 76.

23 DEFILIPPES, F.M. and GUILD, W.R.: Irradiation of solutions of transforming DNA. Rad. Res. 11 (1959) 38.

24 DESAI, I.D., SAWANT, P.L. and TAPPEL, A.L.: Peroxidative and radiation damage to isolated lysosomes. Biochim et Biophys Acta 86 (1964) 277.

25 EKERT, M.B. and GRUNBERG-MANAGO, M.: Effets des rayons γ sur l'efficacité de quelques polyribonucléotides en tant que messagers. C.R. Acad. Sc. Paris 263 (1966) 1762.

26 EKERT, B. and LATARJET, M-F.: Inactivation par les rayons γ des propriétés fonctionelles des RNA de transfert d'E.coli (phenyl-alanine et lysine). Int. J. Radiat. Biol. 20(1971) 521.

27 EKERT, B., MONIER, R. and TORDJMAN, A.: Etude de l'inactivation par les radiation ionisantes des propriétés acceptrices des acides ribonucleiques de transfert. Bull. de la Soc. de Chim. Biol. 50 (1968) 1875.

28 EPHRUSSI-TAYLOR, H. and LATARJET, R.: Inactivation,par les rayons X, d'un facteur transformant du Pneumococque. Biochim et Biophys. Acta 16 (1955) 183.

29 FORAGE, A.J.: The dependence of the oxygen enhancement ratio on the test of damage in irradiated bacteria. Int. J. Radiat. Biol. 20 (1971) 427.

30 GIBBONS, R.A. and HUNTER, G.D.: Nature of the scrapie agent. Nature 215 (1967) 1041

31 HILL, R.F.: A radiation sensitive mutant of Escherichia coli. Biochim. Biophys. Acta 30 (1958) 636.

32 HOWARD-FLANDERS, P.: Physical and chemical mechanisms in the injury of cells by ionizing radiations. Adv. Biol. Med. Physics 6 (1958) 554.

33 HOWARD-FLANDERS, P.: Effect of oxygen on the radiosensitivity of bacteriophage in the presence of sulphydryl compounds. Nature 186 (1960) 485.

34 HOWARD-FLANDERS, P. and MOORE, D.: The time interval after pulsed irradiation within which injury to bacteria can be modified by dissolved oxygen. I. A search for an effect of oxygen 0.02 second after pulsed irradiation. Rad. Res. 9 (1958) 422.

35 HUNTER, G.D.: Scrapie. Progr. med. Virol. 18 (1974) 289.

36 JACOB, F., RYTER, A. and CUZIN, F.: On the association between DNA and membrane in bacteria. Proc. of Roy. Soc. of London (Series B. Biol. Sciences) Vol. 164 (1966) 267.

37 KELLERER, A.M. and ROSSI, H.H.: The theory of dual radiation action. Curr. Topics Radiat. Res. Qtly. 8 (1972) 85.

38 KNIPPERS, R. and STRATLING, W.: The DNA-replicating capacity of isolated E.coli wall-membrane complexes. Nature, Lond. 226 (1970) 713.

39 LATARJET, R., MUEL, B., HAIG, D.A., CLARKE, M.C. and ALPER,T.: Inactivation of the Scrapie agent by near-monochromatic ultraviolet light. Nature 227 (1970) 1341.

40 LEA, D.E.: Actions of Radiations on Living Cells. Cambridge Univ. Press (1946).

41 LEA, D.E. and CATCHESIDE, D.G.: The mechanism of the induction by radiation of chromosome aberrations in Tradescantia. J. Genet. 45 (1942) 216.

42 MICHAEL, B.D., ADAMS, G.E., HEWITT, H.B., JONES, W.B.G. and WATTS, M.E.: A post-effect of oxygen in irradiated bacteria: a submillisecond fast mixing study. Rad. Res. 54 (1973) 239.

43 MILLSON, G.C., HUNTER, G.D. and KIMBERLIN, R.H.: An experimental examination of the scrapie agent in cell-membrane mixtures. II. The association of scrapie activity with membrane fractions. J. Comp. Pathol. 81 (1971) 255.

44 MOORE, J.L.: An induced enzyme in X-irradiated Escherichia coli: Comparison with lethal effects. J. Gen. Microbiol. 41 (1965) 119.

45 MUNSON, R.J., NEARY, G.J., BRIDGES, B.A. and PRESTON, R.J.: The sensitivity of Escherichia coli to ionizing particles of different LET's. Int. J. Radiat. Biol. 13 (1968) 205.

46 NEARY, G.J.: Chromosome aberrations and the theory of RBE. I. General considerations. Int. J. Radiat. Biol. 9 (1965)477.

47 NEARY,G.J., HORGAN, V.U., BANCE, D.A. and STRETCH, A.: Further data on DNA strand breakage by various radiation qualities. Int. J. Radiat. Biol. 22 (1972) 525.

48 SCHAMBRA, P.E. and HUTCHINSON, F.: The action of fast heavy ions on biological material. 2. Effects on T_1 and ØX-174 bacteriophage and double-strand and single-strand DNA. Rad. Res. 23 (1964) 514.

49 SCHOLES, G.: The radiation chemistry of aqueous solutions of nucleic acids and nucleoproteins. Prog. Biophys. and Molec. Biol. 13 (1963) 59.

50 SETLOW, R.B. and CARRIER, W.L.: The disappearance of thymine dimers from DNA: an error correcting mechanism. Proc. Nat. Acad. Sci. 51 (1964) 226.

51 SHENOY, M.A., ASQUITH, J.C., ADAMS, G.E., MICHAEL, B.D. and WATTS, M.E.: Time-resolved oxygen effects in irradiated bacteria and mammalian cells: a rapid-mix study. Rad. Res. 62 (1975) 498.

52 SMITH, D.W., SCHALLER, H.E. and BONHOEFFER, F.J.: DNA synthesis in vitro. Nature, Lond. 226 (1970) 711.

53 SPARVOLI, E., GALLI, M.G., MOSCA, A. and PARIS, G.: Localization of DNA replicator sites near the nuclear membrane in plant cells. Exptal. Cell Res. 97 (1976) 74.

54 TODD, P.W.: Reversible and irreversible effects of ionizing radiation on the reproductive integrity of mammalian cells cultured in vitro. Ph.D. Thesis UCRL 11614 University of California, Berkeley, Calif.

55 TOWN, C.D., SMITH, K.C. and KAPLAN, H.S.: Influence of ultrafast repair processes (independent of DNA polymerase-1) on the yield of DNA single-strand breaks in Escherichia coli K12 X-irradiated in the presence or absence of oxygen. Rad. Res. 52 (1972) 99.

56 VAN der SCHANS, G.P. and VAN der DRIFT, A.C.M.: Comparison of the oxygen-enhancement ratio for γ-ray induced double strand breaks in the DNA of bacteriophage T_7 as determined by two different methods of analysis. Int. J. Radiat. Biol. 27 (1975) 437.

57 WATKINS, D.K.: High o.e.r. for the release of enzymes from isolated mammalian lysosomes after ionizing radiation. Adv. Biol. Med. Physics 13 (1970) 289.

58 WATKINS, D.K. and DEACON, S.: Comparative effects of electron and neutron irradiation on the release of enzymes from isolated rat-spleen lysosomes. Int. J. Radiat. Biol. 23 (1973) 41.

59 WILLS, E.D. and WILKINSON, A.E.: Release of enzymes by irradiation and the relation of lipid peroxide formation to enzyme release. Biochem. J. 99 (1966) 657.

DISCUSSION

PRESSMAN: The single colchicine molecule is supposed to be the cause of a propagating damage of the entire membrane of a susceptible coli cell. Do you think that this would be an interesting system to examine in terms of the effect of radiation and the synergistic effect of oxygen?

ALPER: That is a very interesting point because we discovered that some bacterial mutants, alleged to be sensitive and assumed to be deficient in DNA repair, were, in fact, organisms which carried bacteriocins, but I haven't myself done any specific oxygen experiments except that UV - sensitive mutants are, in general, less protected against ionizing radiation by anoxia than wild type strains. I am quite satisfied that the site of the oxygen effect is membranes. It seems to me that everything points to it. I think that radiation could be a useful tool to study some of the problems you are looking at, particularly since, with membrane damage, you have these very large oxygen effects. I think more and more radiobiologists are becoming interested in the role of membrane damage, and I hope that membrane people will become interested in radiation, even if only as a tool. I think membranes are very important in all radiobiology and therefore in radiotherapy.

BLUMENTHAL: I would like to suggest the introducing of physical chemistry in your system, namely the use of liposomes in your case would be very useful. First of all you talk about the question of sensitivity. The very sensitive assay with liposomes does require the release of a trapped marker. Secondly, the advantage of using the liposome is that one can identify the site of the damage, for instance what happened to the lipid. All the systems that you work with are so gross that one really doesn't know on the molecular level what happened.

ALPER: Well, I would take almost the opposite view. There have been many people working with model membranes and biophysicists who have irradiated them with millions of rads and they have failed to find an effect. When you work with a biological system you are able to apply very delicate tests of damage. As you saw both with the liposome systems and with Bill Cramp's system some biochemical processes must occur before you see the end results. I don't know of any aspect of radiobiology where we have really learned very much from purely physico-chemical models. Maybe we will in time. Maybe we haven't got the right one.

PASSOW: I think there are a lot of things to say about membranes and radiation but I think we don't have that much time. Perhaps a few questions or comments.

ROTHSTEIN: You remind me of some work that was done at Rochester a number of years ago in which it was shown that in red cells, X-irradiation produces disulfide bonds in membrane proteins and in parallel alters the permeability to potassium. Protection against this irradiation effect was observed if the sulfhydryl groups were first reacted with PCMBS, the mercurial that we heard about earlier as an inhibitor of water permeability. The system is oxygen sensitive. Do you have any thoughts about this kind of chemistry as having anything to do with what you are discussing?

ALPER: Let's put it this way: those cells in which we see effects at the lowest doses, and also the effects that matter in any kind of practical context, are always effects on cells whose job it is to divide. For example, liver was known for a long time as a "radio-resistant" organ, but it was found that after hepatectomy, which stimulated the cells to divide, the results of previous irradiation could be seen. It is rather difficult to devise delicate tests, or tests that will work after reasonably low doses of radiation, on cells that never have to divide.

ROTHSTEIN: Yes, but the red cell system loses potassium as a consequence of the irradiation and the same loss of potassium is also found in yeast cells. The latter cells fail to grow because of the loss of potassium. They will grow despite the irradiation damage to the membrane if the extracellular potassium is increased so that a high cellular potassium is maintained. In other words, the inability to divide in this case can be connected specifically to a leakiness to potassium. Perhaps we can discuss this point in more detail later.

LAKOWICZ: Yes, I would like to offer an explanation for why you might see enhancement of oxygen damage in membranes and not in nucleic acids. You had made the comment at the outset that the oxygen needs to get to the point of damage for any effect to be felt. But we have been examining the oxygen permeation rates of a variety of biological materials from membranes to nucleic acid, and we find that the diffusion coefficient of oxygen within membranes is very high, higher than it is in proteins and it is very low within the nucleic acid itself. So, it could be an accessibility phenomenon.

MECHANISMS BY WHICH SMALL MOLECULES ALTER IONIC PERMEABILITY THROUGH LIPID BILAYER MEMBRANES

Gabor Szabo

Department of Physiology and Biophysics
University of Texas Medical Branch
Galveston, Texas 77550

ABSTRACT

Small molecules may be particularly effective cytotoxic agents by way of altering ionic permeability in cell and organelle membranes. The mechanisms by which such molecules may act on membrane permeability are examined here in light of the rather precise understanding of the mechanisms by which ionic permeability is induced and altered in lipid bilayer membranes. In particular, the influence of specific changes in membrane composition on the kinetics of direct and carrier mediated transport of ions are examined in detail and related to the electrical potentials originating from dipolar residues at the membrane surface.

Introduction

An "energized" state, which maintains large differences in the chemical composition and in the electrical potential between internal and external milleu, is essential for the vital function of cells and organelles. Such an energized state is maintained by two functionally different properties of the membrane. First, the membrane acts as an insulator and thereby prevents dissipation of the energized state as a result of unrestricted diffusion. Second, the membrane maintains the energized state by reversibly coupling the electrochemical gradient to the chemical energy sources of the cell, for example by transporting protons (29) or Na and K (38) against their electrochemical gradient using the chemical energy of ATP.

The energized membrane is of course more than a simple insulator. It bears, in addition to the mechanisms which maintain the energized state, specific permeability pathways through which the stored electrochemical energy is utilized to accomplish two different functions. First, the membrane indirectly uses the electrochemical energy to transport solutes into or out of the cell. The Na^+ gradient coupled transport of sugars and amino acids in some mammalian (31) and bacterial (21) cells are examples of such "transport" permeability. Second, the membrane directly uses the electrochemical energy for specific functional purposes. The Na^+ and K^+ permeability of nerve axons, which are responsible for the propagation of electrical signals (8), or the action potential induced Ca^{++} permeability of presynaptic terminals, (20) which result in synaptic transmission, are examples of "functional" permeability.

Insulating as well as selective permeability properties of the membrane are crucial for proper function. Alteration of either of these functions is expected to be toxic. A large number of small molecules is known to alter ionic permeability of cell and organelle membranes. For example, valinomycin increases membrane potassium permeability, thereby uncoupling mitochondria (35), altering sugar transport in intestial brush border membranes (31), and generally injuring most cells by dissipating the internal K^+ concentration. The ionophore X537A and A23187 increase membrane permeability to Ca^{++} and by this mechanism can initiate diverse physiological reactions such as the release of synaptic vesicles from nerve endings (19,17). Quaternary ammonium compounds (3), tetrodotoxin (32) local anesthetics (40), neurotransmitters and their analogues (12) are but some of the more striking examples of small molecules that are toxic by altering membrane permeability. I shall refer to such molecules as "permeatoxins" indicating that they alter cell function by altering the permeability of the membrane.

Permeatoxins such as valinomycin or X537A alter ionic permeability in artifical lipid bilayer membranes as well (30,6). Indeed, the lipid bilayer proves to be a model system well suited for the study of the mechanisms of permeatoxicity, so much so that a good part of the theoretical understanding of the action of permeatoxins originates from the study of this system. For example the carrier mechanism for neutral permeatoxins such as the macrotetralide actins (41) or valinomycin (23) as well as for charged permeatoxins such as weak acid uncouplers (33) or X537A (6); the existance of conductive channels (14); the mechanisms of action of voltage sensitive dyes (46) have all been established using lipid bilayers.

This paper examines the molecular mechanisms by which small molecules alter membrane permeability. The first part classifies and discusses conceivable mechanisms of permeatoxicity, while the second part examines experimentally the kinetics of direct and carrier-mediated membrane permeability as well as the mechanisms by which small molecules can modify these through altering the electrochemical properties of the lipid bilayer.

Mechanisms of Permeatoxicity

Conceptually, permeatoxins can be divided into two basic classes according to their mechanism of action: (1) those which alter membrane permeability by creating new permeability pathways through the membrane and (2) those which alter membrane permeability by modifying permeability pathways already existing in the membrane. Figure 1 summarizes mechanisms of action for permeatoxins within these two broad categories.

Direct permeability, in which a charged permeatoxin penetrates the membrane simply by dissolving in and diffusing across the membrane, is shown in the second column of figure 1. Permeability of an ion depends on its partition into the membrane as well as on its translocation within the membrane. These parameters are determined both by the structure of the ion and by the composition of the membrane. Small, hydrophilic ions (Na, K or Cl, for example) have negligibly small bilayer permeabilities because of the large energy necessary to move these ions from the aqueous phase into the hydrocarbon like membrane interior. Permeability studies of model bilayer and cell membranes show however, that there are a number of ions such as tetraphenylborate (TFB), that can permeate the membrane directly. Hydrophobicity, large ionic size and delocalized charge are those ionic properties which tend to favor direct permeability by favoring partition into the membrane interior. It is possible not only to predict the permeability of a particular ion, but also to "probe" the electrochemical properties of both cell and artificial bilayer membranes (42). For example, oppositely charged hydrophobic cations and anions have been used to assess the sign and the magnitude of the changes in the electrostatic potential induced within a bilayer membrane by altered lipid composition (43).

Hydrophobic ions are permeatoxic agents which are often useful for the study of membrane function since they tend to redistribute themselves according to the electrical potential difference, $v' - v''$, across the two sides (' and ") of the membrane according to the Nernst relation $c'/c'' = \exp(zF[v'-v'']/RT)$ for an ion of valence z (39). Flourescent or colored hydrophobic ions have also been used to follow optically the redistribution of ions in vesicles (39,21), the electrochemistry of ionic adsorption and translocation

FUNCTION ALTERED	IONIC PERMEABILITY OF THE MEMBRANE				
TYPE OF ACTION	INDUCES NEW IONIC PERMEABILITY			ALTERS EXISTING PERMEABILITY	
MECHANISM OF ACTION	DIRECT	MEDIATED		DIRECT	INDIRECT
		CARRIER	CHANNEL		
EXAMPLES	HYDROPHOBIC IONS	NEUTRAL CARRIERS OF IONS		BINDING TO IONOPHORE	SURFACE CHARGE SURFACE DIPOLE MEMBRANE FLUIDITY
	tetraphenyl borate	trinactin valinomycin	gramicidin A alamethicin monazomycin	Tl^+ in gramicidin channel tetrodotoxin in sodium channel	membrane lipid composition; adsorbed charged or dipolar molecules
		CHARGED			
		X-537A A23187			

Figure 1. Mechanism of action of permeatoxic agents

in bilayers (9,46) and the transmembrane potential in natural as well as lipid bilayer membranes (37).

The mid section of figure 1 shows a somewhat more complicated transport process in which a permeatoxin mediates the permeability of an otherwise impermeant ion. The increased permeability may be mediated either by a carrier mechanism (third column) in which the permeatoxic carrier combines with its ionic substrate to form a hydrophobic complex, or by opening a polar pathway or channel across the membrane (fourth column) permitting the flow of small hydrophilic ions.

Carrier mediated transport is similar to direct transport in that the permeant species is a hydrophobic complex. It however differs from direct transport in that the hydrophobic complex is formed by the carrier combining with the carried ion. This allows selective enhancement of permeability. Neutral carriers, which form charged complexes (valinomycin or nonactin with K) as well as charged carriers, which form neutral complexes (X537A with Ca^{++}) are both well known examples of permeatoxic molecules which act by the carrier mechanism. Carrier-mediated transport has two remarkable features: 1) it permits a selective translocation which is governed not by intrinsic lipid solubility but by the rate (or equilibrium) constants of the reaction between the carrier and the transported solute; and 2) it permits a coupling between the flow of the transported molecule and the electrical potential across the membrane whenever the carrier or the complex (or both) are charged.

A number of permeatoxic agents such as the pentadecapeptide gramicidin A (14), alamethicin (10), "excitability inducing material" (4) and the polyene antibiotics nystatin and amphotericin B (26) is believed to act by this mechanism* as evidenced by the quantal fluctuation of ionic conductance seen in many of these systems (28). All of the known channel-forming molecules are antibiotics. Their toxicity appears to derive from their ability to nonselectively increase the ionic permeability of membranes. Indeed, nystatin has been used to alter the internal ionic composition of cells by virtue of creating a rather large but non-selective ionic permeability across the membrane (36).

Small molecules may also alter membrane permeability by altering those permeability pathways which are an integral part of the membrane. As shown in the last two columns of figure 1,

*Postsynaptic conduction in the neuromuscular junction (20) as well as the K and Na conduction in nerve membrane (15) are also believed to be mediated by conductive channels.

this may occur either directly, by specific interaction of the permeatoxin with an existing permeability pathway (fifth column in figure 1) or indirectly, by a nonspecific action of the permeatoxin on an existing permeability pathway through altering the physical and/or chemical properties of the membrane (last column of figure 1). The former mechanism (example of which are blocking of Na^+ channels by tetrodotoxin (32) and of K channels by quaternary ammonium compounds (3) in the nerve membrane) is poorly understood not only for cell membranes but also for model membranes. Thus, specific blocking of the gramicidin A channel by the Tℓ ion is only now beginning to be elucidated (11). In contrast, the latter mechanism (examples of which are effects of a surface charge on the magnitude and kinetics of ionic currents in the nerve membrane (7) and the apparent blocking of Cl^- permeability by a surface potential resulting from adsorption of the highly dipolar phloritin molecule (2) are reasonably well understood not only in model (27) but also in cell membranes. In particular, the effects of a surface potential on carrier-induced permeability have been studied extensively for surface potentials induced by charged (27,42,28) or dipolar residues (42) present at the membrane surface.

Kinetics of Direct and Carrier Mediated Transport

The overall details of direct and carrier mediated transport are well understood and have been reviewed extensively (see for example references 18,24,45,23,22). Figure 2 shows kinetic

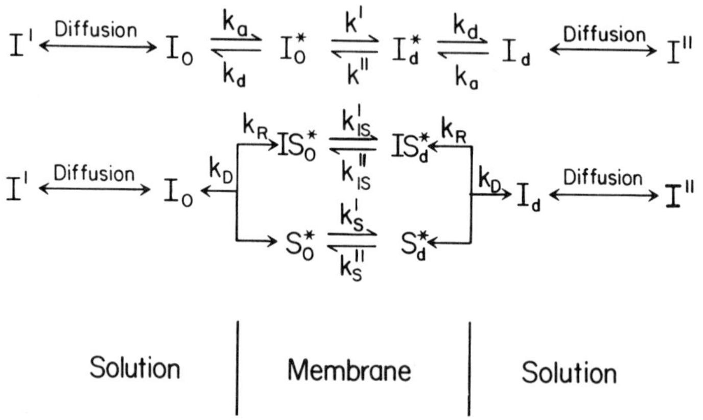

$$I' \xleftrightarrow{\text{Diffusion}} I_o \underset{k_d}{\overset{k_a}{\rightleftharpoons}} I_o^* \underset{k''}{\overset{k'}{\rightleftharpoons}} I_d^* \underset{k_a}{\overset{k_d}{\rightleftharpoons}} I_d \xleftrightarrow{\text{Diffusion}} I''$$

$I' \xleftrightarrow{\text{Diffusion}} I_o$; $I_o \xrightarrow{k_R} IS_o^*$, $IS_o^* \xrightarrow{k_D} I_o$; $IS_o^* \underset{k''_{IS}}{\overset{k'_{IS}}{\rightleftharpoons}} IS_d^*$; $S_o^* \underset{k''_S}{\overset{k'_S}{\rightleftharpoons}} S_d^*$; $I_d \xrightarrow{k_R} IS_d^*$, $IS_d^* \xrightarrow{k_D} I_d$; $I_d \xleftrightarrow{\text{Diffusion}} I''$

Solution | Membrane | Solution

Figure 2: Schematic representation of direct (upper) and carrier-mediated (lower) transport through lipid bilayer membranes in terms of first order kinetic processes. Starred quantities refer to concentration of ion (I) carrier (S) or complex (IS) in the membrane interior. For simplicity, the scheme assumes a symmetrical membrane.

schemes that have been used to describe quite accurately both the steady-state and the kinetic behavior of direct (upper part) and carrier mediated (lower part) transport. These two transport processes are seen to be similar in that a hydrophobic ion, respectively a single species, I, or a carrier complexed ion, IS, moves across the membrane interior. However, in direct transport the hydrophobic ion (I) simply diffuses to and adsorbs at the membrane surface (I*) according to the process:

$$I \underset{k_d}{\overset{k_a}{\rightleftharpoons}} I^* \qquad (1)$$

where k_a is the rate constant for adsorption, k_d is the rate constant for desorption, while in carrier mediated transport the charged complex (IS^+) is formed by the heterogeneous reaction:

$$I^+ + S^* \underset{k_D}{\overset{k_R}{\rightleftharpoons}} IS^* \qquad (2)$$

where k_R is the rate constant of formation and k_D is that of dissociation. Furthermore, the free carrier must also be able to move across the membrane interior (rate constants k_S' and k_S'') in carrier mediated transport.

Techniques for the measurement of kinetic parameters for direct as well as carrier mediated transport have been reviewed in detail (14,22). Direct transport of tetraphenylborate (TFB) (18) and trinactin mediated NH_4^+ transport (5,16) have been particularly well characterized. These transport systems were chosen here, therefore, as model permeability pathways in order to examine the mechanisms by which permeability is altered by dipole potentials at the membrane surface.

A comparative study of cationic and anionic permeability has shown that the electrostatic potential at the surface of monoolein bilayers can be increased by increasing the cholesterol content of the membrane forming lipids (43). This surface potential, which is also seen in monolayers spread at the air-water interface, must originate from oriented dipolar groups since the lipids have no net charge. "Dipolar" potential changes are large (approximately 100mv for membranes of high cholesterol content relative to pure monoolein membranes) and their variation as a function of lipid cholesterol content is well characterized (46). Therefore, these calibrated bilayers were used to assess the effects of a dipole potential on the kinetics of direct and carried mediated transport.

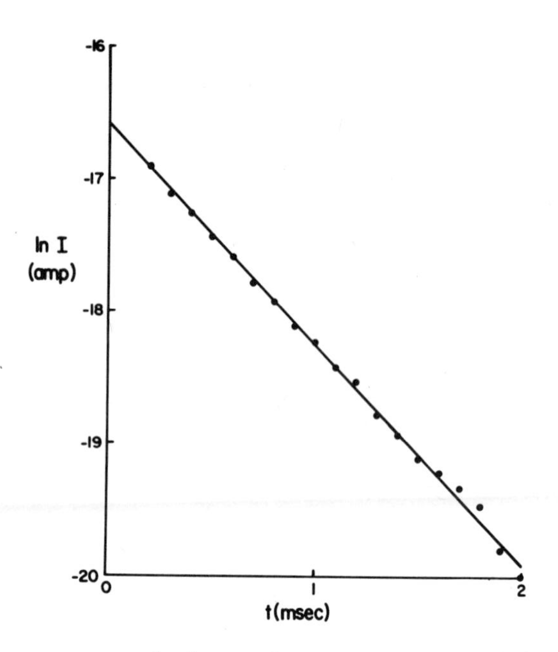

Figure 3: Time course of the membrane current elicited by a 20mv step change in the membrane potential in the presence of $5x10^{-8}$M TFB. Indifferent electrolyte, 1M NaCL. $X_{chol} = 0.83$; membrane area, $6.1x10^{-3}$ cm^2. Filled circles show the natural logarithm of the membrane current read from a photographic record while the solid line shows the regression line ($r^2 = 0.999$) through these points.

Influence of Dipole Potential on the Kinetics of Direct Transport

Kinetic parameters of transport (k', k", and $\beta = k_a/k_d$) were measured for TFB using a step-clamp technique (18) in which the membrane potential is suddenly increased from its zero-current value (0 volts for the experiments described here, for which the aqueous solutions bathing the membrane had identical composition) to some value V and the time course of the resulting membrane current, I (t), is analyzed.

Figure 3 shows the time course of the TFB current in a typical relaxation experiment. The logarithm of the membrane current, plotted as filled circles, is seen to decrease linearly with time. The regression line which connects the data points, has a correlation coefficient $r^2 = 0.999$ indicating that the current decays exponentially with time.* An exponential decay ($r^2 > 0.9$)

*At longer times the TFB current ceases to be exponential and its decay becomes proportional to the inverse square root of time(25).

of the early TFB current is seen for all monoolein bilayers of increasing cholesterol content. The time course of relaxation observed here for cholesterol-monoolein membranes is qualitatively similar to that reported by LeBlanc (26) and Ketter, et al., (18) for phospholipid membranes.

Table I

Influence of Membrane Composition on
The Kinetics of TFB Transport

X_{chol}	P (cm/sec)	τ (msec)	k (sec^{-1})	β $x10^3$ (cm)
0	.072	18	28	2.6
.5	.21	5.6	89	2.3
.75	1.5	.78	640	2.4
.83	2.3	.60	830	1.7

Table 1 shows as a function of increasing cholesterol content initial values of membrane current density I_o (in amperes per cm^2) obtained from the t = o intercept of the I vs. t regression line, as well as time constants τ (in seconds, obtained as the negative inverse of the slope of the I vs. t regression line). At low concentration of TFB ($C_{TFB} > 10^{-6}M$) I_o's are proportional to the aqueous TFB concentration while τ's remain unaltered within the limits of experimental error (45). The TFB permeability, P (in cm/sec) of the membrane can be calculated from the initial membrane current measured for small (<50mv) applied potentials (45):

$$P = \frac{RT}{AF^2} \frac{Io}{CV} \tag{3}$$

where A is the membrane area, V is the potential increment, C is the aqueous concentration of TFB and R,T and F have their usual meaning. P's, shown in the second column of Table I were independent of C_{TFB} for $C_{TFB} < 10^{-6}M$ (45).

The kinetic parameters can be calculated from P's and τ's by the following equations (18,45):

$$k = \frac{1}{2\tau} \qquad \beta = 2\tau P \tag{4}$$

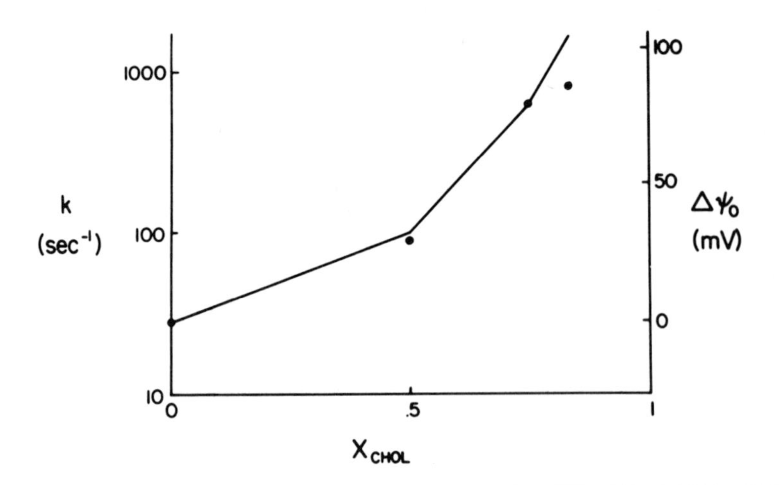

Figure 4: The relation between rates of TFB translocation (k) and changes in the surface potential ($\Delta\psi_o$) as a function of cholesterol more fraction (x_{chol}) in monoolein bilayers. Filled circles show k's plotted from table I using the left hand logarithmic scale. The solid line show $\Delta\psi_o$'s taken from reference 43, and plotted using the right hand scale in such a way that a tenfold change in k corresponds to a 59 mV change in $\Delta\psi_o$.

A single rate constant k suffices to describe the translocation of TFB through the membrane interior since for small applied potentials* $k' \simeq k'' \simeq k$:

$$k' = k \exp\left(\frac{nFV}{RT}\right) \simeq k + \left(\frac{nFV}{RT}\right) k \simeq k \tag{5}$$

$$k'' = k \exp\left(\frac{-nFV}{RT}\right) \simeq k - \left(\frac{nFV}{RT}\right) k \simeq k$$

Individual values of the adsorption (k_a) or desorption (k_d) rate constants can not be obtained from the relaxation experiments since aqueous diffusion of TFB masks these processes. However, the adsorption coeffieient, $\beta = k_a/k_d$, is easily determined from P's and τ's.

---

* The origin of the voltage dependence of k' and k" is considered more in detail in the discussion. Empirically n is found to be near 0.4 for TFB (1,18).

k's and β's are shown in the fourth and fifth columns of table I for various values of the cholestoral mole fraction, x_{chol}, in the membrane forming lipids. It is immediately apparent that x_{chol} greatly alters k's but it has little effect on β's . Thus, nearly all of the increase in the overall TFB permeability results from an increased rate of TFB translocation across the membrane interior.

The increase of the k's is related to the changes of the dipolar surface potential $\Delta\psi_o$, measured by Szabo (43) for the same lipids. This is shown in Figure 4, which plots k's (filled circles, left hand logarithmic scale) as a function of x_{chol} and compares these with $\Delta\psi_o$'s (solid line, right hand scale) drawn so that a 59 mV increase in $\Delta\psi_o$ corresponds to a tenfold increase in k. The agreement between filled circles and solid line suggests that there is a proportionality between the log k and $\Delta\psi_o$. This is expected (see equation 11 of the discussion) only if $\Delta\psi_o$ alters the energy barrier for TFB translocation across the bilayer membrane interior.

The most parsimonious explanation of this observation is that TFB adsorbs at the aqueous side of the dipolar groups from which $\Delta\psi_o$ originate so that $\Delta\psi_o$'s become part of the energy barrier for TFB translocation. Consistent with this finding is the fact that the coefficient of TFB partition, β, is practically unaltered by x_{chol}, that is by $\Delta\psi_o$ (see column 5 of table I).

The Influence of Dipole Potential on The Kinetics of Carrier Mediated Transport

Increasing cholesterol content in monoolein bilayers substantially alters the magnitude and the kinetics of carrier mediated ion transport in a way which is qualitatively similar for all of the well-known carriers of ions such as valinomycin, nonactin or trinactin (44). Specifically, increasing cholesterol content decreases the carrier-induced ion permeability in a way which is consistent with a decreased translocation rate of the complexed cation and, to a smaller extent, a decreased concentration of the uncomplexed carrier at the membrane surface. These effects of cholesterol could be characterized in quantitative detail for the trinactin mediated NH_4^+ transport which, in monoolein membranes, shows well understood kinetics of large amplitude on a time scale which is sufficiently slow for accurate estimates of transport parameters in membranes of high cholesterol content.

Figure 5 shows typical traces for the decay in time of the membrane current resulting from rapid increase of the membrane potential from 0 to 100mV in bilayers formed in 1 M $NH_4C\ell$ solutions from lipids in which trinactin was dissolved at 10^{-5}M concentration.

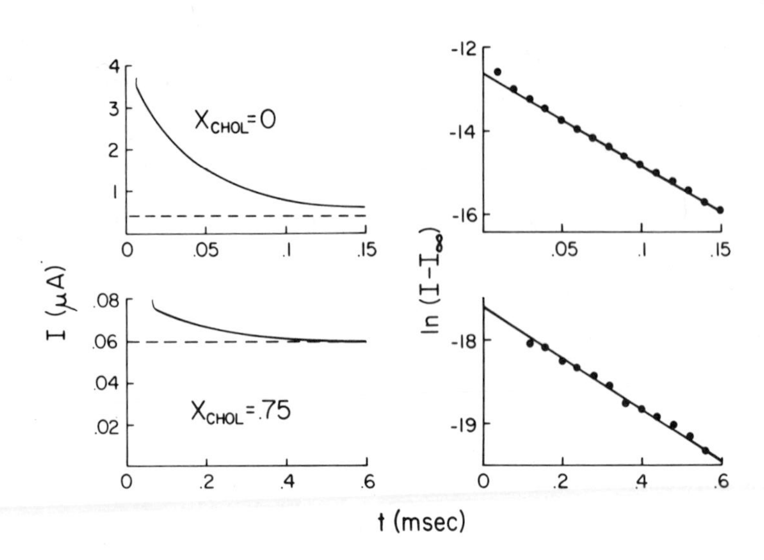

Figure 5: Influence of membrane composition on the magnitude and the time course of the current through lipid bilayers formed from cholesterol monoolein mixtures. 10^{-5}M trinactin in the lipid. 1 M NH_4 Cl solutions, 100mV voltage steps; membrane area: 8.5X 10^{-3} cm^2 for x_{chol} = 0 and 6.1X 10^{-3} cm^2 for x_{chol} = 0.75. Dashed lines show steady state values for the membrane currents. Note the difference in time scales between plots for x_{chol} = 0 and x_{chol} = 0.75.

In contrast to TFB, the membrane current I (t) decays to a large steady-state value (I_∞). Nevertheless, the plot of ℓn ($I_{(t)}$ - I_∞) v.s. time is linear for all of the membrane compositions examined here, confirming and extending Hladky's (16) observation that a single exponential describes accurately the decay of trinactin mediated NH_4^+ current. Figure 5 also shows that the presence of cholesterol in the bilayer decreases the magnitude and slows down the time course of the decay of the membrane current. Relaxation amplitudes $\alpha = (I_o - I_\infty)/I_\infty$, where I_o is the initial current obtained by extrapolation of I(t) - I plots to t = o, and relaxation times calculated from the slopes of I(t) v.s. t plots, are summarized in table II for x_{chol} = 0, 0.5, 0.75, 0.83. Figure 6 shows the influence of cholesterol on the current-voltage relationship of bilayers formed from lipids 10^{-5} M in trinactin in a 1 M HN_4 Cℓ solution. Open circles show steady-state values (I_∞) to which the membrane current relaxes from its initial value (I_o) plotted as filled circles. The shape of the initial current-voltage relationship is not significantly altered by cholesterol. This is indicated by the fact that the I_o's (filled circles) fall on the dashed line drawn to the empirical equation (13)

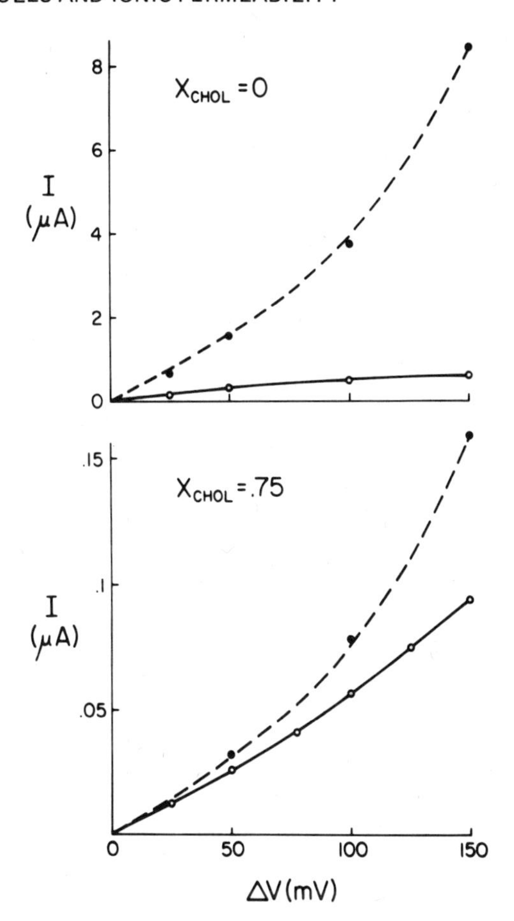

Figure 6: Influence of membrane composition on the shape and magnitude of initial (dashed lines) and steady-state (solid lines) current-voltage relationships. Membranes were the same as those of figure 5.

I_o = constant [sin h (0.36 FV/RT)]. Note however, that the magnitude of the currents are greatly decreased for x_{chol} = 0.75 relative to x_{chol} = 0. This is shown for other values of x_{chol} in table II where initial conductances (G_{oo}) are tabulated. In contrast, open circles show that both the shape and magnitude of the steady-state current-voltage relationship are altered by the presence of cholesterol in the membrane; the steady-state I-V relationship bends toward the V axis for x_{chol} = 0 while it bends toward the I axis for large values of x_{chol}. The behavior of the steady-state I-V curves results from a decreased relaxation amplitude α as x_{chol} increases. For example, the solid lines in Figure 6 were drawn to the empirical equation
$I_\infty = I_o/[1 + \alpha \cosh(0.36\ FV/RT)]$ where α is the relaxation

TABLE II

Influence of Membrane Composition on the Kinetics of TRIN-NH_4^+ Transport

x_{chol}	α	τ (msec)	G_∞ x104 (s/cm^2)	k_{is} (sec^{-1})	γ x104 (sec^{-1})	N^*_{is} (pmole/cm^2)	k_R/k_Dk_s x103
0	2.9	.078	41	4700	2.9	.75	.8
.5	.67	.22	14	920	3.7	1.1	.92
.75	.16	.40	.79	170	4.6	.33	1.8
.83	.11	.72	.15	69	8.0	.19	2.9

amplitude for small applied potentials. These diverse experimentally observed effects of increased membrane cholesterol content can be characterized simply by a decreased initial conductance (G_{oo}), a decreased relaxation amplitude (α) and an increased relaxation time (τ).

The decay of the TRIN-NH_4^+ current is described by a single exponential for all of the experiments reported here, although theoretically the $I_O - I_\infty$ v.s. time plot should be a sum of two exponentials. This is because, as noted by Hladky (16) and confirmed here, the amplitude of one of the exponentials (the faster one) is negligibly small for TRIN-NH_4^+. For this reason it is not possible to deduce the complete set of parameters (k_{is}, k_s, k_R, k_D, see figure 2) for the system under study here. Fortunately, rate constants of interest can nevertheless be calculated in the following way*.

$$k_{is}^{\bullet} = \frac{\alpha}{2\tau\ (\alpha+1)} \tag{6}$$

$$\gamma = \frac{1}{k_D}\ (1 + \frac{k_R a_i}{2\ k_S}) = \tau\ (\alpha+1) \tag{7}$$

$$N^*_{is} = \frac{RT}{0.36F^2}\ \frac{G_\infty}{k_{is}} \tag{8}$$

* These equations are derived (see reference 16) for neutral carrier mediated ion transport provided that the following inequality holds: $k_R + 2k'_{is} >> k''_{is} + 2k_D$. This inequality was found to hold for monoolein bilayers by Hladky (16).

Forward (k'_{is}) and backward (k''_{is}) translocation rates of the TRIN-NH_4^+ complex are nearly equal for small applied potentials (see eq. 5; $k_{is} \simeq k'_{is} \simeq k''_{is}$). N^*_{is} is the equilibrium surface concentration (in units of moles per cm^2) of the complex at the membrane surface.

Table II shows k_{is}, γ and N^*_{is} calculated using eqs. 6,7 and 8, for bilayers of increasing cholesterol content. The last row in table II shows values for the k_R/k_Dk_S ratio calculated for each lipid composition from slopes of γ v. s. NH_4Cl activity plots using data for $C_{NH_4C\ell}$ = 5M, 1M, and .01M.

The data of table II clearly indicate that increasing membrane cholesterol content drastically reduces the rate of translocation of the complexed ion. However, the decrease in k_{is} only partly accounts for the decrease of the initial membrane conductance G_{oo}. Part of the decrease in G_{oo} must be attributed to a decrease of the number of complexed ions (N^*_{is}) adsorbed at the membrane surface. Table II shows N^*_{is} calculated from equation 8. Since $N^*_{is} = (k_R/k_D)\ a_i a_s$, the decrease in N^*_{is} may be due to either a decrease in the equilibrium constant (k_R/k_D) of ion carrier association or to a decreased carrier surface activity (a_s). The last row of Table II indicates that the former is unlikely since k_R/k_Dk_S actually increases with x_{chol} instead of decreasing†. Thus the decrease in N_{is} must be attributed to a decrease of the activity (a_s) of the free carrier in the membrane.

The effect of cholesterol on the kinetics of trinactin mediated NH_4^+ transport are characterized by a large decrease in the translocation rate (k_{is}) of the complex; a smaller decrease in the concentration of the neutral carrier at the membrane surface (a_s) while the association constant k_R/k_D or the translocation rate of the carrier (k_s) appear to remain unaltered.

What is the mechanism by which cholesterol decreases k_{is}? The simplest mechanism would involve an increasingly positive electrostatic potential generated in the membrane interior by the presence of cholesterol, much in the same way as it is involved in increasing the translocation rate of the negative TFB ion. Positive $\Delta\psi_o$ would, however, add to the height of the energy barrier since the TRIN - NH_4^+ complex is positively charged.

Figure 7 shows that this simple mechanism accounts entirely for the observed decrease in k_{is}, which is plotted from table II

---

† It is conceivable but unlikely that k_R/k_D decreases while there is a large decrease in k_S resulting in a nearly constant k_R/k_Dk_S ratio.

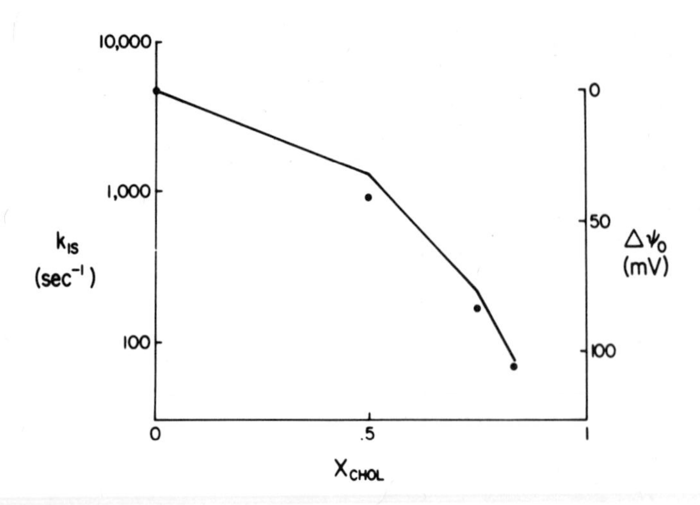

Figure 7: The relationship between rates of translocation for the TRIN-NH_4^+ complex (k_{is}) and changes in the surface potential ($\Delta\psi_0$) as a function of cholesterol mole fraction in monoolein bilayers. Filled circles show k_{is}'s plotted from table II using the left hand logarithmic scale. The solid line shows $\Delta\psi_0$'s replotted from figure 4 using the right hand scale in such a way that a tenfold change in k_{is} corresponds to a 59mV change in $\Delta\psi_0$.

as filled circles using the logarithmic left hand scale. The solid line, drawn to fall on the filled circle at $x_{chol} = 0$, shows the increase of the electrostatic potential $\Delta\psi_0$ in the membrane interior deduced independently using lipophilic cations and anions (46) using the right hand scale which is drawn so that a 59mV change corresponds to a 10 fold change in k_{is}. The agreement between k_{is} and $\Delta\psi_0$ indicates that this is indeed the case.

Discussion

Direct and carrier-mediated transport are both described, at least formally, by a first order rate process for the translocation of a hydrophobic ion (a single, large ion or an ion-carrier complex) through the nonpolar but thin membrane interior. The time-dependent current observed for these systems results from altered surface concentration of the hydrophobic ion by the translocation process.

Details of the translocation process can be understood in terms of a generalized diffusion process (24), which takes into account the energy barrier encountered by an ion during its passage through the hydrocarbon-like membrane interior. Using this formalism, the translocation rate, k, can be expressed as a function of an overall energy barrier ω (x), and a diffusion constant D(x) by the following equation(45):

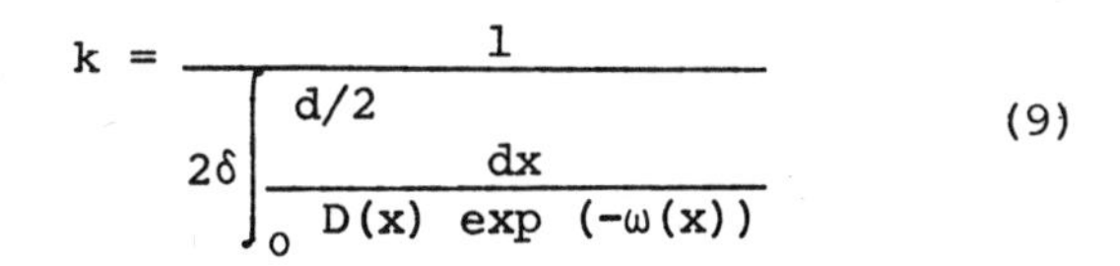

$$k = \frac{1}{2\delta \int_{0}^{d/2} \frac{dx}{D(x)\ \exp\ (-\omega(x))}} \qquad (9)$$

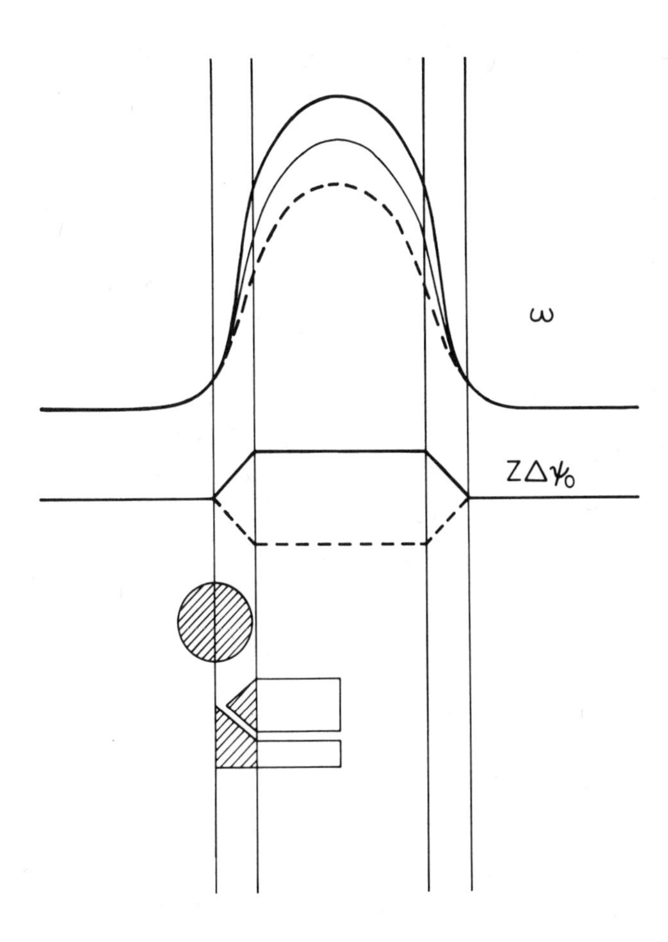

Figure 8: Shematic diagram illustrating the influence of a positive dipole potential ($\Delta\psi_0$) on the energy barrier (ω) for a cation (solid line) or anion (dashed line) of valence z. The lower part shows an arrangement of cholesterol (arrow shape) and monoolein (L-shape) molecules with respect to a hydrophobic ion (shaded disk) adsorbed at the membrane surface. Such an arrangement is consistent with relaxation data for both TFB and TRIN-NH_4^+.

where d is the membrane thickness and δ is the width of the adsorption layer of the hydrophobic ion at the membrane surface. The energy barrier ω (x) is expected to be influenced by any electrical potential difference between the membrane interior and the membrane surface. For a uniform potential ψ_0 in the membrane, $\omega(x) \simeq (\mu o(x)/RT) + (zF\psi_0/RT)$. That is to say, the difference in the electrostatic potential between the membrane surface and the membrane interior is expected to contribute to the height of the energy barrier presented by the membrane interior to the movement of ions. The upper part of figure 8 illustrates this situation schematically for a bilayer membrane. The region of polar head groups is represented by the area between closely spaced vertical lines, while the hydrocarbon-like interior is represented by the area between these. The "intrinsic" energy barrier ω, mainly due to the energy required to move the ion from the aqueous phase into the low dielectric constant membrane interior (13,34), is shown on the top part of the figure as a thin solid line. A difference in the electrostatic potential, which may result from the presence of oriented dipolar head groups, is expected to add (solid lines) or subtract (dashed lines) from the intrinsic energy barrier, depending on the charge of the ion according to the following equation (57):

$$k \simeq \frac{\exp(-zF\Delta\psi_0/RT)}{2\delta \displaystyle\int_0^{d/2} \frac{dx}{D(x)\ \exp\ (\mu o(x)/RT)}} = k_0 \exp(-zF\Delta\psi_0/RT) \quad (10)$$

That is, a proportionality is expected between the logarithm of the translocation rate k, and the change of potential $\Delta\psi_0$, provided that the potential difference arises between the energy minima of adsorption and the membrane interior. For $\Delta\psi_0$ in millivolts and at 22°C.

$$\log k \simeq \log k_0 - 59z\Delta\psi_0 \quad (11)$$

Figure 4 and 7 show that the equation 11 is obeyed for the translocation of the anionic TFB or cationic TRIN-NH_4^+ complex: log k increases for TFB (Z = -1) and log k_{is} decreases for TRIN-NH_4^+ (Z = +1) exactly as expected (solid lines on figures 4 and 7) for the dipole potential that was previously determined by independent measurements. A convenient diagrammatic localization of the dipole potential is shown in the lower section of figure 8, which also shows the head groups (shaded) and hydrocarbon tails (light) of a cholesterol (arrow shape) and of a monoolein (L-shape) molecule in the bilayer membrane.

Direct and carrier mediated transport are similar since in both cases a lipid soluble ion, adsorbed at the membrane surfaces, moves across the interior of the bilayer membrane. The results of Figures 4 and 7 show that this similarity extends to include the location of the permeant ions as well. Both for direct (TFB) and carrier mediated (TRIN-NH_4^+) transport, dipolar potentials in cholesterol-monoolein membranes modify the energy barrier for ion transport across the membrane interior, without altering adsorption to the membrane surface. The most parsimonious explanation of this result locates the adsorbed ions at the aqueous side of the residues from which the dipolar potential originates. Since the dipole potentials are believed to originate from the polar head groups, the adsorption plane for the hydrophobic ions must be located at the aqueous surface of the polar head groups, as shown in the mid section of figure 8.

Lipophilic ions and carriers act as permeatoxic agents not only in lipid bilayers but also in cell and organelle membranes (35). However, the precise dissection of transport steps is much more difficult to achieve in natural membranes not only for technical reasons but also because these membranes often have intrinsic conductive pathways which mask the permeatoxic effect. Nevertheless, choice of suitable systems (large egg cells or tight epithelia, for example) should circumvent these difficulties, yielding eventually a detailed kinetic characterization of permeatoxin induced ionic permeability in cell membranes as well.

Hydrophobic ions and ion carriers are excellent probes of the electrochemical properties of bilayer membranes. Although their use was restricted here to monoolein-cholesterol bilayers, hydrophobic ions are expected to yield similarly useful information on the electrochemical properties of bilayers formed from a variety of lipids (e.g., phospholipids). Furthermore, hydrophobic ions are expected to be extensively useful for investigating the double layer and dipole potentials created in bilayers by the adsorption of biologically active molecules which alter membrane permeability. Such "probing" may be extended directly to at least some types of cell or organelle membranes (such as the aforementioned egg cells or tight epithelia) not only to "sense" the electrochemical properties of the membrane but also to probe for surface potential or fluidity changes induced in the membrane by pharmacological agents which alter permeability. Such direct probing for the effect of these permeatoxic agents should determine whether these act on permeability indirectly, through altering the electrochemical properties of the bilayer (in this case ionic probes should "sense" the effect) or directly through altering specific permeability pathways in the membrane (in this case the probes should sense no effect).

Direct and carrier mediated transport are basically different from channel mediated transport. Permeation occurs in a homogeneous membrane in the former case while permeation occurs in a heterogenous membrane in the latter case. These two types of transport processes appear to represent two extreme cases. In principle, one may conceive transport mechanisms that have properties that lie between these. For example, one may have a selectivity filter, with properties akin to those of a carrier, in series with a highly conductive but non selective channel. Although such intermediate mechanisms have not been reported for model systems, their existence is important since cell membranes, which combine high selectivity and high transport rates, point to the possibility of such intermediate mechanisms in cell and organelle membranes.

Permeatoxic agents which induce ionic permeability in membranes are expected to do so in lipid bilayer membranes so that their mechanism of action can be studied with particular ease. This is the case, for example, for antibiotics such as the macro-tetralide actins, valinomycin, gramicidin A, alamethicin. In contrast, agents that alter already existing permeability pathways may do so either by a nonspecific mechanism (see figure 1), acting indirectly through altering the general electrochemical properties of the membrane, or specifically by interacting directly with a permeability pathway without affecting the general electrochemical properties of the membrane. The former case can be modeled in bilayers by techniques which have been described here (for dipole potentials) or elsewhere(42). In contrast, the latter case is very difficult to model since it requires the specific structure associated with the permeability pathway. Thus, relatively little is known about the molecular mechisms of action for permeatoxic agent which act specifically on a permeability pathway. However, since this is an area of intense pharmacological interest, it is hoped that these specific mechanisms can be understood in terms of more general types of interactions such as those examined in this paper.

Literature References

1 ANDERSEN, O.S., and FUCHS, M.: Biophys. J. 15 (1975) 795.

2 ANDERSEN, S.S., FINKEKSTEIN, A., KATZ, I., and CASS, A.: J. Gen. Physiol. 67 (1976) 749.

3 ARMSTRONG, C.M.: Potassium probes of nerve and muscle membranes, Ch.5, Vol.3, Membranes (Eisenman, G., Ed.). Marcel Dekker Inc. New York (1975).

4 BEAN, R.C.: Protein mediated mechanisms of variable ion conductance in thin lipid membranes, Ch.6, vol. 2, Membranes (Eisenman, G., Ed.). Marcel Dekker Inc., New York (1973).

5 BENZ, R. and STARK, G.: Biochim. Biophys. Acta 382 (1975) 27.

6 CELIS, H., ESTRADA -O.S., and MONTAL, M.: J. Membrane Biol. 18 (1974) 187.

7 CHANDLER, W.K., HODGKIN, A.L. and MEVES, H.: J. Physiol 180 (1965) 821.

8 COLE, K.S.: Membranes, ions and impulses. University of California Press. Berkeley (1968).

9 CONTI, F.: Ann. Rev. Biophys. Bioeng. 4 (1975) 287.

10 EISENBERG, M., HALL, J.D. and MEAD, C.A.: J. Membrane Biol. 14 (1973) 143.

11 EISENMAN, G., SANDBLOM, J. and NEHER, E.: Ionic selectivity, saturation, binding and block of the gramicidin A Channel, In 9th. Jerusalem Symposium: Metal-liquid interactions in organic and biochemistry (Pullman, B. Ed.), Reidel Publ. Co., Dordrecht (1976).

12 GERSCHENFELD, H.M.: Physiologycal Rev. 53 (1973) 1.

13 HALL, J.E., MEAD, C.A., and SZABO, F.: J. Membrane Biol. 11 (1973) 75.

14 HAYDON, D.A. and HLADKY, S.B.: Quart. Rev. Biophys. 5 (1972) 187.

15 HILLE, B.: Ionic selectivity of Na and K channels of nerve membranes, Ch. 4, vol.d, Membranes (Eisenman G., Ed.), Marcel Dekker Inc. New York (1975).

16 HLADKY, S.B.: Biochim. Biophys. Acta 375 (1975) 327.

17 KAO, IL, DRACHMAN, D.B. and PRICE, D.L.: Science 193 (1976) 1256.

18 KETTERER, B., NEUMCKE, B. and LÄUGER, P.: J. Memb. Biol. 5 (1971) 225.

19 KITA, H., and VAN DER KLOOT, W.: Nature (London) 250 (1974) 658.

20 KUFFLER, S.W. and NICHOLLS, J.G.: From neuron to brain Sinauer Associates Inc. Publishers. Sunderland, Mass. (1976).

21 LANYI, J.K., RENTHAL, R. and MCDONALD, R.E.: Biochemistry 15 (1976)1603.

22 LAPRADE, R., CIANI, S., EISENMAN, G., and SZABO, G.: The kinetics of carrier mediated ion permeation in lipid bilayers and its theoretical interpretation, Ch. 2, vol.3, Membranes (Eisenman, G. Ed.), Marcel Dekker Inc. New York (1975).

23 LÄUGER, P.: Science 178 (1972) 24.

24 LAUGER, P., and NEUMCKE, B.: Theoretical analysis of ion conductance in lipid bilayer membranes, Ch.1, vol. 2, Membranes (EISENMAN, G., Ed.), Marcel Dekker Inc. New York (1973).

25 LEBLANC, O.H.: Biochim. Biophys. Acta 193 (1969) 350

26 MARTY, A. and FINKELSTEIN, A.: J. Gen. Physiol. 65 (1975) 515.

27 MCLAUGHLIN, S.G.A., SZABO, G., EISENMAN, G. and CIANI, S.: Proc. Nat. Acad. Sci. U.S. 67 (1970) 1268.

28 MCLAUGHLIN, S., and EISENBERG, M.: Ann. Rev. Biophys. Bioeng. 4 (1975) 335.

29 MITCHELL, P.: Symp. Soc. Gen. Microbol. 20 (1970) 121.

30 MUELLER, P., andRUDIN, D.L.: Curr. Top. Bioenergetics 3 (1969) 157.

31 MURER, H. and HOPFER, U.: Proc. Nat. Acad. Sci. U.S. 71 (1974) 484.

32. NARAHASHI, T.: Federation Proc. 31 (1972) 1124.

33. NEUMCKE, B. and BAMBERG, E.: The action of uncouplers on lipid bilayer membranes, Ch. 3, vol. 3, Membranes (Eisenman, G. Ed.), Marcel Dekker Inc., New York (1975).

34 PARSEGIAN, A.: Nature 221 (1969) 844.

35 PRESSMAN, B.C.: The role of membranes in metabolic regulation (Melham, M.A. and Hanson, R.W. Eds.), Academic Press. New York (1972).

36 RUSSELL, J.M., EATON, D.C. and BRODWICK, M.S.: Biophys. J. 16 (1976) 76a.

37 SALZBERG, B.M. and COHEN, L.B.: Rev. Physiol. Biochem. Pharm. 76 (1976).

38 SKOU, J.C.: Physiol. Rev. 45 (1965) 596.

39 SKOULACHEV, V.P.: Current topics in bioenergetics, vol. 4 (Sandai, D.R. Ed.), Academic Press, New York (1971).

40 STRICHARTZ, G.R.: J. Gen. Physiol. 62 (1973) 37.

41 SZABO, G., EISENMAN, G., and CIANI, S.: J. Membrane Biol. 1 (1979) 346.

42 SZABO, G., EISENMAN, G., MCLAUGHLIN, S., and KRASNE, S.: Ann. N.Y. Acad. Sci. 195 (1972) 273.

43 SZABO, G.: Nature 252 (1974) 47.

44 SZABO, G.: Biophys. J. 15 (1975) 306a.

45 SZABO, G.: The influence of dipole potentials on the magnitude and kinetics of ion transport in liquid bilayer membranes. In Extreme Environment (Heinrich, M.R. Ed.). Academic Press. New York (1976).

46 WAGGONER, A., WANG, C.H., and TOLLES, R.L.: Biophys. J. 16

Supported by USPHS Grant HL 19639. I thank Dr. Simon Lewis for critical reading of the manuscript.

DISCUSSION

WEINER: Do you feel the changes in the monoolein cholesterol ratios would have an effect on the membrane fluidity at all? Is that an important factor particularly to the carrier systems?

SZABO: I think that it is not really an important factor but I have to qualify that. It is not an important factor in the sense that it is not the major influence which would alter ion permeability. This was a surprise to me. I can tell you frankly that what I expected was a loss in the fluidity and no change in the potential and I found the exact opposite. It is quite interesting. We find that it is the potential which is tremendously important for the tenfold change in the membrane conductance.

PRESSMAN: I would like to counter some of the advanced notoriety attributed by Dr. Szabo to ionophores. Valinomycin has been prescribed for treatment of glaucoma, and monensin is being fed to barnyard fowl to cure disease, to cattle to fatten them, and as we will see tomorrow, some of us may ultimately owe our lives to this as a new class of pharmacological agent.

ALLISON: Wouldn't this affect the interaction at the surfaces in the systems that you are talking about.

SZABO: Well, it would affect the hydration energy so it would affect the permeability.

EFFECTS OF NEOMYCIN ON POLYPHOSPHOINOSITIDES IN INNER EAR TISSUES AND MONOMOLECULAR FILMS

J. Schacht, S. Lodhi* and N. D. Weiner*

Kresge Hearing Research Institute and
*College of Pharmacy, University of Michigan
Ann Arbor, Michigan 48109

ABSTRACT

The effect of neomycin on polyphosphoinositides was studied *in vivo* and *in vitro*. *In vivo*, the incorporation of $^{32}P_i$ into phosphatidylinositol phosphate and phosphatidylinositol diphosphate was measured in inner ear tissues. Concentrations of neomycin which decreased the electrophysiological response of the cochlea to sound stimulation also decreased labeling of phosphatidylinositol diphosphate. *In vitro* experiments with brain tissues and polyphosphoinositide extracts indicated a direct interaction between the lipids and neomycin.

Neomycin interacts strongly with monomolecular films of polyphosphoinositides. The interaction appears to be complex and is a function of neomycin concentration in the subphase and surface pressure of the film. Condensation of the polyphosphoinositide film is favored at low neomycin concentrations and low film pressures while expansion of the film is favored at high neomycin concentrations and high film pressures. The interactions of neomycin with other negatively charged films (phosphatidyl inositol and phosphatidyl serine) are much weaker, particularly at low neomycin concentrations.

The metabolic and physiological consequences of the neomycin/polyphosphoinositide interaction are discussed in regard to the ototoxicity of the drug.

INTRODUCTION

Membrane proteins are recognized as the cellular receptors of hormones, neurotransmitters and of a variety of drugs. There is also considerable evidence that some drug actions involve the phospholipids of the plasma membrane. This has been suggested mainly for local anaesthetics, tranquilizers and some CNS active compounds (1,7,17,22,31). In most of these cases the evidence for the proposed mechanism is based on the demonstration of *in vitro* interactions between the drugs and the lipids in artificial membrane systems (monomolecular films, lipid vesicles or bilayers) since it is difficult to obtain evidence for specific interactions in heterogeneous biological membranes. We will describe here the action of the ototoxic antibiotic neomycin on the metabolism of polyphosphoinositides in inner ear tissues and will attempt to correlate these *in vivo* observations with drug/lipid interactions in monomolecular films.

The aminoglycoside antibiotics (neomycin, streptomycin and related compounds) are valuable therapeutic agents by virtue of their antibacterial toxicity. The bactericidal actions are supposed to be mediated by an interference with protein synthesis stemming from drug binding to a protein of the 30 S ribosomal subunit (23). In mammalian systems these drugs can block neuromuscular and ganglionic transmission (24) and they are toxic to the kidney and the inner ear (10).

We have been concerned with the molecular mechanism of drug induced hearing loss, and several characteristics of the antibiotics led us to investigate their influence on phospholipid metabolism. The basic aminoglycosides do not readily penetrate the cell membrane (4); they depress rapidly but reversibly the microphonic output from the lateral line organ (33); and when introduced into the perilymphatic spaces of the guinea pig cochlea, will decrease the cochlear microphonic potentials within minutes (18). This evidence points to an interference with bioelectric events at the membrane. It has been repeatedly suggested, and there is considerable evidence in support, that phospholipids are involved in the control of membrane permeability either through changes of the charge of their polar groups or through their capacity to bind calcium (11,12,16).

While the inner ear, where the physiological effects of neomycin are well documented, seemed specially well suited for *in vivo* studies, other systems were chosen for the investigation of the molecular mechanism of neomycin action: subcellular fractions of brain which is the tissue of choice for the study of polyphosphoinositide metabolism, and monomolecular lipid films.

Monomolecular films represent a relatively simple type of membrane model having a well defined organized structure. They provide a convenient and promising method of studying molecules in a fixed orientation, as well as in a single layer where orientation can be changed by compression of the monomolecular film. As such, they constitute an important model system for the study of many natural phenomena involving surfaces of an oriented array of molecules.

Binding of cations by phospholipid monolayers has been studied extensively. Whereas calcium ions were found to produce little reduction in the area per molecule of phospholipid monolayers possessing no net negative charge (3,32), many workers reported film condensation of phosphatidyl serine in the presence of divalent cations (2,5,15,20,21,25). The extent of the interaction between negatively charged phospholipids and divalent ions is dependent on a number of factors including degree of unsaturation of the fatty acyl chains of the phospholipid, surface pressure, and presence of monovalent ions (15,25,32). Acidic phospholipids have been shown to act as ion exchangers in monolayers (15), and these lipids may play a role in transmembrane cation transport (2) and neuronal excitation (8,11).

It is important to emphasize that whenever acidic phospholipid monolayers have been shown to interact with cations possessing more than one charge and no significant hydrophobic group (e.g., long alkyl chain), the resultant film always shows a reduction in the area per molecule at a given surface pressure of the phospholipid (condensation effect).

METHODS

We have described in detail the procedures of inner ear perfusion, and tissue dissection and lipid analysis (19) and of monolayer measurements. We will only briefly review these methods.

Guinea pigs were prepared surgically for perilymphatic perfusion while under anaesthesia and artificial respiration. Capillaries were implanted into scala tympani and scala vestibuli of the lower turn of the cochlea and connected to a reservoir of "artificial perilymph" for continuous perfusion of the perilymphatic spaces. This technique leaves the middle ear intact so that the ear can be stimulated by sound and cochlear microphonic potentials can be recorded. When required, drugs were added to the perfusion fluid as well as radioactive precursors, e.g., ^{32}P for lipid labeling. The cochlear tissues were then fixed in 10% neutral formaldehyde and dissected under the microscope. Lipids were extracted with an acidified solvent (chloroform - methanol - 2.4 N HCl; 3:2:1 by vol), separated by thin layer chromatography and quantitated by liquid scintillation counting.

The film balance assembly has been described previously (13,14). For each of the film studies, the subphase consisted of 0.1M HEPES buffer at a pH of 7.0. In addition, various concentrations of neomycin were added to the subphase to study the neomycin-phospholipid interactions. In all cases, enough sodium chloride was added so that the final ionic strength was 0.2. The phospholipid spreading solutions were prepared by dissolving the lipids in a hexane/ethanol mixture (90:10, v/v). Compression of

Table I

^{32}P-labeling of inner ear phospholipids

Tissue	Protein	$PhIP_2$	PhIP	PhI+S
	μg	cpm ^{32}P incorporated		
Stria vascularis	8	3360	4155	880
Organ of Corti	10	1520	1780	675

Cochlear perfusions were performed as described in "Methods" with 250 μCi $^{32}P_i$/ml perfusion fluid for 60 min. Values are means of three experiments. PhIP, $PhIP_2$: phosphatidylinositol phosphate, diphosphate; PhI+S: phosphatidyl inositol + phosphatidyl serine.

the film was initiated 45 minutes after speading to allow for equilibration. All experiments were run in duplicate.

RESULTS

Labeling of inner ear lipids

Phospholipids of the inner ear were efficiently labeled by cochlear perfusion with 250 μCi ^{32}P-orthophosphate/ml for 60 min (Table I). The polyphosphoinositides (phosphatidylinositol phosphate and phosphatidylinositol diphosphate) incorporate radioactivity most rapidly, a labeling pattern which had previously been reported for nervous or secretory tissues (26,27,34). The values presented are means of three independent experiments and are expressed as cpm obtained from tissues of a single cochlea.

Effects of neomycin on the inner ear

The toxic effects of neomycin on the function of the inner ear can be assessed through recordings of the cochlear AC potential ("cochlear microphonics") in response to sound. Such recordings showed that the microphonic output is rapidly decreased when neomycin is present in the perfusion fluid at concentrations of 10^{-3}M or 10^{-2}M (18). With 10^{-3}M neomycin, the cochlear microphonic was decreased 32% in 26 min after a delay of 24 min; at 10^{-2}M it was decreased 78% in 50 min after a delay of 10 min. In the presence of 10^{-4}M neomycin or in the absence of this drug, a steady microphonic potential was maintained for at least 60 min.

Parallel to the electrophysiological recordings we measured the drug effect on phospholipid labeling (Table II). In agreement with our earlier findings with chronic systemic application of the

Table II

Effect of neomycin on ^{32}P-phosphoinositides in the inner ear

	Stria Vascularis	Organ of Corti
	cpm $PhIP_2$/cpm PhIP	
No Neomycin	0.64 ± 0.10	0.67 ± 0.12
10^{-4}M Neomycin	0.66 ± 0.11	0.89 ± 0.11
10^{-3}M Neomycin	*0.40 ± 0.04	*0.40 ± 0.06
10^{-2}M Neomycin	*0.35 ± 0.04	*0.48 ± 0.19

Conditions were as in Table I. Values are means of four to ten experiments ± S.D.
*differs from control $p < 0.01$ (One-way ANOVA)

drug (19) we see a decrease of $PhIP_2$ labeling (relative to PhIP) in the presence of neomycin in inner ear tissues (stria vascularis and organ of Corti). It is interesting that 10^{-4}M neomycin does not alter the ^{32}P-incorporation while 10^{-3}M and 10^{-2}M do change both the lipid labeling and the cochlear microphonic potential.

In vitro experiments

Such an *in vivo* observation leaves open several possible explanations: a drug effect on the synthetic enzymes, on the hydrolytic enzymes, or on the substrates of the reactions involved. Although lipid labeling and hydrolysis can be studied *in vitro* with inner ear tissues (19) it is not an optimal approach because a large number of tissues have to be dissected for each experiment. Phosphoinositide metabolism is high in brain and we investigated the neomycin effect on ^{32}P-incorporation and on hydrolysis in this tissue (28). Our studies demonstrated a drug action on both the labeling and the hydrolysis of polyphosphoinositides. These effects seemed to reflect a neomycin action on the lipids themselves rather than on the enzymes of their metabolism. A possible direct drug/lipid interaction was studied by titrating a suspension of polyphosphoinositides with neomycin (Table III). Very low concentrations of the drug increased the turbidity of the lipid dispersion and lowered its pH indicating the formation of a "complex" based on ionic interactions.

Table III

Effects of neomycin on suspensions of polyphosphoinositides

Neomycin µM	Turbidity	pH
0	0.00	6.07
10	0.02	5.97
20	0.40	5.89
30	0.70	5.72
40	1.32	5.63
50	1.34	5.50

Turbidity was measured as O.D. at 520 nm in a buffered (20mM Tris Cl, pH 7.4) suspension of 80% $PhIP_2$ and 20% PhIP (0.40 µmoles lipid-P/ml). pH was measured in an unbuffered aqueous suspension of the same lipids (0.20 µmoles lipid-P/ml).

Monolayer Studies

The monolayer studies strongly support the possibility of a direct drug/lipid interaction. Figure 1 shows the surface pressure vs. percent of trough area curves for monolayers of phosphatidyl inositol (la), phosphatidyl serine (lb) and polyphosphoinositides (lc) with and without the addition of 10^{-4}M neomycin to the subphase. Neomycin causes a condensation of the phosphatidyl inositol film which persists up to a surface pressure of about 18 dynes/cm, after which the two curves continue to be superimposable until the film collapses. Neomycin seems to have very little effect on the π -A curve of phosphatidyl serine and the very small expansion effect seen at very high π values is within experimental error. Whereas the addition of neomycin causes a condensation of the polyphosphoinositide film up to a surface pressure of about 15 dynes/cm, further compression of the film results in a film expansion which persists up to the collapse pressure of the film.

Whereas only the polyphosphoinositide film showed an expansion effect in the presence of 10^{-4}M neomycin, higher concentrations of neomycin resulted in slight film expansion (at higher pressures) for the phosphatidyl inositol film (10^{-3}M neomycin) and for the phosphatidyl serine film (10^{-2}M neomycin). The addition of 10^{-3}M neomycin resulted in film condensation only, in the case of the phosphatidyl serine monolayer. Furthermore, the subphase addition of both 10^{-3}M and 10^{-4}M neomycin to the polyphosphoinositide film yielded superimposable π-A curves.

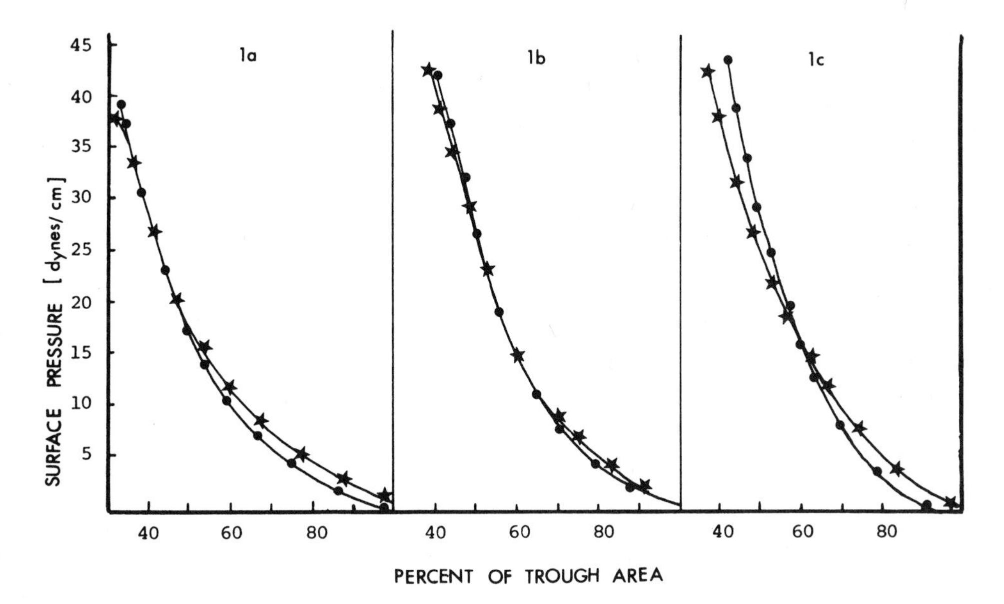

Figure 1. Surface pressure vs. percent trough area curves for phosphatidyl inositol (1a), phosphatidyl serine (1b), and polyphosphoinositides (1c) spread on a subphase consisting of 0.1M HEPES buffer (pH = 7.0, μ = 0.2) (★), and spread on a subphase consisting of 0.1M HEPES buffer (pH = 7.0, μ = 0.2) and 10^{-4}M neomycin (●).

DISCUSSION

The results of the monolayer studies suggest complex ionic equilibria between neomycin and lipids. The nature of the phospholipid, the concentration of neomycin, and the film pressure all exert an influence on the displacement of standard π-A curves. The fact that a soluble penetrating molecule can cause a condensation of a film at low pressures and a significant expansion at high pressures, to the best of our knowledge, has not been reported previously. It appears reasonable to assume that the interaction leading to film condensation must be different than that leading to film expansion.

It has been well documented (20) that acidic phospholipids interact strongly with bivalent metal ions at low concentrations, the interaction being accompanied by a condensation of the monolayer and at an increase in the surface potential. However, the extent of condensation is often a function of the phospholipid (both head group and hydrophobic portion), cation used, and film pressure. Shah and Schulman (32) observed that the π-A curves of dicetyl phosphate were not affected by the presence of divalent metal ions. They postulated a position for calcium in dicetyl phosphate monolayers that would not affect the area of the film but would increase the surface potential. Similarly, we observed almost no condensation effect on addition of 10^{-3}M Ca^{++} to the subphase of phosphatidyl inositol films, whereas 10^{-3} Ca^{++} greatly condensed phosphatidyl serine and polyphosphoinositide films. Although the interaction between acidic phospholipids and divalent ions are electrical in nature, specific "complexes" have been proposed. For example, Papahadjopoulos (20) proposed a model for the "complex" of phosphatidyl serine and calcium whereby the phosphatidyl serine molecules are packed in such a way in the film that the calcium can form six coordination bonds with four phosphatidyl serine molecules to form a linear polymeric arrangement. The fact that phosphatidyl serine shows less condensation effects than the phosphatidyl inositol or polyphosphoinositides in the presence of 10^{-4} neomycin may be due to the difference in unsaturation of the acyl chains or differences in the nature of the polar head groups. Since phosphatidyl serine contains only one unsaturated bond (on the average) in its hydrophobic group, it is possible that the large neomycin molecule, at low concentrations, will have difficulty getting into an area determining position. Even slight changes in the size of the ion can affect condensation and it has been shown that the relative affinity for different bivalent ions on phosphatidyl serine monolayers was calcium $>$ barium $>$ magnesium (20).

When the interaction between an insoluble monolayer and a soluble component of the subphase leads to film expansion, the phenomenon is referred to as monolayer penetration (29). Energetic requirements for penetration were summarized by Schulman and Friend (30). If no interaction between the components of the system occurs, the soluble compound is squeezed out of the surface, and the properties of the mixed film become identical to those of the insoluble film former. If there is a weak interaction, the penetrant dissolves in the monolayer but is gradually "squeezed out" upon compression. Strong interactions yield "complexes" of definite stoichiometry whose π-A characteristics differ from those of either component. In all cases where surface "complexes" were observed, strong interactions were apparent between the polar groups of the penetrant and film former as well as between the hydrophobic groups of the penetrant and film former. The best indication of a strong surface complex was when expansion persisted up to the collapse pressure of the film, and the film collapses as a "unit".

An analysis of Figure 1C shows that in the presence of as little as 10^{-4}M neomycin, expansion of the polyphosphoinositide film persists up to collapse and the film seems to break as a "unit" (the area per molecule is greater in the presence of

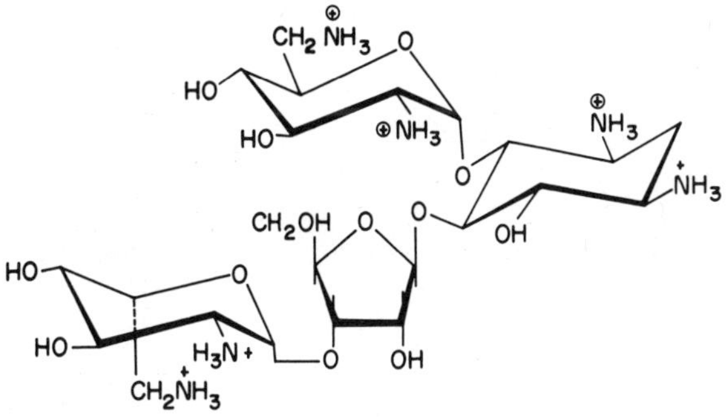

NEOMYCIN B

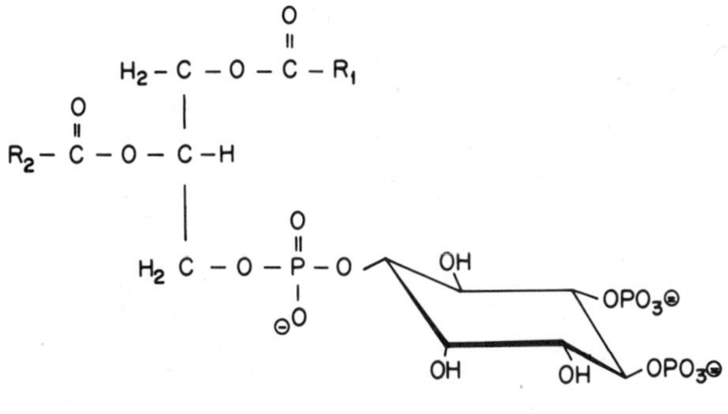

PHOSPHATIDYL INOSITOL DIPHOSPHATE

Figure 2: Structures of neomycin B and phosphatidyl inositol diphosphate. Typical fatty acid composition of the lipid is R_1 = stearoyl, R_2 = arachidonoyl. Phosphate groups and amino groups with circled charges are those pointed out in figure 3.

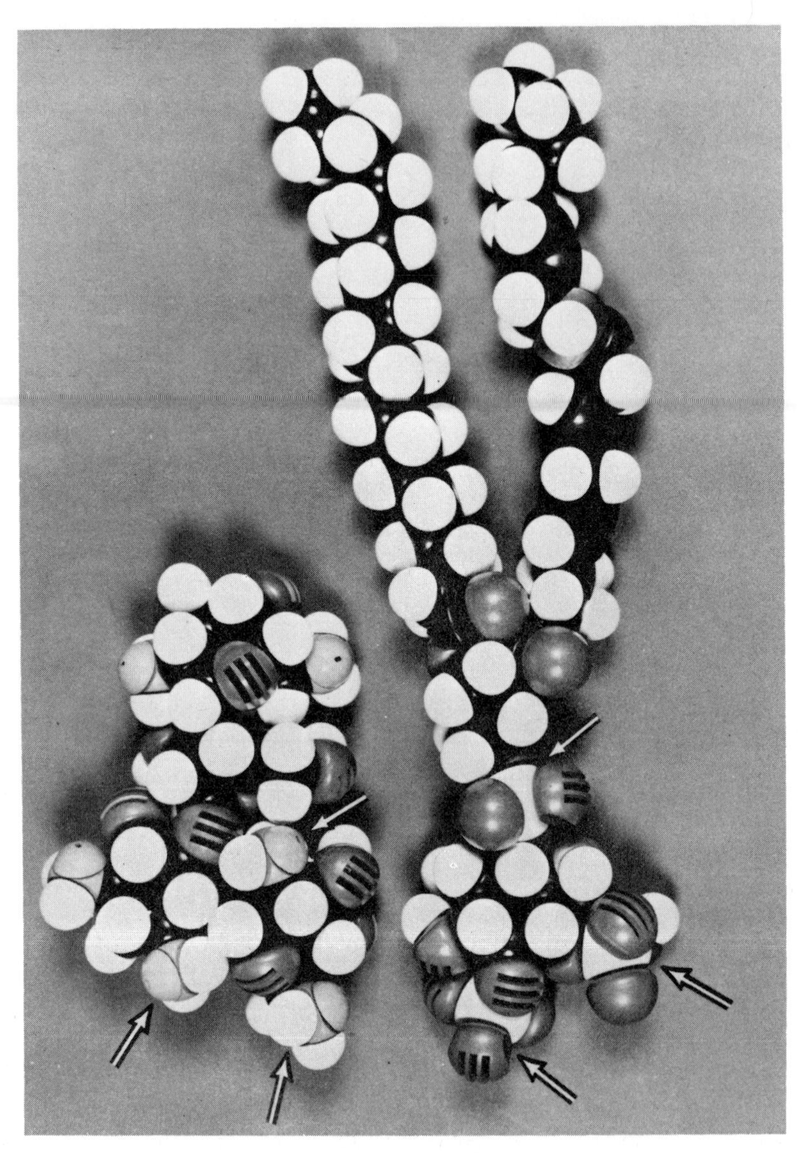

Figure 3: Molecular models of neomycin B (left) and phosphatidylinositol diphosphate (right). Indicated by arrows are the three phosphate groups of the lipid and three amino groups of the drug which are suggested to participate in the "complex" formation.

neomycin). Since it would not be expected that the highly polar neomycin molecule, upon penetration, would interact with the hydrophobic phase, it is surprising that expansion should persist at such high pressures in the absence of the classical hydrophobic penetrant-film former interactions.

An examination of the molecular models of phosphatidyl inositol diphosphate and neomycin (Figure 2 & 3) offers a possible explanation for the expansion effect at high pressures. The three negatively charged groups of the lipid and three positively charged groups of the neamine segment of neomycin produce an excellent 3-point fit. This interaction would fix neomycin in an area determining position in the film which would result in film expansion. Such 3-point attachment is obviously precluded in the case of the phosphatidyl inositol or phosphatidyl serine films, and expansion can only be seen for these films at much higher neomycin concentrations.

The fact that low concentrations of neomycin exert this expansion effect at high surface pressures may have biological significance since it is believed that the lipid packing in biological membranes is comparable with a surface pressure greater than 30 dynes/cm (6).

What would be the biochemical and physiological consequences of such a mechanism of neomycin action? Polyphosphoinositides are labeled by sequential phosphorylation:

(I) Phosphatidylinositol + [γ-^{32}P] ATP $\rightarrow$ phosphatidylinositol [^{32}P]phosphate + ADP

(II) Phosphatidylinositol phosphate + [γ-^{32}P] ATP $\rightarrow$ phosphatidylinositol [^{32}P]diphosphate + ADP.

The formation of a complex with neomycin would remove phosphatidylinositol phosphate as a substrate for the kinase reaction (II) leading to a decreased labeling of phosphatidylinositol diphosphate as we indeed observe *in vivo*.

The correlation of the dose dependency of the neomycin effects on lipid labeling and on cochlear microphonics suggests some connection between the two events. Although our results so far cannot be interpreted as indicating a causal relationship it is interesting to speculate along those lines. The polyphosphoinositides readily form calcium salts (9) and the binding (and release) of calcium in excitable membranes has been ascribed to these lipids (11,12). The release of calcium would be controlled by the hydrolytic reactions:

(III) phosphatidylinositol diphosphate + H_2O $\rightarrow$ phosphatidylinositol phosphate + P_i

(IV) phosphatidylinositol phosphate + H_2O $\rightarrow$ phosphatidylinositol + P_i

in which each product has a lower affinity for calcium than the respective substrate. Thus the reaction sequence I$\rightarrow$IV would constitute a cycle to bind and release calcium in the membrane thereby regulating its permeability to cations.

The formation of a neomycin/lipid complex should first lead to a displacement of calcium. We have indeed observed this with inner ear tissues (19), brain and polyphosphoinositides (13). Secondly, it should block the phosphorylation/dephosphorylation cycle and thus disturb the function of the membrane. Thirdly, it should impose conformational changes-contraction or expansion-which may further influence membrane permeability and function.

A membrane action of neomycin may not be restricted to polyphosphoinositides. We can expect this drug, as a polyamine, to bind to a number of anionic membrane components but we observe this only at high drug concentrations with phosphatidylinositol and phosphatidyl serine. The sum of our *in vivo* and *in vitro* evidence points to the polyphosphoinositides as specific sites involved in the actions of neomycin.

Acknowledgements: This work was supported in part by a grant from the John A. Hartford Foundation, Inc. and by Program Project Grant NS 05785.

REFERENCES

1. ABOOD, L. G. and HOSS, W.: European J. Pharmacol. 32 (1975) 66.

2. ABRAMSON, M. B., KATZMAN, R. and GREGOR, H. P.: J. Biol. Chem. 239 (1964) 70.

3. ALEXANDER, A. E., TEORELL, T. and ABORG, C. G.: Trans. Faraday Soc. 35 (1939) 1200.

4. ANDRE, T.: Acta Radiol. Suppl. 142 (1956) 1.

5. BLAUSTEIN, M. P.: Biochim. Biophys. Acta 135 (1967) 653.

6. DEMEL, R. A., GUERTS VAN KESSEL, W. S. M., ZWAAL, R. F. A., ROELOFSEN, B. and VAN DEENEN, L. L. M.: Biochim. Biophys. Acta. 406 (1975) 97.

7. FEINSTEIN, M. B.: J. Gen. Physiol. 48 (1964) 357.

8. GOLDMAN, D. E.: Biophys. J. 4 (1964) 167.

9. HAUSER, H. and DAWSON, R. M. C.: European J. Biochem. 1 (1967) 61.

10. HAWKINS, J. E. Jr.: Biochemical aspects of ototoxicity, Ch. 24, Biochemical Mechanisms in Hearing and Deafness (Paparella, M.M., ed) Thomas. Springfield, IL (1970).

11. HENDRICKSON, H. S. and REINERTSEN, J. L.: Biochim. Biophys. Res. Comm. 44 (1971) 1258.

12. KAI, M. and HAWTHORNE, J. N.: Ann. N. Y. Acad. Sci. 165 (1969) 761.

13. LODHI, S., WEINER, N. D. and SCHACHT, J.: Biochim. Biophys. Acta 426 (1976) 781.

14. MALICK, A. W., FELMEISTER, A. and WEINER, N. D.: J. Pharm. Sci. 62 (1973) 1871.

15. MALICK, A. W., WEINER, N. D. and FELMEISTER, A.: J. Pharm. Sci. 63 (1974) 398.

16. MICHELL, R. H.,: Biochim. Biophys. Acta.415 (1975) 81.

17. MULE, S. J.: Biochem. Pharmacol. 18 (1969) 339.

18. NUTTALL, A. L., MARQUES, D. M., STOCKHORST, E. and SCHACHT, J.: J. Acoust. Soc. Amer. 57 (1975) S60.

19. ORSULAKOVA, A., STOCKHORST, E. and SCHACHT, J.: J. Neurochem. 26 (1976) 285.

20. PAPAHADJOPOULOS, D.: Biochim. Biophys. Acta 163 (1968) 240.

21. PAPAHADJOPOULOS, D.: Biochim. Biophys. Acta 241 (1971) 254.

22. PAPAHADJOPOULOS, D.: Biochim. Biophys. Acta 265 (1972) 169.

23. PESTKA, S.: Ann. Rev. Biochem. 40 (1971) 697.

24. PITTINGER, C. and ADAMSON, R.: Ann. Rev. Pharmacol. 12 (1972) 169.

25. ROJAS, E. and TOBIAS, J. M.: Biochim. Biophys. Acta 94 (1965) 394.

26. SANTIAGO-CALVO, E., MULÉ, S., REDMAN, C. M., HOKIN, M. R., and HOKIN, L. E.: Biochim. Biophys. Acta 84 (1964) 550.

27. SCHACHT, J. and AGRANOFF, B. W.: J. Biol. Chem. 247 (1972) 771.

28. SCHACHT, J., LODHI, S., and WEINER, N. D.: Trans. Amer. Soc. Neurochem. 7 (1976) 126.

29. SCHULMAN, J. H., and HUGHES, A. H.: Biochem. J. 29 (1935) 1236.

30. SCHULMAN, J. H. and FRIEND, Jr., J. A.: J. Soc. Cosmetic

Chem. 1 (1949) 381.

31. SEEMAN, P.: Pharmacol. Rev. 24 (1972) 583.

32. SHAH, D. O. and SCHULMAN, J. H.: J. Lipid Res. 6 (1965) 341.

33. WERSÄLL, J. and FLOCK, A.: Life Sci. 3 (1964) 1151.

34. WHITE, G. L., SCHELLHASE, H. U., and HAWTHORNE, J. N.: J. Neurochem, 22 (1974) 149.

DISCUSSION

SHAMOO: Please comment about the distribution of phosphoinositides among the various membranes of the cell particularly cells that are sensitive to this drug?

SCHACHT: Neither I nor can anybody else can. There is some evidence from brain cells that they are associated with plasma membranes. I can tell you more about tissue distribution than distribution among cell membranes. That is very interesting, too, in regard to the toxicity of neomycin. Nervous tissue, brain, and kidney are the tissues with the highest polyphosphoinositide content (cf. reference 16, this paper), and these are the tissues that are sensitive to the drug. To demonstrate aminoglycoside effects on brain, one has to by-pass the blood brain barrier.

GENNARO: Is it known whether these dominate in the outer lipid leaflet or the inner lipid leaflet?

SCHACHT: Again, I cannot answer this question because we do not have sufficient information about the subcellular distribution of polyphosphoinositides, let alone their distribution within a membrane. However, enzymes involved in polyphosphoinositide metabolism have been localized on the cytoplasmic surface of the human erythrocyte membrane (GARRETT, R. J. and REDMAN, C. M.: Biochim. Biophys. Acta 382 (1975) 58.)

GENNARO: Have you studied the effect of temperature on the labelling in this system? Can you change the labelling curves downward by cooling?

SCHACHT: We have not checked this.

ALPER: The method you used when you were simultaneously looking at biochemical changes and changes in the cochlear microphonics suggest to me that you would not expect to see an effect of these antibiotics, say on an animal or human being within weeks or months. Is that right? Do you feel that if you can't see these changes immediately then you wouldn't expect to see a longer term effect?

SCHACHT: The experiments that I described were acute experiments where we actually introduced the drug into the cochlea. When you give systemic injections you are slowly building up the drug concentration in the inner ear fluids. I can only tell you that at the time when you can measure first hearing losses following systemic applications the concentration of the drug in the inner ear fluids should be around 10^{-3}M, maybe a little less.

ALPER: It is a very quick reaction.

SCHACHT: Yes, it is a very quick effect, if the concentration of the drug is sufficiently high. The ear accumulates neomycin, by the way, over serum levels, so does the kidney. The delayed toxicity which you see with systemic or oral application is probably due to the slow buildup of the drug concentration in the inner ear.

SHA'AFI: Is it possible that the effect is due to the expansion of one-half of the bilayer with respect to the other?

SCHACHT: This is certainly possible. We have no direct evidence that our suggested mechanism occurs at the plasma membrane. We do know however, from studies with radioactive antibiotics by Andre (ref. 4, this paper), that they penetrate the cell membrane only very slowly. Thus, the plasma membrane is the likely point of attack when you are dealing with such fast reactions as we observe them. Furthermore, we have seen under the conditions of our perfusions a dramatic increase of the permeability of inner ear tissues to perfused amino acids and to phosphate. So, obviously some kind of disturbance of the plasma membrane takes place, but we cannot say by what mechanism.

HOFFMAN: Is there any effect of neomycin on calcium ATPase?

SCHACHT: I don't know of any records about it, but I am willing to bet that there must be one.

REYNOLDS: Does the organ of Corti lose potassium at these levels or can the lesion be prevented with potassium?

SCHACHT: We really cannot tell. We are not directly perfusing the organ of Corti. The organ of Corti is closer to the endolymphatic compartment and we are perfusing the perilymphatic compartments. The neomycin gets to the organ of Corti either through Reissner's membrane, or the basilar membrane.

FEINSTEIN: I presume that at some point the effects of Neomycin on cochlear microphonics become irreversible. When that occurs what is the concentration of the drug in the cochlea and what residual membrane functions are you left with?

SCHACHT: Reversible and irreversible effects have been demonstrated by Wersall und Flock (ref. 33, paper) at the lateral line organ of fish which is a good model for the mammalian ear. Perfusing the lateral line organ with low concentrations of streptomycin led to a reversible decrease of the microphonic potential, whereas higher drug concentrations caused an irreversible loss of the microphonic output. In mammals, the hearing loss following systemic application of antibiotics is usually irreversible. What leads to these irreversible effects we really don't know. You can speculate along two lines. Firstly, since you do alter membrane permeability a prolonged change in membrane permeability may be just lethal to the cell because of the resulting ionic imbalance. Secondly, since neomycin increases membrane permeability, conceivably it can increase permeability to itself. So, after the first damage is being done at the plasma membrane, neomycin could enter the cell a lot more readily and act on intracellular membranes in a similar or even different fashion.

ULLRICH: Doesn't neomycin just act like many other positive-charged macromolecules? Couldn't you just trace the neomycin in these series of positively-charged substances?

SCHACHT: There seem to be a number of similarities between the action of polyamines such as polylysine and of neomycin. From the sum of our *in vivo* studies we feel, however, that the neomycin action is more specifically directed against the polyphosphoinositides. Only high neomycin concentrations cause effects on other lipids *in vitro*, and this may well be the expression of a more general "polyamine effect".

ULLRICH: But, perhaps if you look very carefully, we may find that the polylysine also acts on the polyphosphoinositides.

SCHACHT: Nobody has ever looked at polyphosphoinositides in connection with those drugs. We have done it *in vitro* and we have seen more widespread effects of those drugs on lipid labeling than just on the polyphosphoinositides.

ALLISON: On the subject of environmental pollution I wonder whether young people should not go to pop concerts if they are taking these drugs.

SCHACHT: Well, you can ask should young people go to pop concerts because noise itself is definitely deafing. There are indications that under specific circumstances noise can potentiate the effect of aminoglycoside drugs, or vice versa.

PRESSMAN: Aren't the effects of aminoglycosides on hearing potentiated by diuretics?

SCHACHT: Yes. It has been reported (WEST, B.A., BRUMMETT, R.E. and HIMES, D.L.: Arch. Oto-laryngol. 98 (1973) 32.) that a remarkable potentiation of ototoxicity occurs between aminoglycosides and the diuretic ethacrynic acid. We have only begun to study this phenomenon at the biochemical level. We speculate that the aminoglycoside antibiotics increase the permeability of the inner ear tissues to the diuretic, so that highly toxic concentrations of the latter are attained.

ALLISON: Couldn't both drugs just interfer with polyphosphoinositide turnover?

SCHACHT: Ethacrynic acid has a number of effects on adenylate cyclase, on ATPase, it can uncouple oxydative phosphorylation. We did not observe any action on polyphosphoinositides *in vitro*. So, I can only envision a potentiation the way I just described it.

PASSOW: I think your model experiments with the inositols are very interesting because they demonstrate that, in principle at least, an interaction between meomycin and the inositols is feasible. In the intact cell the inositols constitute only a fraction of the total lipids. I wonder what the interactions would look like if you had done your experiments with a physiological mixture of lipids. In such mixtures basic groups of phosphatidyl ethanolamine could possibly be associated more closely with the acetic inositol groups and thus reduce the effects which you observed in your model experiments.

SCHACHT: Yes, that certainly is a very good point. You probably don't have areas of high polyphosphoinositide concentration although you cannot rule them out. But as I pointed out in the last slide the action on the polyphosphoinositides is thought to be one to one and you would not need even two polyphosphoinositides for a neomycin action to occur. By attachment to one polyphosphoinositide in the membrane you would by virtue of the very high affinity expand the membrane in that particular area.

PASSOW: How did you demonstrate this stoichiometry?

SCHACHT: We haven't yet. It is still speculation.

FEINSTEIN: Well, Ohnishi and Ito (Biochemistry 13:881, 1974) have shown that when phosphatidyl serine and lecithin are present in a mixed lipid membrane, phosphatidyl will tend to segregate out in the presence of calcium into areas which are especially concentrated in phosphatidyl serine, and you might have the same situation. You might have membrane patches which are very rich in phosphoinositide even though they are in low concentration overall.

SCHACHT: Thank you.

HIGH AFFINITY SH-GROUPS ON THE SURFACE OF PANCREAS CELLS INVOLVED IN SECRETIN STIMULATION

I. Schulz and S. Milutinović

Max-Planck-Institut für Biophysik

Kennedyallee 70, 6000 Frankfurt/Main, Germany

ABSTRACT

The effect of p-chloromercuribenzoate (pCMB) on secretin stimulated pancreatic fluid secretion in vivo was investigated and compared with its effect on secretin binding and secretin stimulated adenylate cyclase in isolated pancreatic plasma membranes in vitro. A biphasic effect of pCMB was observed. At low concentrations (10^{-9} - 5×10^{-8}M) pCMB stimulated adenylate cyclase activity, secretin binding and secretin stimulated pancreatic fluid secretion by $\sim$ 50, 25 and 100%, respectively. At higher concentrations (10^{-7} - 10^{-5}M) pCMB inhibited secretin binding by 50%. In the same range of pCMB concentrations secretin stimulated adenylate cyclase was inhibited in a dose dependent fashion. Basal adenylate cyclase activity was much less susceptible to the inhibition by pCMB since about 50 times greater concentration of pCMB is required for half-maximal inhibition (5×10^{-5}M and 10^{-6}M, respectively).

To restrict the effect of the SH group reagent to the outer membrane surface a large Dextran-linked derivative of pCMB was used in in vivo experiments. At 10^{-8}M this compound inhibited secretin induced fluid secretion by 47%. About one half of this inhibition is due to the blocking of SH groups involved in glucose transport since it is abolished by replacing glucose in perfusion fluid by substrates of the Krebs-cycle. The other half of inhibition is directly related to the secretin action since it is abolished by replacing secretin by dibutyryl cAMP and theophylline.

The data show that accessible SH groups located at the cell surface are directly involved in secretin binding and adenylate cyclase stimulation. Since the apparent K_m for secretin stimulation of adenylate cyclase and the K_d for secretin binding were in agreement and since pCMB stimulated and inhibited both secretin binding and secretin stimulated adenylate cyclase activity in the same concentration range it-is suggested that the binding of secretin to its receptor is the rate determining step in the stimulation of adenylate cyclase by this hormone. The biphasic action of pCMB can be best interpreted with the assumption that several categories of SH-groups are present in the plasma membrane.

Introduction

It has been reported that membrane SH groups are essential for the action of the diverse hormones (4, 8, 10, 12, 13, 16, 19). The mechanism, however, by which changes in the SH groups affect the response to these hormones is not well understood. Since Sutherland and collaborators discovered that an adenylate cyclase-adenosine 3'5'-cyclic monophosphate (cAMP) system is involved in the action of most peptide hormones (18), the mechanism by which changes in SH groups affect peptide hormone action could be envisaged to be at least twofold: at the level of hormone-receptor interaction and/or at the level of cyclic AMP formation. In order to determine whether SH-groups in plasma membranes of the pancreas are involved in pancreatic fluid secretion, a large derivative of p-chloromercuribenzoate (pCMB Dextran T 10, molecular weight approx. 10,000) was perfused through an isolated preparation of cat pancreas and its effect compared to that of pCMB on secretin stimulated adenylate cyclase activity and secretin binding to membrane-fractions of pancreatic homogenate. The data suggest that high affinity SH groups on the cell surface of the pancreas control secretin-receptor interaction.

Methods

Perfusion experiments. Parts of this study have been published previously (21). The method of the isolated perfused cat's pancreas is the same as described before (2). Briefly it involves surgically isolation of the organ by ligating all vessels leading to other organs. A Krebs-Henseleit solution pH 7.4 containing 15 mM glucose was perfused through the aorta via the coeliac axis and the superior mesenteric artery by a roller pump at a rate of 5 ml/min at 37°C and equilibrated with 95% O_2 and 5% CO_2. Synthetic secretin which was a generous gift of Prof. E. Wünsch (Max-Planck-Institut für Biochemie, Martinsried b. München, Germany) was delivered by a Braun infusion pump into the arterial cannula at concentrations between 10^{-11} and 10^{-8}M. The main pancreatic duct was cannulated and secretion was measured in 10 min periods. A control period which usually lasted 30 - 60 min, was followed by a period of 40 - 70 min in which pCMB-

Dextran T 10 was added to the perfusate at a concentration of 10^{-8}M. A second control period followed, during which the perfusate contained 10^{-3}M cysteine to reverse the effect of pCMB-Dextran T 10. Cysteine itself had no effect on the flow rate. In some experiments secretin was replaced by dibutyryl cAMP (10^{-3}M) (Boehringer, Mannheim, Germany) and theophylline (5 x 10^{-3}M) (Schuchard, Munich) added directly to the perfusate and glucose was replaced by intermediates of the Krebs-cycle (α-keto-glutarate, fumarate and pyruvate, 5 mM each).

The synthesis of pCMB-Dextran T 10 was carried out by Dr. B. Simon (17). The method used was the same as described by Eldjarn and Jelleun (5) for the preparation of aminoethyl-Dextran T 10 from Dextran T 10 (M.W. 10,000). The coupling of aminoethyl-Dextran T 10 to pCMB was performed by the method of Ohta et al. (12) with some modifications.

Membrane preparation. Membranes were prepared from cat pancreas by cutting the tissue into small pieces in solution containing tris-Cl-buffer 20 mM, pH 7.4, Bacitracin (Sigma, St. Louis, USA) 0.1 mg/ml and soybean trypsin inhibitor (Boehringer, Mannheim, Germany) 0.5 mg/ml. The tissue was homogenized and the homogenate filtered through one layer of gauze and spun at 600 g for 10 min to remove nuclei and large debris. The supernatant was spun at 4,000 g for 10 min and the pellet washed 3x with the same solution by alternate homogenization and centrifugation. The final pellet was resuspended to a protein concentration of about 10 mg/ml in 20 mM tris-Cl pH 7.4 containing 0.5 mg per ml soybean trypsin inhibitor.

Secretin binding experiments (11). Synthetic secretin from Prof. E. Wünsch was iodinated by Dr. Rosselin (Hôpital Saint-Antoine, Paris, France) with ^{125}I by the chloramine method (1). The binding assay consisted of incubating membranes (150-300 µg of protein) at 22°C for 30 min in 0.5 ml of 20 mM tris-Cl-1% albumin buffer pH 7.4 containing 0.1 mg trypsin inhibitor, 0.05 mg Bacitracin, ^{125}I-secretin (0.1 - 0.5 x 10^{-10}M), and when necessary, different amounts (0.5 x 10^{-9}M - 10^{-6}M) of unlabelled secretin. The reaction was started by adding membranes to the preequilibrated assay mixture. After incubation the mixture was filtered without dilution through millipore filters (EHWPO, pore size 0.5 µ). The filters were washed within 15 - 20 sec with 10 ml of ice-cold 20 mM tris-Cl buffer pH 7.4 and transferred to a plastic vial for counting in a γ-counter (Packard). Corrections were made for unspecific binding of ^{125}I-secretin to membranes and to the filters by performing parallel incubations in which excess unlabelled secretin (3 x 10^{-6}M) was added to the membranes together with iodinated secretin. To test pCMB effect on secretin binding, pancreatic membranes (0.5 mg of tissue protein) were preincubated for 10 min at 22°C in 10 ml tris-Cl buffer (40 mM pH 7.35) containing pCMB at different concentrations (10^{-10} - 10^{-3}M). After incubation samples were diluted to 40 ml with 10 mM tris, pH 7.4, and immediately spun down at 10,000 g for 10 min. The obtained pellet was washed once with the same buffer and used for the binding assay with ^{125}I secretin (10^{-11}M) within 15 min.

Adenylate cyclase assay. The adenylate cyclase was assayed by the method of Krishna (9) with modifications described before (11). Briefly, separation of cAMP consists of chromatography on Dowex 50 WX columns, followed by precipitation by $ZnSO_4/Ba(OH)_2$. Incubations were performed at 30°C for 20 min in 0.1 ml of medium containing 0.8 mM ATP α^{32}P, 5 mM $MgCl_2$, 42 mM tris-Cl pH 7.4, 10 mM theophylline, about 100 µg of membrane protein and ATP regenerating system consisting of 5 mM phosphoenol pyruvate and 20 µg pyruvate kinase. Under these conditions linear reaction rates are obtained in the presence and absence of secretin for at least 35 min.

To test the pCMB effect on adenylate cyclase activity membranes obtained by centrifugation in a continuous Ficoll-sucrose gradient (5% Ficoll to 45% sucrose) (manuscript in preparation) were incubated with pCMB in a similar way as for secretin binding experiments.

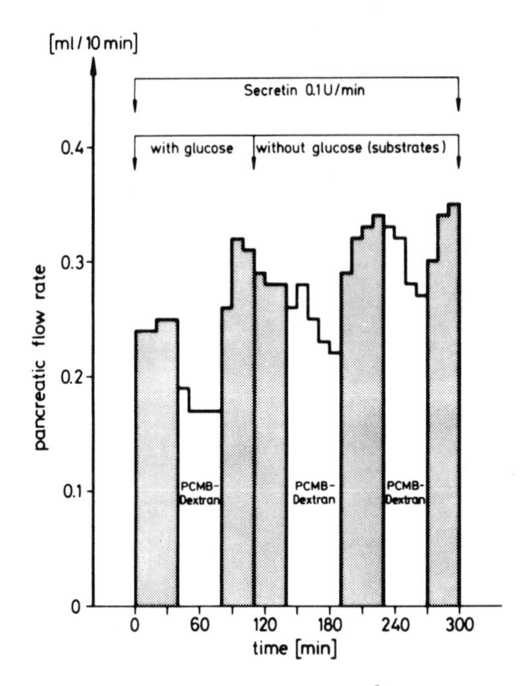

Figure 1. Effect of pCMB-Dextran T 10 (10^{-8}M) on pancreatic secretion rate in the presence and absence of perfusate glucose during secretin stimulation. During the period following perfusion with pCMB-Dextran T 10, the perfusate contained cysteine (10^{-3}M) (21).

Results

Perfusion experiments. In the unstimulated cat pancreas preparation no volume output could be observed. Secretion could be evoked by the addition of secretin in a dose-dependent fashion, half-maximal stimulation being reached at 9 x 10^{-11}M. When pCMB-Dextran T 10 (10^{-8}M) was added to the perfusate pancreatic secretion was inhibited by 47% (S.D. ± 14%, 6 observations in 4 experiments, left panel of Fig. 1). The inhibition could be reversed by a perfusate containing 10^{-3}M cysteine. As shown in Fig. 2, 10^{-3}M dibutyryl cyclic AMP plus 5 x 10^{-3}M theophylline could replace secretin in eliciting pancreatic fluid secretion as has been described before by Case et al. (3). In this case pCMB-Dextran T 10 reversibly inhibited secretion by only 15% (S.D. ± 4.3%, 5 observations in 4 experiments, Fig. 2, left panel).

As glucose or other substrates are necessary to maintain pancreatic fluid secretion (20) the effect of pCMB-Dextran T 10 was checked under conditions in which glucose was replaced by α-keto-

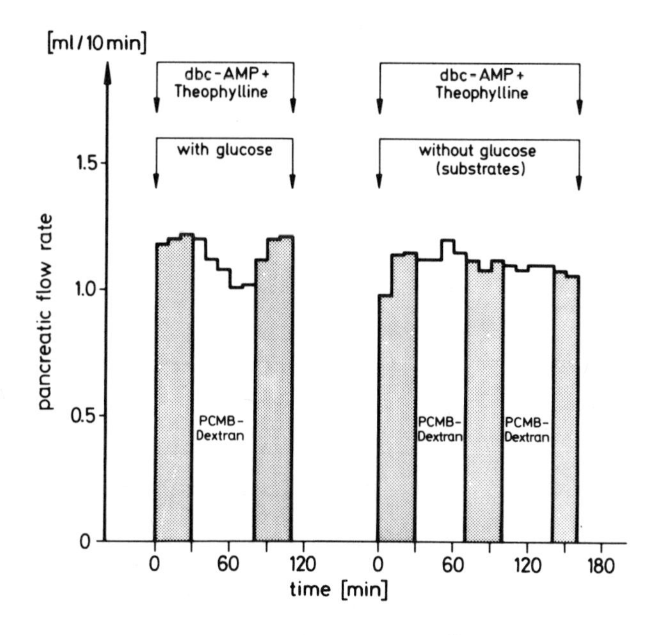

Figure 2. Effect of pCMB-Dextran T 10 (10^{-8}M) on pancreatic flow rate in the presence and absence of perfusate glucose during stimulation with 10^{-3}M dibutyryl cyclic AMP plus 5 x 10^{-3}M theophylline. During the period following perfusion with pCMB-Dextran T 10, the perfusate contained cysteine (10^{-3}M) (21).

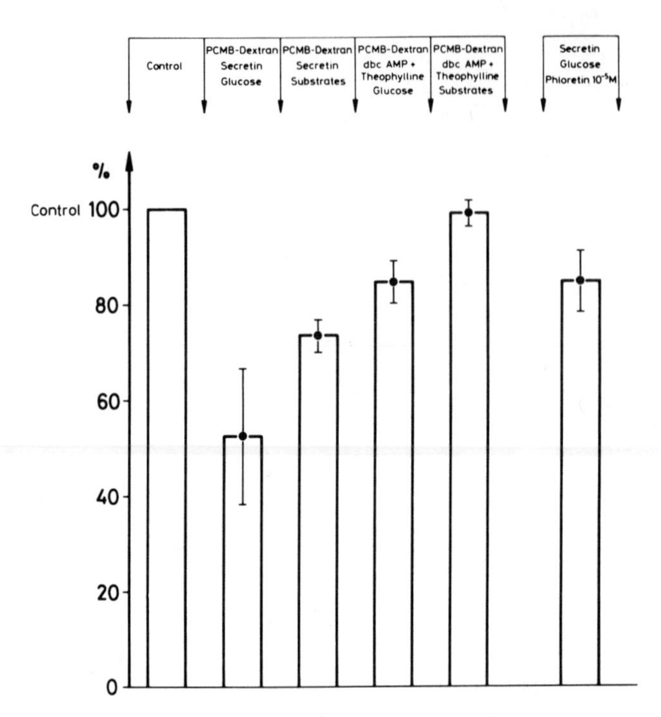

Figure 3. Effect of pCMB-Dextran T 10 (10^{-8}M) on secretin (0.1 unit/min)-stimulated or dibutyryl cyclic AMP (10^{-3}M) plus theophylline (5×10^{-3}M)-stimulated pancreatic volume flow. The substrate was either glucose (15 mM) or α-ketoglutarate, fumarate and pyruvate (5 mM each). Each set of experiments was done with paired controls (without pCMB-Dextran T 10). The mean values ± the S.D. of the mean values are given (21).

glutarate, fumarate and pyruvate (5 mM each). In 7 experiments the secretory flow rate was not changed under those conditions (data not shown). When glucose was replaced by these substrates and fluid secretion stimulated by secretin, pCMB-Dextran T 10 (10^{-8}M) inhibited the secretin stimulated secretion by 26% (S.D. ± 3.4%, 5 observations in 3 experiments, Fig. 1). If both secretin and glucose were replaced by dbcAMP + theophylline and substrates, respectively, pancreatic fluid secretion could not be inhibited by pCMB any more (6 observations in 4 experiments, Fig. 2, right panel).

In a quantitative synopsis of all these data (Fig. 3) it can be seen that the inhibitory effect of pCMB-Dextran T 10 on secretin action and on glucose mediated secretion are additive. If both, se-

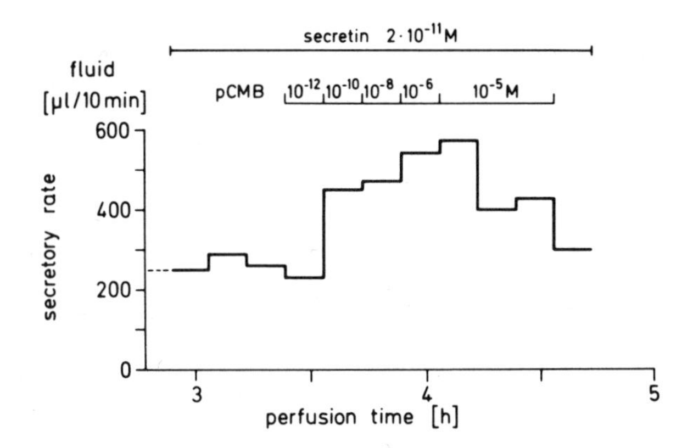

Figure 4. Effect of pCMB on pancreatic fluid secretion.

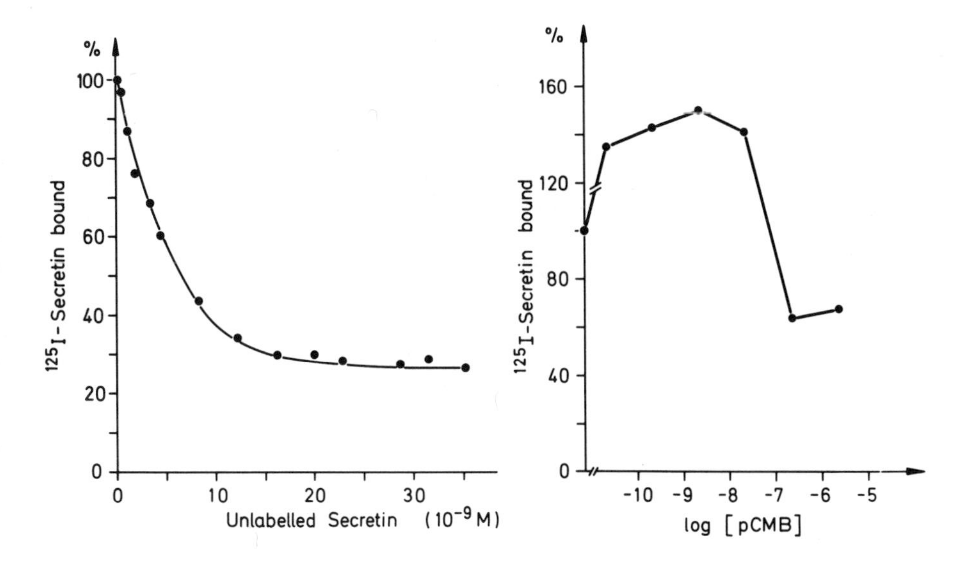

Figure 5 (left). Competitive inhibition of ^{125}I-secretin binding by unlabelled native secretin. Pancreas plasma membranes (300 µg of protein per ml) were incubated at 22°C for 20 min in a final volume of 0.5 ml of the standard incubation medium containing labelled secretin (11 x 10^{-11}M) and the indicated concentrations of unlabelled secretin. Points are mean of triplicate determination. ^{125}I refers strictly to the labelled molecules (11).

Figure 6 (right). Effect of different pCMB concentrations on secretin binding (secretin concentration in the incubation medium was 11 x 10^{-11}M). Each point represents mean of three experiments performed in duplicate.

cretin and glucose are replaced by dbcAMP + theophylline and substrates, respectively, the inhibitory effect of pCMB-Dextran T 10 was abolished. The inhibitor of glucose uptake phloretin inhibited glucose mediated secretion to the same extent as pCMB-Dextran T 10 (right column in Fig. 3).

The effect of pCMB at low concentrations of secretin ($2x10^{-11}$M) which yielded a very small flow rate was checked. It can be seen from Fig. 4 that beginning with 10^{-10}M pCMB the flow rate increases, whereas at 10^{-5}M this stimulated part disappeared. The HCO_3^- output which is stimulated by secretin as well as fluid secretion increased at low pCMB concentrations correspondingly and dropped down when flow rate decreased at higher pCMB-concentrations.

Secretin binding experiments. The binding studies were performed at equilibrium which was reached after 10 min of incubation at 22°C, and at membrane protein concentrations which were on the linear part of a curve when binding of ^{125}I secretin was plotted versus membrane protein concentration (11). Unlabelled secretin, when added simultaneously with ^{125}I labelled secretin to the membranes, reduced tracer binding in a dose dependent fashion with a K_d of 4.1×10^{-9}M (11), indicating that the same sites are involved in the binding of either species (Fig. 5). Preincubation of membranes with pCMB had a marked effect on secretin binding, depending on the dose of this reagent used. At low pCMB concentration (5×10^{-10} - 5×10^{-8}M) an increase in secretin binding up to 50% over the control was observed. Higher concentrations than 10^{-7}M of pCMB inhibited secretin binding by 50% (Fig. 6). A more detailed examination of pCMB inhibition of binding revealed a change in the number of binding sites and in receptor affinity (not shown).

Adenylate cyclase experiments. Secretin stimulates the adenylate cyclase, as tested in the same membrane fraction of the pancreas used in binding experiments with an apparent K_m of $8.4 \pm 0.9 \times 10^{-9}$M, which is in the same range of concentration as the dissociation constant obtained for secretin binding from equilibrium data (11, Table I). Preincubation of pancreatic membranes with pCMB at concentrations as low as 10^{-9}M had a slight stimulatory effect on secretin stimulated adenylate cyclase. Increasing pCMB concentration in the preincubation medium up to 5×10^{-9}M and 10^{-8}M brought about progressive increase in secretin stimulated adenylate cyclase activity by 10 and 20%, respectively (Fig. 7). It should be noticed that the pCMB-induced stimulation is persistent throughout the whole range of secretin concentrations. A reanalysis of these data in the form of double reciprocal plot revealed an increase in V_{max} without change in K_m for secretin (plot not shown). Basal adenylate cyclase activity, i.e. activity in the absence of hormone, did not change in this range of concentrations

TABLE I. Comparison of the kinetic parameters and effect of pCMB. Molar concentrations are given.

	flow rate	secretin stimulated adenylate cyclase activity	secretin binding
apparent K_m	9×10^{-11}	8.4×10^{-9}	4.1×10^{-9}
pCMB			
$10^{-10}-10^{-6}$	stimulates		
$10^{-9}-5x10^{-8}$		stimulates	stimulates
10^{-7}		inhibits	inhibits
10^{-6}	inhibits		
pCMB-Dextran			
10^{-8}	inhibits		

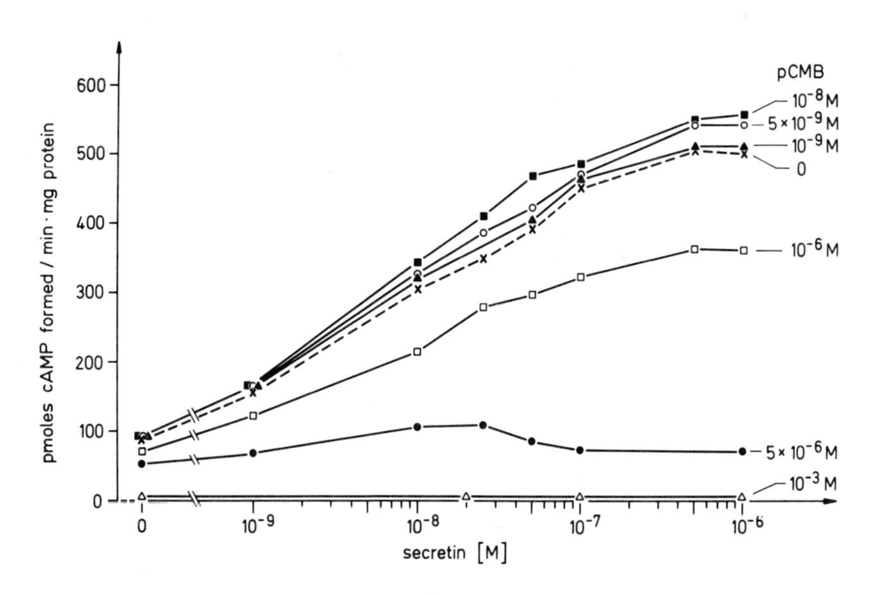

Figure 7. Effect of different pCMB-concentrations on secretin-stimulated adenylate cyclase activity. Points represent mean of two determinations. A representative from five similar experiments is shown.

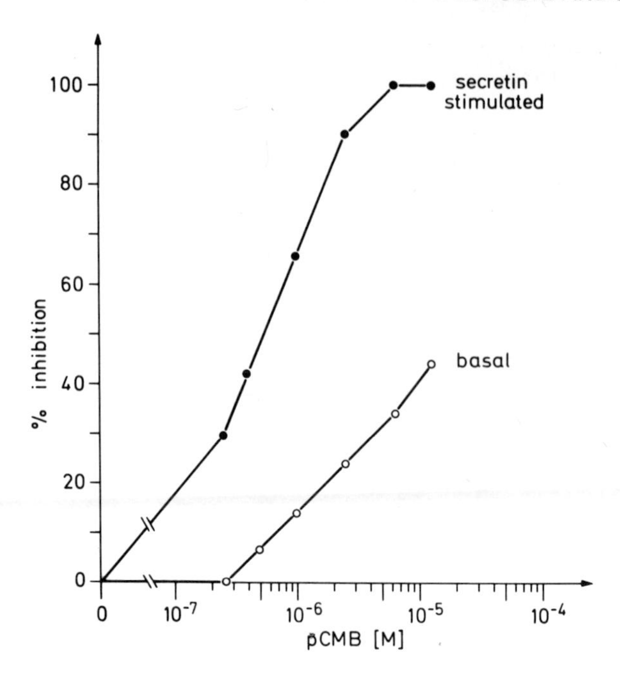

Figure 8. Inhibition of basal and secretin stimulated adenylate cyclase by different concentrations of pCMB. Percent inhibition is shown (% inhibition = 100 x (1 - activity in the presence of pCMB/activity in the absence of pCMB)). Each point represents mean of three determinations.

of pCMB. Increase in concentration of pCMB over 10^{-7}M induced a progressive inhibition of secretin stimulated adenylate cyclase in a dose-dependent fashion. At 6 x 10^{-6}M of pCMB inhibition is complete (Fig. 7,8). Again, re-analysis of the inhibition data (at 10^{-6}M pCMB) in the form of double inverse plot showed inhibition of V_{max} without change in K_m for secretin.

Basal adenylate cyclase activity was much less susceptible to the inhibition by pCMB than secretin stimulated adenylate cyclase. So, at 2.5 x 10^{-7}M pCMB, basal activity remained unchanged, whereas secretin stimulated activity was inhibited by 30% (Fig. 8). At 6 x 10^{-6}M of pCMB, the concentration at which secretin stimulated adenylate cyclase is completely inhibited, basal activity is inhibited only by 30%. Half-maximal inhibitory concentration of pCMB for basal activity is about 50 times greater than that for secretin stimulated adenylate cyclase (5 x 10^{-5} and 10^{-6}M). So, one can clearly distinguish between basal and secretin stimulated activity of pancreatic adenylate cyclase in terms of sensitivity to pCMB inhibition; this implies involvement of two classes of SH groups with different reactivity.

Discussion

In modifying the action of secretin on pancreatic fluid secretion, pCMB could either interfere with functionally important SH groups belonging to the secretin receptor or it could act on SH groups involved in a later step of the chain of events following hormone binding and leading to the physiological response. In order to discriminate between the effect of pCMB on SH groups located at the cell surface and those at intracellular sites involved in secretin stimulation, we used the large derivative of pCMB, pCMB-Dextran T 10 (M.W. $\sim$10,000) which cannot penetrate the cell membrane. In the intact organ pCMB-Dextran T 10 inhibited the secretin stimulated flow rate by 47% when glucose was in the perfusate and by 26% when glucose was replaced by substrates of the Krebs-cycle. When dbcAMP was used to stimulate fluid secretion by bypassing the secretin receptor, and glucose was replaced by substrates, no inhibition of pCMB-Dextran T 10 on fluid secretion could be observed. These data suggest that SH groups involved in glucose uptake and in secretin action are located at the cell surface. As the latter is concerned, however, a distinction between their localization at the hormone binding site or at some other part of the adenylate cyclase system could not be made by these experiments. In order to localize the SH groups responsible for inhibition of fluid secretion by pCMB we examined the effect of pCMB on secretin binding and adenylate cyclase activity in membrane fractions of the pancreas. PCMB has dual effect: Between 5×10^{-10} and 5×10^{-8}M pCMB concentration binding of secretin was enhanced by 40 - 50% over the control value. At higher concentration than 10^{-7}M pCMB, binding was inhibited. A similar effect of pCMB could be observed on secretin stimulated adenylate cyclase activity. The concentration range in which pCMB induces stimulation (between 10^{-9} and 10^{-8}M pCMB) is comparable to that obtained for stimulation of secretin binding, but degree of stimulation is lower (20%). At higher concentrations (between 10^{-7} and 6×10^{-6}M pCMB), pCMB inhibited the secretin stimulated adenylate cyclase. Both stimulatory and inhibitory range of concentrations of pCMB for secretin binding and for secretin stimulated adenylate cyclase are in agreement, suggesting that the same SH groups are involved in binding and adenyl cyclase stimulation. In contrary, basal adenylate cyclase activity was not activated and showed quite different inhibition kinetics by pCMB, indicating that another, less reactive, type of SH group is involved.

The apparent K_m for adenylate cyclase stimulation by secretin and the K_d obtained by secretin binding experiments agree very well

(Table I), which suggests that secretin binding to its receptor is the rate determining step in the stimulation of the adenylate cyclase by this hormone. In the same direction points the dual effect of pCMB: stimulation of secretin binding and the activation of the secretin stimulated adenylate cyclase at low pCMB concentration, inhibition at higher concentration. Furthermore, the finding that pCMB was stimulatory only in the presence of secretin (in vitro and in perfusion experiments) suggests that it modifies SH groups which are involved in the binding of secretin and/or coupling between receptor and catalytic unit and not those involved in catalyzation of ATP to cAMP by the adenylate cyclase system. Otherwise a stimulatory action on the basal activity of adenylate cyclase by pCMB should be expected. In addition, the pCMB concentration required for half-maximal inhibition of basal adenylate cyclase activity is at least 50 times higher than that necessary for half-maximal inhibition of secretin stimulated adenylate cyclase and secretin binding. It seems possible, therefore, to distinguish between SH groups located on secretin receptor and those located on the catalytic part of adenylate cyclase in terms of different reactivity with pCMB. We therefore ascribe the effect of pCMB on secretin stimulated fluid secretion to be located at the receptor site of secretin, which is closely coupled to the adenylate cyclase system.

The results obtained from the perfusion experiments are not directly comparable to the experiments on broken membrane preparations, since the final physiological response to a hormone receptor interaction is not necessarily linearly related to the degree of receptor occupancy and cAMP formation (14). In our studies the half-maximal stimulation of pancreatic fluid secretion in the intact perfused organ was 100 times lower than the dissociation constant for secretin binding and the apparent K_m for adenylate cyclase activation. Also the pCMB evoked changes of binding and adenylate cyclase activity in membranes are not too closely paralleled by the changes evoked by pCMB on fluid secretion in the intact organ. Thus the pCMB inhibition which started on the former event at 10^{-7}M was in the experiments with pCMB-Dextran action on fluid secretion already present at 10^{-8}M, but in the pCMB action on fluid secretion as shown in Fig. 4 only at 10^{-5}M. This discrepancy is not astonishing if one considers that we are dealing with SH groups in membrane protein which have different affinity for pCMB and different accessibility to this reagent. In the binding and adenylate cyclase studies they were in the same status, but might be different from the status in the living cell. In the living cell certainly other factors influence these groups. It was suggested, for example, that intracellular glutathione may maintain at least some of the membrane SH groups in the reduced state, since the susceptibility to oxidation of membrane SH groups critical for Rb^+(6) and amino acid transport (7), markedly increased when the GSH level was reduced

(6, 7). Furthermore it could be envisaged that the large Dextran molecule, to which pCMB is attached could influence its binding and biological effect. However, the dual effect of pCMB: stimulation at low concentration, inhibition at high concentration was observed also in perfusion experiments.

How can we explain the stimulatory and the inhibitory effect of the same substance at different concentrations? Previous studies on the involvement of SH groups in the insulin stimulated transport of glucose into fat cells (4) and insulin stimulated amino acid transport into thymocytes (10) showed that SH group reagents stimulate also these transport processes at low concentrations and inhibit them at higher concentrations. The working model proposed by the authors to explain the role of SH groups on amino acid transport implies that the steady state of two categories of SH groups, lying at or near the external surface of the plasma membrane and buried within the membrane core respectively, regulates the transport rates. It is further assumed that S-S linkage of the buried SH groups is necessary for the active state. S-S linkage between buried and superficial peptide segments are possible but limited. Low levels of pCMB titrate the superficial SH groups allowing a greater proportion of the deeper groups to form S-S linkage. High concentrations of pCMB block also the buried groups. Although the stimulatory effect of SH group reagent on glucose transport into fat cells is different from the stimulation presented in this paper in so far as in the former case SH group reagent mimicks action of insulin in the absence of the hormone, whereas in the latter it is effective only in the presence of hormone, we think that such a model could also hold to explain our data, if we assume that the deeper SH groups are related to the receptor protein and that S-S linkage between these SH groups would bring it into a conformation of high affinity for secretin and/or into conformation suitable for coupling with catalytic unit. Titration of the superficial SH groups (by low pCMB concentrations) would prevent S-S linkages between superficial and buried SH groups, whereas the reaction of pCMB with the deeper class of SH group (at higher concentration of pCMB) would decrease affinity of receptor for secretin and/or prevent coupling of the receptor with catalytic unit, leaving the catalytic unit itself unaffected. The inhibitory effect of pCMB-Dextran T 10 in this model must be explained in that way that the deeper SH groups sometimes emerge so that pCMB attached to Dextran can react with them. In summary our data show that high affinity SH groups are involved in the control of secretin stimulated pancreatic fluid secretion and that they are located in the plasma cell membrane of the target cell. Several categories of SH groups are the most likely interpretation for the stimulatory and inhibitory effects of pCMB, although other possibilities cannot be excluded.

REFERENCES

1 BATAILLE, D., FREYCHET, P., ROSSELIN, G.: Endocrinology 95 (1974) 713.

2 CASE, R.M., HARPER, A.A., SCRATCHERD, T.: J. Physiol. London 196 (1968) 133.

3 CASE, R.M., SCRATCHERD, T.: J. Physiol. London 223 (1971) 649.

4 CZECH, M.P., LAWRENCE, J.C., LYNN, W.S.: Proc. nat. Acad. Sci., Wash. 71 (1974) 4173.

5 ELDJARN, L., JELLEUM, E.: Acta Chem. Scand. 17 (1963) 2610.

6 EPSTEIN, D.L.: Exp. Eye Res. 11 (1971) 351.

7 HEWITT, J., PILLION, D., LEIBACH, F.H.: Biochim. biophys. Acta 363 (1974) 267.

8 JOCELYN, P.C.: Biochemistry of the SH Group. Academic Press. London, New York (1972).

9 KRISHNA, G., WEISS, B., BRODIE, B.B.: J. Pharmacol. exp. Ther. 163 (1968) 379.

10 KWOCK, L., WALLACH, D.F.H., HEFTER, K.: Biochim. biophys. Acta 419 (1976) 93.

11 MILUTINOVIĆ, S., SCHULZ, I., ROSSELIN, G.: Biochim. biophys. Acta, in press.

12 OHTA, H., MATSUMOTO, J., NAGANO, K., FUJITA, M., NAKAO, M.: Biochem. Biophys. Res. Commun. 42 (1971) 1127.

13 ØYE, I., SUTHERLAND, E.W.: Biochim. biophys. Acta 127 (1966) 347.

14 RODBELL, M.: Current Topics in Biochemistry. (Anfinsen, C.B., Goldberger, R.F., Schechter, A.N., Eds.). Academic Press. New York (1972) 187.

15 ROTHSTEIN, A.: Current Topics in Membranes and Transport.(Brenner, F., Kleinzeller, A., Eds.). Academic Press. New York. 1 (1970) 143.

16 SCHRAMM, M., NAIM, E.: J. biol. Chem. 245 (1970) 3225.

17 SIMON, B., ZIMMERSCHIED, G., KINNE-SAFFRAN, E.M., KINNE, R.: J. Membr. Biol. 14 (1973) 85.

18 SUTHERLAND, E.W., ROBISON, G.A., BUTCHER, R.W.: Circulation 37 (1968) 279.

19 TOMASI, V., KORETZ, S., RAY, T.K., DUNNICK, J., Marinetti, G.V.: Biochim. biophys. Acta 211 (1970) 31.

20 WIZEMANN, V., SCHULZ, I.: Pflügers Arch. 339 (1973) 317.

21 WIZEMANN, V., SCHULZ, I., SIMON, B.: Biochim. biophys. Acta 307 (1973) 366.

Acknowledgement

The authors wish to thank Professor K. J. Ullrich for his continuous help and criticism of the manuscript.

DISCUSSION

SHA'AFI: Is it possible that PCMB is acting like calcium?

SCHULZ: Well, first of all calcium has no effect on fluid secretion. What effect do you have in mind?

SHA'AFI: With calcium you usually find that low concentration activates adenylate cyclase and higher concentration inhibits adenylate cyclase.

SCHULZ: Well, in this system calcium only inhibits the adenylate cyclase and is not required, since the presence of EGTA does not influence the activity. This is in contrast to other systems like ACTH stimulated adenylate cyclase. I would doubt that in so different compounds as Ca^{++} and pCMB could act in the same way.

ELDEFRAWI: Have you attempted to titrate the SH-groups in the membranes to see if PCMB is indeed binding to those SH-groups at 10^{-8} or 10^{-9} M? I find this rather intriguing because the affinities we know of for PCMB are usually in the 10^{-5} M range when you bind the PCMB to such groups. Have you made an attempt to look at SH?

SCHULZ: No, we have not measured PCMB binding. I think if it was an effect, at least it should bind shortly and go off again, but it should have done something to that receptor protein because we see the effect.

BLUMENTHAL: Are there other hormone stimulated adenylase systems with SH groups?

SCHULZ: Yes, very many, for instance, F.I. Schramm found that the adenylate cyclase in parotid gland is very sensitive to μmolar concentrations of SH-reagents. Similar observations were made in fat cells and in erythrocytes.

BLUMENTHAL: Does GTP affect binding?

SCHULZ: We checked the effect of GTP on binding of iodinated secretin and found no effect, but Gpp (NH)p, the analogue of GTP, stimulates adenylate cyclase.

BERG: I wonder if you have information on the stoichiometry of available binding sites, such as SH groups, in your preparation to the total available mercurial, in moles per mole. Did you have more sulfhydryl than mercurials?

SCHULZ: We did not measure binding of PCMB and I don't know how much was really bound.

BERG: If you have enough membrane to measure adenylate cyclase in reasonable volumes of 10^{-8} or 10^{-9} molar mercurial it is likely that you had 10 times more available SH than you had mercurial. So, you are possibly dealing with just a distribution of mercurial between binding sites.

SCHULZ: Yes, but we checked all concentration ranges.

BERG: Sure, but you kept your tissue constant.

SCHULZ: Yes.

BERG: So, you are perhaps titrating more and more SH groups.

SCHULZ: Yes. If you incubate longer you will probably get a more pronounced effect on the inhibition than, for example, if you incubate only 10 minutes. The longer you incubate the more SH groups are titrated. But we did not relate PCMB binding to SH groups.

NEUMAN: A very simple test would be to vary the solution to solid ratios. You can't measure the equilibrium concentration. You are measuring only the initial concentration. You might find that some of the disparity between your KMs is not a disparity at all, but it is because you are not really sure of the equilibrium concentration. KMs really should not be initial concentration, but final. By varying the amount of solution to membranes at a fixed concentration you can at least explore whether or not this is a serious problem.

ROTHSTEIN: There may be other factors involved at these low concentration. The SH-mercurial affinity is exceptionally high. Theoretically, therefore, even a small amount of tissue should ultimately soak up all of the mercurial from a large volume of solution. The effects of such small amounts of agents is often seriously modified by factors such as its penetration into the cell and reaction with internal sulfhydryl pools. In your case, with the mercurial in a non-penetrating form that prevents it from reaching the large internal SH pool, I do not think it surprising that you can get effects of very low concentration.

SCHULZ: You mean in membranes?

ROTHSTEIN: Yes, because you have a non-penetrable form of PCMB it does not disappear into the interior of the cell, a problem common to many agents. They are only transiently effective on the membrane because they are "soaked up" by large reservoirs of sulfhydryl groups inside the cell.

ULLRICH: As far as I am aware the amount of pCMB remaining in the 10 ml incubation medium after binding to the membrane proteins (0.5 mg) was not measured. Thus, the actual pCMB concentrations could be much smaller than shown in the graphs.

ROTHSTEIN: Schultz has a large reservoir of agent per unit of tissue even at very low concentrations because of the large volume of solution relative to tissue. Because of the high affinity, almost all available agents will be taken up. For high affinity agents concentrations are not the important thing, but the total amount of agent in the system, because it will all be bound to the membrane.

BLUMENTHAL: Have you any information itself on the production of cyclic AMP and the secretin itself?

SCHULZ: Yes, studies are made on this. If you stimulate with secretin the cyclic AMP level increases in the cell. Then in these membranes you can show that secretin stimulates adenylate cyclase activity. You can potentiate the hormonal effect by phosphoidesterase inhibitors and you can mimic the effect also by analogues of cyclic AMP.

BLUMENTHAL: I was asking more if something was known about the mechanism itself.

SCHULZ: The mechanism is not known, at least not completely. There is a protein kinase which is stimulated by cyclic AMP, and one could speculate that this protein kinase is involved in phosphorylation of membranes which might lead to permeability changes. We have done some studies on phosphorylation by a membrane bound cyclic AMP stimulated protein kinase, but what the function of this phosphorylated protein is in further steps underlying the secretory mechanism, is not known.

SHA'AFI: Do you think the increase of cyclic AMP when you add this hormone is secondary to a possible change in the level of calcium, in other words is the increase a trigger for the secretin of fluid or is it to modulate the amount of calcium that has been released? For example, is calcium or the cyclic AMP the second messenger?

SCHULZ: Increased intracellular calcium is not necessary for fluid secretion since it was shown that secretion persists in Ca^{++} free extracellular medium and the calcium ionophore has no effect, but calcium is necessary for enzyme secretion which is the other part of that organ. We have two main functions in the exocrine pancreas: fluid and bicarbonate secretion and enzyme secretion, which is evoked by another hormone. In the latter calcium is necessary and just like a second messenger or third messenger. But in fluid and bicarbonate secretion calcium doesn't seem to be necessary at all.

FISCHBARG: How would you relate your results to the similar ones that Dr. Gatzy presented yesterday? The second question is how do you visualize the fact that the simulation is followed by an inhibitory effect? Do you think that these actions are extended at two different sites or only one, and if so how?

SCHULZ: Well, I think that two categories of SH groups are involved, and I can only speculate on that and this is a model which is put forward by Kwock et al. (BBA 419: 93, 1976), for example, and other people (Czech et al., Proc. Nat. Acad. Sci., Wash. 71: 4173, 1974), namely that there are SH groups on the surface of the membrane and deeper buried SH groups,and that SH linkage between the deeper buried SH groups are necessary for the active state. Also SH linkage between the superficial SH groups is possible. So if you titrate SH groups on the cell surface then more SH linkage between the deeper groups is possible, which would increase the number of molecules necessary for the active state. This would be an explanation of why low concentrations are working because you titrate the superficial SH groups allowing the buried groups to crosslink with each other. But at higher concentrations of SH group reagents also deeper SH groups are reduced by these agents preventing S-S linkage necessary for the active state.

BERG: If you have inhibition with mercurial bound to dextran, and activation with mercurial not bound to dextran, it would seem that the inhibitory group is out on the surface, but less able to compete with the activating group which is deeper in. You just said it the other way around.

SCHULZ: Right, this was really what we also found hard to

understand, why the PCMB dextran also inhibited the fluid secretion. So, we think that probably if this model really holds, these deeper SH groups have to emerge to the surface and so become accessible to PCMB dextran.

SHAMOO: Where does the calcium ATPase fit in this picture?

SCHULZ: Well, the calcium ATPase probably is involved in enzyme secretion. Each duct ends up into acini where enzymes are produced; by stimulation with acetylcholine a large influx of calcium is produced which triggers enzyme secretion. We think that the calcium ATPase pumps out the calcium again.

Cellular Responses to Toxins

CELL MEMBRANES IN CYTOTOXICITY

A.C. Allison and J. Ferluga

Clinical Research Centre

Harrow, England

ABSTRACT

Silica particles are cytotoxic for macrophages because they damage the membranes around secondary lysosomes in which the particles are engulfed. Hydroxyl groups of silicic acid on the surface of the particles form hydrogen bonds with phosphate ester groups of phospholipids and disrupt a variety of natural and artificial membranes. Asbestos fibers induce secretion of hydrolytic enzymes from cultured macrophages. Magnesium hydroxide groups of chrysotile asbestos interact ionically with ionized sialic acid residues of membrane glycoproteins, increase passive cation flux and produce osmotic lysis. The terminal components of complement (C5b-C9) when inserted into the bilayer structure also increase passive cation flux and produce osmotic lysis. The small complement cleavage product C3a is lytic for several cell types, especially malignant cells. The mechanism by which specifically sensitized thymus-derived (T)-lymphocytes kill tumour cells is discussed. Plasma membranes from effector lymphocytes possess considerable cytolytic potential, which is dependent on the activity of a membrane-associated proteinase.

INTRODUCTION

Of all cellular constituents the plasma membrane is especially important for cytotoxicity. It is the first part of the cell exposed to toxic materials, many of which exert their effects directly on the plasma membrane. Among these are particulate pollutants and several toxins of biological origin. Moreover, certain complement components and immune cells can damage plasma membranes in various ways. Even toxic materials that mediate their effects on cytoplasmic constituents or the nucleus must gain access through the plasma membrane, so that differences in membrane permeability may determine whether a particular cell type is sensitive or resistant.

Among the agents which affect plasma and lysosomal membranes are toxic particles, including silica, asbestos and glass fibers. The chemistry of the interactions with membranes has been studied in some detail as well as the biological consequences of these interactions. Three classes of toxic effects of plasma membranes can be distinguised. Some agents affect specific transport systems, and are discussed by other contributors to the symposium. A second class of agents, including anionic and non-ionic detergents, disrupt membrane structure, producing non-osmotic lysis. The characteristics of this type of lysis are that large molecules such as cytoplasmic proteins or markers attached to them are released from the cells at the same time as potassium, and that the presence in the extracellular medium of non-penetrating solutes does not inhibit the final lytic step. A third class of agents, including several biological toxins, increase the passive permeability of the plasma membrane to ions and other small molecules but not to proteins. Because of the osmotic pressure exerted by the entrapped cytoplasmic proteins, there is a net influx of sodium and water and the cell eventually bursts. The characteristics of such osmotic lysis are that small atoms or molecules, such as potassium or the radioactive analogue ^{89}rubidium, are released from the cells before proteins or the marker ^{51}chromium, and the presence in the extracellular medium of non-penetrating solutes which counterbalance the osmotic effects of cytoplasmic protein inhibits the final lytic step.

The lytic effects of complement components will be reviewed briefly and the mechanism by which immune thymus-derived (T) lymphocytes kill tumour cells will be outlined. When immune lymphocytes kill tumor cells there is prolonged and intimate contact of effector cells and target cell membrane. Plasma membrane fractions of the effector cells can reproduce this cytotoxicity, which facilitates analysis of the underlying mechanism. Thus plasma membranes are not only common targets for cytotoxic agents : the membranes have intrinsic cytolytic potential which can be expressed under appropriate conditions.

INTERACTION OF SILICA WITH MEMBRANES

As a pure mineral silica (silicon dioxide) occurs in three main crystal forms or isomers : quartz, tridymite and cristobalite; other rare variants, produced by high temperatures and pressures and found in meteorite craters, are coesite and stishovite. Although the chemical reactivity of silica is low, most forms are fibrogenic in animals, hemolytic and highly cytotoxic for macrophages in culture (1). The tetrahedral structure of silica appears to be required for the fibrogenic and cytotoxic activities of the mineral. This structure is not found in stishovite which has a configeration similar to the rutile type of titanium dioxide; the latter is also biologically inert.

Shortly after inhalation most particles of silica are ingested by alveolar macrophages, so that their cytotoxic effects on macrophages are responsible for their main biological effects. Some time ago we showed that the cytotoxic effects of silica particles are due to their interaction with the membranes surrounding secondary lysosomes (3). This results in rapid cytotoxic effects in which there is coincident release from cultured macrophages into the surrounding medium of lysosomal and cytoplasmic enzymes (10). This begins about 2 hours after uptake of silica and reaches a maximum (80 to 90%) 6 hours later. Incubation of lysosomal preparations with silica particles also brings about release of enzymes into the medium (12). Silica particles can interact with other membranes, as shown by their well-known hemolytic effect. If silica particles are coated with aluminium hydroxide, phosphatidycholine, protein, or poly-2-vinylpyridine-1-oxide, hemolysis is markedly inhibited (48). Agents chelating calcium or magnesium have no demonstrable effect on silica hemolysis, nor does treatment of the erythrocytes with sialidase inhibit the reaction (2). Silica-induced hemolysis is non-osmotic : K^+ and hemoglobin are released simultaneously, and the presence of sucrose, dextrans or other non-penetrating solutes does not inhibit the release of hemoglobin. We have presented evidence that the main effect of silica is the formation of hydrogen bonds with membrane phospholipids (38).

When silica particles are hydrated the surface consists largely of silicic acid, which is for the most part un-ionized at physiological pH (pKa = 10.8) but which has hydroxyl groups like those of phenol able to form hydrogen bonds with phosphate ester groups of phospholipids, as illustrated in model systems (38). This interaction is also shown by observations that incubation with silica particles increases the permeability of liposomes composed of phosphatidycholine and cholesterol, with no protein (52). Poly-2-vinylpyridine-1-oxide preferentially forms hydrogen bonds with hydroxyl groups of silicic acid, thereby preventing the interaction of the latter with membrane phospholipids (52). This polymer is taken up into secondary lysosomes with silica particles and reduces the fibrogenic effects of silica *in vivo*. The surface of the silica particles has multiple rigidly arranged hydroxyl groups, and the formation of many hydrogen bonds with phospholipids can distort the membranes out of the ordered bilayer configuration and so produce non-osmotic lysis.

INTERACTION OF ASBESTOS WITH MEMBRANES

Inhalation of asbestos fibers can result in fibrogenesis (asbestosis), and the development of two types of cancer (mesotheliomas of the pleura and, especially in smokers, bronchial carcinoma). Around sites of accumulation or injection of asbestos considerable infiltrates of macrophages are found. In other words, asbestos has the capacity to elicit chronic inflammatory reactions. All types of asbestos are carcinogenic and fibrogenic (24). Most experimental work has been done with chrysotile, a fibrous magnesium silicate. When cultures of macrophages are incubated with chrysotile asbestos, there is little cell death (as shown by release of lactate dehydrogenase into the medium and other criteria) but the cells secrete large amounts of hydrolytic enzymes (11). The capacity to induce enzyme secretion from macrophages is shared by many agents which elicit chronic inflammatory reactions, and the two processes may be causally related (10).

Chrysotile asbestos is also hemolytic. The magnesium hydroxide groups on the surface of the fiber are important for the interaction with membranes. Hemolysis by chrysotile asbestos is inhibited by agents chelating magnesium but not by agents selectively chelating calcium (2). Acid extraction of the magnesium hydroxide groups from chrysotile markedly reduces its capacity to hemolyze, induce oxygen secretion from macrophages, and to elicit chronic inflammatory reactions and collagen synthesis _in vivo_ (4). Hence the repeating magnesium hydroxide groups rigidly arranged on the surface of the fiber play a major role in its biological activity. These groups appear to react electrostatically with sialic acid groups of membrane glycoproteins. When erythrocytes are pretreated with sialidase, chrysotile hemolysis is markedly inhibited (24). Chrysotile hemolysis is osmotic: $^{86}Rb^+$ is released before hemoglobin, and release of the latter but not the former is inhibited when the non-penetrating solute, sucrose, is added to the medium. Chrysotile does not induce the release of anions from liposomes consisting of phosphatidylcholine and cholesterol. There is thus an interesting difference between silica, which interacts by hydrogen bonding with membrane phospholipids to disrupt membrane structure and fibrous magnesium silicate, which interacts electrostatically with membrane glycoproteins to increase the passive permeability to cations.

LYSIS BY COMPLEMENT COMPONENTS

Lysis of erythrocytes and other cells by antibody and complement has been extensively studied. It has been suggested that the terminal complement components (C5b to C9) become inserted into the membrane to generate rings which penetrate through the bilayer structure (23). Since the studies of Green and Goldberg (22) it has been recognized that this type of lysis is osmotic. As shown in table I, hemoglobin

TABLE I Effect of non-penetrating solutes on lysis of sheep erythrocytes by rabbit IgM antibody and complement

Additive (30 mOsm)	% Hemolysis	% inhibition of Hemolysis	% inhibition of ^{86}Rb release
-(Control)	62.5	-	-
Dextran T10	76.1	0	0
Dextran T40	15.8	74.7 ($p<0.001$)	0
Sucrose	61.9	1.0 ($p<0.5$)	0

release but not ^{86}Rb release can be prevented by the presence of a non-penetrating solute of molecular weight 40,000 daltons or higher. Dextran of molecular weight 10,000 does not protect, probably because it can penetrate into the cells through the effective "holes" generated by terminal complement components.

We have recently found that the terminal complement components are not required for lysis of tumor cells as a result of complement activation by the classical (antibody-dependent) or alternative (antibody-independent) pathways. The third component of complement C3, is cleaved by proteolytic activity into biologically active products. One of these is a large component, C3b, which we have found to induce enzyme secretion from macrophages in culture (46). Indeed, many agents eliciting chronic inflammatory reactions activate complement by the alternative pathway and induce enzyme secretion from cultured macrophages (45). The other cleavage product, C3a, is a basic polypeptide of molecular weight of 6,900 with a known amino-acid sequence (29). It induces selective degranulation of mast cells (13) and contraction of smooth muscle by means of histamine release as well as direct interaction with the muscle cell plasma membrane (8). We have found that C3a is highly toxic for many tumor cells of human or mouse origin (19, fig. 1). Normal human lymphocytes are relatively resistant, but human lymphocytes transformed by mitogens are sensitive.

The amounts of C3a required for lysis of tumor cells are small (about 1 μg/ml.), less than required for mast cell degranulation. These could readily be obtained in vivo (32). Hence, C3a could be involved in the tumorolytic effects of activated macrophages and in some other types of immunity against tumors.

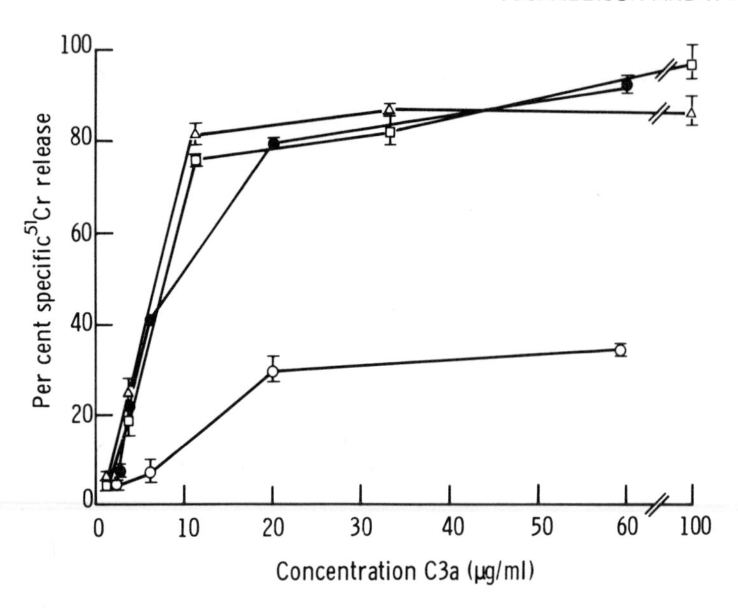

Figure 1. Lysis of human cells by C3a in 6 hours. O-unstimulated lymphocytes; ●-PHA-stimulated lymphocytes; □ -CLA-4 human lymphoblastoid cell line; △ - Chang cells.

THE MECHANISM BY WHICH T-LYMPHOCYTES KILL TUMOR CELLS

The systematic studies of Brunner, Cerottini and their colleagues (34, 9) showed that thymus-derived T-lymphocytes from immunized animals kill allogeneic tumor cells in culture and defined several properties of the system. More recently highly active preparations of lymphocytes have been obtained and various experimental approaches have helped to characterise the mechanism by which lymphocytes kill tumor cells.

Splenic (9) and peritoneal (33) lymphocytes are both highly lytic and activity can be increased by culture with irradiated tumor cells (6) or removal of cells adherent to nylon wool columns (49). Using a ratio of 10 lymphocytes stimulated in vitro to 1 tumor cell, as many as 90% of the tumor cells are lysed within one hour. In many studies (9) the immunosensitive mouse mastocytoma (P815) cells have been employed, but similar results have been obtained with other cell types, including the EL4 lymphoma (6) and a plasmacytoma (49).

The rapidity of killing contrasts with the slower killing of cells by lymphotoxins, which are released into the supernatant when sensitized lymphocytes are incubated with antigen. Even when target cells are made more sensitive by treatment with actinomycin D, several hours incubation is required before killing by lymphotoxin commences (14). Production of lymphotoxins is suppressed by inhibitors of protein synthesis (27) whereas cytotoxicity mediated by T-lymphocytes is only slightly inhibited (34,27). The immunological specificity of lymphocyte killing, and absence of effects on bystander cells (9) are also arguments that contact of lymphocytes and tumor cells is required for killing.

Despite some statemtnts to the contrary, recent experiments support the interpretation that the kinetics of cell killing are those of a first-order reaction. When centrifugation is used to bring lymphocytes and tumor cells together before incubation at 37°, there is linear release of ^{51}Cr without a preliminary lag (15, 44). The slope of the curve is proportional to the ratio of lymphocytes to tumor cells. Thus contact of one lymphocyte with a tumor cell can result in lysis and from the kinetics of the reaction and time-lapse moving pictures (31) it is apparent that the lymphocyte survives and can kill more than one tumor cell.

Several stages in the reaction can be distinguished. The first is movement of the cells so that the plasma membrane of the lymphocyte establishes intimate contact with that of the tumor cell. Agents which depress cell motility, such as cytochalsin B (9 ,31,18) or inhibitors of oxidative phosphorylation or glycolysis (7), prevent the establishment of contacts between lymphocytes and tumor cells and diminish killing; they are ineffective during the final stage of lysis. Agents which increase intracellular cyclic AMP depress tumor cell lysis (28,50) and reduce cell motility (30). In contrast, cholinergic drugs, which raise intracellular levels of cyclic GMP, increase killing (30). Agents which break down cytoplasmic microtubules also inhibit lymphocyte-mediated lysis of tumor cells (51), perhaps because they affect directional motility.

The second stage is the establishment of a firm adhesion between the lymphocyte plasma membrane and tumor cell-membrane. Moving pictures show that control lymphocytes usually pass rapidly over tumor cells whereas immune lymphocytes remain attached to tumor cells for long periods (31). Optimum adhesion of lymphocytes to tumor cells requires Mg^{2+}, although Ca^{2+} provides a less efficient substitute (21,41). The adhesion can be prevented by the presence of dextran in the medium, and established adhesions are disrupted when cells are shaken in the presence of ethyltenediamine-tetraacetic acid (EDTA), which chelates divalent ions (36). Adhesions of lymphocytes and tumor cells are established within one minute of contact at 37°. The third stage requires a living, metabolizing lymphocyte to be in contact with the target cell for a few minutes in the presence of Ca^{2+} (21); during this stage the plasma membrane of the tumor cell is damaged.

Disruption of adhesions of lymphocytes to tumor cells can be used to determine the duration of contact necessary for lysis. When the disruption accurs following 30 minutes of contact, nearly all tumor cells are lysed after incubation for a further 60 minutes (36). Similarly, when the lymphocytes are killed by specific antiserum and complement 45 to 60 minutes after contact, when only a minority of tumor cells are lysed, further incubation of the tumor cells results in lysis of the majority (44,35). Thus the final stages of the lytic reaction can proceed in the absence of an intact lymphocyte, and of Ca^{2+} and Mg^{2+} ions (21).

Since it is the plasma membrane of the lymphocyte which makes contact with the tumor cell, it is of interst that plasma membrane fractions purified from lymphocytes have considerable capcity to lyse tumor cells, whereas fractions enriched in lysosomal and other markers do not (35). Thus the plasma membrane appears to have an inbuilt lytic potential which is manifested after intimate contact with a target cell under appropriate conditions.

The nature of the lesion in the tumor cell following contact with the lymphocyte has been analysed indirectly by observing the rate of release of various markers previously introduced into the tumor cell. It is generally agreed that there is a rapid increase in the rate of loss of the K^+ analogue $^{86}Rb^+$ and of nicotinamide, that loss of protein labelled with amino acids and of ^{51}Cr takes place after some dealy and that release of DNA labelled with ^3H-thymidine or ^{125}I-iododeoxyuridine occurs later still (15, 25, 43, 37,26). Plasmacytoma cells in the presence of immune lymphocytes show nearly complete suppression of thymidine uptake, and considerable inhibition of the uptake of uridine and amino acids, long before ^{51}Cr is released (49).

In general, the observations show that soon after contact with lymphocytes tumor cell plasma membranes become permeable to small markers such as Rb^+ and nicotinamide, but not to cytoplasmic macromolecules. When dextran is added to the medium after contact of lymphocytes and tumor cells has been established, release of ^{86}Rb is not affected but release of ^{51}Cr is strongly inhibited (15). These results suggest that the lysis is osmotic: the lymphocyte plasma membrane changes the tumor cell plasma membrane, which becomes so permeable to small molecules that the ensuing cation fluxes cannot be counterbalanced by their active transport systems. As a consequence, the tumor cells lose K^+ and gain Na^+, Ca^{2+} and water. Increased Ca^{2+} in the cytoplasm could account for the marked cytoplasmic contractions (bubbling movements or zeiosis) which are observed in moving pictures before tumor cells burst (42). The fact that dextran of low molecular weight protects against lymphocyte-mediated lysis (16), whereas dextran of high molecular weight is required for protection against complement lysis, is an argument against the suggestion that the terminal components of the complement system might be involved in lymphocyte-mediated cell killing (39). Later observations from the same authors have shown that antibody-dependent cell killing does not require participation of complement factors C5-C9 (40). This is also evidence against participation of lysophosphatidylcholine generated by phospholipase A, since this type of lysis is non-osmotic and dextran provides no protection[8]. It has been reported (47) that fluorescein is transferred from the cytoplasm of tumor cells to the cytoplasm of lymphocytes. If confirmed, this observation suggests that the lymphocyte establishes a gap junction, or analogous contact permeable to small molecules, with the tumor cell.

The mechanism by which lymphocyte plasma membranes exert these effects is still unknown. Agents such as fluorophosphonates (18) tosyllysylchloromethylketone and tosylphenylanalylchloromethylketone (17), which inhibit proteinases with the formation of covalent bonds, when added to lymphocytes before the reaction strongly depress their capacity to lyse tumor cells. This inhibition is not due to killing of lymphocytes, and can be reversed completely by culturing the lymphocytes.

Cytotoxicity mediated by isolated plasma membranes is similarly but irreversibly inhibited. Thus a plasma membrane proteinase, analogous to the esterases participating in mast-cell degranulation, chemotaxis and other phenomena, (5), may be involved. An inhibitor of phospholipase A has been reported to prevent antibody-dependent cell-mediated cytotoxicity (20), but inhibits lymphocyte-mediated killing of tumor cells only in toxic concentrations (17).

In general, several stages of lymphocyte-mediated killing of tumor cells can be distinguished. The first is active movement of the cells which brings their plasma membranes into intimate contact; this requires living, metabolizing lymphocytes. The second stage is the establishment of a firm adhesion between the lymphocyte and tumor cell, which occurs within a few minutes at room temperature; this occurs most efficiently in the presence of Mg^{2+} but Ca^{2+} is a less efficient substitute. The third stage requires intimate contact of the lymphocyte and tumor cell in the presence of Ca^{2+} for a few minutes. The final stage of lysis of the tumor cell can occur in the absence of the lymphocyte and does not require Ca^{2+} or Mg^{2+}. During this stage the tumor cell membrane shows increased permeability to small molecules but not to macromolecules. As a result of osmotic disequilibrium the tumor cell bursts. Plasma membranes isolated from lymphocytes have considerable cytotoxicity, and studies with inhibitors suggest that a membrane-associated proteinase participates in the reaction.

REFERENCES

1 ALLISON, A.C.: Arch. intern. Med. 128 (1971) 131.

2 ALLISON, A.C.: In Biological effects of asbestos (Bogovski, P., Gilson, J.C., Timbrell, V. and Wagner, J.C. Eds.) IARC publication No. 8 Lyon (1973) p. 89.

3 ALLISON, A.C., HARINGTON, J.S., BIRBECK, M.: J. exp. Med. 124 (1966) 141.

4 ALLISON, A.C., DAVIES, P., MORGAN, A., SCHORLEMMER, H.U. and WAGNER, C.J. Unpublished observations (1976).

5 BECKER, E.L., HENSON, P.M.: Adv. Immunol. 17 (1973) 93.

6 BERKE, G., Amos, DB: Transplant. Rev. 17 (1973) 71.

7 BERKE, G., GABISON, D.: Eur. J. Immunol 5 (1975) 761.

8 BOKISCH, V.A. and MULLER-EBERHARD, H.J.: J. clin. Invest. 49 (1970) 2427.

9 CEROTTINI, J.C., BRUNNER, K.T.: Advanc. Immunol. 18 (1974) 67.

10 DAVIES, P. and ALLISON, A.C.: In The immunobiology of the macrophage (Nelson, D.H. Ed.) Academic Press, New York, p. 427.

11 DAVIES, P., ALLISON, A.C., ACKERMAN, J., BUTTERFIELD, A. and WILLIAMS, S.: Nature 251 (1974) 423.

12 DEHNEN, W. and FETZER, J.: Naturwiss, 54 (1967) 23.

13 DIAS DA SILVA, W., GIELE, J.W. and LEPOW, I.H.: J. exp. Med. 126 (1967) 1027.

14 EIFEL, P.J., WALKER, S.M. and LUCAS, Z.J.: Cell. Immunol. 15 (1975) 208.

15 FERLUGA, J., ALLISON, A.C.: Nature, Lond. 250 (1974) 673.

16 FERLUGA, J., ALLISON, A.C.: Nature, Lond. 255 (1975) 708.

17 FERLUGA, J., ALLISON, A.C.: Unpublished observations.

18 FERLUGA, J., ASHERSON, G.L. and BECKER, E.L.: Immunology 23 (1972) 577.

19 FERLUGA, J., SCHORLEMMER, H.U., BAPTISTA, L.C. and ALLISON, A.C.: Brit. J. Cancer (In the press) (1976).

20 FRYE, L.D., and FRIOU, G.J.: Nature Lond. 258 (1975) 333.

21 GOLDSTEIN, P. and SMITH, E.T.: Eur. J. Immunol. 6 (1976) 34.

22 GREEN, H. and GOLDBERG, P.: Ann. N.Y. Acad. Sci. 87 (1970) 352.

23 HAMMER, C.H., NICHOLSON, A. and MAYER, M.M.: Proc. Nat. Acad. Sci. USA 72 (1975) 5076.

24 HARINGTON, J.S., ALLISON, A.C. and BADAMI, D.V.B.: Adv. Pharmacol. Chemotherapy. 12 (1974) 1.

25 HENNEY, C.S.: J. Immunol.101 (1973) 73.

26 HENNEY, C.S.: Nature, Lond. 249 (1974) 456.

27 HENNEY, C.S., GAFFNEY, J. and BLOOM, B.R.: J. exp. Med. 140 (1974) 837.

28 HENNEY, C.S., LICHTENSTEIN, L.M.: J.Immunol. 107 (1971) 610.

29 HUGLI, T.E.: J. biol. chem. 250 (1975) 8293.

30 JOHNSON, G.S., MORGAN, W.D., PASTAN, I.: Nature, Lond. 235 (1972) 54.

31 KOREN, H.S., AX, W. and FREUND-MOELBERT, E.: Eur. J. Immunol. 3 (1973) 32.

32 LEPOW, I.H., WILLMS-KRETSCHMER, K.? PATRICK, R.A. and ROSEN, F.S. Amer. J. Pathol. 61 (1970) 13.

33 MACDONALD, H.R.: Eur. J. Immunol.43 (1975) 192.

34 MAUEL, J., RUDOLF, H., CHAPUIS, B. and BRUNNER, K.T.: Immunology 18 (1970) 517.

35 MARTZ, E.: J. Immunol. 111 (1973) 1538.

36 MARTZ, E.: J. Immunol. 115 (1975) 261.

37 MARTZ, E., BURAKOFF, S.J., BENACERRAF, B.: Proc. Nat. Acad. Sci. USA 71 (1974) 177.

38 NASH, T., ALLISON, A.C. and HARINGTON, J.S.: Nature 210 (1966) 259.

39 PERLMANN, P., PERLMANN, H., MULLER-EBERHARDT, H.T., MANNI, J.A.: Science 163 (1969) 937.

40 PERLMANN, P., PERLMANN, H., LACHMANN, P.: Scand. J. Immunol. 3 (1974) 77.

41 PLAUT, M., BUBBERS, J.E., HENNEY, C.S.: J. Immunol. 116 (1976) 150.

42 SANDERSON, C.J.: Proc. R. Soc. Lond. B. 192 (1976) 192.

43 SANDERSON, C.J.: Proc. R. Soc. Lond. B. 192 (1976) 221

44 SANDERSON, C.J., TAYLOR, G.A.: Cell Tissue Kinet. 8 (1975) 23.

45 SCHORLEMMER, H.U. and ALLISON, A.C.: Unpublished observations (1976).

46 SCHORLEMMER, H.U., DAVIES, P. and ALLISON, A.C.: Nature 261 (1976) 48.

47 SELLIN, D., WALLACH, D.F.H., FISCHER, H.: Eur. J. Immunol. 1 (1971) 453.

48 STALDER, K. and STOBER, W.: Nature 207 (1965) 874.

49 STEINITZ, M. and WEISS, O.N.: Cell. Immunol. 15 (1975) 403.

50 STROM, T.C., DEISSEROTH, A., MORGANROTH, J., CARPENTER, C.B., MERRILL, J.P.: Proc. Nat. Acad. Sci. USA 69 (1972) 2995.

51 STROM, T.B., GAROVOY, M.R., CARPENTER, C.B. and MERRILL, J.P.: Science 181 (1973) 171.

52 WEISSMAN, G. and RITA, G.A.: Nature New Biol. 240 (1972) 167.

DISCUSSION

LAKOWICZ: I would like to say that I thought the part on the molecular aspects of the toxicity of silica and asbestos was extremely interesting. I was wondering if you would be willing to stick your neck out on the mechanism of carcinogenecity. It is known among European asbestos workers that asbestos is not really an inhalation carcinogen, but is more properly described as being co-carcinogenic with cigarette smoking. Also, as you are probably aware, Dr. Jerina at the National Institutes of Health has shown that polycyclic aromatic hydrocarbons need to be metabolized in order to be carcinogenic. Do you think that points toward a mechanism of carcinogenesis?

ALLISON: You have to distinguish between the two types of tumours. With mesoetheliomas of the pleura and peritoneum there is no evidence of a relationship to cigarette smoking. Injection of different types of asbestos into the pleura of experimental animals produces mesoetheliomas. Apparently, this is a direct effect of asbestos on the mesoethelial cells. In contrast, in the case of bronchial carcinoma there is a multiplicative risk of asbestos exposure and cigarette smoking. One interesting possibility is that the inhalation of asbestos fibers in some way induces metabolic pathways in the lung which are concerned with activation of the hydrocarbons. This is a delicate process where a whole series of enzymes is induced. In the lungs the P-448 cytochrome system is more evident than the P-450 system. There is also the induction of epoxide hydrolases and other enzymes inactivating the activated products of carcinogens. It would be interesting to know whether asbestos can induce these enzymes, but I am not aware of any work on this problem.

ELDEFRAWI: I sit on the other side of the table, and I find the second part of the presentation was more stimulating. You showed in a graph the various aspects of cytotoxicity and the fact that not the whole lymphocyte is needed but that membrane particles could produce that effect. Have you actually done binding studies with these membrane particles to see that they do in fact attach to the T cells?

ALLISON: These microvesicles become attached to the tumor cells, and we can measure this by using radioactive membranes. All of this should be looked at and we plan to do so. In the electron micrograph which I am now showing the lymphocyte plasma membrane microvesicles can be seen attached to the tumor cell plasma membranes. At the point of contact small spherical particles appear, about 25 nanometers in diameter, too small to be vesicles themselves. All the vesicles derived from membranes which are lined by lipid bilayers are in the range of about 50 to 60 nanometers.

BLUMENTHAL: Have you used fluorescent labeled surface components?

ALLISON: We have not. As regards the mechanisms by which asbestos is lytic, evidence is of two types. The first comes from observations that the hemolytic effects of asbestos are markedly diminished by sialidase treatment of the target cells. The second is that if we carry out the reaction in the presence of magnesium-chelating agents or remove magnesium groups from chrysotile asbestos the biological activity is greatly reduced without abolishing the basic fiber structure.

BLUMENTHAL: What was your evidence that this protein esterase was involved in the cytotoxic effect?

ALLISON: This was based partly on studies of isolated membranes where proteinase activity is demonstrable by conventional tests. Moreover, certain proteinase-esterase inhibitors block cytotoxicity efficiently without killing the lymphocytes while other inhibitors block the reaction only at concentrations which limit lymphocyte viability or motility.

BLUMENTHAL: Could this esterase release some component of the effector cell, which then would be cytotoxic for the target cell?

ALLISON: That is possible. All we say is that the enzyme seems to be an essential part of the lytic system. The pattern of inhibition of cytotoxicity mediated by intact lymphocytes is the same as that of the isolated plasma membranes.

BLUMENTHAL: You do, however, know whether the isolated plasma membranes are cytotoxic for the effector cell.

ALLISON: Pretreatment of the effector and then washing abolishes the reaction. Pretreatment of the target cell has no effect. Hence, it seems to work on the effector cell membrane rather than the target cell.

ALPER: I have always thought that this immunological test used for the killing of cancer cells or whatever by T cells was a terribly savage test. What immunologists are interested in, however, is the practical endpoint - that is killing in the sense that I used the word yesterday. Do you really propose that if you didn't have the membrane torn open, with consequent release of protein, the target cells would fail to be killed by contact with the lymphocytes?

ALLISON: In the case of T-lymphocytes, there is no evidence that I am aware of that there is a cytostatic effect. In contrast macrophages activated in certain ways inhibit growth of tumor cells without lysing them.

SIEGEL: I think you showed very nicely in the lysis experiments that the loss of proteins seems to take place about 30 minutes or so after the movement of ions. Is there really a compelling reason to assume that the loss of protein is, in fact, an effect of the ion movement or perhaps both of them are effects of yet some other cause? A technical point is whether or not your cells are incubated in iso-osmotic medium.

ALLISON: Yes, they are incubated in standard medium which is iso-osmotic. The reason why I believe that the two events are related is that in the presence of a non-penetrating solute, such as dextran, rubidium release is still observed but protein release is markedly inhibited. This suggests that the ion leakage and protein loss are normally related, but that when the osmotic problem is overcome by having a non-penetrating solute in the extracellular medium the final stage of lysis can be inhibited.

SIEGEL: If the cells are in an iso-osmotic media, how is it that the cells swell, assuming they swell?

ALLISON: Because the medium is iso-osmotic as far as small ions are concerned. If you have free permeability of the ions in excess of the active transport rate, the protein which is entrapped in high concentration inside the cells has much greater relative osmotic effect than the low concentration of extracellular protein. Under these circumstances water and ions are attracted into the system and eventually the cell membrane is disrupted.

FEINSTEIN: Your very nice freeze-fracture pictures showed clumping of intramembrane particles, which are presumably protein when Sendai virus was inserted in the membrane. But the clumping from the pictures I saw was in areas remote from the virus. Do you have any idea why this happens?

ALLISON: Dr. T. Bachi has shown that once the virus gets into the membrane the clustered intramembranous particles appear first around the virus and then further away from the virus. The clusters contain virus antigens as shown by immunoferritin labelling. The virus penetrates into the membrane and its antigen interacts with the protein of the membrane, clusters them, and the clusters float away as rafts from the virus.

FEINSTEIN: What is the evidence that that clumping is not related, say, to the other effects; that is the ion exchange effect and the osmotic stress, and not directly an effect of virus insertion into the membrane? In other words, it may occur as a result of exchange of potassium and sodium, or osmotic stress on the membrane structure, causing this clumping of protein.

ALLISON: In the presence of sucrose where the actual hemolytic event is prevented the same changes are seen. So, they don't appear to follow from hemolysis. But, of course, we do believe that the clustering of the particles is related to the increased ion permeability across the membrane. We further believe that this process of clustering is related to the capacity of Sendai virus to produce membrane fusion. There are two theories about how membrane fusion occurs. According to one, which I subscribe to, there is interdigitation of protein or glycoprotein particles on apposed membranes and then later the phospholipid areas come together. Lucy and others believe that the fusion takes place in the areas of the phospholipid denuded of proteins (Lucy, J.A., In: Lysosomes in Biology and Pathology, Vol. 2, Ch. II, [1969] pp. 313-341).

ULLRICH: Does calcium play any role in the position of the membrane particles?

ALLISON: Calcium is essential for the interaction of Sendai virus with the red cell membrane, which I have discussed, and also for the membrane fusion which is induced by Sendai virus. Calcium is also required for other membrane fusion reactions (Poste, G. and Allison, A. C., Biochim. Biophys. Acta 300: [1973] 421-465).

SCHACHT: You ruled out lipid asbestos interactions by experiments with liposomes. Were those again the lecithin cholesterol liposomes? What about possible interactions with negatively charged liposomes?

ALLISON: We didn't do those experiments because we thought they would be uninterpretable. Some interactions would certainly be observed. The reason for believing that the main interaction is with glycolipids is that treatment of cells with sialidase markedly diminishes cytolysis mediated by asbestos. This enzyme does not affect phospholipids.

STANLEY: Does the protease activity in your membrane preparation increase after mitogenic stimulation?

ALLISON: It does.

STANLEY: How much?

ALLISON: Three or four-fold.

STANLEY: What if you take the lymphocytes and culture them; can you get the same activity from the culture medium?

ALLISON: We have not tried that. The problem is our cells don't like being cultured in the absence of serum, which has proteinase and inhibitors.

ALTERED DRUG PERMEABILITY IN MAMMALIAN CELL MUTANTS

V. Ling, S.A. Carlsen and Y.P. See[1]

The Ontario Cancer Institute, and the Department of

Medical Biophysics, University of Toronto, Toronto, Canada

ABSTRACT

The properties of colchicine uptake into Chinese hamster ovary cells were examined and found to be consistent with an unmediated diffusion mode. This uptake was stimulated several fold by metabolic inhibitors. The activation energy of colchicine uptake was found to be 19 kcal per mole; a similar value was obtained in cells stimulated by cyanide. Drug resistant mutants with greatly reduced colchicine permeability have been isolated. They displayed a pleiotropic phenotype, being cross-resistant to a variety of unrelated compounds. The basis of this pleiotropy was due also to reduced drug permeability. Examination of the lipids and fatty acids of parental and mutant cell membranes revealed no major differences. However, a 170,000 dalton surface glycoprotein was observed to be associated with colchicine resistance. This glycoprotein was postulated to be a modulator of drug permeability. All these data are consistent with the concept that mammalian cells are able to regulate the permeation of drugs entering by an unmediated diffusion process.

INTRODUCTION

The control of membrane permeability to a diversity of compounds is an essential function of eukaryotic cells. The transport of natural metabolites such as sugars, amino acids, vitamins, nucleosides, and their analogues has been actively investigated in recent years, and properties of their uptake are consistent with the presence of membrane carrier molecules responsible for their transport (6,7,9,13,18,19). Conceptually, the uptake of these substances is thought to be modulated via the appropriate transport molecules. In contrast, our understanding of mechanisms (or lack of mechanisms) regulating the permeation of compounds which enter cells via an unmediated diffusion mode is relatively meagre. It seems likely that a very large variety of substances including many complex chemotherapeutic drugs permeate mammalian cells in this latter manner.

As an approach towards understanding how certain cytotoxic drug molecules enter mammalian cells, we have investigated the properties of colchicine uptake into Chinese hamster ovary (CHO) tissue culture cells. We have characterized also a number of independent colchicine resistant mutants with reduced drug permeability. Studies with these mutants have enabled us to gain an insight into aspects of drug uptake into mammalian cells.

RESULTS

Colchicine Uptake. To determine whether colchicine permeates CHO cells via a facilitated or an unmediated mode, we examined the kinetics of colchicine uptake. As shown in Fig. 1, the initial rates of colchicine permeation increased linearly as a function of drug concentration up to 1×10^{-4}M. These non-saturation kinetics data are consistent with an unmediated mechanism of colchicine uptake. Carlsen et al (3) have demonstrated also that this uptake was not competed by colcemid a close structual analogue of colchicine, nor was it affected by the presence of sulphydryl reagents such as p-chloromercuribenzoic acid and p-chloromercuriphenylsulfonic acid which are potent inhibitors of a number of mediated systems. On the other hand, non-lytic concentrations of membrane active compounds such as non-ionic detergents and local anesthetics stimulated the uptake rates of colchicine (3,15). All these data suggest strongly that colchicine enters CHO cells via an unmediated diffusion mode.

Stimulation of colchicine uptake was observed also in the presence of metabolic inhibitors (20). As illustrated in Fig. 2, in the presence of potassium cyanide, or other metabolic inhibitors (20) and in the absence of glucose, the rate of colchicine uptake is stimulated several fold. This increased rate can be restored to normal by the addition of glucose (Fig. 2) or other metabolizable sugars (20). Recent studies employing rotenone to maintain the internal ATP concentration of the cells at various levels have indicated that the increased rate of colchicine uptake is related to the reduced level of ATP inside the cells (Carlsen unpublished observation). Thus it appears that colchicine

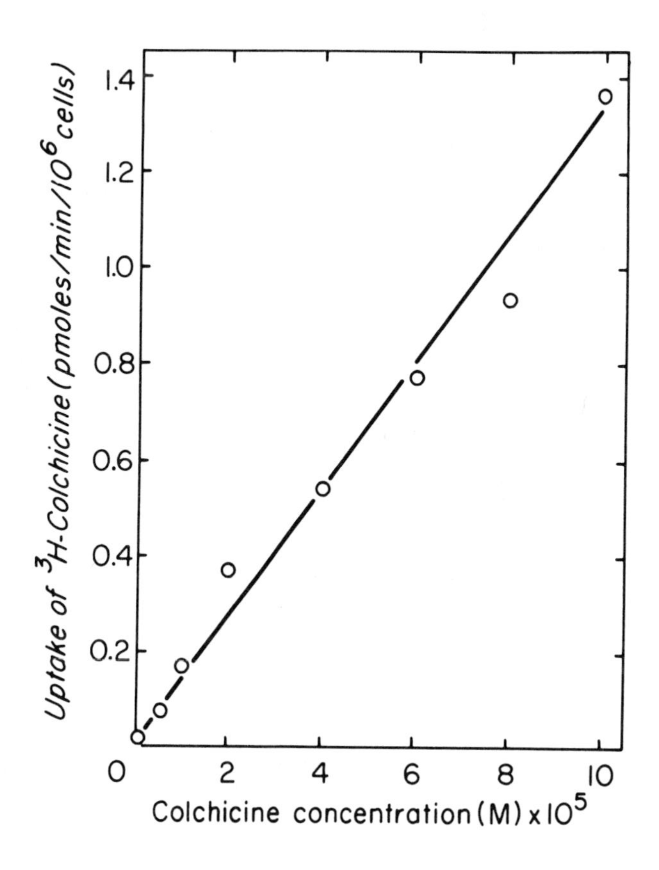

Figure 1: Rate of colchicine uptake as a function of colchicine concentration. The initial rates of colchicine uptake into CHO cells were measured as previously described (20).

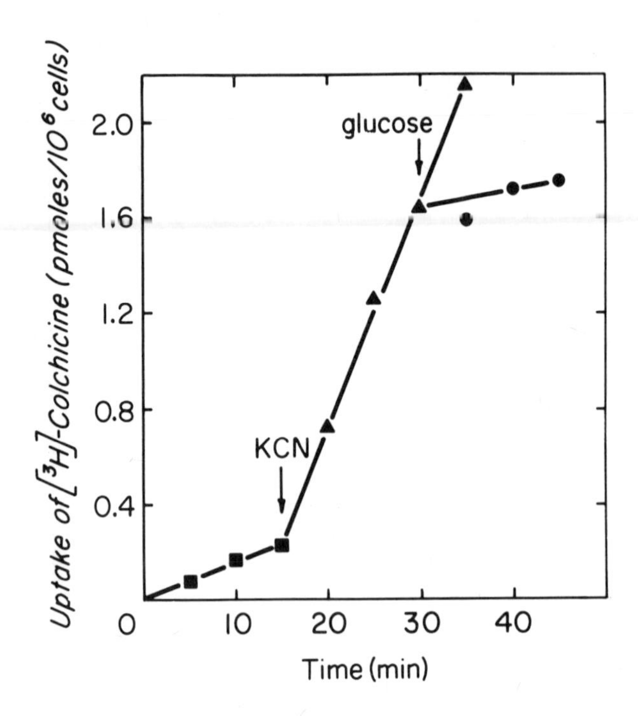

Figure 2: Effect of potassium cyanide and glucose on colchicine uptake. The uptake of labelled colchicine was measured as described previously (20). Incubation of drug sensitive cells, Aux B1 with labelled colchicine was initiated in phosphate buffered saline. At times indicated, KCN (▲-▲) to a final concentration of 2 mM and glucose (●-●) to a final concentration of 15 mM were added.

permeation into CHO cells is modulated by an energy dependent process, the mechanism of which is not yet understood.

The possibility that the cyanide stimulation of drug uptake is due to the inactivation of an energy-dependent carrier mediated drug efflux system has been raised by some investigators (4,6), however, this mechanism seems unlikely in our system as discussed previously (20). Moreover, the activation energies of colchicine uptake in cyanide stimulated and non-treated cells are similar (Fig. 3), indicating that colchicine enters the cells via the same mechanism under the stimulated or normal condition. It should be noted also that the activation energy obtained for colchicine permeation (19 kcal/mole) (Fig. 3) is similar for that obtained for the passive diffusion of glycerol and erythritol into _Acholeplasma laidlawii_ B cells and their liposomes (16). This lends further support for an unmediated mode of colchicine uptake into CHO cells.

Isolation of colchicine resistant mutants. The isolation of a number of independent colchicine resistant (CH^R) mutants of CHO cells has been described (14,15). The initial expectation was that mutants with different mechanisms of resistance would be obtained. Surprisingly, of more than 30 clones isolated under a variety of conditions, all of them exhibited reduced colchicine permeability (14,15,20). In the highly resistant cells, the initial rates of colchicine permeation were reduced by more than 20 fold when compared to that of the parental cells (15,20) There was also a strong correlation between the increased resistance, as detected by colony-forming ability of the mutant cells in the presence of drug, and the reduced drug permeability (15). These results indicate that in CHO cells at least, mutations involving membrane alterations leading to colchicine resistance are relatively common compared to other possible mechanisms of resistance.

The rate of colchicine uptake into mutant cells could also be stimulated by cyanide (c.f. Fig. 2)(20) and "detergent-like" compounds (3) such that the rate of drug permeation in the stimulated mutant cells was similar to that of stimulated parental cells. Although the mutant phenotype associated with reduced drug permeability was not evident in stimulated cells, the response of the mutant lines to the stimulatory agents was somewhat different (3,20); for example, increased amounts of non-ionic detergents were required to stimulate the mutant cells as compared to the parental line (3). Thus it appears that the alteration(s) conferring reduced drug permeability to mutant cells may also affect the ability of those cells to respond to detergents and metabolic inhibitors.

One characteristic feature of the CH^R cells was their unexpec-

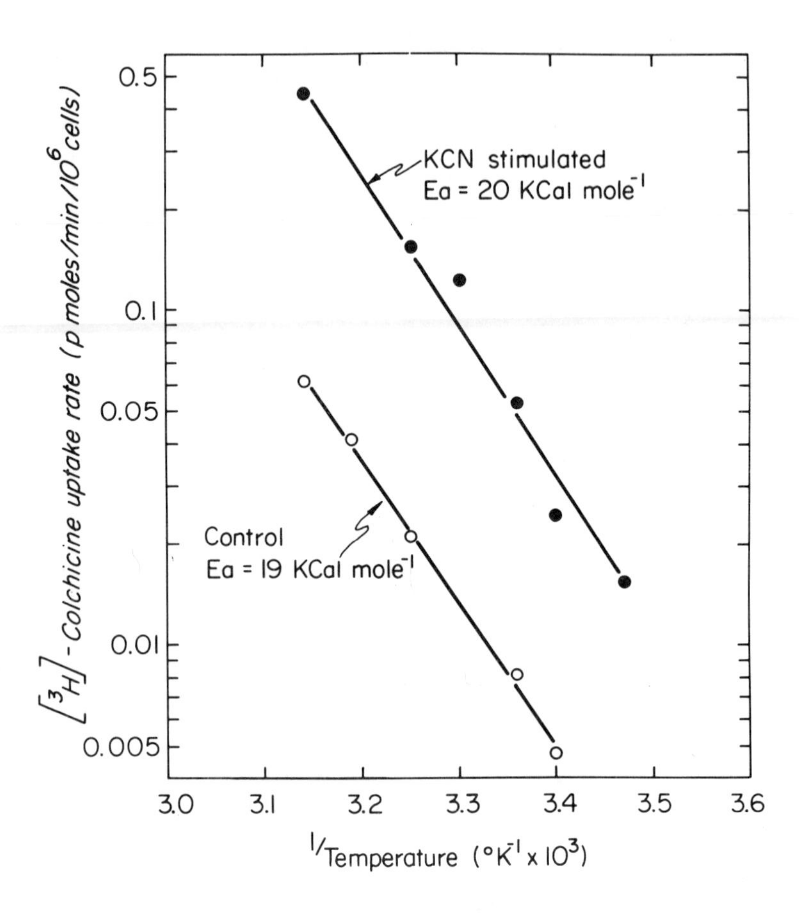

Figure 3: Arrhenius plot of colchicine uptake. Drug sensitive Aux B1 cells were used in this experiment. Initial rates of uptake were measured as in Figure 1. Colchicine permeability was stimulated with 2 mM potassium cyanide (●-●). Activation energies (Ea) were calculated from the slopes of the curves.

TABLE I. Drug cross-resistance of a colchicine resistant mutant.

Drug	Relative resistance*	
	Aux B1 (parental line)	CH^RC5 (mutant line)
Colchicine	1	184
Colcemid	1	16
Puromycin	1	105
Daunomycin	1	76
Adriamycin	1	25
Vinblastine	1	29
Gramicidin D	1	144
Emetine	1	29
Ethidium bromide	1	11
Cytochalasin B	1	11

* Relative resistance was determined by a relative growth assay of the mutant line in the presence of varying concentrations of drug compared to that of the parental line. This data was compiled from those presented by Bech-Hansen et al (1) by permission.

ted pleiotropic response to a wide variety of compounds (1,14,15). As can be seen in Table 1, the colchicine resistant mutant, CH^RC5 is cross-resistant to a plethora of structually unrelated compounds; furthermore, this mutant possessed increased sensitivity to a variety of membrane active agents such as some non-ionic detergents, local anesthetics and certain steroid hormones (1). While the mechanism of this increased sensitivity is not presently understood, the cross-resistance to other drugs is likely also the result of reduced permeability. We have observed for example, that the mutant cells possessed reduced permeability to labelled puromycin, actinomycin D and colcemid (20, Carlsen and Ling unpublished observations).

Two pieces of genetic evidence have confirmed the hypothesis that the pleiotropic phenotype results from the same mutation that confers colchicine resistance. First, analysis of hybrid cells obtained from cell-cell fusion of colchicine-sensitive and colchicine resistant mutants, indicated that colchicine resistance was expressed as an incompletely dominant phenotype. The pleiotropic phenotype (cross resistance and collateral sensitivity) was also expressed as an incompletely dominant phenotype in these hybrid cells (V. Ling and R.M. Baker, manuscript in preparation). Second, revertants have been isolated in which the mutants have lost most of their colchicine resistance and at the same time, the pleiotropic phenotype had reverted towards the wild-type phenotype (14).

The fact that a single or a limited number of mutations concurrently affected the permeability of a number of different compounds (Table 1) suggests that all these compounds have a similar mechanism of uptake. This lends further support to the hypothesis that colchicine permeates the cells via an unmediated mode since it would seem unlikely that a transport molecule facilating the influx (or efflux) of colchicine would have such a broad specificity as reflected by the cross-resistance of the colchicine mutants to a variety of compounds. On the other hand, alterations affecting an unmediated mode of transport might be expected to have rather wide ranging effects on membrane functions.

Biochemical characterization. As an approach towards elucidating the molecular mechanism of reduced drug permeability in the colchicine resistant mutants, we investigated initially the lipid composition of these cells and that of their plasma membranes. _A priori_ several observations suggested that the site of alteration in the mutant cells might be localized in the membrane lipids. For example, as documented above, colchicine permeates CHO cells via an unmediated process and presumably diffuses through the lipid bilayer of the plasma membrane. The fact that the colchicine resistant mutants display a wide ranging pleiotropy to a variety of amphipathic drugs is consistent with some global alteration(s) in the hydrophobic portion of the plasma membrane. Moreover, Bech-Hansen _et al_ (1) have shown that there was a correlation between the relative distribution coefficient of a number of different drugs in an octanol:buffer system and the relative resistance of the mutant lines to these compounds. In systems such as _Escherichia coli_, _Acholeplasma laidlawii_, red blood cells and in artificial membranes where the lipid composition of the membranes could be manipulated, it was observed that the permeability to passively diffusing compounds was dependent on the lipid composition of the membranes (5,8,16). Such factors as the cholesterol:phospholipid ratio and the relative amounts of saturated to unsaturated fatty acids were found to be important. Thus these parameters for the parental line and one colchicine resistant mutant and their plasma membranes were examined (Table II and Table III). It can be seen from Table II that the lipid composition of the parental line (Aux B1) is similar to that of a highly resistant mutant, $CH^{R}C4$. As expected, the major lipid components found in the plasma membranes of these cells were cholesterol and phospholipid. No significant difference was observed between mutant and parental line in their cholesterol to phospholipid ratios. In comparison, the cholesterol content in red cells had to be reduced by more than 30 percent before signi-

TABLE II. Lipid composition of CHO cells and surface membranes

Lipid class	Whole cells		Surface membranes	
	Aux B1	$CHR^{R}C4$	Aux B1	$CH^{R}C4$
Cholesterol	19.9 ± 0.5	20.2 ± 0.6	132.4	136.9
Cholesterol ester	16.0 ± 0.7	12.3 ± 0.7	Trace	Trace
Neutral glyceride	9.4 ± 2.1	9.4 ± 1.2	Trace	Trace
Free fatty acid	4.3 ± 0.5	4.5 ± 0.3	Trace	Trace
Phospholipid	125.4 ±13.2	110.7 ±10.3	351.1	384.5
Cholesterol to phospholipid molar ratio	0.32± 0.04	0.36± 0.04	0.71	0.74

Note: Aux B1 and $CH^{R}C4$ are respectively the drug sensitive parental line and a highly colchicine resistant line. Plasma membranes were prepared by the procedure of Brunette and Till (2). The preparation and analysis of lipids were performed as previously described (21,22). Values are expressed as means ± S.D. in μg/mg protein. Values for whole cells were obtained from four different determinations; for membranes from two different determinations.

TABLE III. Fatty acid composition of CHO cells and surface membranes

Fatty acids	Whole cells		Surface membranes	
	Aux B_1	CH^RC_4	Aux B_1	CH^RC_4
14:0	2.0	2.0	1.7	2.5
15:0	1.7	1.8	3.9	3.3
16:0	21.7	18.7	22.4	19.7
16:1	5.2	4.9	2.7	2.5
17:0	2.0	2.2	3.5	2.8
18:0	20.2	20.2	28.7	30.8
18:1	30.7	32.6	23.8	26.6
18:2	4.2	5.1	3.5	3.9
18:3	Trace	Trace	Trace	Trace
20:4	12.3	12.7	9.8	8.0
Unsaturated fatty acids %	52	55	40	41

Note: Cells and membranes are as described in Table II. Fatty acid methyl ester for GLC analysis of fatty acids were performed as described by Morrison and Smith (17). Values are averages from two separate determinations and expressed as percent of total fatty acids.

ficant alterations in glycerol permeability were observed (8). Analysis of the fatty acids of the mutant and parental cells also revealed that these lines are similar (Table III). The relative amounts of unsaturated fatty acids in parental and mutant cell surface membranes were 40 and 41 percent respectively.

These data clearly indicate that, in contrast to what might be anticipated from studies with other systems (5,8,16), the main lipid components found in the plasma membranes of mutant and parental CHO cells are similar. This argues against a mechanism of reduced drug permeability in mutant cells resulting from major changes in the lipid composition.

Examination of membrane proteins and carbohydrates revealed the presence of a glycoprotein of about 170,000 daltons in the membrane of a colchicine resistant line while at the same time this protein was present in both the parental and revertant line in considerably lower amounts (peak III of Fig. 4) (10,11). Further analyses of a number of independent mutant and revertant clones indicated that the amount of this 170,000 dalton glycoprotein, as determined

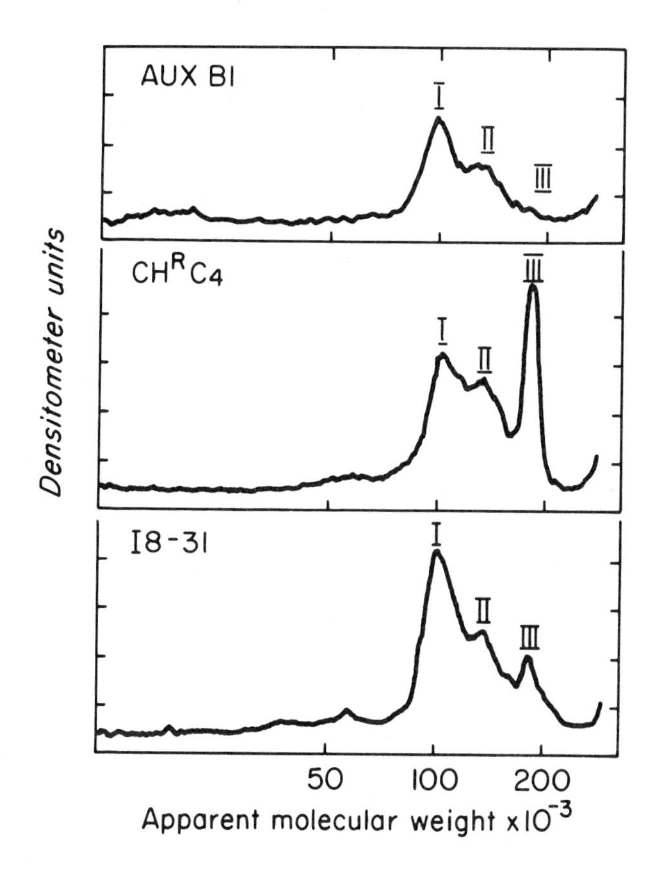

Figure 4: Separation of surface labelled membrane glycoproteins by polyacrylamide gel electrophoresis. Labelling was performed by the galactose oxidase-^{3}H-borohydride technique. Separated proteins were visualized by radioautography and quantitated by scanning with a densitometer. Details are described by Juliano et al (11). Aux B1 is the drug sensitive parental line; CH^{R}C4 is a colchicine resistant mutant; 18-31 is a drug sensitive revertant isolated from CH^{R}C4 (14). Peak III represents the 170,000 dalton surface component associated with colchicine resistance. (Figure adapted from Juliano et al (11) with permission).

by surface labelling, correlated with the degree of colchicine resistance (reduced colchicine permeability) (10). Because this glycoprotein seemed to be associated with drug permeability, we have designated it as the P surface glycoprotein.

To determine whether or not the P glycoprotein was associated only with cells selected for colchicine resistance in CHO cells, we have examined the surface glycoproteins of an actinomycin D resistant hamster line isolated in another laboratory (12) and a daunomycin resistant CHO line from our laboratory. These lines also displayed a pleiotropic phenotype and are thought to be altered in their drug permeability. In both cases, a P-like cell surface glycoprotein was observed (10; Shales, Bech-Hansen and Ling, unpublished observation). Thus the presence of this type of glycoprotein(s) appears to be associated with different drug resistant lines.

DISCUSSION

In this paper we have presented a number of different observations which are completely consistent with the hypothesis that colchicine permeates CHO cells via an unmediated diffusion mode. We have also made two major observations which support the concept that this mode of drug permeation is regulated in CHO cells. First, as shown in Fig. 2, the permeation of colchicine is greatly enhanced in the presence of cyanide; yet, the activation energies of uptake in the stimulated and normal state are similar (Fig. 3). These data indicate that the basic mechanism of colchicine permeation is the same in the presence or absence of metabolic inhibitors, and suggest that under normal conditions, in the absence of inhibitor, an energy dependent process operates to modulate the entry of colchicine and other drugs into CHO cells (20). While the mechanism of this process is not known, it does display selectivity. For example, presumed passively diffusing compounds such as erythritol and thiourea are not affected by this process (20). Second, our characterization of CHO cell mutants with reduced colchicine permeability, revealed the presence of an unique surface glycoprotein(s), the P glycoprotein, in these cells (Fig. 4). Furthermore, the amount of this glycoprotein correlated with the degree of reduced drug permeability (10). Thus these data suggest that the P glycoprotein in mutant cells has the remarkable property of limiting the permeability of a compound (colchicine) which enters cells by an unmediated diffusion mode. The permeability of other structually unrelated compounds are also affected, as reflected by the pleiotropic phenotype of the mutant cells (Table I). It seems unlikely that the P glycoprotein *per se* is a drug carrier protein but rather, it may affect some general property of the plasma membrane leading to reduced drug permeability. The mechanism by which this reduced permeability is generated is not known, however it has been speculated that the P glycoprotein may in fact modulate the fluidity of the cell membrane (14). If this is so, its sphere of influence is likely to be restricted to only certain domains of the membrane since uptake of erythritol, thiourea, α-aminoisobutyric acid, nucleosides, and 2-deoxyglucose appears not to be altered in mutant cells (20; Carlsen unpublished observation; Bech-Hansen unpublished observation).

It seems clear from these studies that the mammalian cell membrane is not functioning merely as a passive barrier to compounds which permeate the cell by an unmediated diffusion mode. Investigation into mechanisms through which the cell are able to modulate the entry or exit of these compounds may hold many surprises.

Our study of the colchicine resistant cells may be of general clinical significance. As has been noted previously (14), a number of different drug resistant cultured cell lines and tumours display cross-resistance to a variety of drugs. It seems likely that the mechanism of resistance in these systems results from altered drug permeability also. The possibility that similar types of alterations may occur in human tumour cells has been raised (14). Thus the understanding of the regulation of colchicine permeability may provide important new concepts in chemotherapy (3,14).

ACKNOWLEDGEMENT

We wish to thank Michael Naik and Alice Chase for excellent technical assistance. We thank also N.T. Bech-Hansen, J.E. Till, R.L. Juliano, and J.E. Aubin for helpful discussions. This work was supported by grants from the Medical Research Council of Canada, and the National Cancer Institute of Canada and by a contract from the National Institutes of Health of the United States. One of us (SAC) was supported by a Medical Research Council of Canada Studentship.

[1] Present address: Dept. of Anatomy, University of Ottawa, Ottawa, Canada K1N 9A9.

REFERENCES

1. BECH-HANSEN, N.T., TILL, J.E. and LING, V.: Pleiotropic phenotype of colchicine-resistant CHO cells: cross-resistance and collateral sensitivity. J. Cell. Physiol. 88 (1976) 23.

2. BRUNETTE, D. and TILL, J.E.: J. Membrane Biol. 5 (1971) 215.

3. CARLSEN, S.A., TILL, J.E. and LING, V.: Modulation of membrane drug permeability in Chinese hamster ovary cells (submitted for publication).

4. DANØ, K.: Active outward transport of daunomycin in resistant Ehrlich ascites tumor cells. Biochim. Biophys. Acta. 323 (1973) 466.

5. DAVIS, M-T.B. and SILBERT, D.: Changes in cell permeability following a marked reduction of saturated fatty acid content by *Escherchia coli* K-12. Biochim. Biophys. Acta. 373 (1974) 224.

6. GOLDMAN, I.D.: Transport energetics of the folic acid analogues, methotrexate, in L1210 leukemia cells. J. Biol. Chem. 244 (1969) 3779.

7. GOLDMAN, I.D.: The characteristics of the membrane transport of amethopterin and the naturally occurring folates. Ann. N.Y. Acad. Sci. 186 (1971) 400.

8. GRUNZE, M. and DEUTICKE, B.: Changes of membrane permeability due to extensive cholesterol depletion in mammalian erythrocytes. Biochim. Biophys. Acta. 356 (1974) 125.

9. HEINZ, E.: Coupling and energy transfer in active amino acid transport. Current topics in membranes and transport 5 (1974) 137.

10. JULIANO, R.L. and LING, V.: A surface glycoprotein modulating drug permeability in Chinese hamster ovary cell mutants. (submitted for publication).

11. JULIANO, R., LING, V. and GRAVES, J.: Drug-resistant mutants of Chinese hamster ovary cells possess an altered cell surface carbohydrate component. J. Supramolec. Struct. 4 (1976) 521.

12. LANGIER, Y., SIMARD, R. and BRAILOVSKY, C.: Mechanism of actinomycin resistance in SV40 transformed hamster cells. Differentiation 2 (1974) 261.

13. LE FEVRE, P.G.: Sugar transport in the red blood cell: structure-activity relationships in substrates and antagonises. Pharmacol. Rev. 13 (1961) 39.

14. LING, V.: Drug resistance and membrane alteration in mutants of mammalian cells. Can. J. Genetics and Cytology 17 (1975) 503.

15. LING, V. and THOMPSON, L.H.: Reduced permeability in CHO cells as a mechanism of resistance to colchicine. J. Cell. Physiol. 83 (1974) 103.

16. MCELHANEY, R.N., DEGIER, J. and VAN DER NEUT-KOK, E.C.M.: The effect of alterations in fatty acid composition and cholesterol content on the nonelectrolyte permeability of Acholeplasma laidlawii B cells and derived liposomes. Biochim. Biophys. Acta. 298 (1973) 500.

17. MORRISON, W.R. and SMITH, L.M.: J. Lipid. Res. 5 (1964) 600.

18. OLIVER, J.M. and PATERSON A.R.P.: Nucleoside transport. I. A mediated process in human erythrocytes. Can. J. Biochem. 49 (1971) 262.

19. PLAGEMANN, P.G.W. and ERBE, J.: Nucleotide pools of Novikoff rat hepatoma cells growing in suspension culture. IV. Nucleoside transport in cells depleted of nucleotides by treatment with KCN. J. Cell Physiol. 81 (1973) 101.

20. SEE, Y.P., CARLSEN, S.A. TILL, J.E. and LING, V.: Increased drug permeability in Chinese hamster ovary cells in the presence of cyanide. Biochim. Biophys. Acta. 373 (1974) 242.

21. WEINSTEIN, D.B., MARSH, J.B., GLICK, M.C. and WARREN, L.: J. Biol. Chem. 244 (1969) 4103.

22. WEINSTEIN, D.B., MARSH, J.B., GLICK, M.C. and WARREN, L.: J. Biol. Chem. 245 (1970) 3928.

ALPER: If you are using the same carrier for colchicine and other things, would you not expect to find competition if you look, say, at colchicine uptake and then use actinomycin at the same time?

LING: We looked at this closely and could find no competition between colchicine and puromycin, for example.

WEINER: Did you attempt to do any release experiments in which you loaded the cells and then looked at the kinetics of release rather than uptake?

LING: It is difficult with colchicine because it binds very tightly to the microtubule protein. But we really haven't done any good efflux experiments.

GENNARO: A difference in the pattern of "general" protein synthesis and the synthesis rate of tubulin could account for a lot of this and could even produce a different protein on the surface of the cell. How do these cells differ in the amount of tubulin they synthesize compared to the general protein synthetic activity?

LING: Before I answer that question, could you explain to me first how differences in the rate of tubulin synthesis could give you a different cross-resistance pattern?

GENNARO: All of the compounds that you mention either affect general protein synthesis, or have as their target microtubule protein. You may think you are looking at the surface of the cell, but part of your effect may be due to something that is a metabolic target of the compound.

LING: Okay, in answer to your question the amount of tubulin that is inside a cell varies somewhat, but there is no consistent pattern. The more resistant cell doesn't necessarily have more or less tubulin. There is no correlation between the amount of tubulin and the resistance.

LAKOWICZ: Would you comment on the possible metabolism of these compounds?

LING: We have looked at that fairly carefully, because as you well know this could be another mechanism of resistance. For example, a resistant cell could have an increased ability to degrade or modify the drug in some way. To test this possibility, we have examined by thin-layer chromatography the radioactive colchicine taken up by wild-type and mutant cells. We found that the labelled material migrated identically with non-labelled

LING: marker colchicine. Furthermore, we have incubated resistant cells with colchicine and after some time determined whether or not the medium was still cytotoxic to sensitive cells. In other words, questioning whether or not mutant cells inactivated the drug in the medium. In fact the potency of the medium was almost the same as medium made up with fresh colchicine.

LAKOWICZ: Changes in the relative amounts of protein on the surface or glycoprotein and lipid would affect the amount of area that is available for passive diffusions. Do you think this might have any effect?

LING: They may very well have. The size of the cell is not any different but we don't know anything about the area of the membrane.

GOLDMAN: I am not sure I understand what you mean by a mediated passive diffusion.

LING: I meant unmediated passive diffusion.

GOLDMAN: Could you not have a carrier system with a very high Km?

LING: One could. You would also have to postulate that this carrier is located on the membrane and that this carrier has specificity for a wide variety of compounds.

FEINSTEIN: Do you have any evidence about whether or not the P protein exists in very few copies even in normal cells? Also, in some cells some glycoproteins are accessible to proteolytic digestion. Have you tried this with your cells and can it reverse the permeability effect?

LING: In our wild type cells, if we label with glucosamine or amino acids, we see a very small peak of glycoprotein in the region where the P protein is normally located. So, there may be a small amount of the P protein in wild-type cells. With regard to your second question - we were able to treat these cells with trypsin and chymotrypsin and cleave off a lot of surface glycopeptides. However, the permeability in these treated cells was apparently not greatly altered.

STANLEY: Well, my question is based on inside information. You can reverse the permeability barrier by using low concentrations of detergents.

LING: That is right.

STANLEY: Does that alter the presence of the "new" peak of glycoprotein in the mutant?

LING: I can't answer that question, because we really don't know. But we were able to reverse the permeability of mutant cells by glucose starvation or by treatment with cyanide, similar to the stimulation that you see with the parenal line. In one experiment it appeared that the P glycoprotein was still there under those stimulated conditions.

NARAHASHI: I am still puzzled with the effect of potassium cyanide and local anesthetics. I understand that both stimulate the uptake of colchicine. Have you ever tried the effect of high potassium solution which might increase general permeability?

LING: We haven't tried high potassium. We have done an experiment with ouabain and it appeared to have no effect.

NARAHASHI: How do you interpret the effects of local anesthetics?

LING: Well, one interpretation is that it increases the fluidity of the membrane.

NARAHASHI: Local anesthetics generally decrease the permeability.

LING: Only if it is mediated transport.

SHA'AFI: Local anesthetic does increase the fluidity of membranes. We have shown that most of the agents you have used increase significantly the fluidity of the membrane presumably through interacting with the lipid region. This would account for the increase of the permeability to colchicine.

LING: Thank you.

LECTIN RECEPTORS AND LECTIN RESISTANCE IN CHINESE HAMSTER OVARY CELLS

Pamela Stanley and Jeremy P. Carver

Department of Medical Genetics, University of Toronto

Medical Sciences Bldg., Toronto, Ontario, Canada M5S 1A8

ABSTRACT

Chinese hamster ovary (CHO) cells previously selected in a single-step for resistance to one or two different lectins and assigned to individual phenotypic groups on the basis of their unique patterns of lectin resistance, have been examined for their lectin-binding abilities. The lectin-binding parameters of CHO cells were shown to be very complex in a detailed study of the binding of ^{125}I-WGA to wild-type (WT) cells. On the basis of these results, standard assay conditions were established and comparative binding studies between the twenty-two WT and lectin-resistant (Lec^{R}) clones were performed. A general correlation of lectin resistance with decreased lectin-binding ability and of lectin sensitivity with increased lectin-binding ability was found, although several exceptions to this trend were observed.

INTRODUCTION

Lectins bind to specific sugar sequences which are present in the carbohydrate moieties of glycoproteins (and glycolipids) of the cell membrane. Therefore lectin-binding sites (or receptors) may be distributed amongst many membrane components of grossly different size and/or function. Although lectin cytotoxicity has been extensively investigated in relation to its potential therapeutic use against cancer, the detailed mechanism(s) by which lectins exert their cytotoxicity are poorly understood (14, 15) and in only one case has the molecular mechanism of cytotoxicity been described. Ricin, the toxin from Ricinus communis has been shown to consist of two subunits, one of which binds specifically to the cell membrane and another which, after endocytosis of the molecule, inhibits protein synthesis via a direct interaction with cellular ribosomes (19).

Chinese hamster ovary (CHO) cells resistant to the cytotoxicity of a variety of plant lectins have been isolated and partially characterized in our laboratory (10,22-25). Each lectin-resistant (Lec^R) line was selected in a single-step by isolating surviving colonies from a population of CHO cells incubated for 8-9 days in the presence of a particular lectin. Some Lec^R lines were subsequently subjected to a second single-step selection with a different lectin. All of these Lec^R phenotypes have been shown to possess unique patterns of lectin resistance which differ from WT for more than one lectin and often for the following five lectins: PHA (the phytohemagglutinin from _Phaseolus vulgaris_); CON A (concanavalin A); WGA (wheat germ agglutinin); RIC (the toxin from _Ricinus communis_) and LCA (the agglutinin from _Lens culinaris_).

In one particular Lec^R isolate a biochemical mechanism of lectin resistance has been described. CHO cells selected for resistance to PHA were found to have acquired a high degree of resistance to three other lectins — WGA, RIC and LCA — as well as a 4- to 6-fold increase in sensitivity to the cytotoxicity of CON A compared with parental or wild-type (WT) CHO cells. These PHA-, WGA-, RIC- and LCA-resistant cells (designated $Pha^{R}I$, $Wga^{R}I$, $Ric^{R}I$ or $Lca^{R}I$ depending on the selective lectin) were shown to lack a glycosyltransferase enzyme activity specific for the transfer of N-acetylglucosamine (GlcNAc) to terminal α-mannosyl (Man) residues via a β-linkage (24). Cells which have lost this N-acetylglucosaminyltransferase activity (GlcNAc-T) should possess glycoproteins containing carbohydrate side chains which terminate in Man and therefore lack the terminal sugar sequence GlcNAc-Galactose-Sialic acid. Evidence in support of this hypothesis has been obtained in a variety of ways in our laboratory (10,22,24) and also by Gottlieb _et al._ (5,6) and Meager _et al._ (12,13) who observed the same phenotype in RIC-resistant CHO and BHK cells, respectively. Therefore, it may be concluded that resistance to the cytotoxicity of PHA, WGA, RIC and LCA arises in cells which possess a reduced ability to interact with these lectins by virtue of their inability to complete the terminal sugar sequence on many carbohydrate side chains which constitute the binding sites for these lectins.

The study of lectin receptors, and the nature of the membrane changes associated with lectin resistance may also be approached by measuring lectin-binding abilities. In the case of the $Pha^{R}I$ phenotype discussed earlier, such studies showed that lectin resistance correlated with a decreased ability to bind that lectin at the cell surface while increased lectin sensitivity correlated with increased lectin binding (22,24). It was of interest to determine whether the mechanisms of lectin resistance in our other Lec^R isolates could also be correlated with changes in binding characteristics. In this report, we have compared the abilities of the Lec^R CHO cell lines to bind iodinated lectins at the cell surface and have attempted to correlate lectin resistance with lectin-binding ability for each of the lectins PHA, CON A, WGA, RIC and LCA.

MATERIALS AND METHODS

Cell Lines (Nomenclature). The selection and growth characteristics of Lec^R CHO cells have been described in detail previously (22, 23, 25). The nomenclature of these cell lines, given in Tables I and II, may be briefly summarized. Each Lec^R line was isolated from a parental clone auxotrophic for either proline (designated Pro^-) or glycine-, adenosine-, and thymidine (designated Gat^-), and named for the lectin(s) with which it was selected (Pha^R, Con A^R, Wga^R, Ric^R or Lca^R). Different phenotypes were defined by their distinct patterns of lectin resistance (23). Those phenotypes which differ but which were selected by the same lectin are designated types I, II or III. For example, $Pro^-5W^{II}4B$ describes clone B which was isolated at 34° from colony 4 which arose following a selection for WGA resistance from Pro^-5 CHO cells. (Cell lines cloned at 37° are labeled numerically while those cloned at 34° are labeled alphabetically.) As described in the Introduction, the same Lec^R phenotype was isolated independently in eight selections with four different lectins. Since each of these lines exhibits a loss of a specific GlcNAc-T activity, they are designated type I for the selective lectin ($Pha^{R}I$, $Wga^{R}I$, $Ric^{R}I$ or $Lca^{R}I$).

Iodination of Lectins. For binding assays, the five lectins PHA (Burroughs Wellcome, England, purified for mitogenic activity), CON A (Pharmacia, Sweden), WGA (Sigma Chemical Co., U.S.A.), RIC (prepared by a modification of the method of Nicolson et al. (16)) and LCA (prepared by the method of Young et al. (26)) were iodinated by the lactoperoxidase method as previously described (22). Briefly, 1 mg lectin was incubated with 500 μCi carrier-free ^{125}I (The Radiochemical Centre, England), 250 μg lactoperoxidase (Calbiochem, U.S.A.) and 8.8 mM hydrogen peroxide in phosphate buffered saline (PBS) pH 7.0 to a volume of 0.5 ml. Incubations were carried out at 37° for 30 min with shaking. The reaction was stopped by transferring the tubes to ice and by the addition of 0.5 ml 10^{-4} M potassium iodide. Unbound iodide was removed by dialysis against 0.05 M phosphate buffer pH 7.0 at 4°. The iodinated lectins were adjusted to contain 0.1% bovine serum albumen (BSA; Sigma Chemical Co., U.S.A., Fraction V) and 0.02% sodium azide and stored at 4° where they remained stable for up to 3 months. More than 95% of the label in these preparations was precipitable by cold 10% trichloracetic acid. The efficiency of iodination (amount of total ^{125}I which became covalently linked to lectin) varied from 28% (PHA) to 84% (CON A). For WFA, RIC and LCA it was ~ 70-80%. Recoveries following dialysis were approximately 80% giving specific activities for the final preparations of ~ 0.05 μ Ci/μg (PHA), 0.2 μCi/μg (CON A, WGA, RIC and LCA). Samples were counted in a Nuclear Chicago gamma counter with 41% efficiency.

To determine the saturation binding characteristics of WGA to WT CHO cells, ^{125}I-WGA of high-specific activity (8) was required. WGA (30 μg) was reacted with 500 μCi carrier-free ^{125}I in the presence of 2.5 μg chloramine-T (British Drug Houses, England), 10% dimethylsulfoxide (to inhibit secondary oxidations (21)) in a final volume of 50 μl. After 15 min at room temperature the reaction was stopped by the addition of 5 μg potassium metabisulfite. Unbound iodide was removed by chromatography on a 9x0.5 cm column of Sephadex G-25 equilibriated with PBS and 5% BSA. The ^{125}I-WGA eluted in PBS in fractions 5-11 (150 μl per fraction) was pooled, the volume adjusted to 1.0 ml with 0.1% BSA in PBS and stored at 4°. Rechromatography of 1 μl of the pooled fractions showed that 96% of the ^{125}I present was bound to WGA. To show that this degree of iodination did not significantly alter certain properties of WGA, 2 mg WGA were iodinated under identical conditions with potassium iodide trace-labeled with carrier-free ^{125}I. Approximately 70% WGA was recovered as ^{125}I-WGA, giving a specific activity of ~ 0.2 μCi/mg. The cytotoxicity and circular dichroism characteristics of this preparation were essentially identical to those of unlabeled WGA indicating that this level of iodination had not significantly affected the properties of the WGA.

Based on the recovery of ~70%, the ^{125}I-WGA prepared from 30 μg WGA possessed a specific activity of ~ 18 μCi/μg, equivalent to ~ 1 atom ^{125}I per 6 molecules WGA.

<u>Lectin Binding Assays.</u> The abilities of Lec^R CHO cells to bind ^{125}I-lectins were determined using the same preparations of iodinated lectins for all experiments. Cell lines classified as being in the same phenotypic group in Table I were prepared for assay on the same day. The cells from 900 ml cultures were washed 3 times with 50 ml 0.85% saline at 4° to remove serum, diluted to 5x10^6 cells per ml with PBS, pH 7.2, containing 1 mM Mg^{++} and 1 mM Ca^{++} and brought to room temperature just prior to being used in the binding assay. A 0.1 ml aliquot of washed cells in PBS at 5x10^6 cells/ml was added to a glass test tube containing 0.1 ml ^{125}I-lectin in PBS 20% BSA. The final concentration of each ^{125}I-lectin in these assay mixtures was calculated from the specific activities given above as 20 μg/ml (PHA), 7.5 μg/ml (CON A, RIC, LCA) and 12.5 μg/ml (WGA). This amount of each lectin contained ~ 200,000-300,000 cpm and, therefore, could be used for many weeks without serious loss of activity due to isotope decay.

The incubation mixtures were allowed to stand one hour at room temperature which ensured that equilibrium binding conditions had been reached. To remove unbound ^{125}I-lectin each reaction mixture was gently resuspended with a pasteur pipette and layered over 2 ml of 5% BSA. A 0.2 ml PBS wash of the reaction tube was also layered onto the 5% BSA. The tubes were spun at 2500 rpm for 15 min in a MSE Mistral centrifuge (rotor 62303) at room temperature. The supernatant (containing the unbound ^{125}I-lectin) was removed by aspiration with a bent pasteur pipette and the ^{125}I-lectin bound to the cell pellet was counted directly in the reaction tube. Control experiments showed that more than 90% of ^{125}I-lectin binding was inhibited by including an excess (1 mg /ml) of unlabeled lectin in the reaction mixture indicating that iodination of the lectins had not altered their ability to compete with unlabeled lectin and that non-specific cell-associated label represented approximately 10% of that which was specifically bound. The per cent ^{125}I-lectin bound was determined by calculating the cpm in the cell pellet and dividing by the total cpm present in the original reaction mixture less the cpm remaining in the washed reaction tube ($<$ 2% of the total cpm present) on the assumption that these latter cpm were not available to bind during the reaction. Specific binding abilities were calculated as per cent cpm bound per mg cell protein. Protein determinations were performed in triplicate on 0.1 ml aliquots of cells at 5×10^6 cells per ml which had been stored at -20°. These samples were dried down, redissolved in H_2O containing 1% sodium dodecyl sulphate (SDS) and assayed for protein by the method of Lowry _et al._ (11). Standard curves were performed using BSA in the same H_2O-SDS diluent.

Saturation binding experiments using ^{125}I-WGA at high-specific activity were adapted to accommodate the large range of WGA concentrations required to cover saturation binding. Pro^{-}5 CHO cells cultured at 34° to between 2 and 6×10^5 cells/ml were washed in PBS and resuspended at 10^7 cells/ml. A 50 µl aliquot of cells was added to 0.15 ml PBS containing 30% BSA and ^{125}I-WGA of different specific activities. Variations in specific activity were achieved by diluting the ^{125}I-WGA (18 µCi/µg) with a stock solution of unlabeled WGA (4 mg/ml). To show that labeled and unlabeled WGA were behaving as competitive inhibitors, binding activities were determined at the extremes of each log concentration using overlapping ^{125}I-WGA concentrations of different specific activities. Non-specific binding was determined by including a 10- to 40-fold excess of unlabeled WGA in an identical set of control reaction tubes. After 1 hr at room temperature, each reaction mixture was diluted with ~ 3 ml cold PBS and washed onto a GF/C glass fibre filter (previously soaked for at least 2 hr at room temperature in 10% BSA) with three, 10 ml volumes of cold PBS. The filters and empty reaction tubes were counted and, after correcting for non-specific binding and for radioactivity bound to the reaction tubes, the specific cpm bound were calculated.

RESULTS AND DISCUSSION

Lectin Resistance in CHO Cells

The previous assignment (23) of Lec^R clones to particular phenotypes was based mainly on a comparison of their lectin-resistance properties (D_{10} values; Table I). The D_{10} value is defined as the dose of a given lectin which reduces

TABLE I. Lectin resistance in Lec^R CHO cells

Lec^R Phenotype	Cell Line	LECTINS				
		PHA	CON A	WGA	RIC	LCA
Pha^{R}II	Pro$^-$5P^{II}12-2	$>10^5$	74	150	50	69
	Pro$^-$5P^{II}12-3	$>10^5$	68	190	50	69
Wga^{R}II	Pro$^-$5W^{II}4B	40	74	460	10	44
	Gat$^-$2W^{II}4C	48	110	1500	25	42
Ric^{R}II	Pro$^-$5R^{II}3C	34	100	67	2000	160
	Gat$^-$2R^{II}2A	30	88	47	1400	150
ConA^{R}I	Pro$^-$C^{I}B211	460	170	170	650	370
	Gat$^-$C^{I}D11A	850	130	180	150	510
Pha^{R}IConA^{R}II	Gat$^-$2P^{I}C^{II}2A	$>10^5$	310	>3900	1000	$>10^5$
	Gat$^-$2P^{I}C^{II}3B	ND	ND	ND	ND	ND
Wga^{R}IIPha^{R}III	Gat$^-$2W^{II}P^{III}5B	$>10^5$	18	>4700	40	$>10^5$
Wga^{R}IIRic^{R}III	Gat$^-$2W^{II}R^{III}4B	630	47	>4100	100	7
Pha^{R}I	Pro$^-$5P^{I}3B	$>10^5$	16	2300	1600	$>10^5$
	Gat$^-$2P^{I}9A	$>10^5$	13	3100	1800	$>10^5$
Wga^{R}I	Pro$^-$5W^{I}3C	$>10^5$	24	2900	1800	$>10^5$
	Gat$^-$2W^{I}1N	$>10^5$	12	2900	1200	$>10^5$
Lca^{R}I	Pro$^-$5L^{I}2C	$>10^5$	16	2400	1200	$>10^5$
	Gat$^-$2L^{I}4C	$>10^5$	15	1300	1300	$>10^5$
Ric^{R}I	Pro$^-$5R^{I}1C	$>10^5$	16	2100	1100	$>10^5$
	Gat$^-$2R^{I}4C	$>10^5$	15	2400	1000	$>10^5$

A summary of the Lec^R CHO phenotypes isolated in our laboratory. The doses of different lectins which reduce cell survival to 10% (D_{10} value) have been compared for WT and Lec^R lines. The D_{10} values previously published by Stanley et al. (23) have been expressed as a percentage of the D_{10} for the appropriate parental WT cell line. The cell line nomenclature is as described in Materials and Methods, except for the abbreviations: P = PHA; C = CON A; W = WGA; R = RIC; L = LCA.

cell survival in a particular line to 10%. The reproducibility of these assays is apparent from the similarity of the D_{10} values for the $Pha^{R}II$ clones which do not represent independent isolates but clones which were both derived from the $Pha^{R}II$ colony 12. A standard error for D_{10} values may be estimated for at least three lectins (CON A, WGA, RIC) from the values obtained for the eight independently-isolated clones ($Pha^{R}I$, $Wga^{R}I$, $Ric^{R}I$, $Lca^{R}I$) which possess the same Lec^{R} phenotype by a number of criteria (23). For CON A the error is ±28%, for WGA ±28% and for RIC ±22%. In Table I we present the relative lectin resistances of Lec^{R} compared with WT CHO cells determined by calculating the ratio of the D_{10} for Lec^{R} cells to the D_{10} for the appropriate parental line and expressing this as a percentage for each lectin. These values may be used to locate individual Lec^{R} clones in Figure 2A-E. Some of the Lec^{R} lines assigned to the same phenotypic group because of their very similar lectin-resistance properties possess D_{10} values for certain lectins which differ from each other by more than 30% (e.g., RIC resistance of $Wga^{R}II$ clones). Therefore, it was of interest to define these Lec^{R} phenotypes according to another parameter - their ability to bind iodinated lectins at the cell surface.

Lectin Binding to WT and Lec^{R} CHO Cells

In order to choose assay conditions under which the comparative binding abilities of the Lec^{R} lines could be determined, we have examined the binding of ^{125}I-WGA to WT CHO cells in detail (Fig. 1). The data have been plotted according to the method of Scatchard for two main reasons: 1) Scatchard plots exhibit curvature when any cooperativity and/or heterogeneity occurs in the affinities of binding sites; 2) the range of concentrations which must be assayed to cover 20-80% saturation binding may be determined. This range of saturation has been shown to be sufficient for accurate estimation of the binding parameters (4). In those cases in which there exist several classes of receptors with different affinities, this criterion means that binding should be determined from high ligand concentrations where 80% or more of the lowest affinity sites are saturated to low concentrations at which 20% or less of the highest affinity sites are saturated. For the binding of ^{125}I-WGA to WT CHO cells, it is clear that to achieve 20-80% saturation, a 5-6 log range of ^{125}I-WGA concentrations must be examined (Fig. 1). At low lectin concentrations, the Scatchard plot is curvilinear with a 'concave downward' shape indicating that there is a class of high affinity WGA receptors which bind WGA with positive cooperativity. At higher degrees of saturation, the curve is 'concave upward' indicating either negative cooperativity amongst the same high affinity receptors or a new class of negatively cooperative lower affinity receptors or the existence of several classes of receptors with different affinities for WGA.

Cuatrecasas has reported similar binding behaviour in his study of the interaction of WGA and CON A with fat cells (3). Other investigators have examin-

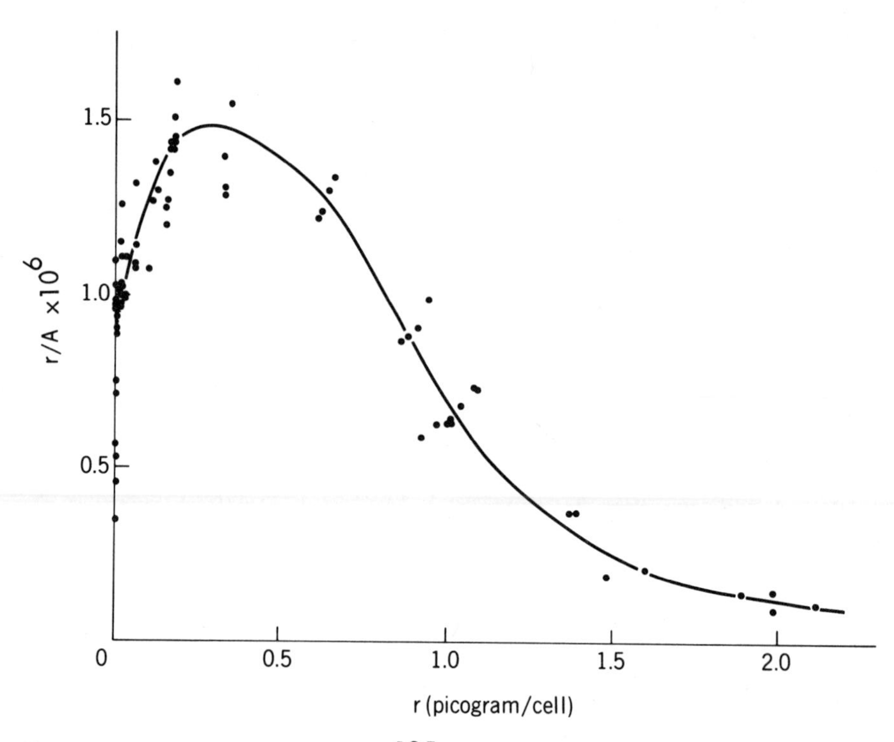

Figure 1. Scatchard plot of ^{125}I-WGA binding to Pro$^-$5 (WT) CHO cells. r is the amount of ^{125}I-WGA specifically bound per cell, A is the amount of free ^{125}I-WGA. The data were obtained from three separate experiments, each performed in duplicate.

ed much narrower concentration ranges, corresponding to partial saturation of the low affinity sites only. In some cases, nonlinearity was found (1, 20) and interpreted as corresponding to two classes of receptors of different affinity (1). In other cases, simple linear Scatchard plots were obtained (2, 27), from which it was concluded that the lectin bound to a single class of receptors of identical affinity.

In determining the lectin-binding phenotypes of all the Lec^R CHO cell lines, standard conditions which would be sensitive to as many of the binding parameters as possible were chosen. Therefore, lectin concentrations which corresponded to approximately 50% saturation of the lower affinity sites for WGA were used. Preliminary results showed that similar concentrations used for the four other lectins also corresponded to partial saturation of the low affinity sites. A further reason for using these conditions was the observation that for the five lectins the differences in binding capability between WT and Pha^{R}I CHO cells could be demonstrated clearly at similar concentrations (22,24).

The specific binding activities of each Lec^R line for the five lectins are presented in Table II. Normalization was achieved by determining the per cent ^{125}I-lectin bound per µg cell protein thus accounting for the progressive decay of the isotope, variations in counting efficiency and cell number. The errors in these measurements may again be calculated by comparing the normalized specific binding abilities of the eight Pha^{R}I-type independent clones and was found to be ±37% for PHA, ±30% for CON A, ±22% for WGA, ±20% for RIC and ±25% for LCA.

Because of the complexity observed for the binding of WGA to WT CHO cells (see Fig. 1), caution must be exercised in interpreting differences between the binding capabilities of the Lec^R cell lines. In particular, a reduction in the amount of lectin bound at a given cell number and lectin concentration may mean that one, some, or all of the following events have occurred: (i) the number of receptors of all affinity classes may have been reduced; (ii) receptors of a particular class may have been reduced or eliminated; (iii) the affinity of all the receptors of one particular class may have been reduced; (iv) the extent of positive cooperativity at low degrees of saturation may have been reduced; or (v) the extent of the negative cooperativity at high degrees of saturation may have increased. In any given case, determination of which of these possibilities has occurred requires detailed analysis of the appropriate Scatchard plot. What is clear, however, is that all these possibilities require a modification of some membrane component, whether it be an alteration of oligosaccharide side chain structure, glycoprotein amino acid sequence or an altered membrane component responsible for controlling receptor mobility (and thereby, perhaps, cooperativity).

Comparisons between Lectin Resistance and Lectin Binding Abilities

To investigate the relationships between lectin-binding abilities and lectin resistance, the normalized binding data for Lec^R cells were compared to WT CHO binding by expressing the amount bound to the Lec^R cells as a percentage of that bound to the appropriate WT for each lectin. These binding activities were compared to lectin resistance in the same cell line by plotting them (Fig. 2A-E) against the Lec^R D_{10}, expressed as a percentage of WT D_{10} (see Table I).

Within the experimental errors of approximately ±30%, a number of general correlations are apparent. In all cases, there appears to be a correlation between resistance to the selective lectin and a decreased ability to bind that lec-

TABLE II. Binding of iodinated lectins to Lec^R CHO cells

Lec^R Phenotype	Cell Line	LECTINS				
		PHA	CON A	WGA	RIC	LCA
WT	Pro^-5	7.6	10	31	5.8	8.7
	Gat^-2	7.8	12	32	7.4	12
Pha^RII	$Pro^-5P^{II}12\text{-}2$	1.7	21	36	16	15
	$Pro^-5P^{II}12\text{-}3$	1.5	20	31	17	15
Wga^RII	$Pro^-5W^{II}4B$	7.7	10	23	17	9.4
	$Gat^-2W^{II}4C$	8.1	11	11	23	12
Ric^RII	$Pro^-5R^{II}3C$	8.1	6.6	23	3.4	5.2
	$Gat^-2R^{II}2A$	7.0	6.5	24	3.8	6.1
$ConA^RI$	Pro^-C^IB211	4.8	4.0	21	3.2	4.8
	Gat^-C^ID11A	7.5	9.7	26	10	12
Pha^RIConA^RII	$Gat^-2P^IC^{II}2A$	0.56	12	16	0.77	4.1
	$Gat^-2P^IC^{II}3B$	0.39	10	15	0.62	5.4
Wga^RIIPha^RIII	$Gat^-2W^{II}P^{III}5B$	0.4	30	5.4	5.9	1.4
Wga^RIIRic^RIII	$Gat^-2W^{II}R^{III}4B$	3.9	23	22	4.9	17
Pha^RI	Pro^-5P^I3B	0.25	15	6.7	0.77	1.1
	Gat^-2P^I9A	0.35	16	10	1.2	1.3
Wga^RI	Pro^-5W^I3C	0.46	33	13	0.91	1.8
	Gat^-2W^I1N	0.75	32	15	0.97	2.0
Lca^RI	Pro^-5L^I2C	0.80	33	14	1.1	1.9
	Gat^-2L^I4C	0.69	29	13	0.80	1.8
Ric^RI	Pro^-5R^I1C	0.50	29	10	1.2	1.5
	Gat^-2R^I4C	0.47	27	12	1.0	1.4

The ability of each Lec^R CHO cell line to bind ^{125}I-lectins was determined as described in Materials and Methods and is expressed as per cent cpm bound of the total cpm present per μg protein normalized to the average observed value of 70 μg protein per 5×10^5 washed cells.

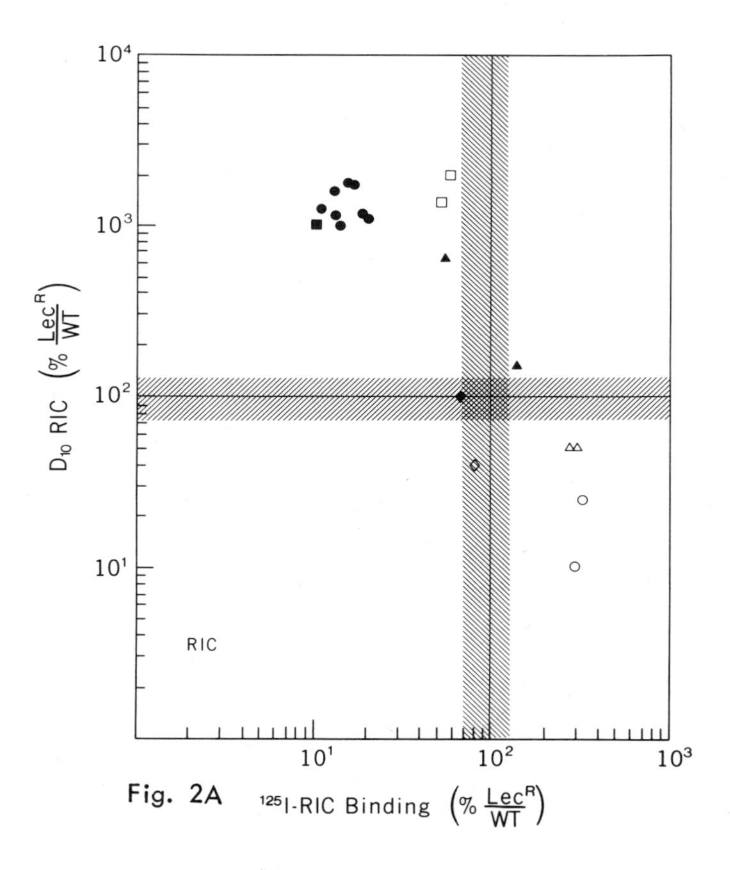

Fig. 2A

Figure 2. Relationships between lectin-resistance and lectin-binding abilities in Lec^R CHO cells. The comparative lectin resistances of Lec^R compared with WT CHO cells (Table I) were plotted against the comparative lectin-binding abilities of these cell lines (cpm per μg protein bound to Lec^R as a percentage of that bound to WT CHO cells). The intersection at 100% on both scales represents the position of WT CHO cells. These comparisons have been made separately for each of the five lectins RIC (A), PHA (B), LCA (C), WGA (D) and CON A (E). The Lec^R phenotypes in Tables I and II are represented by the following symbols:

Pha^RI, Wga^RI, Ric^RI and Lca^RI (●);
Pha^RII (△);
Wga^RII (○);
Ric^RII (□);
Con A^RI (▲);
Pha^RIConA^{R}II (■);
Wga^RIIPha^{R}III (◇);
Wga^RIIRic^{R}III (◆).

It must be noted that a number of points on these graphs represent minimum values for lectin resistance (see Table I). Points included in the hatched areas are within experimental error (±30%) of the WT values.

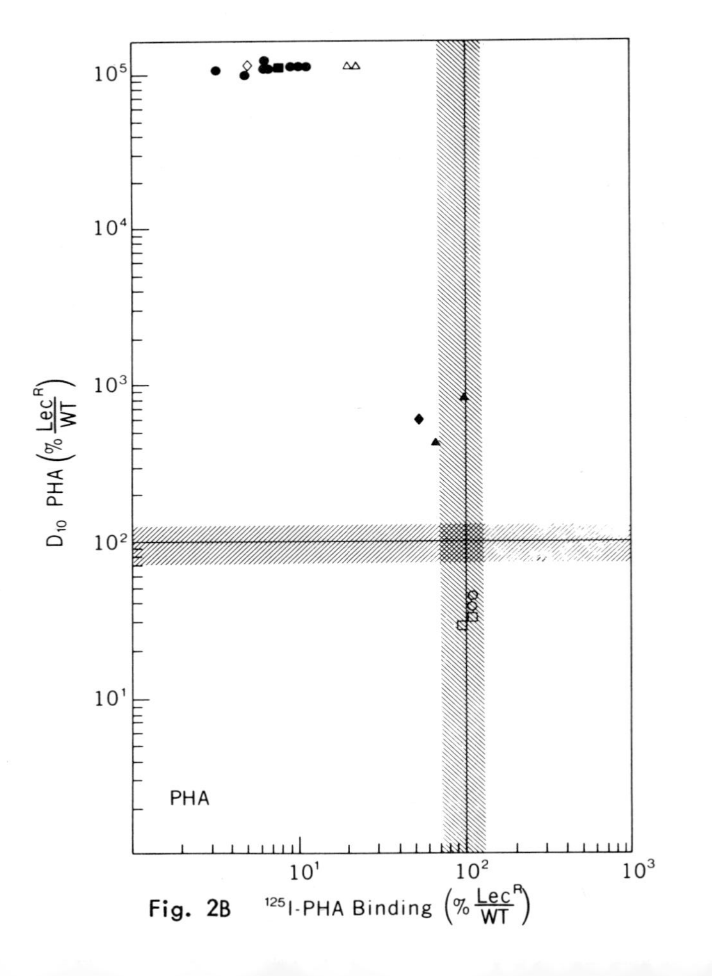
10^5
10^4
10^3
10^2
10^1
D_{10} PHA (% $\frac{Lec^R}{WT}$)
PHA
10^1
10^2
10^3
^{125}I-PHA Binding (% $\frac{Lec^R}{WT}$)

Fig. 2B

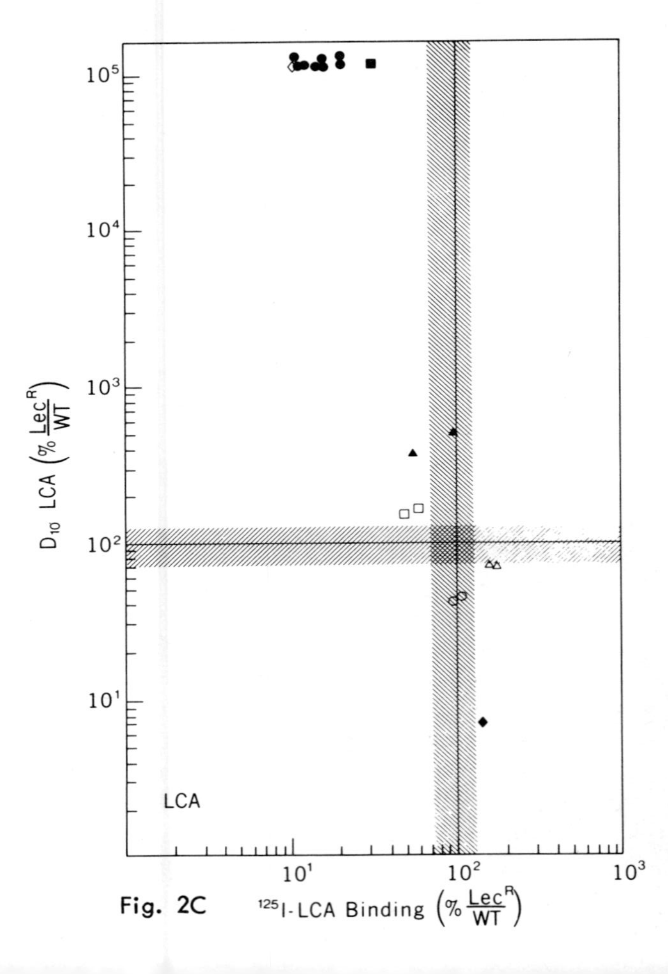
10^5
10^4
10^3
10^2
10^1
D_{10} LCA (% $\frac{Lec^R}{WT}$)
LCA
10^1
10^2
10^3
^{125}I-LCA Binding (% $\frac{Lec^R}{WT}$)

Fig. 2C

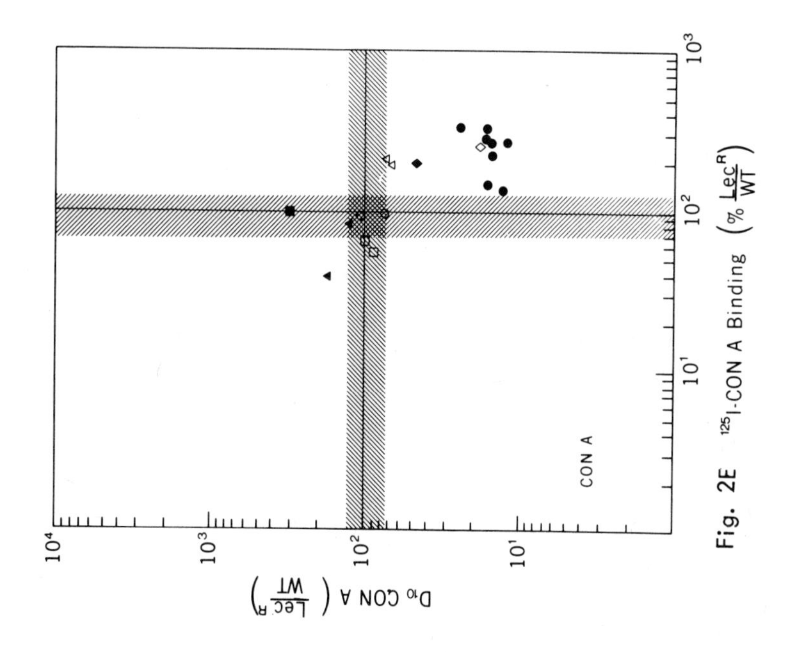
CON A
D_{10} CON A ($\frac{Lec^R}{WT}$)
^{125}I-CON A Binding (% $\frac{Lec^R}{WT}$)

Fig. 2E

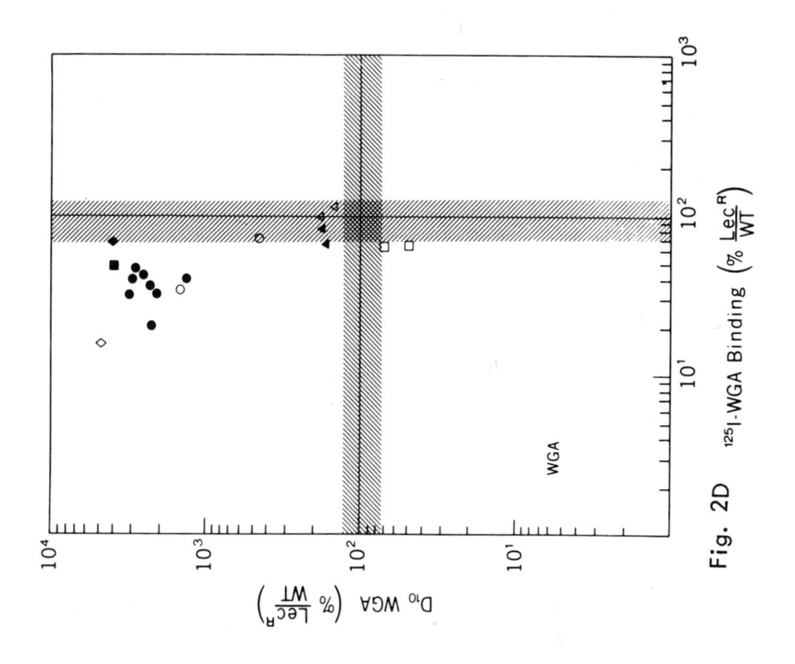
WGA
D_{10} WGA (% $\frac{Lec^R}{WT}$)
^{125}I-WGA Binding (% $\frac{Lec^R}{WT}$)

Fig. 2D

tin. Secondly, lectin sensitivity appears to be correlated with increased lectin-binding ability for CON A, RIC and LCA, although for PHA and WGA this correlation has not been observed.

A number of more specific conclusions may also be drawn from the graphs for each lectin. For RIC, resistance appears to be generally correlated with decreased binding except for two phenotypes which differ in RIC binding but not in RIC-resistance (Fig. 2A). Also, some Lec^R lines which exhibit widely different degrees of RIC sensitivity possess very similar RIC binding abilities and a few clones exhibit phenotypes differing from the general trend (e.g., the W^{II}P^{III} clone which is sensitive to RIC but binds less ^{125}I-RIC than WT CHO cells). For PHA, there appear to be at least three classes of Pha^R CHO cells - those exhibiting a high degree of PHA-resistance and low PHA-binding, those with medium resistance and intermediate binding and a class which is more sensitive to PHA than WT but apparently possesses unchanged PHA-binding ability (Fig. 2B). For LCA, there seems to be a general correlation between lectin resistance and lectin binding in the various classes of Lec^R cells altered in resistance to LCA (Fig. 2C). However, the relationship is quite qualitative and there are exceptions, such as cell lines which are either resistant or sensitive to LCA but bind WT amounts of ^{125}I-LCA at the cell membrane. For WGA, resistance to the lectin appears to be correlated with decreased binding but the range of binding abilities of highly WGA-resistant clones compared with WT CHO cells varied from 16-70% (Fig. 2D). The number of Wga^R classes among the Lec^R phenotypes appears to be at least four, including cells which are more sensitive to WGA than WT but bind less ^{125}I-WGA than WT cells (e.g., Ric^{R}II). Con A^R CHO cells seem to belong to a number of classes including cells which possess WT lectin-resistance with decreased lectin-binding or high lectin-resistance with WT lectin binding (e.g., P^{I}C^{II}2A), although it must be remembered that the latter clone was derived from a Pha^{R}I clone which exhibited markedly greater CON A binding than WT cells (23,25). Although CON A resistance does not always appear to correlate with a decrease in CON A-binding ability, it seems that CON A sensitivity is, so far, always correlated with an increased ability to bind CON A (Fig. 2E).

From Figure 2 it is apparent that the lectin-binding characteristics of the Lec^R CHO cell lines have provided an additional parameter with which to differentiate their phenotypes. While most phenotypic groups remained unchanged, the results suggest that the clones previously classified as Wga^{R}II (4B and 4C) do not possess identical phenotypes and that the Con A^{R}I clones (B211 and D11A are also different. The Wga^{R}II clones are widely separated for RIC and WGA (Fig. 2A, D) while the Con A^{R}I clones do not fall together for RIC, LCA and CON A (Fig. 2A, C, E). Further evidence that supports these distinctions has been observed. For example, compared with Pro$^-$5W^{II}4B, the frequency of Sendai virus induced fusion of Gat$^-$2W^{II}4C is very low (Stanley and Siminovitch, manuscript in preparation). Secondly, the Con A^{RI} lines appear to differ in their relative sensitivity to temperature in that $Pro^{-}C^{I}B211$ exhibits a more marked temperature dependence for growth than $Gat^{-}C^{I}D11A$ (Baker and Cifone, manuscript in preparation; Stanley, personal observation).

Another parameter to consider in comparing lectin binding by cells and resistance to the cytotoxic effects of lectins, is the degree of saturation of the binding sites required to produce efficient killing. A comparison of D_{10} values with saturation levels of lectin binding is complicated by two factors: 1) by the fact that the two experiments are performed with vastly different numbers of cells and that lectin binding shows cell density effects (Stanley and Carver, unpublished results); 2) that D_{10} values are determined in the presence of serum which contains lectin-binding glycoproteins thereby decreasing the effective lectin concentrations. However, it can be shown from the data for WGA that, as has been found for CON A-induced agglutination (20), only a small fraction (corresponding largely to the high affinity positively cooperative sites) of the total number of sites need be occupied to produce efficient killing. It appears, therefore, that the cytotoxic effect of WGA may require interaction of the lectin with receptors which are specific mediators of its toxicity.

Other investigators have examined the relationship between lectin resistance and lectin binding and have reported similar correlations to those we have observed in Lec^{R} CHO cells (5, 6, 9, 12, 13, 17). However, some cell lines have been isolated which are highly lectin resistant but which do not appear to exhibit decreased binding of the selective lectin. Meager et al. (13) have observed RIC-resistant BHK cells which bind more ^{125}I-RIC than WT BHK cells and SV403T3 (18) and MOPC 173 (7, 27) cells selected for CON A resistance have been shown to exhibit WT CON A binding abilities. The CON A-resistant MOPC 173 cells appear to have a reduced ability to mobilize CON A membrane receptors and this has been postulated as the basis of their CON A resistance (7).

A major aim of further characterizing the phenotypes of Lec^{R} CHO cells is to obtain clues as to the mechanisms of lectin cytotoxicity. In all the Lec^{R} CHO cells the extent of binding of the lectin used to select the particular Lec^{R} cell line was found to be significantly reduced and for the majority of resistance phenotypes this correlation was found to extend to the other lectins. Since it appears that only a small fraction of the lectin receptors need be occupied to produce cell death, this correlation might not be expected. One possible explanation is that an enzyme defect (e.g., a glycosyltransferase) which leads to incompleted carbohydrate structures for the receptors involved in the cytotoxicity of a particular lectin, may also be required for the biosynthesis of the majority of the cell-surface receptors specific for that lectin. Therefore, a generalized loss of lectin binding might be expected to occur despite the involvement of only a small fraction of the total lectin receptors in lectin-mediated cytotoxicity.

In summary, two mechanisms of lectin resistance have so far been described in mammalian cells - one due to the lack of a specific GlcNAc-T activity which results in markedly reduced lectin-binding abilities (6, 12, 24) and the second due to reduced lectin receptor mobility which results in no change in lectin-binding ability (7). It seems clear from the relationships between lectin resistance and lectin binding observed in the experiments presented here (Fig. 2A-E) that there are many other possible mechanisms of lectin resistance which may cover the entire spectrum of lectin-binding abilities. Thus, as the biochemistry of lectin resistance is better understood, the mechanisms of lectin toxicity should become clearer.

Acknowledgements

The authors gratefully acknowledge the critical comments on the manuscript of Drs. L. Siminovitch and M. Pearson, the excellent technical assistance of Ms. Nancy Stokoe and Ms. Wendy MacDougall and the financial support of the Medical Research Council of Canada (MA-3732 to JPC; MT-4732 to L. S.); the National Cancer Institute of Canada and the NIH to L. S.

REFERENCES

1. ADAIR, W. L. and KORNFELD, S.: Isolation of the receptors for wheat germ agglutinin and the *Ricinus communis* lectins from human erythrocytes using affinity chromatography. J. Biol. Chem. 249 (1974) 4696.

2. BETEL, I. and van den BERG, K. J.: Interaction of concanavalin A with rat lymphocytes. Europ. J. Biochem. 30 (1972) 571.

3. CUATRECASAS, P.: Interaction of wheat germ agglutinin and concanavalin A with isolated fat cells. Biochem. 12 (1973) 1312.

4. DERANLEAU, D. A.: Theory of the measurement of weak molecular complexes. I. General considerations. J. Am. Chem. Soc. 91 (1969) 4044.

5. GOTTLIEB, C., SKINNER, A. M. and KORNFELD, S.: Isolation of a clone of Chinese hamster ovary cells deficient in plant lectin-binding sites. Proc. Nat. Acad. Sci. (U. S.) 71 (1974) 1078.

6. GOTTLIEB, C., BAENZIGER, J. and KORNFELD, S.: Deficient uridine diphosphate-N-acetylglucosamine glycoprotein N-acetylglucosaminyltransferase activity in a clone of Chinese hamster ovary cells with altered surface glycoproteins. J. Biol. Chem. 250 (1975) 3303.

7. GUÉRIN, C., ZACHOWSKI, A., PRIGENT, B., PARAF, A., DUNIA, I.,

DIAWARA, M.-A. and BENEDETTI, E. L.: Correlation between the mobility of inner plasma membrane structure and agglutination by concanavalin A in two cell lines of MOPC 173 plasmacytoma cells. Proc. Nat. Acad. Sci. (U.S.) 71 (1974) 114.

8. HUNTER, W.M. and GREENWOOD, F.C.: Preparation of Iodine-131 labelled human growth hormone of high specific activity. Nature 194 (1962) 495.

9. HYMAN, R., LACORBIÈRE, M., STAVAREK, S. and NICOLSON, G.: Derivation of lymphoma variants with reduced sensitivity to plant lectins. J. Nat. Cancer Inst. 52 (1974) 963.

10. JULIANO, R. L. and STANLEY, P.: Altered cell surface glycoproteins in phytohemagglutinin-resistant mutants of Chinese hamster ovary cells. Biochim. Biophys. Acta 389 (1975) 401.

11. LOWRY, O. H., ROSEBROUGH, N. J., FARR, A. L. and RANDALL, R. J.: Protein measurements with the folin phenol reagent. J. Biol. Chem. 193 (1951) 265.

12. MEAGER, A., UNGKITCHANUKIT, A. and NAIRN, R.: Ricin resistance in baby hamster kidney cells. Nature 257 (1975) 137.

13. MEAGER, A., UNGKITCHANUKIT, A. and HUGHES, R. C.: Variants of hamster fibroblasts resistant to Ricinus communis toxin (Ricin). Biochem. J. 154 (1976) 113.

14. NASPITZ, Ch. K. and RICHTER, M.: The action of phytohemagglutinin in vivo and in vitro, a review. Progr. Allergy 12 (1968) 1.

15. NICOLSON, G. L.: The interactions of lectins with animal cell surfaces. Internat. Rev. Cytol. 39 (1974) 89.

16. NICOLSON, G. L., BLAUSTEIN, J. and ETZLER, M. E.: Characterization of two plant lectins from Ricinus communis and their quantitative interaction with a murine lymphoma. Biochem. 13 (1974) 196.

17. NICOLSON, G. L., ROBBINS, J.C. and HYMAN, R.: Cell surface receptors and their dynamics on toxin-treated malignant cells. J. Supramolec. Struct. 4 (1976) 15.

18. OZANNE, B.: Variants of Simian virus-40 transformed 3T3 cells that are resistant to concanavalin A. J. Virol. 12 (1973) 79.

19. REFSNES, K., OLSNES, S. and PIHL, A.: On the toxic proteins abrin and ricin. Studies of their binding to and entry into Ehrlich ascites cells. J. Biol. Chem. 249 (1974) 3557.

20. SCHNEBLI, H.P. and BACHI, T.: Reaction of lectins with human erythrocytes. Exptl. Cell Res. 91 (1975) 175.

21. STAGG, B.H., TEMPERLEY, John M. and ROCHMAN, H.: Iodination and the biological activity of gastrin. Nature 228 (1970) 58.

22. STANLEY, P., CAILLIBOT, V. and SIMINOVITCH, L.: Stable alterations at the cell membrane of Chinese hamster ovary cells resistant to the cytotoxicity of phytohemagglutinin. Somatic Cell Genetics 1 (1975) 3.

23. STANLEY, P., CAILLIBOT, V. and SIMINOVITCH, L.: Selection and characterization of eight phenotypically distinct lines of lectin-resistant Chinese hamster ovary cells. Cell 6 (1975) 121.

24. STANLEY, P., NARASIMHAN, S., SIMINOVITCH, L. and SCHACHTER, H.: Chinese hamster ovary cells selected for resistance to the cytotoxicity of phytohemagglutinin are deficient in a UDP-N-acetylglucosamine:glycoprotein N-acetylglucosaminyltransferase activity. Proc. Nat. Acad. Sci. (U.S.) 72 (1975) 3323.

25. STANLEY, P. and SIMINOVITCH, L.: Selection and characterization of Chinese hamster ovary cells resistant to the cytotoxicity of lectins. In Vitro 12 (1976) 208.

26. YOUNG, N.M., LEON, M.A., TAKAHASHI, T., HOWARD, I.K. and SAGE, H.J.: Studies on a phytohemagglutinin from the lentil III. Reaction of Lens culinaris hemagglutinin with polysaccharides, glycoproteins, and lymphocytes. J. Biol. Chem. 246 (1971) 1596.

27. ZACHOWSKI, A., PRIGENT, B., MONSIGNY, M. and PARAF, A.: Susceptible and non-susceptible phenotypes of MOPC 173 plasmocytoma to the killing effect of concanavalin A: Study on the concanavalin A sites. Biochimie 56 (1974). 1621.

DISCUSSIONS

ROTHSTEIN: Is the toxicity of all lectins through the same kind of mechanism, that is, a subunit getting into the cell and stopping protein synthesis?

STANLEY: No. The only toxic mechanism that is really understood at a molecular level appears to be for ricin. This molecule is composed of two subunits, one is the binding subunit, the other appears to be the toxic subunit which interacts with ribosomes to inhibit protein synthesis with the result that the cells die. For all the other lectins no one seems to know exactly what the mechanism is and, in fact, some of them have been tested in in vitro protein synthesis assays and no inhibitory effect uncovered (Nicolson, G.L., Int. Rev. Cytol. 39: 89-190, 1974, The Interactions of Lectins with Animal Cell Surfaces).

ROTHSTEIN: So, you have independent mechanisms of cytotoxicity and of binding?

STANLEY: Yes, I would think so if lectin binding is the essential first step. These cells are highly resistant to ricin presumably because they have lost ricin receptors whereas if one were to look at in vitro protein synthesis, I am sure they would not be resistant to ricin.

ROTHSTEIN: Chemically are there two subunits in all of the lectins or is that just for ricin?

STANLEY: No, ricin and abrin appear to be the only ones with two subunits like that.

PHA could be but its mechanism of toxicity is not well characterized. It seems to have two glycoprotein subunits, but it is not sure whether one or both have the necessary requirements for its biological activities.

LAKOWICZ: With regard to the presence of capping in these cells which is dependent on the tubulin protein; are there any differences in the tubulin content?

STANLEY: We have just started to look at capping; for example, the cells that have lost the glycosyl transferase enzyme do cap perfectly well with Con A. They presumably wouldn't cap with PHA because they don't bind PHA. There are people who have isolated Con A-resistant plasmocytoma cells, and in that case they feel that they have a mobility mutant because using freeze fracture techniques they find that the intramembranous particles do not aggregate in the presence of Con A, and yet the binding parameters

STANLEY: of this mutant don't seem to have changed (Guerin, C., Zachowski, A., Prigent, B., Paraf, A., Dunia, I., Diawara, M-A., Benedetti, E.L. PNAS 71: 114-117, 1974, Correlation between the mobility of inner plasma membrane structure and agglutination by concanavalin A in two cell lines of MOPC[173] plasmocytoma cells).

CRAMP: Are the receptor sites randomly spaced?

STANLEY: I think the current idea is that the receptor sites are at certain points along carbohydrate chains which may be much more complex than simple sugars. Thus, even along the same carbohydrate chain one could bind different lectins though not necessarily together depending on what happens in terms of their competition. The carbohydrate side chains are distributed all over the place. Glycoproteins can have more than one type of carbohydrate side chain and many glycoproteins have the same carbohydrate side chains. In many ways it is very non-specific in terms of lectin receptors because the glycoprotein molecules with receptors have extremely diverse functions. In another way lectin receptors are probably very specific with respect to the carbohydrate sequence that they interact even though this sequence is represented all over the place.

LING: It seems to me that with different kinds of mutants you have a variety of cells with altered carbohydrates on the cell surface. Have you any feeling as to what the function of these carbohydrates might be?

STANLEY: It is hard to say that it is very important because the cells that have lost the glycosyl transferase activity have lost a lot of their surface carbohydrate and still function perfectly well. The only thing that we have noticed that seems to have changed is their ability to adhere to glass or plastic. They also seem to bind more to each other. So, in terms of looking at cell aggregation, it might be an interesting system in the sense of artificially being able to design adhesion experiments once we have characterized other parameters of the phenotype. But otherwise they seem to divide and grow very well.

BIOLOGICAL APPLICATIONS AND EVOLUTIONARY ORIGINS OF IONOPHORES

Berton C. Pressman and Norberto T. deGuzman

Department of Pharmacology, University of Miami

School of Medicine, Miami, Florida 33152

ABSTRACT

The structure and physical chemical properties are reviewed of various ionophores, small molecules which transport cations across biological membranes. The subclass of carboxylic ionophores has dramatic pharmacological properties, dilating the coronary arteries and increasing cardiac contractility, which make them especially interesting. They share related structures containing a terminal carboxyl group hydrogen bonded to the opposite end of the molecule in a cyclic conformation. A variety of oxygen atoms, in heterocyclic rings or linear ether configurations, hydroxyls, and/or ketonic carbonyls as well as carboxyls, focus upon a sphere which can ligand to appropriately fitting cations via ion-dipole interaction. Extreme dependency of liganding energy on fit confers striking ion selectivity upon ionophores. It is speculated that carboxylic ionophores originated as prosthetic groups of ion carriers and represent a natural experiment by the streptomyces genus in the construction of non-proteinaceous binding sites. Ultimately production of these prosthetic groups hypertrophied and subsequently they were elaborated as antibiotics to provide their producers with a survival advantage over competitors. Analogous ionophores have now been synthesized. Unique applications for these compounds are emerging, based either on their selective toxicity or ability to modify physiological processes by altering membrane gradients.

Since the recognition of their remarkable transport-inducing properties on isolated mitochondria, (1, 2) ionophores have been studied intensively, both as novel tools for perturbing biological systems and for the underlying molecular basis of their unique biological properties. This paper will be concerned with the applications of ionophores to intact organisms and the question of their evolutionary origin.

As this conference is multidisciplinary, I will begin by reviewing briefly the structures and physiochemical properties of ionophores. A compendium of representative ionophores is given in Fig. 1. Ionophores form complexes with cations by focusing an array of oxygen atoms about a cavity in space into which various cations fit more or less snugly. Once in place the cation ligands to these oxygens via induced ion-dipole interactions. In the diagramatic representation of Fig. 1 the liganding oxygen atoms, as

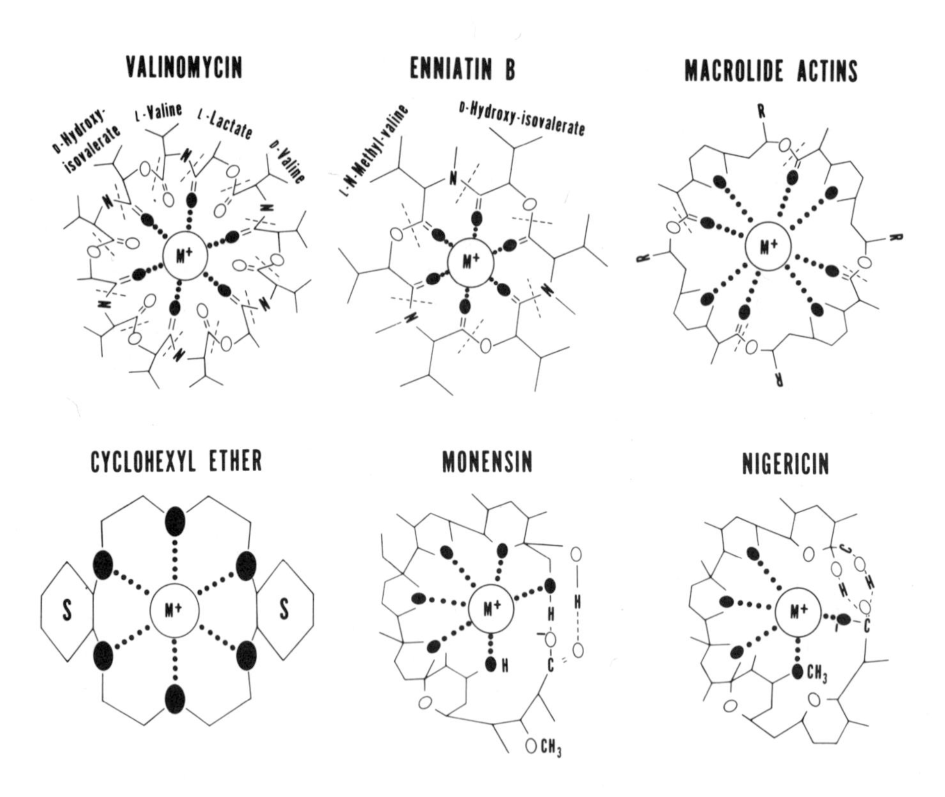

Figure 1. Structures of representative ionophores. The filled in osygens are those known by X-ray crystallography to ligand to complexed cations.

identified by X-ray crystallography of solid complexes, by various investigators are filled in black. Although the strength of any one ligand is moderate, the backbone of ionophores is especially designed to bring a multitude of oxygens into close proximity with the cation. Consequently the total free energy of complexation becomes formidable enough to replace the polar solvation shell of the cation with ionophore and confer on the cation the lipophyllicity of the outwardly oriented hydrocarbon groups of the ionophore, thereby rendering the complex soluble in low polarity organic solvents as well as the lipid interior of biological membranes. Since ion-dipole interaction has a high order distance dependency, even slight constraints of ligand focusing by the ionophore backbone can affect the balance between the free energies of complexation and desolvation, consequently each ionophore structure determines a unique complexation spectrum which is an extremely sensitive function of the cationic crystal radii.

The transport properties of ionophores arise from the lipid solubility of ionophore-cation complexes and the rapidity with which ionophores can reversibly replace the cation solvation shell at high dielectric-low dielectric interfaces such as exist at the surface of a membrane. The rapid exchange kinetics derive from the structure of ionophore backbone, most marked in the naturally occurring ones, which reduces the energy of activation of the complexation-decomplexation reaction sequence.

The wide variations among ionophores are illustrated by the individual structures of Fig. 1. Valinomycin, the first recognized ionophore (1, 2) has a 36 atom ring which convolutes into a "bracelet" conformation capable of encaging cations in three dimensions. It shows a striking ionic discrimination between the biologically significant monovalent ions, preferring K^+ over Na^+ by 10,000:1 and not reacting appreciably with the divalent cations Ca^{2+} or Mg^{2+}. Its depsipeptide structure, alternating α-amino and α-hydroxy acids, is reminiscent of polypeptides. The alteration of LL and DD optical configurations is required for the proper folding of the chain for the cage-like conformation of the molecule.

Enniatin B, like valinomycin a cyclic depsipeptide, has only 18 ring atoms and consequently ensnares cations in a more planer conformation. This produces a less critical fit and hence less K^+:Na^+ discrimination than observed with valinomycin.

Nonactin, the subunits of which are held together exclusively by ester bonds is patently non-proteinaceous. In addition to the ester carbonyls it also introduces heterocyclic ether oxygen atoms into the eight membered ligand cage.

The polyether dicyclohexyl-18-crown-6 represents an early

synthetic ionophore. Its simple structure ligands exclusively through ether oxygens and lacks hydrolyzable bonds or asymmetric carbons. Although extremely stable chemically it is not as efficient a transport agent as the natural ionophores and has found use in introducing ionic catalysts (e.g., KOH) into organic reaction solvents.

Nigericin and monensin employ heterocyclic ether oxygens, as well as other types, for liganding cations. This class of ionophores does not have a covalently linked ring but rather relies on a terminal carboxyl which hydrogen bonds to one or two hydroxyl groups of the opposite end of the molecule. This weak cyclizing ligand is re-enforced considerably by the conformation constraints contributed by the heterocyclic rings and asymmetric centers. The charge of the complexed ion is offset by the change on the dissociated carboxyl so that the resultant complex is an electrically neutral zwitterion. Nigericin and monensin illustrate two options within this ionophore subclass; one ligand of nigericin is ionic while the carboxyl of monensin is too distant from the cation to form an ionic ligand. By way of contrast the ionophores we have previously described are devoid of ionizable groups, hence their complexes acquire the positive charge of the complexed cation. This distinction accounts for radically different biological properties: the neutral ionophores carry net charges across membranes while the carboxylic ionophores cannot, being obliged to serve as neutral, exchange-diffusion type carriers. Accordingly the ionophores are divided generically into two subclasses, neutral and carboxylic. Further general information and details about ionophore structure, behavior and applications can be found in recent reviews (3, 4).

The neutral ionophores have *in vitro* antibiotic activity and the Russian workers explored the antibiotic spectra of natural and synthetic depsipeptides (3). Unfortunately these compounds are exceedingly toxic and consequently have never been clinically useful as systemic antibiotics. We are able to demonstrate with Myobacterium phlei that the ion transporting properties of ionophores were the likely basis of their antibiotic properties (5). In Fig. 2, the addition of valinomycin, after a transient artifact detected by the ion selective electrode, due to the ethanol solvent, causes an increase in the rate of K^+ uptake as well as a stimulation of respiration. As the preparation goes anaerobic K^+ uptake ceases. Restoration of O_2 with H_2O_2 (broken down by bacterial catalase) causes a resumption of K^+ uptake. The experiment was terminated by the addition of the carboxyl ionophore dianemycin which permitted the release of energy accumulated K^+ in exchange for H^+ uptake (upward deflection). It is significant that M. phlei is one of the organisms most susceptable to depsipeptide ionophores (3).

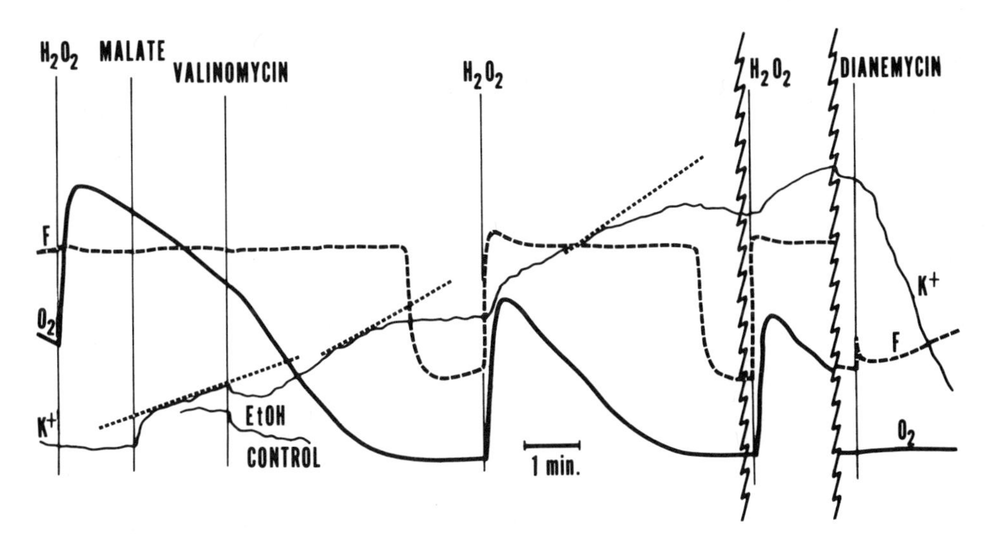

Figure 2. Effect of valinomycin and dianemycin on M. phlei. The tracing F represents NADH fluorescence, upward deflection indicating oxidation to NAD. The O_2 trace represents oxygen content of the medium as detected by a Clark electrode. K^+ represents K^+ of the medium detected by a glass ion selective electrode.

Ionophores have been used intensively as tools for studying the metabolism of energy-transducing systems (mitochondria, chloroplasts, etc.) as well as serving as a focus of intensive physical and chemical investigations as model carriers (3, 4). The biological applications of ionophores were extended considerably by recognition of the ability of certain ones to transport the key biological control ions, catecholamines (X-537A) (6, 7) and Ca^{2+} (X-537A and A-23187) (6-8). This information prompted us to test the effects of X-537A on cardiovascular preparations.

When we perfused isolated guinea pig hearts with X-537A our most hopeful expectations were fulfilled when we observed increases in both heart rate (chronotropy) and isometric tension development (inotropy) (Fig. 3). The amplitude of the electrical signal obtained from surface electrodes (EKG) also increased. As infusion continued we ultimately observed an increase in resting tension (incipient contracture) indicative of overwhelming the relaxation system responsible for removing Ca^{2+} from the contracted myofibrils. Upon washout of the ionophore the preparation returned to control activity. Similar observations have been reported for the isolated rabbit heart (6, 7).

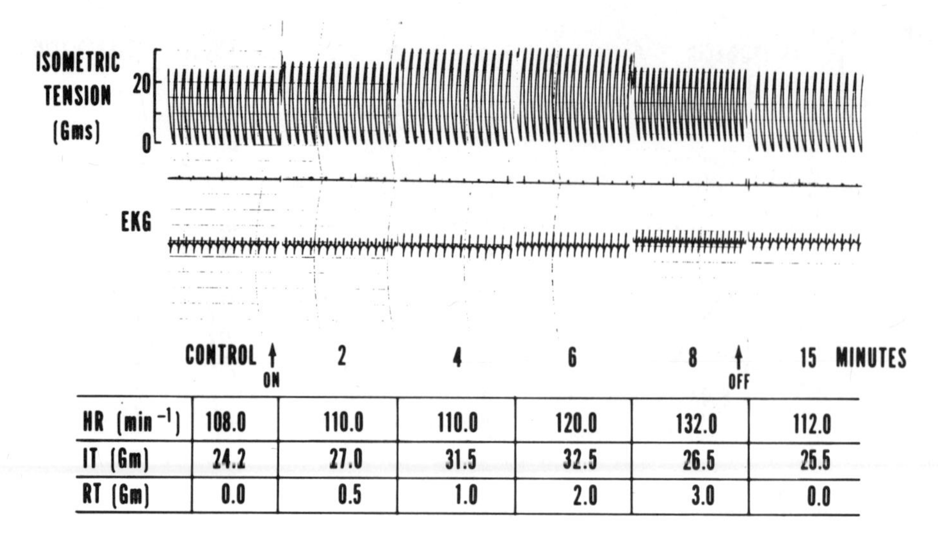

	CONTROL	2	4	6	8	15 MINUTES
HR (min^{-1})	108.0	110.0	110.0	120.0	132.0	112.0
IT (Gm)	24.2	27.0	31.5	32.5	26.5	25.5
RT (Gm)	0.0	0.5	1.0	2.0	3.0	0.0

Figure 3. Effect of X-537A on the spontaneously beating perfused guinea pig heart. At the designation on 10^{-5}M X-537A was added to the perfusing Tyrode's solution. Electrocardiographic tracings (EKG) were obtained from electrodes on the heart surface. HR indicates heart rate, IT, peak isometric tension and RT resting or diastolic tension.

Our guiding premise that direct Ca^{2+} and catecholamine mobilization were the basis of X-537A action nevertheless proved incorrect as we ultimately discovered that all other carboxylic ionophores (except the Ca^{2+} selective A-23187) had even more potent cardiovascular effects than X-537A despite their relative inability to transport Ca^{2+} and catecholamines (9). The relative inotropic potencies of several carboxylic ionophores are given in Table 1. The choice of the most promising ionophore for therapeutic evaluations is based on several factors besides its potency. For our detailed studies of the cardiovascular properties of ionophores on intact, anesthesized, appropriately instrumented dogs (10) we eventually chose to standardize on monensin which has six times the inotropic activity of X-537A but only 0.003 and 0.0001 of the ability of the latter to transport norepinephrine and Ca^{2+} respectively *in vitro*. (4, 9)

When a single dose of monensin (0.05 mg/kg) is injected intravenously into anesthesized mongrel dogs the mean blood flow of the left anterior descending coronary artery rises several fold without any response in aortic pressure or the rate of pressure development in the left ventricle, taken as a measure of cardiac contractility (Fig. 4). Previous studies have confirmed that the response of the latter function parallels the direct measurement

Table 1

Relative Inotropic Activities of Carboxylic Ionophores on the Heart of the Anesthesized Dog

Ionophore	Relative Inotropic Activity
X-537A	1.0
Lysocellin	1.0
Nigericin	3.1
Dianemycin	3.1
Monensin	5.5
Salinomycin	12.3
X-206	12.9
A-204	14.2

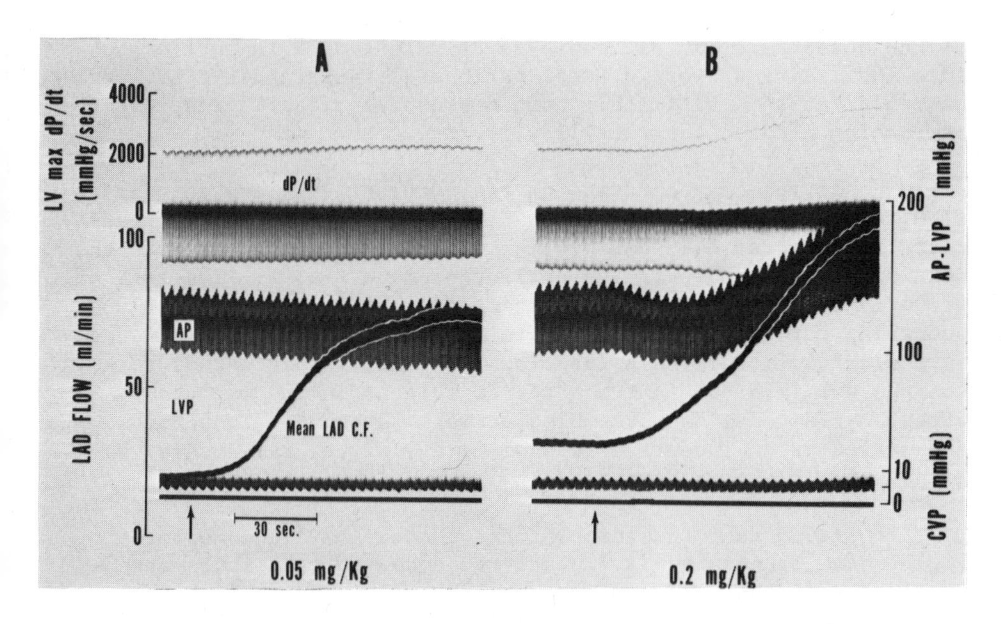

Figure 4. Response of anesthesized dog to monensin. In experiment A 0.05 mg/kg monensin dissolved in alcohol was injected. Mean LAD C.F. indicates mean blood flow in the left anterior descending coronary artery obtained from a magnetic flow probe. AP indicates arterial pressure recorded from a transducer tipped catheter inserted in the aorta. dP/dt is the derivative of the pressure measured from a transducer tipped catheter located in the left ventricle. The upper limit of this function, LV max dP/dt is an index of the contractility of the heart.

of tension development obtained from a strain gauge sutured directly onto the surface of the heart (10). After an interval of an hour to permit recovery a second, larger dose of X-537A was administered (0.2 mg/kg). This dose usually doubled dP/dt but larger doses (0.5 mg/kg), when tolerated, can evoke as much as a five fold increase in dP/dt. The inotropic response would be expected to increase the perfusion pressure on the coronary vessels, and indeed the mean flow through the left anterior descending coronary vessel rises eight fold, however, most of this rise proceeds the increase in dP/dt and therefore indicates a direct effect, i.e., a primary dilatation of the coronary vessels. A corresponding but lesser effect must also occur at the peripheral vasculature because at the very time dP/dt begins its increase arterial pressure (AP) (measured directly in the aorta) actually falls prior to rising in response to the increasing force of cardiac contractility. This effect is also seen with X-537A but is less marked (10). This experiments reveals a sequence of discrete cardiovascular responses to carboxylic ionophores. If the dose is low enough the response may be limited to an increase in coronary blood flow, i.e., a dilation of the coronary arteries. At higher doses this is followed by a transient drop in aortic pressure indicating a drop in total peripheral resistance, i.e., peripheral dilation. Ultimately aortic pressure rises in response to increasing cardiac contractility (dP/dt).

The degree of separation of the contractility and coronary flow responses, each with its own therapeutic potential, appears characteristic of each carboxylic ionophore, being greater in the case of monensin than in the earlier studied X-537A. The task of properly evaluating the multitude of presently known naturally occurring carboxylic ionophores to the degree of meaningful statistical significance is a laborious one, however it serves to sharpen our interest in the patterns evident in the way the streptomyces genus has assembled various carboxylic ionophores and serves as an incentive to explore any structural generalizations which might be discerned.

In Fig. 5 the structures of the two carboxylic ionophores of Fig. 1 have been laid out linearly in order to facilitate structural comparisons. Although rings A-E are virtually identical, monensin has three liganding oxygens in the A ring (filled in black) while nigericin has none. The extra linear chain length of monensin in place of the F ring of nigericin serves to displace the carboxyl out of the complexing configuration. The spirane carbon linked to two oxygen atoms between the D and E rings is a recurrent feature of most of the natural carboxylic ionophores.

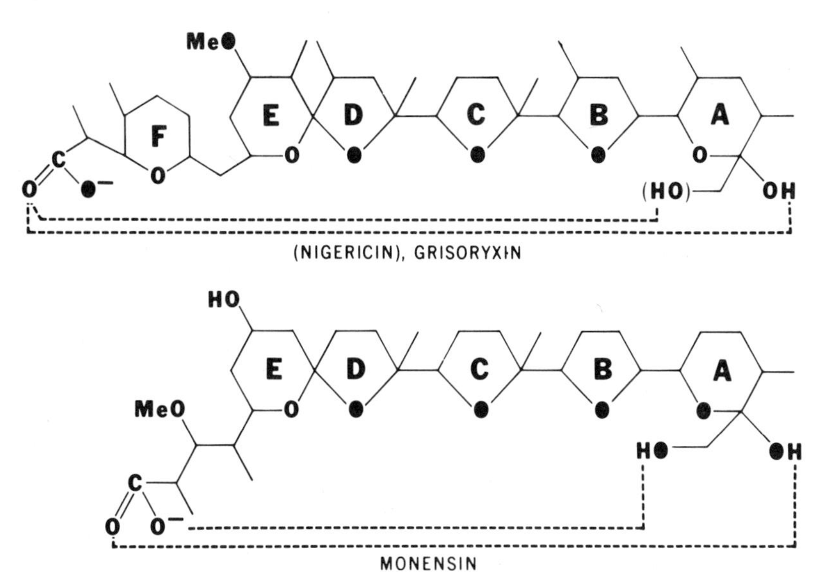

Figure 5. Structures of nigericin and monensin. The structure of grisoryxin is closely related to nigericin except that the hydroxyl in parentheses is absent.

In Fig. 6 three ionophores are depicted having a "S" ring on a side chain affixed to the main backbone via an ether linkage. The complexing configurations of all three ionophores are distinct. Dianemycin utilizes neither the S ring nor its carboxyl for liganding, raising the question of the function of the S ring. Note also the ketonic carboxyl of dianemycin which makes it amenable to conformational study by circular dichroic measurements. The side chain is displaced to the F ring in A-204. It still does not partake in complexation, however both carboxyl oxygens do. The evolutionary origin of the S ring becomes evident in Ro21-6150 where it contributes two liganding oxygens. Note the striking similarity of structure between dianemycin and Ro21-6150, differing in that the S ring joins the C ring of the former and the E ring of the latter. We speculate that the enzyme systems responsible for attaching the S ring evolved to complete the complexation configuration of Ro21-6150 or an analagous ionophore, and once the enzymatic capability for attaching S rings was established, it was retained vestegially in such structures as dianemycin and A-204A. It is also possible that the S rings play some conformational role in the latter two ionophores.

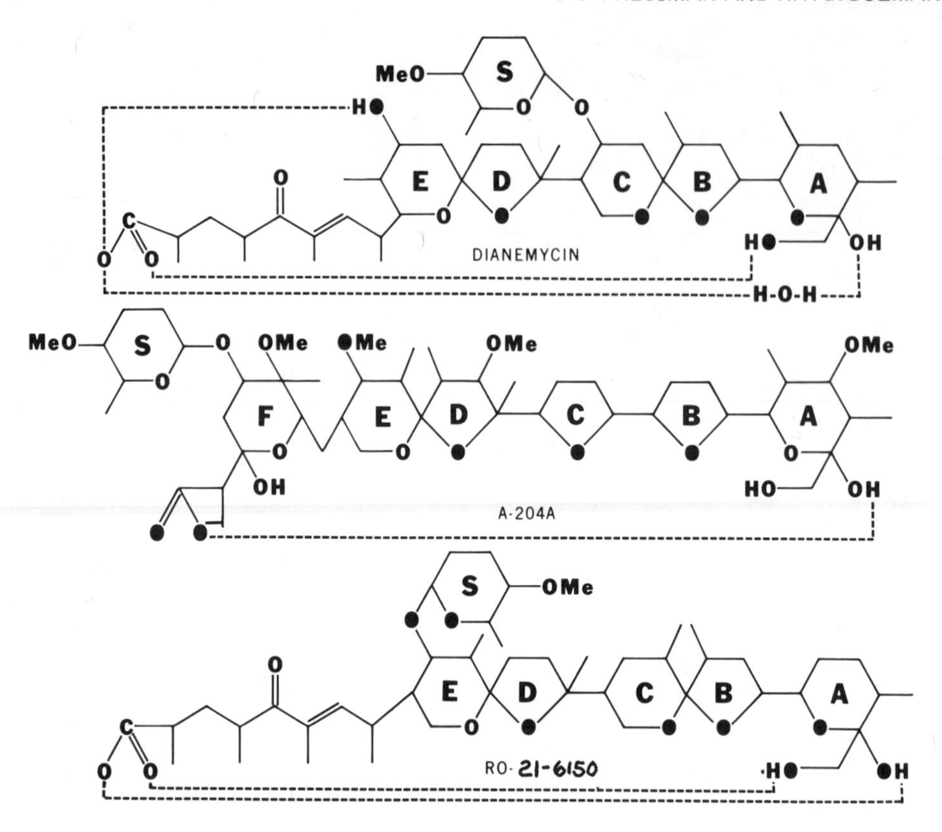

Figure 6. Structure of carboxylic ionophores having side chains or "S" rings.

Figure 7 contrasts the six-ringed X-206, devoid of spirane carbons, with the two adjacent spirane rings of the B, C and D rings of salinomycin. Preliminary tests indicate that salinomycin might be even more interesting pharmacologically than monensin, however supplies available to us are limited at present.

Figure 8 shows structural similarities between lysocellin, the simplest of the natural ionophores, with X-537A. The A and B heterocyclic rings are similar, as are the neighboring (left to right) liganding ketonic and hydroxyl oxygens. Lysocellin has a typical heterocyclic C ring while the C ring of X-537A is aromatic, in fact a substituted salicylate. The aromatic ring is a fluorophore (excitation 318 nm, emission 420 nm) which offers an exceptionally convenient means of measuring complexation spectra. Each cation complexation species has its own characteristic

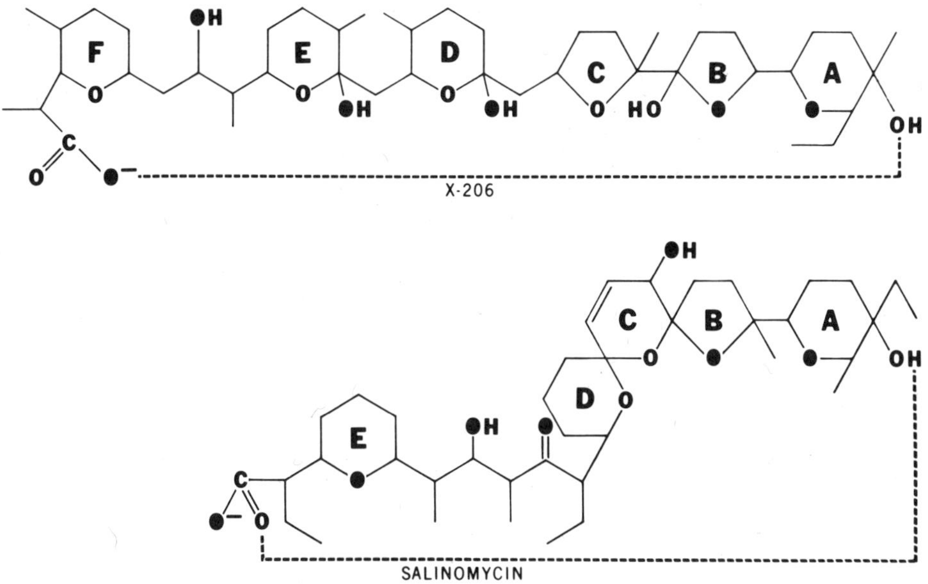

X-206

SALINOMYCIN

Figure 7. Comparison of structures of X-206 having no spirane carbons with salinomycin with two adjacent rings linked by spirane carbons.

fluorescence quantum yield, and so complexation affinities in homogenous solvent systems can be determined by simple fluorometric titration (11, 12). X-537A complexes virtually all cations of all valances including the key biological control agents Ca^{2+} and the biogenic amines; the only cations of low affinity are hindered tertiary and quaternary amines (10). Since X-537A contains several CD bands arising from the aromatic carboxyl and hydroxyl as well as the ketonic carboxyl, it is particularly amenable to conformational changes undergone during complexation by CD spectroscopy (12, 13). As lysocellin does not possess a constituative fluorescent indicator, its complexation spectrum has not been explored as well.

A-23187 shares with X-537A the ability to complex and transport Ca^{2+} but complexes Na^{+} and K^{+} extremely poorly. (8, 14) For many purposes this makes A-23197 an extremely useful tool for selectively transporting Ca^{2+} (and to some extent Mg^{2+}) in biological systems. The polyvalent selectivity may arise from the nitrogen atoms which alone, among the known carboxylic ionophores, appear to be part of the cation liganding system. The nitrogens

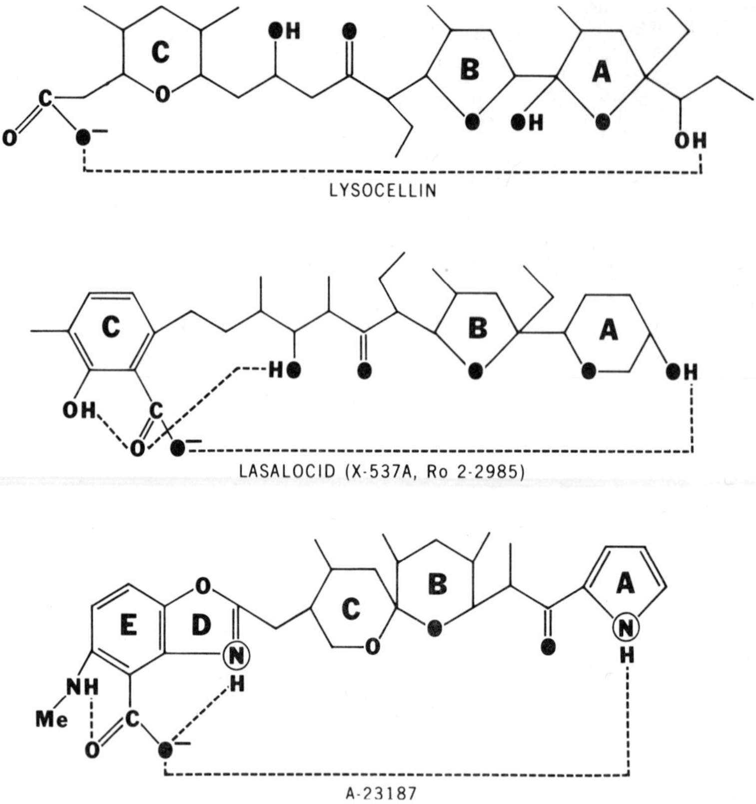

LYSOCELLIN

LASALOCID (X-537A, Ro 2-2985)

A-23187

Figure 8. Three novel carboxylic ionophores. Note that ring C of X-537A and rings A and D-E of A-23187 are both chromophores and fluorophores.

are less electro-negative than oxygens and might be expected to interact more efficiently with the increased charge density of polyvalent cations or via coordination with the d electron orbitals of transitional cations. A-23187 has other features which establish its relationship to the non-introgenous carboxylic ionophores: The B and C rings are joined by a spirane carbon liganded to oxygen atoms and it is cyclized via a head-to-tail hydrogen bond, although in this case the hydrogen donor is an amine group. It is highly fluorescent and amenable to measurement of cation affinities by fluorometric titration (11).

The common features of the carboxylic ionophores offer a

basis for speculation over their origin in various strains of streptomyces. One possibility is suggested by the strategy formulated by Cram in synthesizing a series of synthetic carboxylic ionophores with marked molecular chirality (i.e., asymmetry) which display markedly different affinities for the ennantiomorphs of a given amino acid (15). Cram reasoned that the polyether configuration of the synthetic crown polyether (c.f. Fig. 1) would be a more chemically stable way of deploying liganding oxygens about a cavity of critical size than the more hydrolyzable peptide and ester bonds of valinomycin. Asymmetric recognition centers were added with the binaphthyl group (Fig. 9). In three dimensions the naphthyl groups overlap providing either a right-handed or left-handed twist relative to the plane of the polyether ring. Furthermore additional groups can be attached to the naphthyl rings to permit carboxyls to take positions above and/or below the plane of the polyether ring to produce a three dimensional cage reminiscent of the cage formed by valinomycin and nonactin (Fig. 1). In this cage however, one or two ligands would be ionic, so that the resultant ionophore would be a member of the carboxylic subclass (16).

The speculation we favor about the genesis of the ionophores is that they originate as the prosthetic substrate-binding moieties of membrane-bound carrier systems. Ultimately the ligands confining the ionophore moieties may have been lost (it would be essential for translocases controlling membrane permeability to have an "anchor" moiety, otherwise the ionophore moiety would diffuse into all the lipid pools of the host, as well as neighbor-

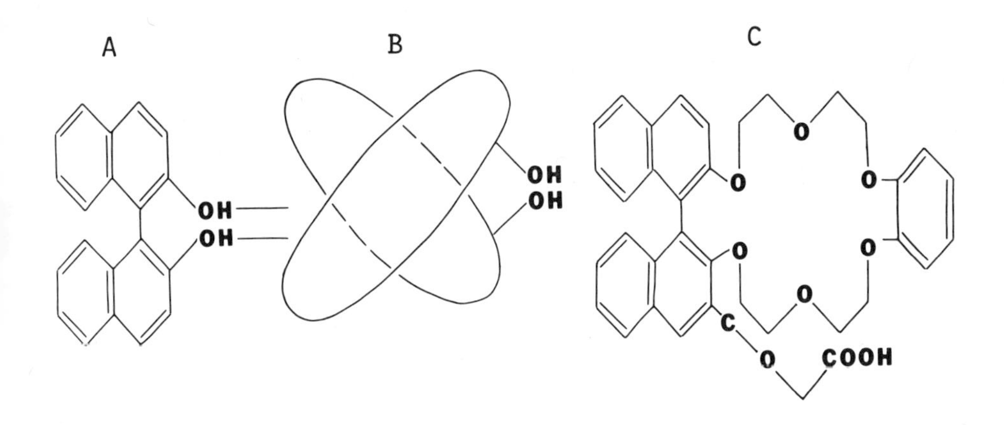

Figure 9. Synthetic polyether carboxylic ionophore.
A. Hydroxylated binaphthyl group. B. Steric equivalent of A.
C. Complete ionophore.

ing organisms, conferring identical permeabilities on all accessible membranes). The naked ionophore moiety may have then propliperated and subsequently become capable of sallying forth from the producing organism to wage bacterial warfare on all competing organisms in its environment not able to tolerate the disruption of transmembrane gradients brought about by the absorption of foreign ionophores.

Because of the stability of genetic coding, higher organisms appear to have lost the knack of incorporating D configurations into proteins c.f. valinomycin and ennistin,(Fig. 1). The D configurations facilitate the compact folding of the ionophore backbone required for optimal focusing of the liganding oxygens, but there is no *a priori* reason that equivalent focusing could not be achieved with all L configurations in proteins at the expense of wider backbone interspacings. The cyclic decapeptide ionophore antanamide from the mushroom Aminita phaloides contains all L amino acids. The protein-like features of valinomycin and enniatin B are self-evident in Fig. 1, however the structures of the carboxylic ionophores diverge markedly from those of the proteins. We would like to speculate that the streptomyces genus anticipated the strategy of Cram by some eons and that carboxylic ionophores represent a natural experiment with a non-proteinaceous system of form-fitting, selective liganding systems. They may have originated as prosthetic groups of translocases and then, much like their less bizarre counterparts, the depsipeptides, once they took on an antibiotic function, various permutations and combinations of the recurrent groupings were selectively programmed for on the basis of the ability of the resultant compounds to disrupt the ion gradients of competing organisms. It is intriguing to note that the Cram group has not only constructed polyether ionophores, but by placing an appropriately nucleophyllic SH group on the binaphthyl moiety, they have even fashioned catalytic ester hydrolases (18). Thus the carboxylic ionophores may be relics of a phylogenetic experiment to supplement proteinaceous enzymes with functionally equivalent polyether structures. The experiment was abandoned in higher organisms presumably because of the advantages of a simpler, more universal system of L polypeptide form-fitting catalysts, but the carboxylic ionophores remain as a relic of an ancient experiment to synthesize functional alternatives to proteinaceous catalysts. This hypothesis is supported by the finding that a mutant of Streptomyces griseus, which has lost the capacity to synthesize ionophores, has impaired ability to accumulate K^+ compared to its ionophore-synthesizing parent strain (19). One wonders whether the development of the polyether ionophores was also accompanied by an equally unorthodox template system.

In summary then the ion transporting ionophores are unique and powerful biological agents which are now finding a multitude of applications. They may have arisen as an early phylogenetic natural experiment. Their wide spectrum of ionic selectivity has provided valuable tools for biological experimentation. Their antibiotic properties have found an economically important application as a feed additive to control coccidiosis in poultry (20). As food additives they alter the flora of the rumen to make rumenents more efficient in converting fodder into meat (21). The tetramethyl homologue of nonactin is an effective mitocide (22).

The current interest of our laboratory in these remarkable compounds, the ionophores, focuses on their unique pharmacological properties. While we find that the neutral ionophores as a group are strongly cardiotoxic, the carboxylic ionophores are well tolerated and have a variety of cardiovascular effects which hopefully will find therapeutic applications. They increase cardiac contractility and coronary blood flow, decrease peripheral resistance, and promote diuresis, effects which may find application for the treatment of congestive heart failure, myocardial hypoxia resulting from infarcts, and shock of various etiologies. In the future new natural and systematically synthesized ionophores, particularly of the carboxylic subclass, may further extend their applications, particularly in the therapeutic area.

Acknowledgement. I wish to thank Dr. Walter Hempfling for collaboration with the M. phlei studies and Dr. N. T. deGuzman for collaboration with the cardiovascular studies. This work was supported in part by grants from NIH (HL-16117), the Florida Heart Association, and American Heart Association (75692)

REFERENCES

1 MOORE, C. and PRESSMAN, B. C.: Biochem. Biophys. Res. Commun. 15 (1964) 562.

2 PRESSMAN, B. C.: Proc. Natl. Acad. Sci. U.S.A. 53 (1965) 1076.

3 OVCHINNIKOV, YU., IVANOV, V. T. and SHKROB, A. M.: Membrane Active Complexones, B.B.A. Library, Vol. 12, Elsevier, New York, (1974).

4 PRESSMAN, B. C.: Am. Rev. Biochem. 45 (1976) 501.

5 PRESSMAN, B. C.: Ion transport induction by valinomycin and related antibiotics, Wirkungsmechanismen von Fungiziden and Antibiotics. Academie-Verlag. Berlin. (1967) p. 3.

6 PRESSMAN, B.C.: Carboxylic ionophores as mobile carriers for divalent ions. The Role of Membranes in Metabolic Regulation (Mehlman, M. A., Hanson, R.W., Eds.) Academic. New York (1972) p. 149.

7 PRESSMAN, B. C.: Fed. Proc. 32 (1973) 1698.

8 REED, P. W. and LARDY, H. A.: Antiobiotic A23187 as a probe for the study of calcium and magnesium function in biological systems, The Role of Membranes in Metabolic Regulation (Mehlman, M. A., Hanson, R. W. Eds.) Academic. New York (1972) p. 111.

9 PRESSMAN, B. C. and deGUZMAN, N. T.: Ann. N. Y. Acad. Sci. 264 (1975) 373.

10 deGUZMAN, N. T. and PRESSMAN, B. C.: Circulation 49 (1974) 1072.

11 CASWELL, A. H. and PRESSMAN, B. C.: Biochem. Biophys. Res. Commun. 49 (1972) 292.

12 DEGANI, H., FRIEDMAN, H.L., NAVON, G. and KOSSOWER, E. M.: J. Chem. Soc. Chem. Commun. 49 (1973) 431.

13 PRESSMAN, B. C. and deGUZMAN, N. T.: Ann. N. Y. Acad. Sci. 227 (1974) 380.

14 PFEIFFER, D. R., REED, P. W. and LARDY, H. A.: Biochemistry 19 (1974) 4007.

15 HELGESON, R. C., KOGA, K., TIMKO, J. M. and CRAM, D. J.: J. Am. Chem. Soc. 95 (1973) 3021.

16 CRAM, D. J.: Synthetic host-guest chemistry, Applications of Biochemical Systems in Organic Chemistry (Jones, J. B., Ed.) Wiley. Interscience New York (1976).

17 WIELAND, T., FAULSTICH, H., BURGERMEISTER, W., OTTING, W., SHEMYAKIN, M. M., OVCHINNIKOV, YU. A., and MALENKOY, G.G.: FEBS Lett. 9 (1970) 89.

18 CHAO, Y. and CRAM, D. J.: J. Am. Chem. Soc. 98 (1976) 1015.

19 KANNE, R. and ZAHNER, H.: Z. Naturforsch. 31 (1976) 115.

20 SHUMARD, R. F. and COLLENDER, M.E.: Antimicrobial Agents and Chemotherapy - 1967 (1968) 369.

21 RAUN, A. P., COOLEY, C. O., POTTER, E. L., RICHARDSON, L. F., RATHMACHER, R. P. and KENNEDY, R. W.: J. Anim. Sci. 39 (1974) 250.

22 SAGAWA, T., HIRANO, S., TAKAHASHI, H., TANAKA, N., OISHI, H., ANDO, K., and TOGASHI, K.: J. Econ. Entomol. 65 (1972) 372.

Penicillin-Binding Proteins of Bacteria

John W. Kozarich, Christine E. Buchanan, Susan J. Curtis, Sven Hammarström and Jack L. Strominger

Harvard University, The Biological Laboratories
16 Divinity Avenue, Cambridge, Massachusetts 02138

ABSTRACT

The penicillin-binding components of bacteria are presumably enzymes of cell wall peptidoglycan synthesis, and one or more of these is a presumed killing site for penicillins. All organisms which have so far been examined contain multiple penicillin-binding components, e.g., there are five in *Bacillus subtilis*, six in *Escherichia coli*, and four in *Staphylococcus aureus*. Progress in studying these components includes: 1) The demonstration that the hydroxylamine-induced release of penicillin G from several penicillin-binding components is enzymatically catalyzed. In addition, the effect of sulfhydryl reagents on the catalytic and penicillin binding activities of *E. coli* carboxypeptidase IA suggests that a thiol group is involved in a deacylation reaction of the enzyme. 2) One of the *S. aureus* binding components can be isolated by affinity chromatography based on the reversibility of its penicillin binding. Under appropriate conditions it can catalyze transpeptidase, carboxypeptidase and penicillinase activities. 3) Altered penicillin-binding components are found in mutants of *B. subtilis* which have been isolated as step-wise penicillin-resistant organisms. 4) The penicilloyl residue of the penicillin-binding components is released at a slow rate in a novel, enzymatically catalyzed degradation. The products of this degradation in the case of the *B. stearothermophilus* D-alanine carboxypeptidase have been identified as phenylacetylglycine and 5,5-dimethylthiazoline carboxylic acid.

REPORT

The extensive research which resulted in the identification of the enzymatic cross-linking of peptidoglycan (transpeptidation), the final step in bacterial cell wall synthesis, as the reaction inhibited by penicillin has been thoroughly reviewed (6). It was postulated (24) that the inhibitory action of penicillin is due to the structural similarity of penicillins to the D-alanyl-D-alanine terminus of the peptidoglycan chain. Recognition of the penicillin by the transpeptidase as a peptidoglycan terminus would ultimately result in the cleavage of the β-lactam ring of the penicillin. The resulting penicilloyl-enzyme complex would be inert to nucleophilic attack by the natural peptidoglycan acceptor or to hydrolysis and, therefore, active transpeptidase could not be regenerated. Cell lysis and death results from the subsequent weakening of the cell wall.

The search for the bacterial transpeptidase was complicated by the discovery of multiple membrane-bound penicillin-binding components (PBCs) for a wide variety of bacteria (5,25). In addition, some of these components are not essential for cell viability. Recently, covalent affinity chromatography techniques have permitted the rapid purification of detergent-solubilized PBCs(4). Isolation of these components by this and other procedures has enabled the study of a number of *in vitro* reactions catalyzed by purified enzymes (4,12,15) Our current work reviewed in the present paper involves the elucidation of four aspects of PBC function. They are 1) the chemical and enzymatic nature of penicillin binding and release; 2) the analysis of PBCs for *in vitro* activities related to possible *in vivo* functions (e.g., carboxypeptidase, transpeptidase, and penicillinase activities); 3) the determination of potential penicillin killing sites by the study of penicillin-resistant mutants; and 4) penicillin degradation by PBCs. Ultimately, these studies may help to clarify the relationships between the action of penicillins on and the reactions of bacterial cell wall synthesis by these proteins.

1. The Nature of Penicillin Binding and Release

The ability of neutral hydroxylamine to release covalently bound penicillin from PBCs (17,18) under conditions where normal amide and ester linkages were not cleaved suggested that the penicilloyl-protein complex involved a thiolester linkage. Although this was based on purely chemical considerations, later work suggested that this release may involve enzymatic activation of the complex (8).

We have found that the hydroxylamine-induced release of penicillin G from the PBCs of *B. subtilis* and *S. aureus* is an enzymatically catalyzed reaction (16). The penicillin G-protein com-

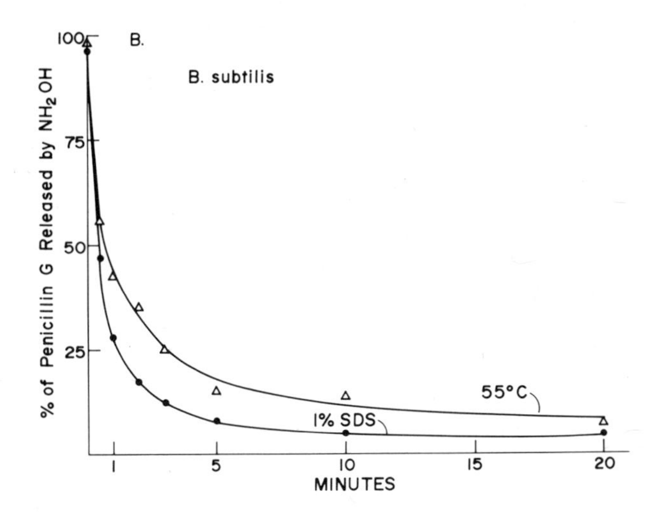

Figure 1. The rate of inhibition of neutral hydroxylamine release of ^{14}C-penicillin G bound to B. subtilis carboxypeptidase by denaturation. Purified penG-carboxypeptidase was pretreated by incubation at 55°C (Δ-Δ) or with 1% SDS at 25°C (●-●) for various time intervals. The sample was then treated with 0.7 M neutral hydroxylamine for 1 hr. Bound penG was measured by paper chromatography (16).

plexes were isolated by treatment of the PBCs with ^{14}C-penicillin G followed by gel filtration. Treatment of the complex with neutral hydroxylamine resulted in the quantitative release of penicillin G from the protein as the penicilloyl hydroxamate. Penicillin G was not released from purified B. subtilis carboxypeptidase by neutral hydroxylamine treatment after pretreatment of the penicilloyl-enzyme complex at 55°C or with 1% SDS at 25°C (Fig. 1). The decrease in release as a function of denaturation time strongly suggests that the reactivity of the penicilloyl-enzyme bound to hydroxylamine is dependent upon the native conformation of the complex. Similar results have been obtained with S. aureus PBCs and evidence suggests similar phenomena in the PBCs of other organisms. In addition, the mild conditions of denaturation makes the transfer of the penicilloyl group highly unlikely.

On the basis of these results, the binding of penicillin via a thiol group appears improbable for these PBCs. However, the enzymatic requirements for the release is reminiscent of the hydroxylamine-induced release of acetyl as the hydroxamic acid from acetylchymotrypsin (11). Upon denaturation the acetyl group was no longer reactive to nucleophilic attack by the hydroxylamine. Later work established the attachment of the acetyl group *via* a serine residue (1). This possibility is currently under investigation.

The use of thiol inhibitors, however, has provided valuable information about the penicillin binding site of carboxypeptidase IA of Escherichia coli. There are six major penicillin-binding components in E. coli (20,21). Although the specific biological function of components 1 through 4 is not known, recently evidence has been obtained which indicates that components 5 and 6 are the subunits of the enzyme, carboxypeptidase IA (21). Carboxypeptidase IA of E. coli was first purified by Tamura, Imae, and Strominger (23). This enzyme is membrane-bound and its activity is severely inhibited by penicillin. It has also been found that carboxypeptidase IA binds penicillin covalently and reversibly (23).

Initially, Tamura et al. investigated the ability of p-chloromercuribenzoate (pCMB) to inhibit the penicillin binding and the carboxypeptidase activity of IA. This reagent reacts with thiol groups specifically. It was found that pCMB severely inhibited the carboxypeptidase activity of IA but did not inhibit its penicillin G binding activity. An interpretation of this result is that a thiol group is important to the carboxypeptidase activity of the enzyme, but this thiol groups does not bind penicillin G. Tamura et al. suggested that this thiol group was not involved in the formation of the acyl-enzyme intermediate postulated to be produced during the carboxypeptidase reaction, but did function in the deacylation reaction of the enzyme.

TABLE I--Effect of pCMB, NEM, and DTNB on Penicillin G Binding and Carboxypeptidase Activity of Enzyme IA*

Inhibitor	Penicillin G Binding		Carboxypeptidase Activity	
	cpm	% Inhibition	cpm	% Inhibition
none	3463	0	10390	0
0.5mM pCMB	3483	0	-	-
0.4mM pCMB	-	-	2351	77
12.5mM NEM	3602	0	7400	29
10 mM DTNB	3176	9	1725	83

*Penicillin G binding assays and carboxypeptidase assays were performed as described by Tamura et al.(23). An assay was started by adding enzyme (5 μg of protein) to an assay mixture containing all other components of the assay including the inhibitor. ^{14}C-penicillin G was present in the assays at a concentration of 18.5 μM and UDPMAGP-^{14}C-ala-^{14}C-ala, at a concentration of 2.4 μM.

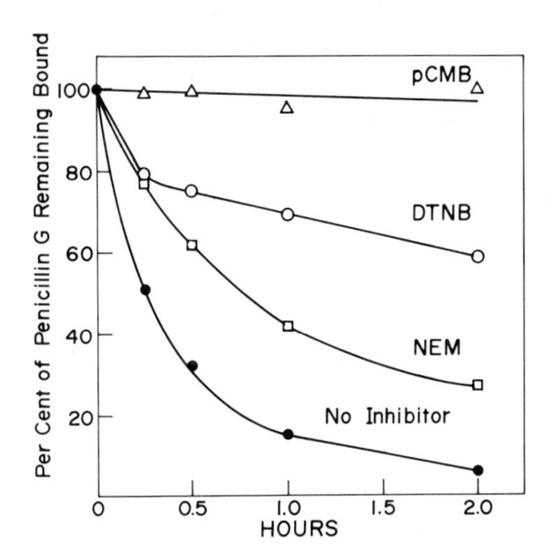

Figure 2. Inhibition of the release of bound penicillin G from carboxypeptidase IA. ^{14}C-Penicillin G was allowed to bind to carboxypeptidase IA for 5 minutes at 25°C in the presence of pCMB (Δ), DTNB (O), NEM (□), or no inhibitor (●). Then unbound penicillin G was degraded and the release of bound penicillin at 25°C measured as described by Tamura _et al_. (23).

The effects of reagents which react with thiol groups on carboxypeptidase IA have been more extensively investigated. Neither pCMB, nor 5,5'-dithiobis-(2-nitrobenzoic acid) (DTNB), nor N-ethylmaleimide (NEM) inhibited the release of penicillin G greater than 95%. DTNB and NEM also inhibited this release approximately 40% and 30% respectively. Thus all three reagents did not inhibit binding of penicillin G to carboxypeptidase IA, but significantly inhibited both carboxypeptidase activity and the release of bound penicillin G. The data presented suggest that there are at least two important chemical groups important in the catalytic site of carboxypeptidase IA. One of these groups, a thiol group, is important to the release of bound pencillin from the enzyme, and thus may be involved in the deacylation reaction of the carboxypeptidase. The other chemical group is important for the binding of penicillin to the enzyme and may also be involved in the acylation reaction which produces an acyl-enzyme intermediate.

To further elucidate the properties of the enzyme site of carboxypeptidase IA which binds penicillin, the inhibition of penicillin binding by the carboxypeptidase substrates UDPMAGDAA and DALAA was investigated. The data in Table II indicate that both these substrates can severely inhibit penicillin G binding. Since pCMB inhibits carboxypeptidase IA activity when UDPMAGDAA or

TABLE II.--Inhibition of Penicillin G Binding by Carboxypeptidase IA Substrates*

Additions	Penicillin G Bound (cpm)	% Inhibition
None	2509	0
1 mM pCMB	2720	0
0.51mM UDPMAGDAA	1689	33
0.51mM UDPMAGDAA + 1mM pCMB	1406	44
1.02mM UDPMAGDAA	559	78
1.02mM UDPMAGDAA + 1mM pCMB	836	67
None	2525	0
1 mM pCMB	2243	11
4.7mM DALAA	1433	43
4.7mM DALAA + 1mM pCMB	1557	38
18.8mM DALAA	613	76
18.8mM DALAA + 1mM pCMB	932	63

*The penicillin G binding assay was started by adding carboxypeptidase IA (5μg protein) to an assay mixture containing all other assay components including, where indicated, pCMB, UDPMAGDAA, or DALAA. ^{14}C-penicillin G was allowed to bind to the enzyme for 30 sec. Details of the assay have been previously described by Tamura et al. (23).

diacetyl-L-lys-D-ala-D-ala (DALAA) is the enzyme substrate (23), the ability of pCMB to prevent inhibition of penicillin binding by these substrates was investigated. When pCMB was present in the penicillin binding assay, it did not significantly reduce the inhibition of binding by the carboxypeptidase substrates (Table II). This result suggests that when the thiol group involved in the carboxypeptidase deacylation reaction is blocked by pCMB, the carboxypeptidase substrates can still compete with penicillin G for binding to the chemical group involved in the acylation reaction. Thus pCMB can not prevent inhibition of penicillin G binding by DUPMAGDAA and DALAA.

The data presented suggest that the acylation and deacylation reactions involved in carboxypeptidase activity are not catalyzed by the same chemical group in the active site of the enzyme. Penicillin may inhibit carboxypeptidase activity be reacting with the group(s) involved in the acylation reaction. Isolation and amino acid analysis of peptide fragments of carboxypeptidase IA to which either penicillin or a carboxypeptidase substrate is bound would help to determine if this hypothesis is correct. These experiments are currently in progress.

2. Isolation of a Carboxypeptidase-Transpeptidase-Penicillinase from S. aureus.

Studies of in vitro transpeptidation by exocellular and membrane-bound transpeptidases have been reviewed (12, 19). We have studied the PBCs of S. aureus strain H purified by affinity chromatography for transpeptidase activity (15). S. aureus H was chosen since it contains no major carboxypeptidase or penicillinase activities. Moreover, initial studies of penicillin G binding to a particulate membrane fraction of S. aureus suggested the presence of only two PBCs, thus, simplifying the possibility of establishing the killing site of penicillin for this organism (5). SDS polyacrylamide gel electrophoresis of affinity chromatography elutions revealed four major PBCs (Fig. 3). The 115,000 MW protein and the two 100,000 MW proteins bind penicillin "irreversibly" ($t_{1/2}$ for penicillin release at 37° is 3 hours). The 46,000 MW protein did not bind detectable amounts of penicillin G by the usual binding procedures (5). Analysis of the three elution fractions for transpeptidase activity using diacetyl-L-lysyl-D-alanyl-D-alanine (DALAA) as the donor and glycine as the acceptor revealed transpeptidation (formation of diacetyl-L-lysyl-D-alanyl-glycine (DALAG)) to be proportional to the amount of the 46,000 MW protein present. Based on its reversible binding to penicillin, this protein could be purified by prebinding membranes with penicillin G prior to the affinity chromatography. Those components which "irreversibly"

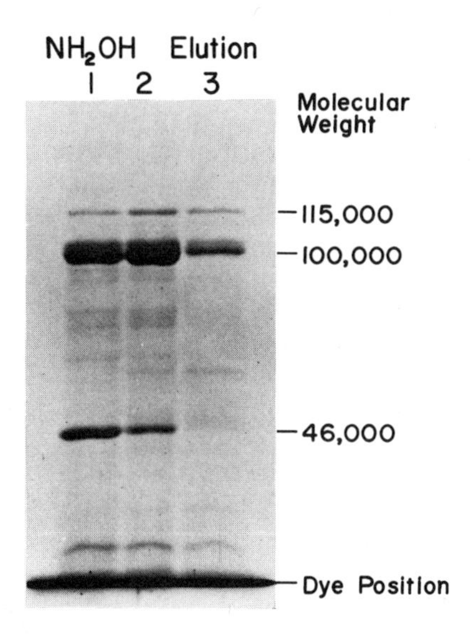

Figure 3. SDS polyacrylamide gel electrophoresis of the penicillin binding components of S. aureus H released from the affinity column by three successive washes with hydroxylamine (4,15).

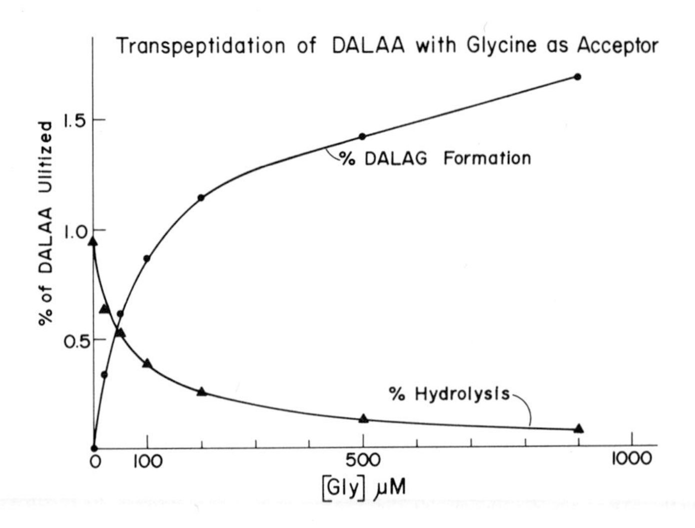

Figure 4. Transpeptidation of DALAA to DALAG (●-●) and inhibition of hydrolysis to DALA (▲-▲) by glycine. Di-^{14}C-acetyl-L-lys-D-ala-D-ala (1mM) was incubated at 37°C with the purified transpeptidase (5μg) in the presence of increasing concentrations of glycine for 2 hr. Products were isolated by paper electrophoresis(15).

bound the penicillin would be excluded from the column enabling selective purification of the 46,000 MW protein. The purified protein retained full transpeptidase activity. The Km for DALAA was 100 mM and the Vmax was 400 nm/min/mg protein. In the absence of acceptor, the protein catalyzed an hydrolysis of DALAA to diacetyl-L-lysyl-D-alanine (DALA) occurring at 50% the transpeptidation rate.

At a concentration of 1mM DALAA, 40μM glycine resulted in equal amounts of DALAG and DALA formed (Fig. 4). When hydroxylamine was used as the acceptor, 7mM of this acceptor was required in order to obtain equal amounts of DALA and DALA hydroxamate. The ability of glycine, the weaker nucleophile, to more efficiently compete with hydrolysis than hydroxylamine implies that the transpeptidation reaction is not merely a chemical partitioning of a carboxypeptidase intermediate between water and acceptor. For 50% inhibition of DALAA hydrolysis, 40μM D-alanine was required while >10mM L-alanine produced the same effect. The stereochemical requirement of the acceptor is apparent from this experiment.

Inhibition of transpeptidation by penicillin G occurred only at high concentrations (>25μg/ml) of the antibiotic. Incubation of the protein with penicillin G resulted in the formation of penicilloic acid, indicating a penicillinase activity. Kinetic studies of this activity revealed a Km of 20μM and a Vmax of 500 nm/min/mg protein. The similarity of the Vmax for the penicillinase activities is suggestive of a similar rate determining step in each reaction.

The coincidence of all three activities on a single enzyme is interesting from an evolutionary standpoint. If penicillin, as an analog of D-alanyl-D-alanine, results in an irreversible penicilloylation of the enzyme, it is conceivable that certain mutations might facilitate the hydrolysis of the penicilloyl-enzyme bond. Since a D-alanyl-enzyme bond presumably formed by the initial reaction of DALAA with the transpeptidase might be expected to have the same lability as the corresponding penicilloyl-enzyme bond, hydrolysis of this intermediate could also be facilitated by the enzyme. Therefore, all three activities could occur at the same active site.

Along these lines, isolation of a penicilloyl-enzyme intermediate and a D-alanyl-enzyme intermediate would provide direct evidence of the identity of the active site(s) of these reactions. Attempts to isolate a penicilloyl-enzyme complex were successful if the complex was rapidly denatured by acetone precipitation. Other methods allow sufficient time for complete hydrolysis of the intermediate to yield free penicilloic acid. The stability of the denatured complex is similar to those described above. Current work is being directed toward the isolation of a substrate-enzyme complex to compare the active sites of both intermediates.

3. Analysis of Penicillin Resistant Mutants

Bacillus subtilis membranes contain five penicillin binding components (5). The major component, PBC V, is the D-alanine carboxypeptidase, which has been shown not to be vital for the cell (24). No enzymatic activity has been identified yet for any of the other PBCs. Unlike PBC V, three of the proteins, PBCs I,II, and IV, bind penicillins much as would be predicted for the killing site (5). They therefore are candidates for the penicillin killing site, and indeed, one of them might be the transpeptidase. The question is, then, which (if any) of the different PBCs corresponds to the lethal target of penicillin? Our approach to solving this problem was to determine which components became more resistant to penicillin upon mutation of the organism to greater penicillin resistance (9).

In order to isolate highly penicillin resistant mutants it was necessary to do a stepwise selection (Fig. 5) since a single step penicillin resistant mutant shows only a small increase in penicillin resistance (10). Cloxacillin was chosen as the selective penicillin since it is not susceptible to the β-lactamase of *Bacillus* species. The cloxacillin resistant mutants showed a

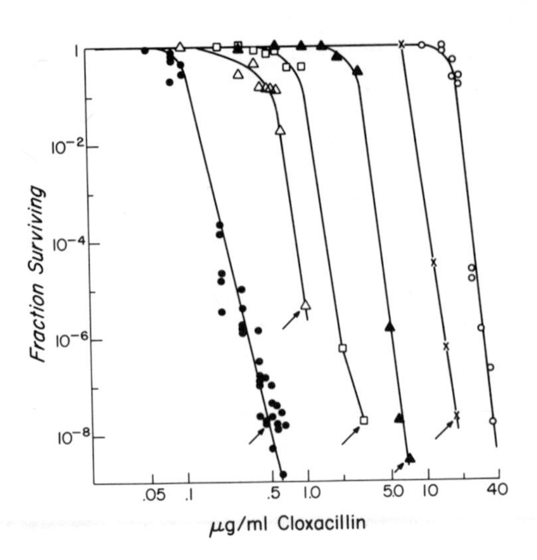

Figure 5. Survival curves of B. subtilis and the five mutants. The curves (from left to right) correspond to survival of the wild type, mutants 1, 2, 3, 4, and 5. The arrows indicate where a survivor was picked to be used for isolation of the next step mutant.

parallel increase in resistance to oxacillin and dicloxacillin, but no substantial change in their sensitivities to 6-aminopenicillanic acid (6-APA) or penicillin G.

The PBCs of B. subtilis strain Porton wild type and the five mutants were isolated by covalent affinity chromatography essentially as described (7). Two interesting observations were evident in stained SDS polyacrylamide slab gels (Fig. 6). In mutant 5, PBC I was not present. Either it did not bind to the 6-APA of the affinity column, perhaps because it was insensitive to penicillin, or else the protein was absent in the mutant. This protein could also not be detected in the membrane when ^{14}C-penicillin G was used as a label. When wild type PBC I is purified, antibody to it can be prepared and used to search for cross-reacting antigenic material in the mutant. In any case, it is clear that the wild type protein is not essential for normal growth, since mutant 5 has normal morphology and grows, sporulates, and germinates normally. In addition, it is likely that PBC I is not a penicillin killing site since mutant 5 is still quite sensitive to 6-APA and penicillin G which have been shown not to bind to PBC I.

Another PBC alteration can also be seen on the stained gel (Fig. 6). An examination of PBC II in the series of mutants suggests there were several discrete changes in the mobility of PBC II, one evident in mutants 4 and 5 as compared to mutant 3. Another slight alteration in PBC II may have occurred in mutant 1. Such small changes in the electrophoretic mobility could be the result of a single amino acid change affecting the net charge of the protein. Indeed, preliminary results with isoelectric focusing of the PBCs do indicate that the isoelectric points of the mutant 5 and wild type PBC II are different.

The ability of the individual PBCs to bind cloxacillin was measured indirectly by competition for subsequent binding of labeled penicillin G. The results (Table III) show that PBC I in mutants 1-4 and PBC IV in mutants 1-5 retained essentially the same sensitivity to cloxacillin as the wild type. PBC II, on the other hand, appeared to undergo three discrete changes in its resistance to cloxacillin, i.e., at mutants 1, 4 and 5. These changes were roughly similar to the changes in the resistance levels of the organisms (Fig. 5). As described above, the change was accompanied by a change in electrophoretic mobility in mutant 4, and possibly in mutant 1. However, direct proof of the relationship between the altered PBC II and increased penicillin resistance would require further genetic studies.

There was no apparent change in PBC II or any other PBC in mutants 2 and 3. Some other change must have occurred in these

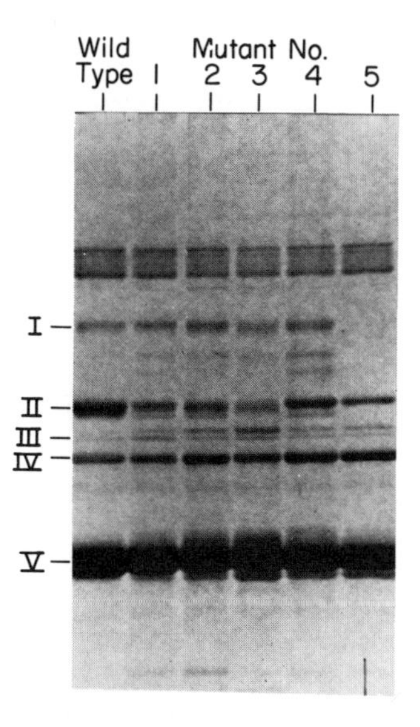

Figure 6. SDS polyacrylamide slab gel of PBCs isolated by affinity chromatography from B. subtilis and five cloxacillin resistant mutants.

TABLE III. Cloxacillin resistance of the PBCs.*

	Wild Type	Mutants				
		1	2	3	4	5
PBC I	0.54	1.08	1.55	1.04	0.66	----
PBC II	1.75	3.8	2.65	2.18	11.4	72
PBC IV	0.8	0.92	1.62	1.18	0.76	1.43

*Resistance is defined as the amount of cloxacillin (μg/ml x minutes) required to inhibit ^{14}C-penicillin G binding by 50%.

mutants, possibly a membrane change reflected in altered permeability rather than in an altered PBC. Increased production of a penicillinase was ruled out.

The affinity of the mutant PBCs for penicillin G was also determined by the competition binding assay. Interestingly, there as no significant change in the sensitivity of PBC II (or any other PBC) to penicillin G despite the obvious alteration of the protein's electrophoretic mobility. This corresponds with the fact that the mutants themselves are not more resistant to penicillin G. This indicates that the active sites of penicillin sensitive proteins may be modified in their sensitivity to one β-lactam antibiotic without any alteration in sensitivity to others. This fact may have implications for therapy with β-lactam antibiotics and it is paralleled by wide differences in the sensitivity of PBCs in wild type organisms to different β-lactam antibiotics. A striking example of this is the extreme sensitivity of PBC II in E. coli to formamidino penicillin as compared to all other penicillins which have been examined (20).

PBCs III and V have previously been eliminated as candidates for the penicillin killing site (3,5). Here it has been shown that PBC I is probably not the killing site either, since the protein, if it exists at all, has no affinity for certain penicillins which are still lethal for the organism. PBC IV also cannot be the lethal target of cloxacillin, at least, since it did not change in its affinity for cloxacillin in any of the five cloxacillin resistant mutants. Thus, all the evidence points to PBC II as the likely target for killing of B. subtilis by β-lactam antibiotics.

4. Spontaneous Release of Penicillin Degradation Products by Penicillin-Binding Components.

In a number of cases the bound penicilloyl moiety of the penicillin binding components is released relatively rapidly, e.g.,

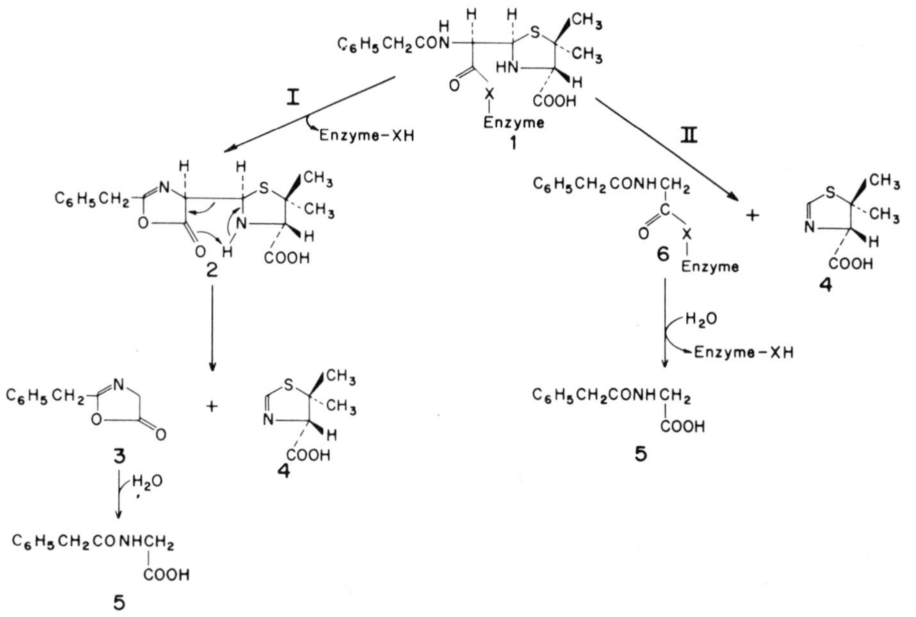

Figure 7. Possible mechanisms for the formation of phenylacetylglycine, 5, and 5,5-dimethylthiazoline carboxylic acid, 4, from benzylpenicilloyl-D-alanine carboxypeptidase, 1.

in the case of the *E. coli* carboxypeptidase IA and the 46,000 MW binding component of *S. aureus* discussed above. In both of these cases the product of release is penicilloic acid, i.e., these binding components have a penicillinase activity. In other cases the bound penicilloyl moieties released at slower rates. For example, PBC 5 of *B. subtilis* and the high MW PBCs of *S. aureus* release the penicillin degradation products with a half-time at 37° of about 3 hr, while the D-alanine carboxypeptidase of *Bacillus stearothermophilus* releases them with a half-time of 10 min at 55°, its temperature optimum. It was, in fact, the relatively rapid release from the thermophilic D-alanine carboxypeptidase at 55° that initially led to the detection of these compounds. The products of release from the *B. stearothermophilus* enzyme have been identified as phenylacetylglycine, 5, and 5,5-dimethylthiazoline carboxylic acid, 4 (13,14) and the products released from the *B. subtilis* enzyme appear to be identical. This degradation almost certainly proceeds through the mechanism outlined in Fig. 7 which is similar to the mechanism of degradation in anhydrous trifluoroacetic acid (2). This degradation would be promoted by protonation of the thiazoline ring nitrogen either in the chemical degradation or in the enzymatic degradation, and it may provide clues to the mechanism of the late steps in the enzymatically catalyzed reaction. The release of dimethylthiazoline carboxylate is analogous to the release of D-alanine from substrate and it could be facilitated by protonation of the amino group of the D-alanine residue. A similar degradation has been observed for a D-alanine carboxypeptidase from *Streptomyces* but a different mechanism has been proposed (26).

Thus considerable progress has been made in the understanding of the penicillin-binding components of several microorganisms and of the reactions which they may catalyze. However, the complexity of the problem is illustrated by the fact that all bacteria which have been studied so far contain multiple penicillin-binding components and that the functions of these in normal bacterial growth, presumably in some aspect of cell wall biosynthesis, remains unknown. A great deal more work will be required in order to obtain a complete understanding of the mechanism by which penicillin kills bacteria.

This work was supported by research grants from the National Institutes of Health (AI-09152) and National Science Foundation (BMS71-01120).

REFERENCES

1 Anderson, B.M., Cordes, E.H., and Jencks, W.P.: J. Biol. Chem. 236 (1961) 455.

2 Bell, M.R., Carlson, J.A. and Oesterlin, R.: J. Org. Chem. 37 (1972) 2733.

3 Blumberg, P.M. and Strominger, J.L.: Proc. Nat. Acad. Sci. USA 68 (1971) 2814.

4 Blumberg, P.M. and Strominger, J.L.: Proc. Nat. Acad. Sci. USA 69 (1972) 3751.

5 Blumberg, P.M. and Strominger, J.L.: J. Biol. Chem. 247 (1972) 8107.

6 Blumberg, P.M. and Strominger, J.L.: Bacteriol. Rev. 38 (1974) 291.

7 Blumberg, P.M. and Strominger, J.L.: Methods Enzymol. 34 (1974) 401.

8 Blumberg, P.M., Yocum, R.R., Willoughby, E., and Strominger, J.L.: J. Biol. Chem. 249 (1974) 6828.

9 Buchanan, C.E. and Strominger, J.L.: Proc. Nat. Acad. Sci. USA in press.

10 Demerec, M.: J. Bacteriol. 56 (1948) 63.

11 Dixon, G.H., Dreyer, W.J., and Neurath, H.: J. Amer. Chem. Soc. 78 (1956) 4810.

12 Ghuysen, J.M. et al: Bulletin De L'Institut Pasteur 73 (1975) 101.

13 Hammarstrom, S. and Strominger, J.L.: Proc. Nat. Acad. Sci. USA 72 (1975) 3463.

14 Hammarstrom, S. and Strominger, J.L.: submitted for publication.

15 Kozarich, J.W. and Strominger, J.L.: submitted for publication.

16 Kozarich, J.W., Willoughby, E., and Strominger, J.L.: submitted for publication.

17 Lawrence, P.J. and Strominger, J.L.: J. Biol. Chem. 245 (1970) 3653.

18 Lawrence, P.J. and Strominger, J.L.: J. Biol. Chem. 245 (1970) 3660.

19 Mirelman, D. and Sharon, N.: Biochem. Biophys. Res. Commun. 46 (1972) 1909.

20 Spratt, B.G. and Pardee, A.B.: Nature 254 (1975) 516.

21 Spratt, B.G. and Strominger, J.L.: submitted for publication.

22 Strominger, J.L., Willoughby, E., Kamiryo, T., Blumberg, P.M. and Yocum, R.R.: Ann. N.Y. Acad. Sci. 235 (1974) 210.

23 Tamura, T., Imae, Y., and Strominger, J.L. J. Biol. Chem. 251 (1976) 414.

24 Tipper, D.J. and Strominger, J.L.: Proc. Nat. Acad. Sci USA 54 (1965) 1133.

25. Yocum, R.R., Blumberg, P.M., and Strominger, J.L.: J. Biol. Chem. 249 (1974) 4863.

26. Frere, J-M., et al. Nature 260 (1976) 451-454.

DISCUSSION

PASSOW: You showed two bands and demonstrated that the fluorescence could be diluted out of one of the bands; I would very much like to understand the technique.

KOZARICH: The method is fluorography (W.M. Bonner and R.A. Laskey, Enr. J. Biochem. 46: 83, 1974) which involves counting of gels; essentially, after running the SDS polyacrylamide gel you take the gel, treat it with the solution of DMSO and PPO to impregnate the gel with PPO, and then quench the impregnation by adding water. One obtains a gel which is completely impregnated with PPO. You dry this gel down and then expose the gel to film. We usually keep these gels at minus 70°, and it is sensitive down to 100 DPM (ibid.).

PASSOW: You said that you obtained your results by some dilution technique. Could you explain the technique?

KOZARICH: Yes. We wanted to establish that if the component was actually behaving as a penicillinase activity. We know that, at least for some components, that the penicilloyl-enzyme bond is chemically rather stable since if you denature it is no longer reactive to hydroxylamine. We suspected that, perhaps, for a large majority of enzymes, even those that are highly reactive such as penicillinases, that this reactivity isn't due *per se* to the increased chemical reactivity of the penicilloyl-enzyme bond but it is merely an enzymatic activation. So, we thought that if we could rapidly denature the proteins by acetone precipitation then we could isolate a chemically stable penicilloyl-enzyme complex even for penicillinases. The experiment specifically was taking labeled penicillin, adding it to a mixture of the components, and then adding acetone to precipitate. The dark bond represented the low molecular weight protein. In the second slot, hot penicillin was added but then, prior to acetone precipitation, cold penicillin was added to dilute in which case only a dilution in the smaller molecular weight component is obtained. The other two components which bind penicillin very tightly, in essence irreversibly, were not diluted.

Toxic Chemicals as Molecular Probes of Membrane Structure and Function

TOXIC CHEMICAL AGENTS AS PROBES FOR PERMEATION SYSTEMS OF THE RED BLOOD CELL

A. Rothstein and P.A. Knauf

The Research Institute

The Hospital for Sick Children, Toronto, Ontario, Canada

ABSTRACT

Chemical agents with different capacities to penetrate into the membrane and with different chemical reactivities can be used to gain information concerning the location of transport sites in the membrane structure and the particular functional ligands. If the agents are highly specific in their interactions and if their inhibitory effects are irreversible, they can also be used as probes to identify the transport components. Several examples are cited using the human red blood cell as a model. The anion transport system in particular has been studied by the use of non-penetrating irreversible inhibitors, and more recently with a photoaffinity probe, NAP-taurine. In the dark the latter is transported in competition with the normal inorganic anions but after exposure to light, it becomes fixed in an irreversible bond that allows identification of the sites of its transport. It is proposed that anion transport involves a transmembrane protein of about 90,000 daltons that forms a channel through the lipid bilayer. The exchange of anions occurs via a gating mechanism containing a specific anion-binding site. Transport of water, cations and sugars may also involve similar transmembrane protein channels.

Introduction

Much of our insight into the molecular mechanisms that underlie biological phenomena has been achieved by the use of toxic agents capable of perturbing functional macromolecules in chemically defined ways. The cell membrane is particularly accessible for investigation by such agents because of its direct contact with the external environment. In the case of non-penetrating agents, for example, interactions are entirely restricted to externally exposed membrane ligands. Other agents may penetrate the membrane by particular pathways determined by their chemical and physical properties, thereby reaching sub-populations of reactive sites within the membrane or within the cell.

In order to be useful as a membrane probe, a chemical agent must have two characteristic features. It must be able to reach the functional ligands in question wherever they may be located in the membrane structure, and it must have the capacity to react with those ligands after access has been achieved. Thus the agent-function relationship is dependent on a diffusibility factor and on a chemical reactivity factor. Experimental assessment of these factors for a variety of agents allows several kinds of information to be obtained concerning the transport systems, including: (a) the specific ligands that may be functionally involved, (b) their location in the membrane, (c) the molecular nature of the transport system, (d) the identity of the membrane components that are involved, and (e) the arrangement of the components in the membrane.

Chemical probes can be most productively used in membrane systems in which the components, particularly proteins, and the functions are well described. One of the best systems for such studies is the red blood cell. It has the additional advantage that it contains no internal membranes to complicate interpretations. Furthermore, its membrane can be readily isolated in a functional state in the form of resealed ghosts or small vesicles into which probes can be incorporated. In this presentation, the red cell membrane will, therefore, be used as a model to illustrate a number of ways in which toxic chemical agents can be used to gain insight into the molecular nature of transport and permeation mechanisms.

Specificity of agent-membrane interactions

If an agent is to be an effective probe for a particular membrane transport system, it should inhibit that system, but not others; that is, it should be <u>function specific</u>. If, furthermore, the agent is to be used to designate the membrane components associated with the function, it should react only with those components; that is, it should be component specific.

Specificity of agent-membrane interactions is determined by a number of factors. The most obvious one is chemical reactivity. For example, mercurials such as PCMBS, because of their exceptionally high affinity for sulfhydryl groups (40), will in low concentrations be highly specific for those ligands. If a function is perturbed, then some role for sulfhydryl groups in that particular function can be assumed. This form of specificity can be called ligand or group specificity. Another class of agents that has been used extensively for probing membranes reacts with amino groups. This includes 1-fluoro-2,4-dinitrobenzene (DNFB)(28,36,60), diazosulfanilic acid (DASA)(4),2.4,6-trinitrobenzene sulfonic acid (TNBS)(1,28),1-isothiocyano-4-benzene sulfonic acid (ITA) (19),2-methoxy-5-nitrotropone (MNT)(28,34),4,4′-diisothiocyano-2,2′-stilbene disulfonic acid (DIDS)(9,10,30) and pyridoxal phosphate (PDP)(7). Although their reactions are usually attributed to amino groups, many of them can react with other ligands as well. Pyridoxal phosphate does form a highly specific Schiff-base with amino groups (12), but DNFB, on the other hand, may also react with phenolic hydroxyl, sulfhydryl, guanidino and imidazole groups (52). Attribution of effects of many of these agents to amino groups must be considered tentative unless direct chemical evidence is available.

Almost all proteins possess reactive ligands such as sulfhydryl, amino, imidazole, tyrosyl and carboxyl, and because many transport functions may depend in some way on such ligands, ligand specificity would not be expected to confer a high degree of specificity in terms of either particular membrane components or particular membrane functions. In actual practice, however, ligand specific agents sometimes turn out to possess an unexpectedly high degree of function and component specificity. The explanation for this unusual behavior resides in the factor of accesibility. Only certain proteins of the membrane are exposed on its outer surface directly accessible to external chemical agents. Other proteins are less accessible because they are inserted into the lipid bilayer or are located on the inner surface of the membrane (45,46).

In the red blood cell, the arrangement of many proteins in the membrane has been assessed (5,17,21,50). It has been demonstrated, for example, that the proteins exposed on the outside are anchored in the lipid bilayer by hydrophobic interactions, and many of them "span" the membrane so they are exposed on the inner face of the membrane as well (figure 1). These hydrophobic proteins are not water soluble unless they are extracted from the membrane by detergents (50,59). Because of their firm association with the membrane the term "intrinsic" has been applied to such proteins (45). The term "extrinsic" has been applied to the many proteins associated with the inner face of the membrane that are associated largely by ionic interactions. These can be dissociated by gentle perturbations, such as changes in pH and ionic strength, and they are soluble in water (50). Because of their internal location, they are protected from interaction with non-penetrating agents and to some degree from slowly penetrating ones.

Protein ligands exposed on the outer face of the membrane can be readily distinguished from those that are "hidden" within the membrane on the basis of their direct accessibility to external agents. For example, p-chloromercuribenzene sulfonic acid (PCMBS) can react directly from the outside of the intact cell with only a small superficial population of membrane sulfhydryl groups amounting to less than 2% of the total, whereas in the leaky membrane of the red cell ghost in which internal membrane proteins are accessible, the same agent is capable of reacting with over 40% of the total sulfhydryl groups (28,40). Because PCMBS can penetrate the intact membrane only at a slow rate, in short term experiments its interactions are restricted to the superficial groups and its functional effects are limited to a specific inhibition of the sugar transport system (40,51). In the leaky ghost, however, sulfhydryl agents are capable of inhibiting the $Na^+ - K^+$ - ATPase involved in cation transport (55). The latter system in the intact cell is protected in the short term against the action of PCMBS by its inaccessibility. In longer experiments, however, the agent can eventually reach internal populations of sulfhydryl groups and inhibition of the cation transport system results (38). The functional specificity, in this case, is time

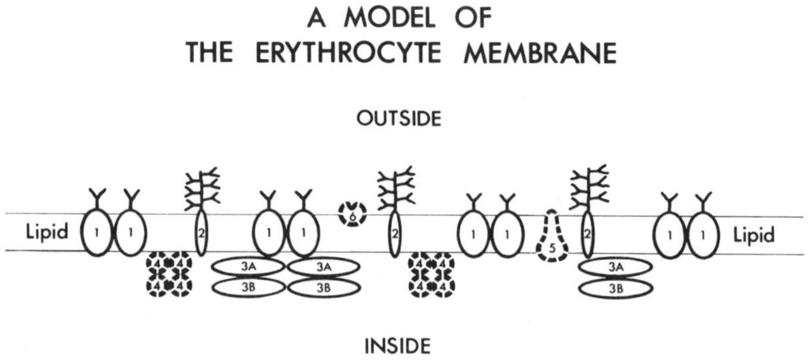

A MODEL OF
THE ERYTHROCYTE MEMBRANE

OUTSIDE

Lipid

Lipid

INSIDE

Figure 1. Proposed arrangement of proteins in the red blood cell membrane (21).

dependent and is clearly determined by accessibility factors. On the other hand, the failure of PCMBS to affect anion permeability can probably be attributed to chemical specificity since non-penetrating amino-reactive agents are capable of inhibiting this transport system (10,28).

Amino-reactive agents provide another example in which the functional specificity depends largely on accessibility factors rather than on chemical specificity per se. The uncharged agent DNFB is relatively non-specific in every sense (28,34,36,60). It penetrates the membrane rapidly, reacting with many sites in many membrane proteins, and perturbing many transport systems (including sugar, cation permeability, and anion transport). Other amino-reactive agents that cannot penetrate the membrane, or that do so slowly, are more specific. For example, MNT has a large immediate effect on anion permeability, but only a relatively small effect on cation permeability (28,34). The differential in effects can be largely eliminated by the addition of alcohol which presumably allows access of the agent to the cation controlling sites (34). Its effect on cation permeability is increased, whereas that on anion permeability is not. In the case of the non-penetrating agents, 4-acetamido-4′-isothiocyano-2,2′-stilbene disulfonic acid (SITS) or DIDS, the effects are more specific, with anion permeability being the only function that is known to be perturbed (10,28). With DIDS, the dependence of its specificity of interaction on accessibility factors can be dramatically illustrated. In the intact cell, DIDS is almost entirely associated with a single membrane protein component known as band 3 (figure 2). Thus, it is not only functionally specific, but component specific as well. In a ghost preparation in which the permeability barrier is destroyed (leaky ghost), on the other hand, DIDS is capable of reacting with virtually every protein component and its component specificity is very low (10).

A high degree of functional specificity and of ligand specificity does not necessarily correlate with a high degree of component specificity. Thus PDP is chemically specific for amino groups due to the formation of a Schiff's base (12). Like DIDS, its action can be restricted to the outer surface of the cell and its effects to the anion transport system (7). Yet it reacts not only with band 3, but with three other surface components (glycoproteins) as well (figure 3). Thus DIDS, an agent that is not particularly specific in a chemical sense (it can react with amino, sulfhydryl and guanidino (9,10)), is more specific in its action on surface proteins than is PDP, an agent that is highly specific in its chemical interactions. The difference is presumably due to accessibility factors, even though all the proteins involved are exposed on the outer surface. The accessibility in this case

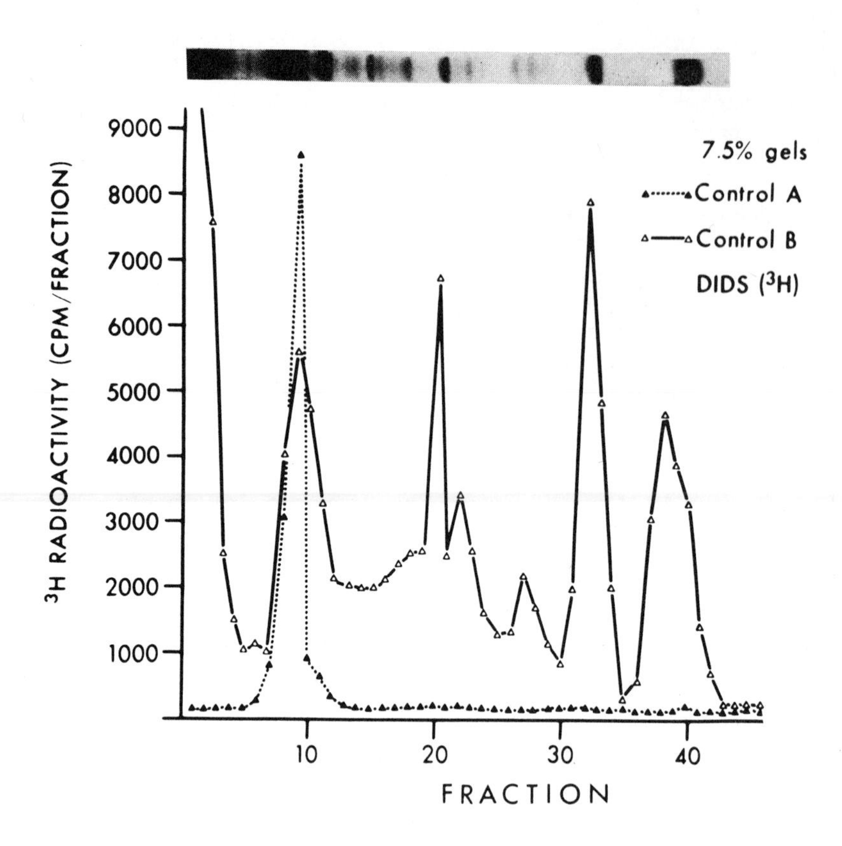

Figure 2. (^{3}H)DIDS binding to membrane proteins of intact cells and leaky ghosts as determined by SDS-acrylamide gel electrophoresis (10). The gel at the top is stained for protein with Coomassie Blue.

may be determined by charge and perhaps steric factors. The glycoproteins are rich in carboxyl groups of sialic acid which may act as a negative screen that prevents the larger bivalent anion, DIDS, from reaching reactive ligands.

Another important form of specificity can be called site specificity. Specific transport involves an interaction of the transported ion or molecule with a membrane site often called the carrier. The specificity of the interaction, like that of a substrate-enzyme interaction, may depend on the presence in the functional site of several ligands that contribute to the substrate binding. Chemical agents that resemble the substrates can, therefore,

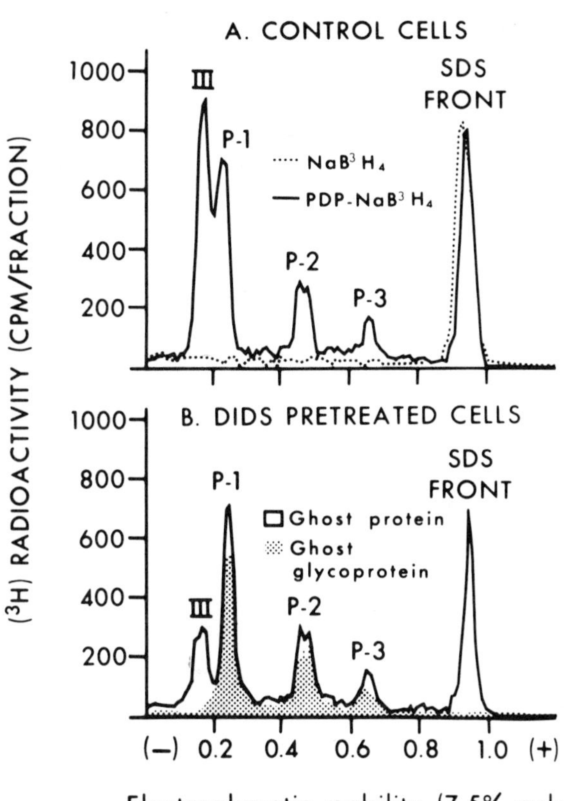

Figure 3. PDP binding to membrane proteins from normal cells and DIDS-treated cells, as determined by SDS-polyacrylamide gel electrophoresis (7). The stippled portion of the lower segment represents the PDP binding to glycoproteins isolated from the membranes. The upper gel is stained for protein with Coomassie Blue and the lower gel for carbohydrate with periodic acid-Schiff reagent.

be expected to show a high degree of site specificity. This factor can account for the specificity and potency, for example, of sugar, amino acid and nucleoside analogs in perturbing the corresponding transport systems. It may also account for the exceptional potency of cytochalasin B as an inhibitor of sugar transport (31).

In the case of ion transport, site specificity would not be expected to be as important a factor as in the case of transport of organic molecules. Nevertheless, it may exist to some degree. For example, the anion transport system can be inhibited by many anionic substances, with a wide range of potencies. A comparison of analogs of the disulfonic stilbenes and other organic anions (9) suggests that the inhibitory site contains multiple positively charged groups (probably 3) and an adjacent hydrophobic center (figure 4). The complex nature of the site can account for its exceptionally high affinity for DIDS, intermediate affinity for monosulfonic compounds, and may account for its ability to bind transportable inorganic anions.

Location of transport sites

In the previous section it has been noted that the accessibility of reactive ligands may be an important determining factor in their perturbation by chemical agents. It is therefore possible to use agents with different capacities to penetrate into the membrane to

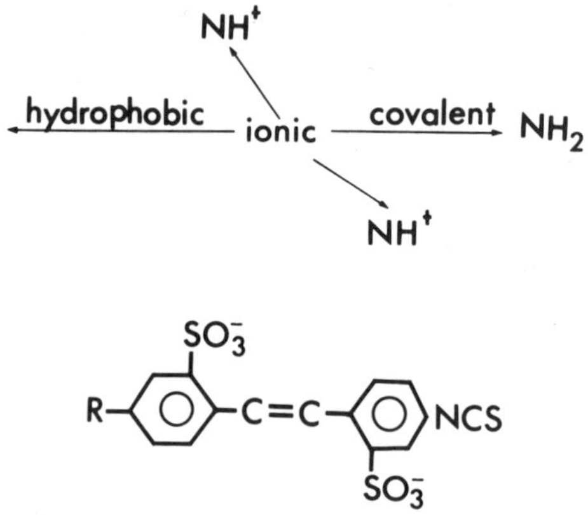

Figure 4. A schematic map of the inhibitory site of anion transport based on the potency of stilbene disulfonate analogs and other organic anions (9).

gain information concerning the location of functional sites. For example, sugar transport can be inhibited by PCMBS under conditions of minimal penetration (57). It can therefore be concluded that superficial sulfhydryl groups are involved. Similarly, because non-penetrating agents such as the disulfonic stilbenes can inhibit anion transport, it can be concluded that these inhibitory sites are also superficial (9,10,28). In contrast, the effects of a number of agents on cation permeability and on cation transport suggest that the reactive sites of these systems are in the interior of the membrane or on its inner face (29,38,40,53). For example, the non-penetrating amino reactive agents, DIDS or SITS, have no effect on cation permeability, whereas penetrating agents with a similar chemical reactivity can increase the cation permeability (28).

Slowly penetrating agents can be used to distinguish superficial from internal transport systems by the sequence of their effects. Thus the effect of PCMBS on the superficially located sugar transport sites is immediate (57), whereas that on the internally located K^+ permeabilty (53), or Na^+, K^+ active transport sites is delayed (38). In parallel, after removing external agent, the reversal of the effect is rapid for sugar transport, but slow for effects on cation permeability or transport. The sulfhydryl groups associated with cation permeability and transport are located in an internal membrane compartment into which and out of which PCMBS slowly diffuses. If its rate of permeation is increased by raising the temperature, then the delay is reduced (53).

The pathways by which chemical probes penetrate the membrane, as well as their rates of penetration, may determine their capacity to interact with particular functional sites (29,43). PCMBS and p-chloromercuribenzoic acid (PCMB) can both react rapidly with the same number of sulfhydryl groups in isolated "leaky" ghosts. With the intact cell, however, their behavior is quite different. PCMB diffuses more rapidly (up to 100 times) into the cell, particularly at low pH, due presumably to the rapid flow of its uncharged form through lipid. PCMBS, on the other hand, can only penetrate slowly via ion permeation channels. Because the sulfhydryl sites controlling cation permeability seem to be located in these pathways, PCMBS, the slowly penetrating agent, is a considerably more effective probe for the cation permeation system than PCMB, a rapidly penetrating agent. At equal concentrations, the effect of PCMBS is large and prolonged, whereas that of PCMB is small and transient. The main pathway for penetration of PCMBS, an anion, is via the anion transport system, for its flux is inhibitable to a large extent by agents such as SITS or DIDS that are specific inhibitors of anion transport. The Y-intercept of figure 5 represents

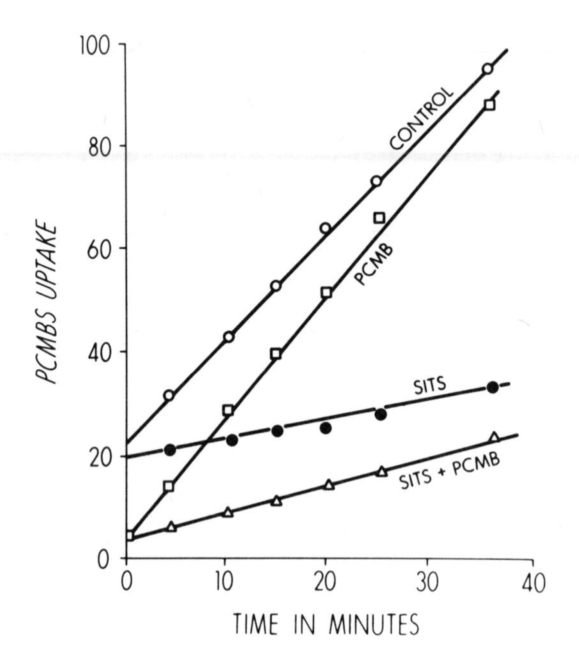

Figure 5. The effect of SITS and of PCMB on the surface binding (Y-intercept) and uptake (slow component) of PCMBS (43).

surface binding by sulfhydryl groups. It is displaced by other sulfhydryl agents such as PCMB but not by SITS. The slow component represents penetration. It is not blocked by PCMB, but is substantially inhibited by SITS.

Another method of determining the location of functional sites is to directly compare the effect of non-penetrating agents on the outer face with their effect on the inner face. The outer face can be studied in intact cells and the inner face using inside-out vesicles prepared from ghosts. By such studies it was demonstrated that ouabain inhibits cation transport only on the outside face of the membrane (35). Recently, a similar "sidedness" has been demonstrated for the effect of disulfonic stilbenes on anion transport. In resealed ghosts, the agent was inhibitory from the outside, but not the inside (60).

From studies such as those outlined above, a crude map of the membrane can be drawn (figure 6), with various sites located at the inner or outer faces of the membrane, or within internal membrane compartments. Certain pathways by which agents reach the internal sites can also be proposed. It must be emphasized, however, that all definitions of locations, of compartments and of pathways are operational. Thus sites on the outer surface are defined as being accessible to non-penetrating agents, and pathways are defined in terms of their sensitivity to agents.

It is of considerable interest that the operational map based on accessibility of various functions places certain functional ligands at the outer surface, others within the membrane, and others at the inner face of the membrane (figure 6). The biochemically determined location of membrane proteins (figure 1) is entirely consistent with the functional map. In particular the evidence for a sulfhydryl pool within an interior compartment of the membrane (28,40,43) suggests that some proteins span the membrane,as recently proposed in the fluid-mosaic model (46).

Identification of transport components

In order to understand the mechanisms of transport in molecular terms it is necessary to identify the transport components and to determine their arrangement in the membrane. Identification in some cases is relatively easy because the isolated transport systems still possess inherent measurable properties such as specific substrate binding sites or characteristic enzyme properties. For example, the cation transport system, when isolated, still possesses ATPase activities that are dependent on specific cation binding. The anion transport system, on the other hand, has no useful built in "markers". Its affinity for its substrates, Cl^- and HCO_3^-, is relatively low (13,18), and

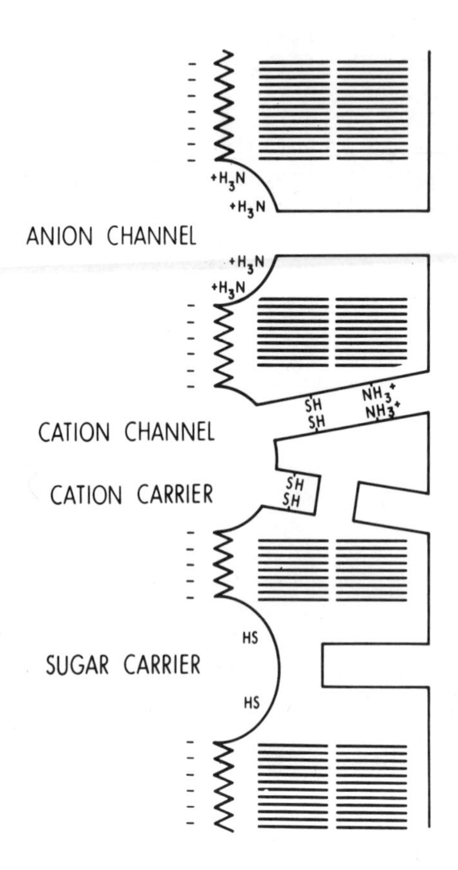

Figure 6. A schematic diagram of the location of functional ligands in the membrane (43).

it has no associated enzyme activities. Most studies of this system have therefore involved the use of specific inhibitory chemical probes that can react irreversibly to mark the components permanently so that they can be identified after various isolation and purification procedures. They include DNFB (60), TNBS (1), DASA (22), ITA (19), DIDS (9,10), PDP (7) and N-(4-azido-2-nitrophenyl)-2-aminoethyl sulfonic acid (NAP-taurine) (8,42).

The reactive groups of these agents include fluoro, diazo, isothiocyano, aldehyde, and nitrene. Many, but not all, are anions. All can react with amino groups, but specificity varies from that of PDP (very high (12)) to that of the nitrene (very low) which can even displace H from C-H bonds (47). They also vary considerably in functional and component specificity. DIDS has been one of the most useful agents (9,10). It specifically affects anion permeability, it is exceptionally potent, and in the intact cell it is highly localized in one component, a protein of about 95,000 daltons known as band 3 (figure 2).

It has already been pointed out that the component specificity of DIDS is largely attributable to accessibility factors. It has also been suggested that its high inhibitory potency is due to the structure of the inhibitory site (figure 4). The two sulfonic acid groups of DIDS presumably are chelated by two + charged groups in the site, with perhaps some contribution of hydrophobic interactions (9). The covalent reaction of the isothiocyano group occurs after the reversible ionic binding to the site (9,30). The latter is responsible for the inhibitory effect, because disulfonic stilbenes having no covalent binding group are highly potent.

Most of the DIDS is localized in band 3 with a 1 to 1 correspondence between binding and inhibition (10,30). This finding would support the idea that band 3 is the transport component, assuming that the inhibition involves a first order reaction (one DIDS reacts with only one transport molecule). But, because the mechanism of inhibition (and therefore the nature of the inhibition-binding relationship) is not known, and because other minor DIDS binding components may also have a linear relationship, the conclusion concerning band 3 is not established in an absolute sense.

The question of minor components is a difficult one to resolve. Although DIDS is largely bound to band 3, a small fraction (about 6%) is found in the major sialoglycoprotein (11) and it has also been reported in a "satellite" band (30). The use of the proteolytic enzyme pronase allows some clarification. When intact cells are treated with pronase, the DIDS associated with the glycoprotein is liberated into the medium without any effect on the anion transport system or its inhibition by DIDS (11).

This major sialoglycoprotein can therefore be eliminated as the site of DIDS inhibition. Other minor components, such as the "satellite" band or those that cannot be resolved from the general background of DIDS distribution on SDS-acrylamide gels, cannot be rigorously excluded as potential sites for anion transport. It has been pointed out, however, that such minor components would have to possess an unusually high rate of turnover to account for the Cl^- fluxes, approaching or exceeding that of the fastest enzymes such as catalase (19,44). This consideration would favor the abundant protein, band 3, as the anion transport protein. Pronase treatment of the intact cell also results in the splitting of band 3 into two segments of 65,000 and 35,000 daltons (see gels at top of figure 7), again with no impairment of anion transport activities (11). Most of the DIDS is associated with the 65,000 dalton segment (figure 7) and it is this component

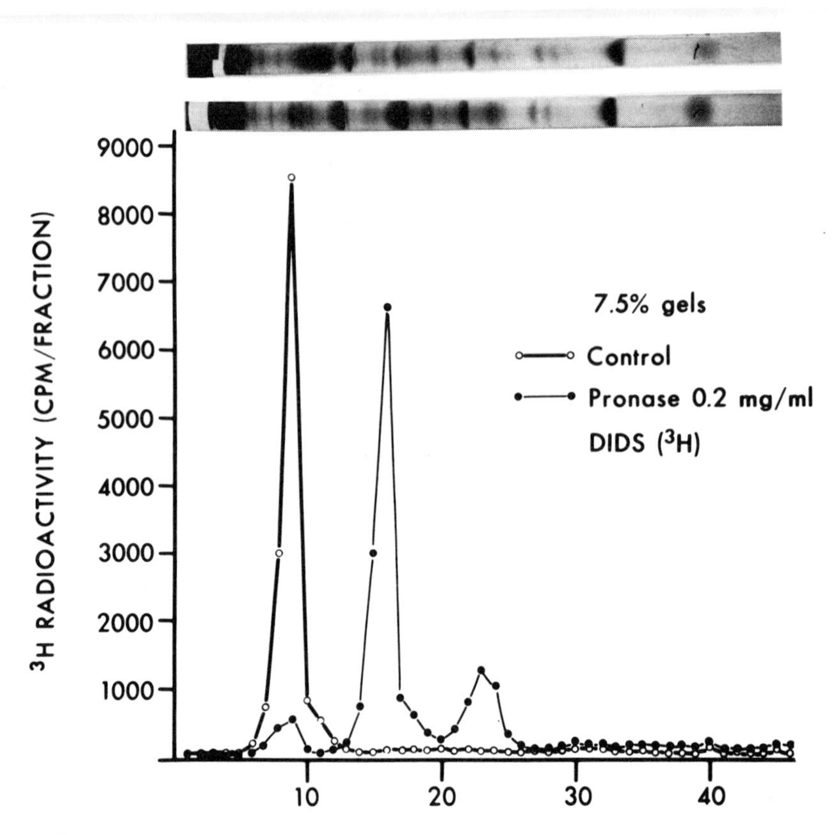

Figure 7. The effect of pronase treatment of the intact cell on DIDS binding to membrane proteins as determined by SDS-acrylamide gel electrophoresis (11). The gels at the top are stained for protein with Coomassie Blue. The upper one is the control and the lower from pronase treated cells.

that seems to be the one involved in anion transport. This conclusion is supported by the finding that Triton-X-100 extracts, enriched in either the 95,000 dalton component or in the 65,000 dalton component can, when reconstituted in lecithin vesicles, markedly increase the anion fluxes (41). On the other hand, reconstitution of the same proteins extracted from DIDS pretreated cells causes no change in anion fluxes.

Another question that can be raised is the nature of the inhibition by DIDS. The site, a cluster of positively charged groups (figure 4), would be an appropriate binding site for anions as part of an anion transport system. DIDS, also an anion, might therefore inhibit by interacting with the substrate binding site of the transport system. This possibility is attractive, but it is not proven because DIDS might act indirectly at a distance from the transport system by a steric effect or by causing some configurational charge. This possibility must be seriously considered because many non-ionic substances can also inhibit anion transport, presumably by mechanisms that do not involve the positively charged membrane ligands (15).

Several studies throw some light on the question of the relationship of the inhibitory site and the transport site. The reconstitution in lecithin vesicles of anion transport with extracts enriched with the 95,000 dalton, or the pronase resistant 65,000 dalton segment (mentioned above) indicates that these proteins probably contain transport as well as inhibitory sites (41). The conclusion is, however, limited by the fact that the extracts used in reconstitution contained small amounts of other proteins. Another approach is to use probes that are substrates for, as well as inhibitors of, the anion transport system. For example, PDP can penetrate the cell slowly and its flow is inhibited by DIDS, a specific inhibitor of anion transport (7). It is also a reversible inhibitor of sulfate transport. Because PDP seems to be a substrate for the anion transport system, it presumably interacts directly with the anion binding site and might therefore inhibit competitively by displacing substrate from the site.

On addition of (^{3}H)NaBH$_4$, the reversible complex of PDP with amino groups is converted to an irreversible bond so its binding sites can be identified by the inserted ^{3}H. The inhibitory effect also becomes irreversible, but its extent is almost the same as the reversible inhibition before addition of NaBH$_4$. Low concentrations of PDP are highly localized in band 3, but with higher concentrations it is also found in three glycoprotein bands (figure 3). The PDP-sites in band 3 are largely common to DIDS, so that pretreatment with either agent largely eliminates the binding of the other.

The most definitive studies have been carried out with NAP-taurine, a photoreactive probe (47,49). In the dark, NAP-taurine is an unreactive organic anion, but on exposure to light, its azido group is converted to a reactive nitrene capable of non-specific irreversible interaction with most organic ligands (since it even displaces H from CH (47,49)). It therefore has the potential of being transported by the anion system in the dark and of interacting with binding sites, regardless of their chemical structure, after exposure to light.

In the dark at $0^{o}C$, NAP-taurine penetrates very slowly (47,49), but at higher temperatures it passes rapidly into the cell (8,42,48). Its flux is blocked by two chemically different inhibitors of inorganic anion transport, DIDS and dipyridamole, suggesting that NAP-taurine and the inorganic anions permeate by the same mechanism (8). Furthermore, the degree of inhibition of anion fluxes by NAP-taurine is dependent on the concentration of the inorganic anions (8,25,42). The kinetics of inhibition are quantitatively consistent with competition for a common site (figure 8). In addition, the effectiveness of the physiological anions in displacing NAP-taurine is consistent with their relative affinities for transport as expressed by their Km's, with $HCO_3^- > Cl^-$ These results suggest that NAP-taurine is not only transported by the same system as the inorganic anions, but that it reversibly inhibits their fluxes by competing for the binding site of the transport system.

Among the sites that are occupied by NAP-taurine in the dark are those with which DIDS reacts to produce its irreversible inhibition. This conclusion is based on the observation that reversibly bound NAP-taurine can protect the cells against the irreversible inhibition by DIDS and against interaction of DIDS with band 3 protein (Table 1). The experiment also suggests that the inhibition of anion fluxes by DIDS is a consequence of its reaction with the substrate binding site of the transport system (8,42).

When cells are exposed to light in the presence of NAP-taurine, the agent is irreversibly bound, mainly to band 3 protein (8,42,47,49) (figure 9). The sites are largely in common with the binding sites for DIDS (pretreatment with DIDS markedly reduces the binding of NAP-taurine to band 3 and vice versa (8). In addition, after pronase treatment, most of the NAP-taurine is found in the 65,000 dalton segment (as is the case for DIDS, figure 7). If the cells are exposed for a longer period of time so that the agent reaches the inside of the cell, many additional protein bands associated with the inner aspect of the membrane are labelled (bottom panel of figure 9).

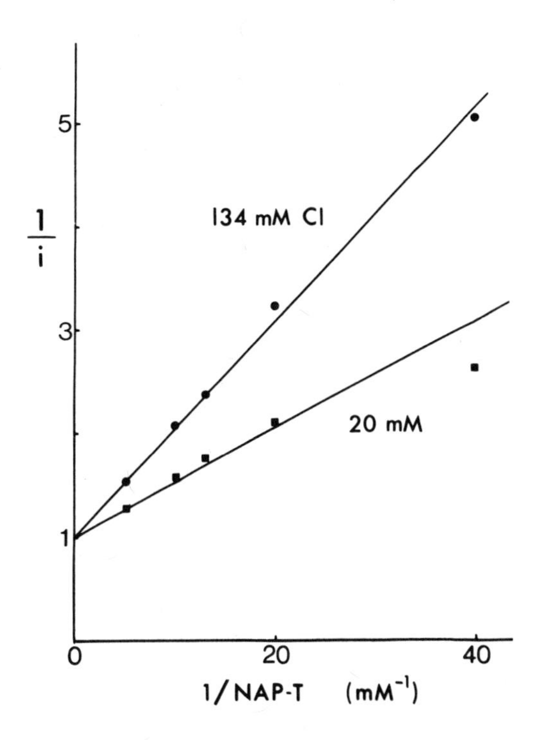

Figure 8. The kinetics of NAP-taurine inhibition of sulfate fluxes in the presence of low and high Cl^- (25). The reciprocal of the fractional inhibition ($^1/i$) is plotted against the reciprocal of the inhibitor concentration. Straight lines of increasing slope (with increasing Cl^-) are consistent with competition of Cl^- and NAP-taurine for the transport site.

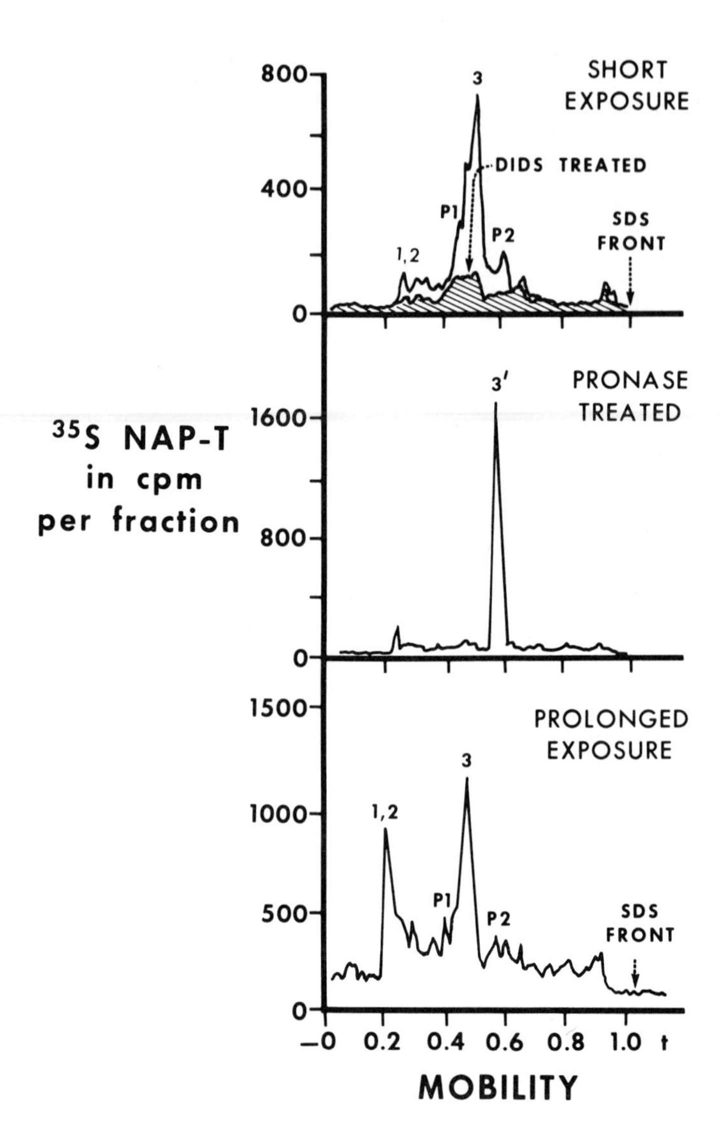

Figure 9. NAP-taurine binding to membrane proteins from normal, DIDS-treated and pronase treated cells as indicated by SDS-acrylamide gel electrophoresis (8).

Table I: The protective effect of reversibly bound NAP-taurine on the inhibition of sulfate efflux by DIDS (8)

	Inhibition in % of control	
	8μM DIDS	4μM DIDS
	50μM NAP-taurine	100μM NAP-taurine
DIDS only	88, 92	72, 83
NAP-T (dark) + DIDS, then washed	63, 58	22, 25
NAP-T (dark) then washed, + DIDS	95, -	66, 78

Cells (10% hematocrit) preloaded with (^{35}S) sulfate (5mM) were exposed to DIDS for 5 minutes at 10°C in the presence or absence of NAP-taurine. They were then washed twice with albumin (0.5%) containing buffer and twice with buffer. In some aliquots of cells, NAP-taurine, 50 or 100μM, was present during DIDS-treatment; in others it was added and then washed out prior to DIDS treatment. Control samples of blood cells were subjected to the same washing procedures. Fluxes were measured at 37°C at 5% hematocrit after the washing procedure. All procedures were carried out in semi-darkness to avoid photolysis of the NAP-taurine.

In order to equate the irreversible binding sites of NAP-taurine in band 3 with its reversible binding to the transport sites, it is essential to determine whether the distribution of the probe after exposure to light reflects its distribution in the dark, with no significant degree of redistribution.

The behavior of the NAP-taurine-anion system was therefore examined by a number of criteria that can be used to define a photoaffinity probe (26). In each case, the evidence supported the conclusion that the irreversible NAP-taurine binding reflects its reversible interactions in the dark. The evidence includes:

(1) The effects of NAP-taurine on anion transport are reversible in the dark and irreversible after exposure to light.

(2) The reversible inhibition in the dark corresponds closely with the irreversible inhibition after exposure to light.

(3) Completely photolyzed NAP-taurine is not inhibitory.

(4) The presence of scavengers in the medium capable of reaction with the nitrene has no effect on the irreversible inhibition.

These findings indicate that the binding sites revealed by NAP-taurine include the substrate binding sites of the transport system, and suggests that they are also the sites of action of PDP and DIDS.

NAP-taurine may, in addition to reacting with transport sites, interact with other positively charged sites that are not functionally significant. In order to differentiate transport sites from non-specific sites, experiments were carried out with different concentrations of Cl^- under conditions arranged so that the NAP-taurine inhibition (in the dark) was substantial in one case and was almost completely reversed in the other. After exposure to light, the amount of binding of NAP-taurine to band 3 or to the 65,000 dalton segment was substantially reduced in parallel with reversal of inhibition (figure 10). This experiment indicates that a substantial fraction of the band 3 binding sites are common to Cl^- and to NAP-taurine (42). It supports the view that band 3 contains a binding site for anions essential to the transport.

Proteins in band 3 may be involved in other functions as well. A sulfhydryl reactive agent (6) that inhibits water and non-electrolyte permeation is covalently associated with band 3. Sugar transport has been localized in this protein component by use of probes and other techniques (24,31,54). Band 3 also contains acetyl choline esterase (3), the phosphorylated intermediate of the Na^+-K^+-ATPase involved in Na^+-K^+ transport (2,27) and it is the preferred substrate for phosphorylation by the protein phosphokinase (39). It is functionally heterogeneous and presumably is also chemically heterogeneous. Other evidence of heterogeneity includes its susceptibility to pronase (11), its behaviour in isoelectric focusing (Morrison, M., personal communication), and its capacity to bind the lectin concanavalin A (16). The total number of monomers of band 3 is of the order of 1.0 to 1.5 million per cell (50). Most of them are probably involved in anion transport, a substantial number in sugar transport, and small numbers in the other functions (42).

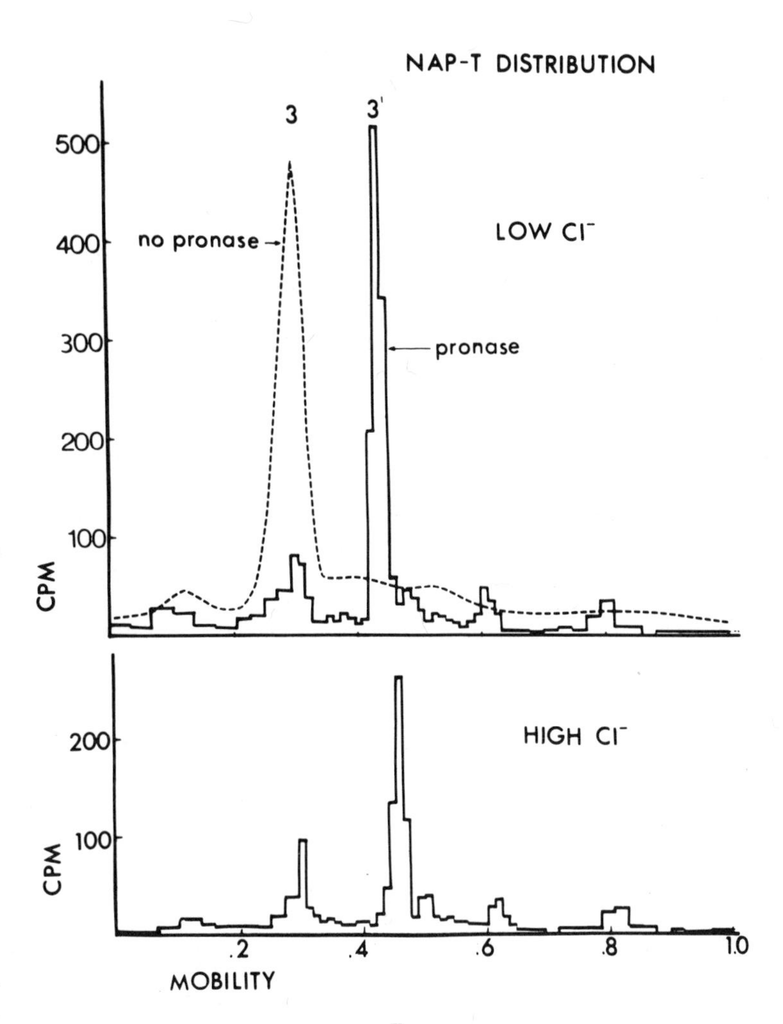

Figure 10. The effect of high Cl^- concentrations on NAP-taurine binding to membrane proteins from pronase treated cells as indicated by SDS-acrylamide gel electrophoresis (25). The dotted line represents the distribution of NAP-taurine in proteins from normal cells.

Arrangement of the transport protein in the membrane and the mechanism of transport

Chemical agents can not only be used to identify components involved in transport, but they can also help to determine how those components are arranged in the membrane and how they function in transport. It has already been pointed out that non-penetrating agents such as DIDS can react with band 3 from the outside of the cell (9, 10), confirming other studies (using proteolytic enzymes, other chemical probes and iodination by lactoperoxidase (5,17,32,37,50,56)) that demonstrate the exposure of part of this protein on the outer surface of the cell. It has also been pointed out that pronase cleavage from the outside splits band 3 into two segments of 65,000 and 35,000 daltons and that most of the DIDS is located in the 65,000 dalton segment (11) (figure 7). Another important finding is that band 3 protein "spans" the membrane and is accessible from the inside as well as the outside. This was first proposed on the basis of comparisons of interactions of the non-penetrating agent formylmethionyl methyl phosphate (FMMP with the itact cell and with the leaky ghost (5). In the intact cell, only those peptides exposed to the outside could react, but in the ghost, the agent could reach additional residues exposed to the inside. The conclusion that band 3 "spans" the membrane has since been confirmed by a number of procedures using inside out vesicles, resealed ghosts, and intact cells. The probes have included proteolytic enzymes,lactoperoxidase catalysed iodination,PDP and NAP-taurine (5,7,17,32,37,48,50).

Band 3 protein is not symmetrically arranged across the bi-layer. Thus the sugar moieties (8% of the mass is carbohydrate) are exposed to the outside but not to the inside (50,59). Proteolytic attack with pronase from the outside largely splits the protein into segments of 35,000 and 65,000 daltons (11), whereas attack from the inner face leads to considerable degradation (23,56). From the outside band 3 is not susceptible to tryptic attack, but from the inside it is split into segments of 48,000 and 58,000 daltons(20,51). The sulfhydryl groups which can be cross-linked by oxidizing agents to produce dimers of band 3 are localized at the inside face (51). At the inner face, band 3 is specifically associated with other membrane proteins including glyceraldehyde-3-phosphate dehydrogenase (58).

As would be expected of a protein inserted through the membrane band 3 is a highly hydrophobic intrinsic protein that is closely associated with the fatty acid chains of the phospholipids. Like other hydrophobic proteins, it is not soluble in water, but it can be extracted by detergents (50,59). These and other data have led to the suggestion that band 3 is an S shaped protein with one loop exposed on the outside, another on the inside of the membrane and with three segments inserted into the membrane (20).

Hydrophobic proteins would be expected to be visible as intramembranous particles (IMP) by the freeze fracture technique of electron microscopy. It had been pointed out that band 3 might be a component of such particles on the basis that the other hydrophobic components (the sialoglycoproteins) could only account for about 15% of their bulk (5,17). More recently, direct evidence linking band 3 to the particles has been presented (14,42).

In considering the molecular mechanism of transport, the role of band 3 (a membrane spanning, hydrophobic portein) must be taken into account and the kinetic characteristics of transport must also be accommodated. The kinetics in the case of anion transport can be summarized briefly (13,18,44):

1. The process is "saturable" by high concentrations of substrate and it displays competition between anions when they are present together. These characteristics suggest that in the process of transport, anions must bind to a site.

2. The transport system displays specificity as indicated by different Km's + Vm's for different anions. The normal substrates Cl^- and HCO_3^- have relatively high Km's and high Vm's. Organic anions have lower Km's and much lower Vm's. Thus the binding site must display a range of affinities and the transport a range of operating rates.

3. All but a small fraction of the inorganic anion flux involves an electrically silent, obligatory one-for-one exchange.

4. Inorganic anions can compete with each other for transport, but they also inhibit each other by a non-competitive mechanism. A variety of other compounds can inhibit including organic anions, amino-reactive agents, and non-ionic substances. Presumably, several mechanisms of inhibition exist.

Two general mechanisms of transport have been proposed to account for the kinetics, one involving interaction of the anions with a "mobile" membrane carrier (18), and the other with "fixed" positive charged groups (33). The two mechanisms are difficult to distinguish on the basis of kinetic information.

Band 3 contains a specific binding site for transported anions, but the whole protein cannot act as a "mobile" carrier. It is a large protein, probably present in the membrane as a dimer, held in place by many hydrophobic and perhaps hydrophilic interactions with lipids and other proteins, and with certain segments always exposed to the outside. It is therefore highly improbable that the whole protein could act as a mobile carrier because of energetic considerations. Also, the rate of anion exchange is almost as rapid as the fastest enzyme reactions (19,44), much too fast for migrations or rotations of macromolecules. It furthermore seems unlikely on kinetic grounds that a segment of the protein could carry the anion through the lipid. The turnover of mobile lipid soluble peptide cation carriers, such as valinomycin (30,44), is lower by several orders of magnitude than that of the anion carrier. It seems more reasonable that the band 3 protein provides an aqueous channel through the membrane through which the transport can occur (42) (figure 11). Although band 3 is relatively rich in hydrophobic amino acid residues (17, 19,59), over 60% of them are hydrophilic. In the membrane, the protein must be arranged with hydrophobic residues turned outward toward the lipids and hydrophilic residues turned in toward each other, forming an aqueous channel. Band 3 may be S shaped so that 3 segments of the peptide may be side by side in the lipid (20). Furthermore, the protein is probably a dimer (50,59) so that 6 segments might be available to form a channel.

The protein channel could not be open for free diffusion of anions because the conductive fluxes are low, whereas the exchange fluxes are high (44). Thus a diffusion barrier must be present, perhaps a hydrophobic zone. With this model in mind, the transport kinetics would largely reflect the characteristics of an exchange mechanism operating across this barrier with the binding site perhaps acting as a gate that can move back and forth, carrying anions (42). The obligatory exchange can be explained in two ways. The gate may have 2 binding sites, one facing inward and the other out. It could flip, or rotate,only if both sides were loaded with anions. Such a mechanism would involve a "simultaneous" exchange. The alternative is a gate with one binding site that could move from outward to inward facing only if loaded with an anion. Such a mechanism would involve a "sequential" exchange. In either case, the mobile element of the protein would move only a small distance, with no insurmountable energy barriers to be overcome. Such a configurational perturbation would have the large temperature coefficient that is characteristic of the anion transport system and could be influenced by any agents that perturbed the protein or lipid-protein structure. An alternate type of mechanism would utilize a site that does not move, but in which access to the site from inside or outside would be altered by alternately shifting the location of the diffusion barrier.

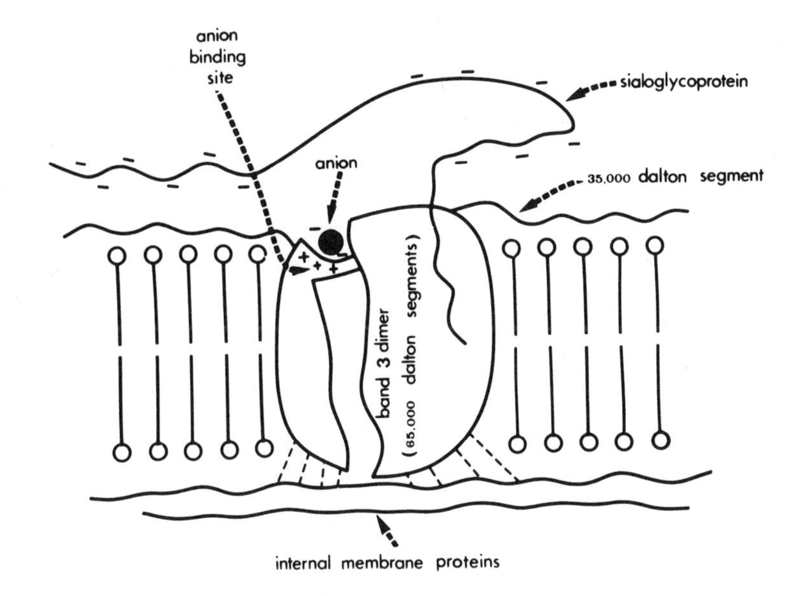

Figure 11. A schematic diagram of band 3 as a protein channel for anion transfer through the membrane (42). Reprinted from FEDERATION PROCEEDINGS 35: 3-10, 1976.

Kinetically, the models proposed above would behave like a classical mobile carrier, except that a small segment of a protein is the moving element, and that it acts as a gate in an aqueous protein channel, rather than by solubilizing the substrate in the lipid bilayer.

Chemical agents can potentially be used to test the idea that some element of the protein is alternately exposed to the inside or outside. A preliminary test is illustrated in Table 2 (42). The cells were loaded with PDP, a high affinity anion that flows through the membrane very slowly (7). Its flow was further reduced by the presence of the non-ionic anion transport inhibitor, dipyridamole (15), and the external PDP was then washed away. Under such conditions, the carrier site should be distributed with most of it facing inside, in equilibrium with the internally located high affinity anion. The number of sites facing the outside when titrated with the non-penetrating agent, DIDS, was reduced by about 50% (Table 2). If the dipyridamole was then washed out so that the turnover of the carrier increased, external DIDS was able to "capture" over 90% of the sites. In appropriate controls, the presence of dipyridamole or of traces of external PDP not

Table II: PDP present on the inside of the membrane reduces the (^{3}H)DIDS binding to the outside (42)

Location of PDP	(^{3}H)DIDS binding to cells	
	Dipyridamole present	Dipyridamole absent
None	100	100
Outside	12	11
Outside then washed	85	83
Inside	55	89

For "PDP outside", cells (30%Hct) were exposed to PDP (10 mM) for 10 minutes before addition of DIDS. For "outside then washed", the exposure to PDP was followed by two washes with buffered saline containing albumin (0.5%) and two with buffered saline. Exposure to (^{3}H)DIDS was for 10 minutes. For "inside" the cells were exposed to PDP for 20 hours at 37°C at pH 7.0 with the medium replaced every 5 hours. The washing and treatment with (^{3}H)DIDS were the same as above. All data are averages for three experiments, given as percent binding relative to controls with no PDP. Statistical analysis indicates that the value for "inside", in the presence of dipyridamole (55%) is significantly lower than the control with no dipyridamole (89%) at a level of 0.01.

washed away caused only a 10% reduction in the number of DIDS binding sites. The trans effect of internal PDP in reducing the number of sites titratable from the outside is consistent with a sequential type of mobile site model as proposed.

Because components of band 3 may also be involved in transport of water, non-electrolytes (6), Na^+ and K^+ (2,27), and sugars (24,31,54), it is suggested that protein channels may provide a general mechanism for controlled transfer of hydrophilic solutes through the membrane.

Acknowledgement:

This study was supported in part by grants from the Medical Research Council of Canada, Grant MT4665 and Grant MT4836

REFERENCES

1. ARROTTI, J.J. and GARVIN, J.E.: Biochem. Biophys. Res. Commun. 49 (1972) 205.

2. AVRUCH, J. and FAIRBANKS, G .: Proc. Nat. Acad.Sci. U.S.A. 69 (1972) 1216.

3. BELLHORN, M.B., BLUMENFELD, O.O. and GALLOR, P.M.: Biochem. Biophys. Res. Commun. 39 (1970) 267.

4. BERG, H.C.: Biochim. Biophys. Acta 183 (1969) 65.

5. BRETSCHER, M.S.: Science 181 (1973) 622.

6. BROWN, P.A., FEINSTEIN, M.B. and SHA'AFI, R.I.: Nature, London, 254 (1975) 523.

7. CABANTCHIK, I.Z., BALSHIN, M., BREUER, W. and ROTHSTEIN, A.: J. Biol. Chem. 250 (1975) 5130.

8. CABANTCHIK, I.Z., KNAUF, P.A., OSTWALD, T., MARKUS, H., DAVIDSON, L., BREUER, W. and ROTHSTEIN, A.: Biochim. Biophys. Acta (in press).

9. CABANTCHIK, I.Z. and ROTHSTEIN, A. : J. Membrane Biol. 10 (1972) 311.

10. CABANTCHIK, I.Z. and ROTHSTEIN, A.: J. Membrane Biol. 15 (1974) 207.

11. CABANTCHIK, I.Z. and ROTHSTEIN, A.: J. Membrane Biol. 15 (1974) 227.

12. CHURCHICH, J.E.: Biochemistry 4 (1965) 1405.

13. DALMARK, M.: J. Gen. Physiol. 67(1976) 223.

14. DA SILVA, P. and NICOLSON, G.L.: Biochim. Biophys. Acta 363 (1974) 311.

15. DEUTICKE, B.: Naturwiss. 57 (1970) 172.

16. FINDLAY, F.: J. Biol. Chem. 249 (1974) 4398.

17. GUIDOTTI, G.: Ann. Rev. Biochem. 41 (1972) 731.

18. GUNN, R.B., DALMARK, M., TOSTESON, D.C. and WIETH, J.O.: J. Gen. Physiol. 61 (1973) 185.

19. HO, M.K. and GUIDOTTI, G.: J.Biol. Chem. 250 (1975) 675.

20. JENKINS, R.E. and TANNER, M.J.A.: Biochem.J. 147 (1975) 393.

21. JULIANO, R.L.: Biochim. Biophys. Acta 300 (1973) 341.

22. JULIANO, R.L. and ROTHSTEIN,A.: Biochim. Biophys. Acta 249 (1971) 227.

23. JUNG, C.Y. and CARLSON, L.M.: J. Biol. Chem. 250 (1975) 3217.

24. KAHLENBERG, A.: J.Biol. Chem. (in press).

25. KNAUF, P.A., DAVIDSON, L., BREUER, W. and ROTHSTEIN, A.: Manuscript in preparation.

26. KNAUF, P.A., DAVIDSON, L., BREUER, W., OSTWALD, T. and ROTHSTEIN, A.: Manuscript in preparation.

27. KNAUF, P.A., PROVERBIO, F. and HOFFMAN, J.F.: J. Gen. Physiol. 63 (1974) 305.

28. KNAUF, P.A. and ROTHSTEIN, A.: J. Gen.Physiol. 58 (1971) 190.

29. KNAUF, P.A. and ROTHSTEIN, A.: J. Gen. Physiol. 58 (1971) 211.

30. LEPKE, S., FASOLD, H., PRING, M. and PASSOW, H.: J. Membrane Biol. (in press).

31. LIN, S. and SPUDICH, J.A.: Biochem. Biophys. Res. Commun. 61 (1974) 1471.

32. MUELLER, T.J. and MORRISON, M.: J.Biol. Chem. 249 (1974) 7568.

33. PASSOW, H.: Progress in Biophysics (Butler, J.A.V. and Noble, D. Eds.) Pergamon. New York. 19 (1969) 425.

34. PASSOW, H. and SCHNELL, K.F.: Experientia 25 (1964) 460.

35. PERRONE, J.R. and BLOSTEIN, R.: Biochim. Biophys. Acta 291 (1973) 680.

36. POENSGEN, J. and PASSOW,H.: J. Membrane Biol. 6 (1971) 210.

37. REICHSTEIN, E. and BLOSTEIN, R.: J. Biol. Chem. 250 (1975) 6256.

38. REGA, A.F., ROTHSTEIN, A. and WEED, R.I.: J. Cell. Physiol. 70 (1967) 45.

39. ROSES, A.D. and APPEL, S.H.: J. Membrane Biol. 20 (1975) 51.

40. ROTHSTEIN, A.: Current Topics in Membranes and Transport (Bronner, F. and Kleinzenner, A. Eds.).Academic Press, New York, 1 (1970) 135.

41. ROTHSTEIN, A., CABANTCHIK, I.Z., BALSHIN, M. and JULIANO, R.: Biochem. Biophys. Res. Commun. 64 (1975) 144.

42. ROTHSTEIN, A., CABANTCHIK, I.Z. and KNAUF, P.: Fed. Proc. 35 (1976) 3.

43. ROTHSTEIN,A., TAKESHITA, M. and KNAUF, P.A.: Biomembranes v.3, Passive Permeability of Cell Membranes, (Kreuzer, F. and Slegers, J.F.G. Eds.). Plenum. New York (1971) 393.

44. SACHS,J.R., KNAUF, P.A. and DUNHAM, P.B.: Transport through red cell membranes, Ch.15, The Red Blood Cell, v.II, 2nd. Ed. (Surgenor, D.M. Ed.). Academic Press, New York (1975) 613.

45. SINGER, S.J.: Ann. Rev. Biochem. 43 (1974) 805.

46. SINGER, S.J. and NICOLSON, G.L.: Science 175 (1972) 720.

47. STAROS, J.V., HALEY, B.E. and RICHARDS, F.M.: J. Biol. Chem. 249 (1974) 5004.

48. STAROS, J.V., HALEY, B.E. and RICHARDS,F.M.: J. Biol. Chem. (in press).

49. STAROS, J.V. and RICHARDS, M.: Biochemistry 13 (1974) 2720.

50. STECK, T.L.: J. Cell. Biol. 62 (1974) 1.

51. STECK, T.L., RAMOS, B. and STRAPAZON, E.: Biochemistry 15 (1976) 1154.

52. STEIN, W.D.: In "The structure and Activity of Enzymes (Goodwin, T.W, Hartley, B.S. and Harris, J.I. Eds.). Academic Press. New York (1964) 133.

53. SUTHERLAND, R.M., ROTHSTEIN, A. and WEED, R.I.: J. Cell. Physiol. 69 (1967) 185.

54. TAVERNA, R.D. and LANGDON, R.G.: Biochim. Biophys. Acta 323 (1973) 207.

55. TOSTESON, D.C.: Ann. N.Y. Acad. Sci. 137 (1966) 577.

56. TRIPLETT, R.B. and CARRAWAY, K.L. Biochemistry 11 (1972) 2897.

57. VAN STEVENINCK, J., WEED, R.I. and ROTHSTEIN, A.: J. Gen. 48 (1965) 617.

58. YU, J., FISHMAN, D.A. and STECK,T.L.: J. Biol. Chem. 250 (1975) 9176.

59. YU, J. and STECK, T.L.: J. Biol. Chem. 250 (1975) 9170.

60. ZAKI, L., FASOLD, H., SCHUHMANN, B. and PASSOW, H.: J. Cell. Physiol. 96 (1975) 471.

DISCUSSION

HOFFMAN: In your chloride protected system how did you substitute or how did you raise chloride?

ROTHSTEIN: In the data presented it was sucrose.

HOFFMAN: Will that not change the membrane potential?

ROTHSTEIN: Right, when we substitute sucrose for anions we do change the potential across the membrane and the binding and inhibitory effects of NAP-taurine could be thereby alterred. But the reversal of inhibition by chloride and nitrate were substantially different. Because the chloride and nitrate substitutions were identical, the potentials should be the same for both. Therefore the binding affinity for nitrate must be higher than for chloride and this nitrate-chloride difference in protecting against NAP-taurine inhibition cannot be accounted for by any difference in potentials. We are now doing a series of experiments in which the anion concentrations are varied but in which the potential is always zero. (Since the meeting, these experiments have been carried out. The protective effect of chloride is independent of the potential and it can be attributed to a competition of Cl and NAP-taurine for the same binding sites).

HOFFMAN: Could we see that slide again? That is the one

that has 1/i on the ordinate?

ROTHSTEIN: Yes. It is figure 8.

HOFFMAN: What does the i stand for?

ROTHSTEIN: The fractional inhibition.

BRODSKY: Does this mean that the more chloride you add the more you protect against the NAP-taurine effect?

ROTHSTEIN: Yes. High chloride competes for the binding site, protecting against the binding of NAP-taurine and against its inhibitory effects.

PRESSMAN: How does that channel explain the high exchange diffusion rate of chloride ions as opposed to the slow chloride mediated permeability of charge?

ROTHSTEIN: Let us put the model on (Figure 11). In the model we assume an aqueous channel but it cannot be an open channel allowing free diffusion of anions, because, as you point out, the conductive permeability to chloride is very low compared to the exchange process. Thus the channel has to be blocked by a diffusion barrier. We also conclude that a barrier must be present because the behaviour of the anion exchange kinetics is typical of a carrier system. Such behaviour implies that an anion binding site is an essential part of the transfer system. We have identified binding sites with DIDS and with NAP-taurine that are related to their inhibitory effects. We think that the same binding sites are associated with transport. We suggest that they can exist in two conformations, one facing out and one facing in, and that they can shift from one to the other when loaded with anions, acting like a gate that swings back and forth with anions allowing them to pass the diffusion barrier. Because the gate cannot return unless loaded with another anion, it operates as a one-for-one exchange system. The diffusion barrier might be a local hydrophobic region within the protein channel. To support the concept outlined above, it has been demonstrated that the binding sites accessible from the outside to the non-penetrating agent DIDS, are reduced in number when the high affinity anion, pyrixodal phosphate is present only inside the cell (reference 42). This experiment suggests that the sites are mobile and can therefore be recruited from an outside-facing to an inside-facing conformation.

For those of you who do not believe in abstractions, I would show you the anion binding sites directly. DIDS is a bifunctional reagent, with an isothiocyano group on each end. When it reacts with the membrane sites it reacts, at least initially, at one end, the other end being free to react with added substances. If ferritin is added to DIDS-treated membranes it reacts with the free end of the DIDS. The binding sites can then be seen as a dark spot on electromicrographs (reference 42). If the membranes are treated at low pH before they are fixed, the sites are clustered rather than dispersed. The particles seen by freeze fracture (intramembraneous particles) under the same conditions are also clustered rather than solitary. Independent evidence also suggests that band 3 is largely associated with these particles. Thus the ferritin particles mark the sites that are associated with transmembrane proteins, and with the inhibitory effects of DIDS on anion transport.

BLUMENTHAL: Does the number of particles per square micron correlate with the number of transport sites?

ROTHSTEIN: There has been come controversy about the number of transport sites. Dr. Passow in his presentation indicates that there may be about 1.5 to 1.8 million DIDS molecules bound to band 3 protein in each cell. Our most recent number is 1.0 to 1.1 million with DIDS or with NAP-taurine. But only about half of the NAP-taurine sites are reversible by chloride (see for example, figure 10 and reference 42), so we believe that the maximum number of anions transport sites may be about 600,000. There are about 1 to 1.2 million monomers of the band 3. Because it seems to be dimeric in the membrane, there are also about 5 to 6 hundred thousand dimers, or about 1 transport site per dimer. The labelling with ferritin amounted to about 90,000 sites per cell, or about 15% of the dimer sites.

SHAMOO: Since band 3 is the major protein in red blood cells, would it constitute 60 percent of the protein?

ROTHSTEIN: Twenty-five percent.

SHAMOO: But you would agree that it is a major protein.

ROTHSTEIN: It is one of the major components.

SHAMOO: So, if you probably label it to anything it will be by simple mass action low and highly visible. Have you reconstituted at 65 thousand plus or minus five thousand molecular weight protein intervesicles and treated them with Dig. to see if you get the same anion permeability data you could get with this protein?

ROTHSTEIN: I showed you a slide from reference 41 which illustrated the reconstitution of anion transport with protein fractions enriched with band 3 or with the 65,000 dalton protein derived from band 3 (over 90% of the stainable protein). Further reconstitution studies have been put aside until band 3 is successfully repurified. Some progress has recently been made in this direction (since the conference, Woloshin and Cabantchik (1976, FEBS Symposium on Biochemistry of Membrane Transport ed. G. Semenza and E. Carafoli, Springer-Verlag, (in press) have reported the preparation of vesicles capable of specific anion transport, whose protein content is up to 98% band 3).

SHAMOO: When you incorporated the 65,000 dalton segment of band three and liposomes 65K protein into vesicles, did you look at the freeze fraction and see whether it corresponds to intramembrane particle?

ROTHSTEIN: Yes, we did and there are particles of about the same size as those seen in red cell membranes together with some clumped or clustered material.

CHEMICAL AND ENZYMATIC MODIFICATION OF MEMBRANE PROTEINS AND ANION TRANSPORT IN HUMAN RED BLOOD CELLS

H.Passow, H.Fasold, S.Lepke, M.Pring, B.Schuhmann

Max-Planck-Institut für Biophysik and Biochemisches Institut der Universität
Frankfurt/M, Germany

ABSTRACT

The paper is introduced by a review of the developments which lead to the suggestion of an involvement of the protein in band 3 (nomenclature of Steck, ref.2) in anion transport across the red cell membrane. Subsequently, it is shown that DIDS and its dihydroderivative H_2DIDS, which both played an essential role in the reviewed work, not only combine with the protein in band 3 but also with other membrane constituents. At maximal inhibition 1.1-1.3 molecules of DIDS or H_2DIDS are bound per molecule of protein in band 3. Combined treatment with $^3H_2$DIDS and esternal chymotrypsin, pronase or papain demonstrates the existence of peptides in the protein in band 3 which differ with respect to their accessibility or susceptibility to proteolysis. Each enzyme affects the protein differently and produces different changes of anion transport. In contrast to external trypsin which has neither an effect on the protein in band 3 nor on anion transport, internal trypsin splits the protein in band 3 completely. Fragments of 58,000 and 48,000 Daltons remain attached to the membrane while other products of hydrolysis are released into the medium. Anion transport is partially inhibited but continues to exhibit the essential features seen in the intact cell. The described results are compatible with an involvement of some component of the protein in band 3 in anion transport. They show that additional evidence is required to provide more definitive proof of such involvement.

I. Anion Transport and The Protein in Band 3.

The present paper summarizes recent attempts to identify the membrane constituents involved in anion transport across the red cell membrane. These attempts have lead to the suggestion that the protein in band 3 of SDS polyacrylamide gel electropherograms is one of these constituents. A more detailed discussion shows, however, that although the suggestion is a useful guide for further work, the problem is not yet solved. Thus the emphasis of this paper does not rest on the provision of definitive conclusions, but on the presentation of an example of the problems which arise in connection with the identification of membrane constituents which participate in passive ion transport. It will become apparent that it may be more difficult to be successful with the study of systems of passive rather than active transport. This is largely due to the fact that the anion transport system in the red cell membrane does not seem to exhibit the same degree of specificity with respect to the ion to be transported as the active transport system for K + Na, or Ca; that there is no really specific inhibitor, like ouabain in K + Na transport; and that there is no coupling to ATP hydrolysis which permits, in a suspension of broken membranes, to determine the activity of the transport system by means of the activation produced by the addition of the ions to be transported.

The discussion of anion transport will concentrate on two special topics: the selectivity of inhibitors used to label membrane constituents and the use of pro-

Abbreviations used in the text:

DIDS = 4,4'-diisothiocyanato stilbene-2,2'-disulfonic acid. H_2DIDS = dihydro DIDS. SITS = 4-acetamido 4'isothiocyanato stilbene-2,2'-disulfonic acid. DAS = 4,4'-diacetamido stilbene-2,2'-disulfonic acid. APMB = 2-(4'-aminophenyl)-6-methylbenzenethiazol-3',7-disulfonic acid. DNFB = 1-fluoro-2,4-dinitrobenzene. I_2DIDS = 4,4'-diisothiocyanato diiodo stilbene-2-2'-disulfonic acid. TNBS = 1,3,5-trinitrobenzene sulfonic acid. MNT = 2-methoxy-5-nitrotropone. SDS PAGE = sodium dodecyl sulfonate polyacrylamide gel electrophoresis. K = kilo (e.g. 106 K = 106,000).

teolytic enzymes in the study of the uniformity of the protein in band 3. To introduce these special subjects we shall recall some pertinent information on the disposition of the membrane proteins, recapitulate briefly some of the more important properties of the protein in band 3, and review the developments which eventually lead to the suggestion that it plays an important role in anion transport across the red cell membrane.

On SDS gel electropherograms, 8 - 1o "major" membrane proteins have been identified as Coomassie blue stainable bands. Using enzymes and irreversibly binding, penetrating and non-penetrating chemical reagents it has been shown that all except two of these major membrane proteins are located at the inner membrane surface. The two exceptions are proteins which span the membrane. Many additional proteins are attached to the membrane in amounts too small to be detectable by the usual staining procedure. These protein species include, for example, acetylcholinesterase which is located at the outer membrane surface (13,14), and the transport ATPases for Ca^{++} and alkali ions, which span the membrane. In fact, using ^{35}S diazo sulfanilic acid as a label for the proteins in the outer membrane surface, it has been shown that in this surface there exist proteins with a whole spectrum of molecular weights, ranging from more than 1oo,ooo to less than 1o,ooo Daltons (1).

Band 3 in the nomenclature of Steck (2) or band E in that of Tanner and Boxer (3) is one of the most prominent Coomassie blue stainable bands in SDS gel electropherograms of the red cell membrane. It is somewhat diffuse and contains a glycoprotein with a carbohydrate content of about 7% (3,4). The estimates of the molecular weight obtained in a number of laboratories range from 88,ooo to 1o5,ooo Daltons (5,6,7). The protein comprises about 2o - 3o% of the total protein content of the red cell membrane, or about o.9 - 1.o x $1o^6$ molecules/cell (2,8). It spans the membrane (9,11) and can be cross-linked in situ via disulfide bridges to form dimers (10). Although definite proof is still lacking, it has been repeatedly suggested that in the intact membrane the protein forms aggregates with glycophorin, another major (glyco-)protein which also traverses the lipid bilayer, and that such aggregates represent the essential constituents of the electron-microscopically visible "membrane particles" (11, 12). The fibrous protein spectrin which extends as a meshwork at the inner membrane surface seems to be involved in fixing the positions of these particles within the membrane (15). It has been proposed to call the membrane particles "Permeaphors" (16). This is based on reports that the protein in band 3 mediates the transport of water (17,18) and sugars (19,20), controls passive cation movements (21) and participates in anion exchange (22,23). Although it is fascinating to think of electronmicroscopically visible packages of membrane proteins which control the flow of water and many solutes, additional evidence is required to justify such terminology. Nevertheless, at least so far as anion exchange is concerned there are good reasons to suspect that the protein plays indeed a role.

The suggestion that the protein in band 3 participates in anion transport emerged from studies on the chemical and enzymatic modification of anion transport and membrane constituents. In previous work from this laboratory it was shown that a variety of amino reactive reagents including DNFB (24, 25), MNT (26) and TNBS (27) as well as a number of enzymes including pronase (28) and papain (29) are powerful inhibitors of anion transport in the human red cell. This work was confirmed by others (30). It was shown in addition that the inhibitory effect of pronase could not be related to esterolytic effects on the principal membrane lipids (28) and it was demonstrated that inhibition by maleylation took place without measurable modification of amino lipids (31). It was suggested therefore that membrane proteins control the anion exchange across the red blood cell membrane. At this stage, Knauf and Rothstein reported that SITS, a non-penetrating isothiocyanate of a stilbene disulfonic acid first used for membrane studies by Maddy (32) inhibits anion transport (30). The diisothiocyanate derivatives I_2DIDS and DIDS were found to be even more effective than SITS. While I_2DIDS reacted with a variety of membrane proteins (33), DIDS was bound primarily to the protein in band 3 (22) and it was concluded that this protein mediates anion transport (22). At the same time the observation was reported that two different inhibitors of anion transport, SITS (or DAS) and DNFB have common binding sites only at the protein in band 3, and it was pointed out that this suggests a participation of this protein in the control of anion transport (23,34). The presence of common binding sites on the protein in band 3 was confirmed with another pair of inhibitors (pyridoxal phosphate and DIDS (35)) and attempts were made to use p-isothiocyanatobenzene sulfonic acid in place of DIDS as a label for the identification of the anion transport protein (36).

Attempts to isolate the anion transport system followed two different avenues. In our laboratory we degraded most of the membrane proteins by incorporation of trypsin into red cell ghosts. This lead to the formation of vesicles which, in many respects, are similar to liposomes. SITS and DNFB continued to inhibit anion transport in these vesicles. The only common binding sites for the two agents were those on two peptides derived from the protein in band 3 (23). This work will be discussed below in more detail. Rothstein and his associates, on the other hand, isolated a protein fraction from the red cell membrane which consisted of about 80% of the protein in band 3. They used this fraction for recombination with liposomes in order to reconstitute the anion transport system. The results were encouraging (37). It is obvious, however, that both approaches need further refinement.

The described experiments, notably the more enthusiastic reports of the Toronto Group, were widely accepted as sufficient proof for the participation of the protein in band 3 in anion transport. In our own publications, we clearly suggested such involvement but we placed much emphasis on still unresolved problems (23, 34,42). We pointed out that all of the described attempts to identify the anion transport protein involve the use of isothiocyanates. These compounds are capable of reacting with amino groups and possibly other functional groups of protein molecules, regardless of whether these groups are constituents of the anion transport system or not. The actual occurence of reaction with groups other than those on the protein in band 3 has been shown in white ghosts where the proteins at the inner membrane surface are accessible to DIDS. Virtually all of these proteins reacted with the agent (ref.22,fig.7). Moreover, it has also been shown that many of the amino groups in human serum albumin also combine with H_2DIDS (38). It seems, therefore, that in contrast to the specificity of inhibition of active cation transport by ouabain, the specificity of inhibition of anion transport by DIDS or H_2DIDS is primarily (although perhaps not exclusively) due to limited accessibility rather than susceptibility of the binding sites. This conclusion is important for two reasons. Firstly, it cannot be excluded with certainty that isothiocyanate binding sites other than those in band 3 are related to inhibition, but escaped detection as a discrete band because of their small numbers. Secondly, the isothiocyanate binding sites in band 3 may not represent a population of functionally identical sites and include populations not involved in anion transport.

Before entering the discussion of these two points, it should be mentioned that the binding sites for isothiocyanates like SITS, DIDS etc. do not seem to be the only ones which are involved in anion transport. This was inferred from work with resealed ghosts which showed that certain inhibitors like phlorizin (39) or DAS exert their effect at the outer membrane surface while others like APMB acted at either surface (23,34). Moreover,even among phlorizin and DAS a striking difference was observed. Although both agents affect transport with about equal strength, only DAS has common binding sites with DNFB on the protein in band 3 while phlorizin has not (34). One may suspect, therefore, that there participate in the control of anion transport three different sets of sites (i) fixed sites at the outer membrane surface for DNFB, DAS, etc., (ii) other fixed sites in that surface for phlorizin, and (iii) common sites for APMB and DNFB which are accessible at both surfaces and may belong to a mobile carrier. It is unknown whether or not all three sets of binding sites are associated with the protein in band 3. The participation of at least two types of sites - transfer sites and modifier sites - has also been inferred from studies on the concentration and pH dependence of anion exchange (4o, 41). A discussion of the possible functions of the various sites in anion translocation across the membrane is outside the scope of the present paper and will be presented elsewhere.

II. The Relationship Between Inhibition of Anion Transport and The Binding of DIDS or Its Dihydroderivative H_2DIDS to The Proteins in The Red Cell Membrane.

In view of the significance of the work on DIDS binding for the identification of the anion transport protein, it would seem useful to determine the relationship between DIDS binding and inhibition of anion transport, and to estimate the number of DIDS molecules bound at maximal inhibition. Such attempt was made by Cabantchik and Rothstein (22). In their work, for the determination of DIDS binding, tritiated DIDS was used as a radioactive label for DIDS. However, during the tritiation procedure, the double bond of the stilbene molecule is reduced and a diethane derivative, $^3H_2$DIDS, is formed. It was discovered that the reactivity of this compound towards the proteins of the red cell membrane is considerably lower than that of DIDS, and that for this reason the previous results which were obtained with mixtures of DIDS and $^{3}H_2$DIDS were erroneous. Separate studies with DIDS and H_2DIDS yielded the following picture (42):

On SDS polyacrylamide gel electropherograms we found, in addition to the previously observed labeling of band 3 and glycophorin, another distinct labeled band corresponding to a molecular weight of about 70,000 Daltons. Altogether about 30% of the total radioactivity bound could not be recovered on the protein in band 3. This fraction was virtually independent of the total amount of $^{3}H_2$DIDS attached to the membrane. The relationship between binding and inhibition was linear, regardless of whether total binding, binding to the protein in band 3, or the difference between the two was plotted against anion flux. Inhibition was complete when total binding was about 1.88×10^{6} molecules H_2DIDS/cell. This value is almost one order of magnitude higher than the previous estimate of 0.3×10^{6} molecules/cell (22). The corresponding values for binding to the protein in band 3 and proteins other than those in that band were 1.2×10^{6} and 0.6×10^{6} molecules/cell, respectively (fig.1). Measurements of DIDS binding yielded essentially similar results. Binding at complete inhibition was, however, about 10% lower than the corresponding values for H_2DIDS and the proportion of DIDS bound to proteins other than glycophorin and those in band 3 was somewhat smaller than for H_2DIDS. These findings are in accord with a participation of the binding sites on the protein in band 3 in anion transport. They are also compatible, however, with the assumption that binding sites on small numbers of other membrane constitutents are involved.

The new estimates permit the calculation of turnover numbers for anion transport per H_2DIDS binding site on the protein in band 3. Using recent estimates of maximal rate and apparent activation enthalpy for chloride transport (41) we obtain a value of $2\text{-}3 \times 10^{4}$/sec at 25°C which is close to the turnover number observed for valinomycin mediated potassium transport across artificial lipid bilayers at that temperature (43). Such turnover number is unlikely to reflect a rotation in toto of molecules of a molecular weight of 100,000 Daltons. Therefore, if the molecules in band 3 should in fact be involved in anion transloca-

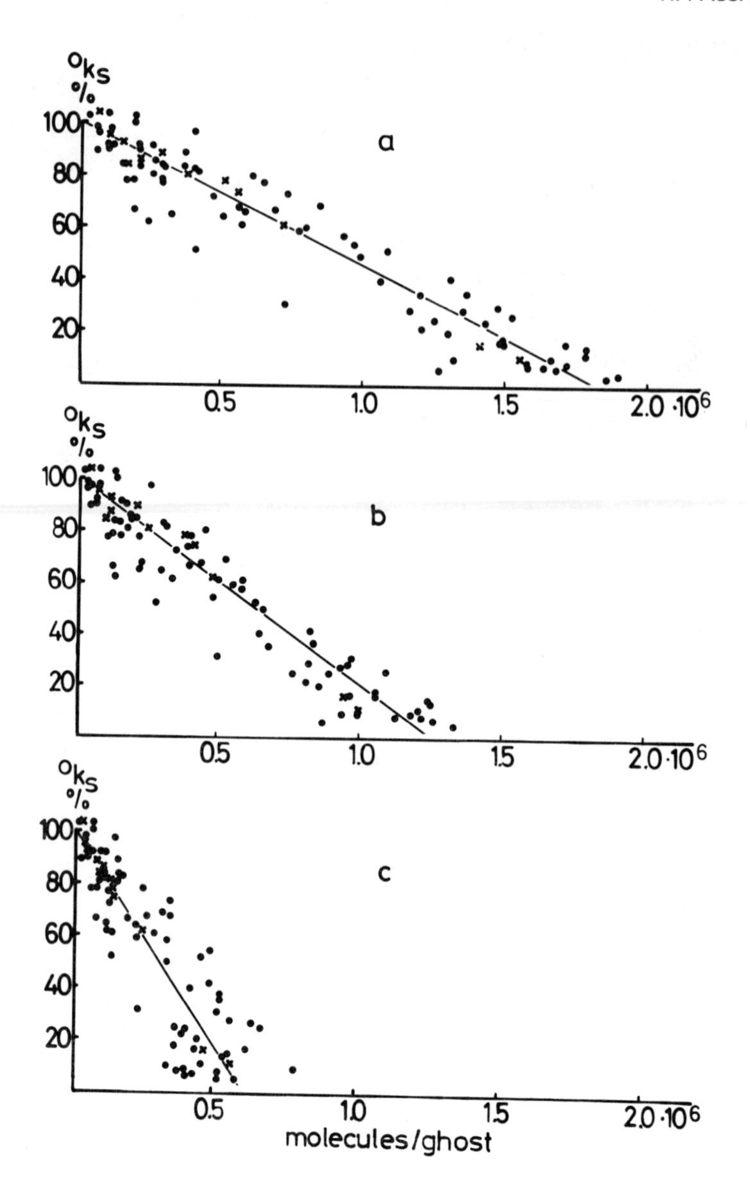

Fig. 1: Relationship between sulfate equilibrium exchange and H_2DIDS binding to (a) the red cell membrane as a whole (b) the protein in band 3 (c) membrane constituents other than those in band 3 (difference between (a) and (b)). Ordinate: rate constant, as a percent of control without H_2DIDS. Abscissa: binding, molecules per ghosts. Same data as in ref. (42).

tion across the membrane, it would seem necessary to postulate that this is accomplished by a rotation of side chains rather than of the molecules as a whole. The rate of such rotation is unknown but the possibility cannot be ruled out that it is at least one order of magnitude higher than that of valinomycin-mediated transport. This would imply that smaller numbers of H_2DIDS binding sites than the total number determined in band 3 could suffice to accomplish anion transport at the observed rate.

Our new figures on the binding of DIDS or H_2DIDS also permit a calculation of the stoichiometrical ratio for the number of binding sites per protein molecule in band 3 and yields values of 1.2 to 1.3. This raises the question whether or not all binding sites are functionally equivalent and involved in anion transport. Previous work showed that not all of the common binding sites for SITS and DNFB can be related to anion transport and suggests that these sites consist of two functionally distinct populations (34).

In summary, the results of our reinvestigation of the relationship between binding and effect of H_2DIDS or DIDS are compatible with the assumption of a participation of at least a fraction of the binding sites on the protein in band 3 in anion transport. However, they do not rule out the possibility that small numbers of other binding sites instead of those in band 3 are involved. Experiments with internally trypsinized ghosts described below would support the view that some component of band 3 plays an important role.

III. Enzymatic Modification.

<u>A. Externally applied enzymes.</u>The enzymatic modification of anion transport (28,29,30) and membrane proteins (1, 5, 7, 45, 46), including the membrane proteins of DIDS-labeled cells (44), has been investigated in the past. So far, however, no attempt has been made to study the effects on membrane proteins, H_2DIDS binding, and anion transport systematically under identical conditions. Below the results of experiments performed under such conditions will be briefly summarized. It will be shown that each of a number of different proteolytic enzymes produces a different effect on the membrane proteins, including the protein in band 3, and that these differences are accompanied by characteristic

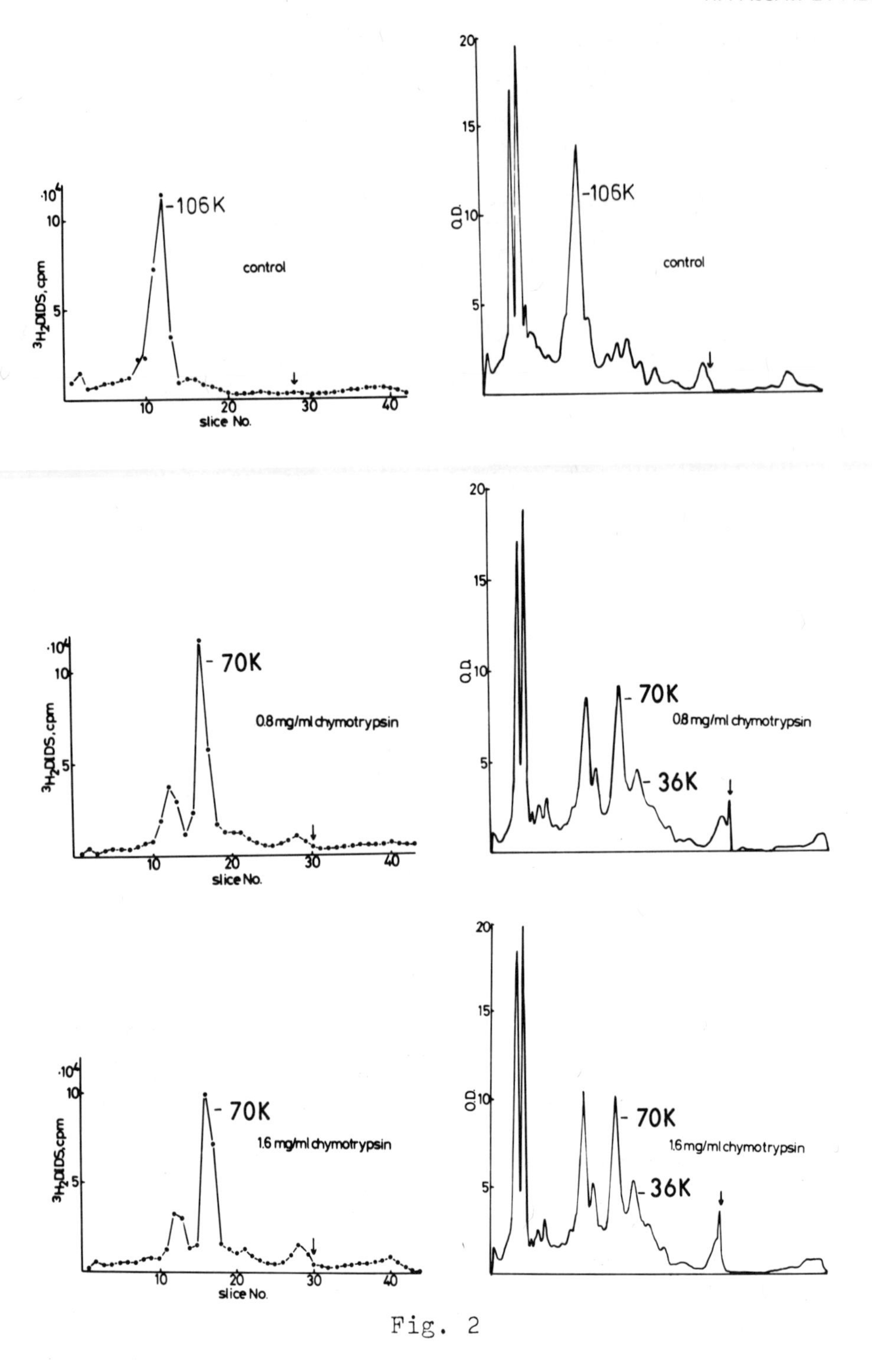

·10⁴
10
5
³H₂DIDS, cpm
-106K
control
10
20
30
40
slice No.
20
15
10
5
O.D.
-106K
control
·10⁴
10
5
³H₂DIDS, cpm
- 70K
0.8 mg/ml chymotrypsin
10
20
30
40
slice No.
O.D.
- 70K
0.8 mg/ml chymotrypsin
- 36K
·10⁴
10
5
³H₂DIDS, cpm
- 70K
1.6 mg/ml chymotrypsin
10
20
30
40
slice No.
O.D.
- 70K
1.6 mg/ml chymotrypsin
- 36K

Fig. 2

Fig. 2: Effect of chymotrypsin on proteins of $^3H_2$DIDS labeled red cell membranes. Intact cells were first labeled by incubation at 37°C for 9o min. at a $^3H_2$DIDS concentration of 25 µM (hematocrit 1o%), washed, and then subjected to enzymatic treatment at 37°C for 6o min. at the enzyme concentrations indicated in the figure. Subsequently, the membranes were isolated, dissolved in SDS (o.5%), and subjected to SDS-PAGE in 5% gels. The gels were run in duplicates. One of each pair was sliced and used for the determination of radioactivity (left), the other was stained with Coomassie blue and scanned in a Gilford spectrophotometer (right). 7o K refers to the major of the 2 fragments of the 1o6 K protein. On the gels, this band is located at the same position as plasma albumine, i.e. at about 67.ooo Daltons. However, since it is not clear whether this is significantly lower than 7o.ooo Daltons as observed for the major split products of pronase or papain, it is designated in the figures and in the text by 7o K.

differences of the effects on anion transport. It will further be demonstrated that the resistance of the protein in band 3 to enzymatic hydrolysis is not uniform, indicating that there exist differences of accessibility or susceptibility to enzymatic degradation. These observations support the previously expressed suspicion that the protein in band 3 is not only heterogeneous with respect to its carbohydrate moiety but that there may also exist conformational or chemical differences in the peptide chains. The observations are also compatible with the view that the population of binding sites on the protein in band 3 for the isothiocyanates of stilbene disulfonic acids and related compounds is not homogeneous (34).

The non-uniform response of the protein in band 3 to enzymatic hydrolysis is illustrated in fig. 2. In the experiments represented there, red cells which had first been labeled at a saturating concentration of $^3H_2$DIDS were treated with various concentrations of chymotrypsin. The gel electropherograms show that part of the protein in band 3 is split into two moieties with molecular weights of 70,000 and 36,000 Daltons and that a fraction of the protein is resistant to enzymatic hydrolysis. This is more clearly demonstrated in fig.3 where the number of $^3H_2$DIDS molecules associated with the newly formed 70,000 Dalton component and that associated with the undigested 106,000 Dalton protein are plotted against the employed chymotrypsin concentrations. Evidently, part of the protein in band 3 is much more resistant to chymotryptic digestion in situ than the rest. This indicates differences of accessibility and susceptibility among the individual protein molecules in band 3 (c.f. ref. 45, p.1155).

The experiments in fig. 4 convey a similar message. In these experiments, the red cells were first exposed to the enzymes indicated in the figure (chymotrypsin, papain, pronase), washed, and subdivided into two batches. One of these batches was treated with a saturating concentration of H_2DIDS to determine ("titrate") the number of $^3H_2$DIDS binding sites in the enzymatically modified membranes (i.e. the sum of surviving sites and newly exposed sites). The other was used for determinations of sulfate equilibrium exchange. This allows to establish the correlation between H_2DIDS binding sites and anion equilibrium exchange in the enzymatically modified cells.

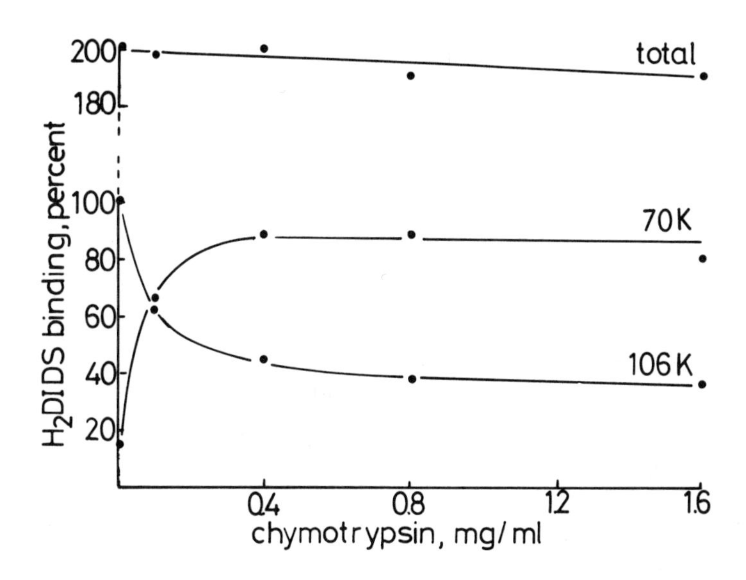

Fig. 3: Effect of chymotrypsin on $^{3}H_2$DIDS labeled red cell membranes. Same experimental conditions as in fig.2. Ordinate: $^{3}H_2$DIDS as a percent of $^{3}H_2$DIDS on the protein in band 3 in untreated cells. Abscissa: concentration of chymotrypsin. "total", "70 K" and "106 K" refer to $^{3}H_2$DIDS associated with the outer membrane surface as a whole, the 70 K fragment, and the protein at the original location of band 3, respectively. 100 percent corresponds to about 1.23 x 10^{6} H_2DIDS molecules/cell.

In spite of hydrolysis by chymotyrpsin of part of the protein in band 3 into 70,000 and 36,000 Dalton fragments, there is very little inhibition of anion transport (figs. 4, 5). A comparison with the results obtained in the experiments in fig. 3 shows that chymotryptic cleavage neither seems to expose new H_2DIDS binding sites nor to remove previously existing sites. The non-uniform response of the protein to the enzymatic attack is again clearly exhibited. The experiments suggest that either the surviving 1o6,ooo Dalton peptide +), or the newly formed peptides or both continue to mediate anion transport nearly as effectively as in the intact cell.

P r o n a s e , at low concentrations, where the inhibition of anion transport is insignificant, converts part of the protein in band 3 into a peptide of a molecular weight of about 7o,ooo Daltons, and into smaller fragments. The rest of the 1o6,ooo Dalton protein is fairly resistant to degradation. This action of pronase is similar to that of chymotrypsin, and supports the conclusions drawn above concerning the heterogeneity of the protein in band 3 and the transport of anions in the modified cells. In contrast to chymotrypsin, however, with increasing concentrations of pronase, both, the surviving 1o6,ooo Dalton peptide and the 7o,ooo Dalton peptide are further degraded (fig. 4) and anion transport is inhibited (figs. 4,5). The decrease of the number of $^3H_2$DIDS binding sites on both the 1o6,ooo Dalton and the 7o,ooo Dalton peptide is linearly related to the inhibition of anion transport (fig. 5). It is not possible, therefore, to decide which of the two peptides could be involved in anion transport. In contrast to chymotrypsin, pronase exposes as well as releases $^3H_2$DIDS binding sites on the membrane as a whole as well as on the protein in

+) Footnote: In all experiments, except the double labeling experiments with trypsinized ghosts described in section III B, molecular weights (MW) were estimated by running on the same gel the unknown sample and ^{14}C DNP derivatives of proteins of known MW. In the trypsin experiments (performed in 1973/74, ref. 23) standard proteins were run on separate gels. According to ref. 45, the MW's of the major fragments obtained after digestion by trypsin, chymotrypsin, papain and pronase are about equal. We report a MW for the major fragment of digestion by trypsin which is somewhat lower than that produced by the other three enzymes. This does not necessarily indicate a factual difference but may reflect a difference of our former and present method of measuring MW's.

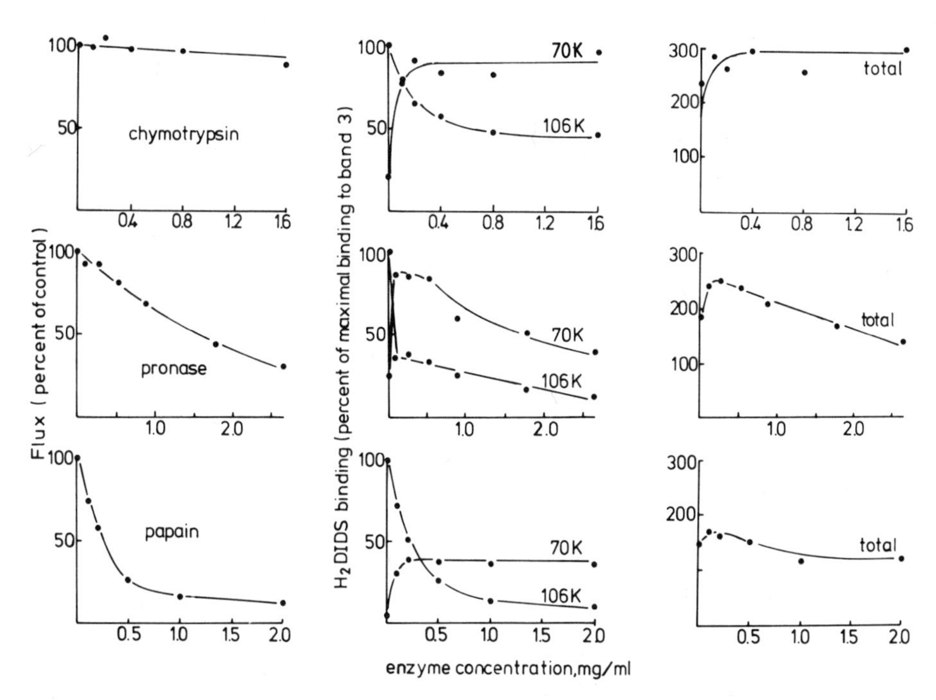

Fig. 4: Effects of increasing concentrations of proteolytic enzymes on anion exchange (left) and the capacity to bind H_2DIDS to the outher membrane surface as a whole (right) or to the isolated protein in band 3 (1o6 K) and the major peptides (7o K) derived from that band by enzymatic hydrolysis (middle). Red cells incubated at 37°C, pH 7.4, 1o% hematocrit, for 6o min. in phosphate buffered (2o mM) saline containing 5 mM Na_2SO_4 and the enzymes at the concentrations indicated on the abscissa. After removal of the enzymes by washing, the cells were subdivided for separate measurements of sulfate equilibrium exchange and $^3H_2$DIDS binding. The latter was measured after 9o min. of incubation at 37° C in the medium described above containing 25 µM $^3H_2$DIDS. Bindind is expressed as a percent of binding to band 3 in the absence of enzymatic treatment. 1oo% corresponds to about 1.23 x $1o^6$ molecules H_2DIDS/cell.

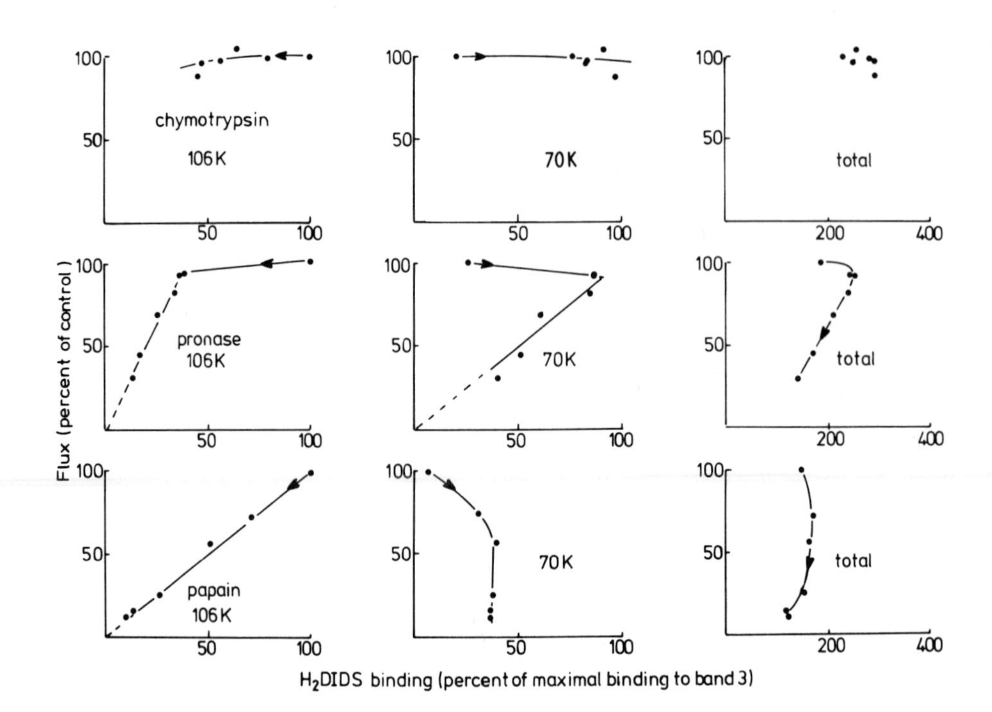

Fig. 5: Relationship between sulfate flux and number of $^3H_2$DIDS binding sites on enzymatically modified cells. Intact cells are first exposed to the desired enzyme and subsequently subdivided into two batches. One is used for the determination of $^3H_2$DIDS binding in the presence of a large excess of $^3H_2$DIDS in the medium, the other for flux measurements in the absence of $^3H_2$DIDS. This procedure was performed at a number of different enzyme concentrations. Corresponding values of binding and flux obtained at the various enzyme concentrations are plotted against each other . Same data as in fig. 4. The arrows indicate the direction of increasing enzyme concentrations. The curves denoted 7o K do not start at the origin. The reason is that in intact, untreated cells we always find some H_2DIDS binding at a distinct band representing a molecular weight of 7o,ooo. The number of the sites is somewhat variable in different samples of blood. Binding is expressed as a percent of binding to band 3 in the absence of enzymatic treatment. 1oo% corresponds about 1.23 x $1o^6$ molecules H_2DIDS/cell.

band 3. The latter is evident if one forms the sum of $^3H_2$DIDS binding to the 106,000 and 70,000 Dalton fragments in fig. 4.

P a p a i n , like pronase, inhibits anion transport (fig. 4). The enzyme also converts part of the protein in band 3 into several peptides including a peptide with a molecular weight of about 70,000 Daltons which is fairly resistant to further degradation. The remaining part of the 106,000 Dalton protein is hydrolyzed into fragments all of which have molecular weights of less than 70,000 Daltons. The emergence of different types of products of enzymatic hydrolysis again suggests a heterogeneity of the protein in band 3. There exists a linear relationship between anion transport and the number of $^3H_2$DIDS binding sites that can be titrated on the 106,000 Dalton peptide. In contrast, over a wide range of papain concentrations which induce large variations of anion transport, the number of titrable $^3H_2$DIDS binding sites on the 70,000 Daltons peptide remain essentially constant (fig. 5). Papain, like pronase, exposes new H_2DIDS binding sites and releases previously existing ones (fig. 4).

The observation that the effects of pronase and papain on transport are similar while those on the protein in band 3 are different, does not necessarily contradict the assumption that this protein is involved in anion transport. In spite of the fact that a 70 K fragment is formed by each of the two enzymes, it is unlikely that these fragments are identical. One enzyme may split the transport protein closer to the active center than the other and this may account for the differences of the effects.

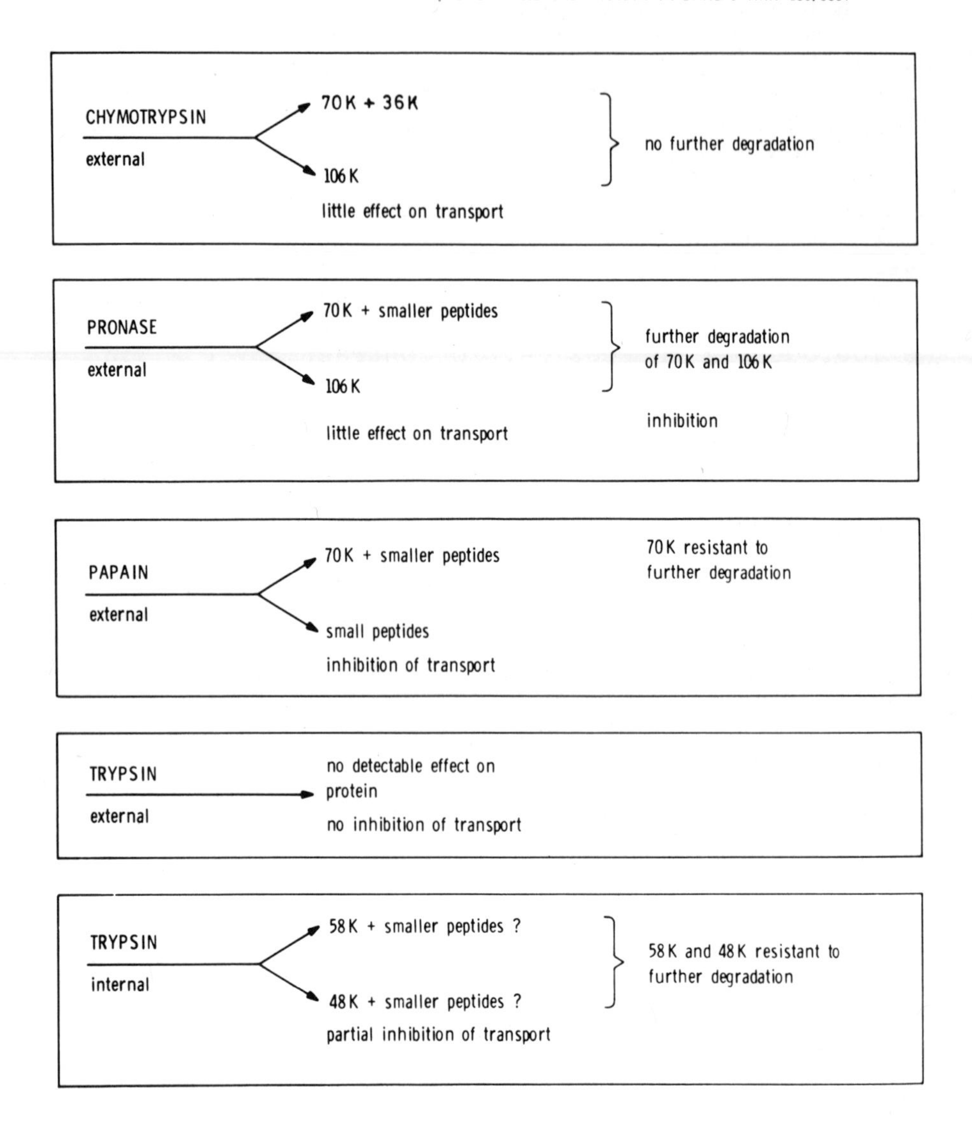

Fig. 6: Summary of conclusions drawn from the data in Figs. 4 and 5.

B. Internally applied trypsin. The results described above were obtained after exposure of intact cells to the desired enzyme in the external medium. Below we shall supplement this information by a brief summary of experiments with an internally applied enzyme, trypsin. As has already been mentioned, external trypsin neither affects the protein in band 3 nor does it produce inhibition of anion transport. Internally applied trypsin behaves quite differently. Using previously described methods (47) it is possible to incorporate into red cell ghosts varying concentrations of the enzyme, to reseal them, and to study the ensuing changes of membrane proteins and anion transport. Without producing much hemolysis, internal trypsin leads to a fragmentation of the ghosts into large and small vesicles. Spectrin is digested first, followed by the protein in band 3 and the other bands of SDS polyacrylamide gel electropherograms. After removal of about 60% of the total membrane protein, enzymatic degradation comes to an end. Under these conditions we usually find 3 bands representing molecular weights of 58,000, 48,000 and 34,500 Daltons. Sometimes we observe a fourth band at 13,000 Daltons (fig. 7). The two bands at 58,000 and 48,000 Daltons are derived from the protein in band 3. This is demonstrated in fig. 8 where it is shown that the common binding sites

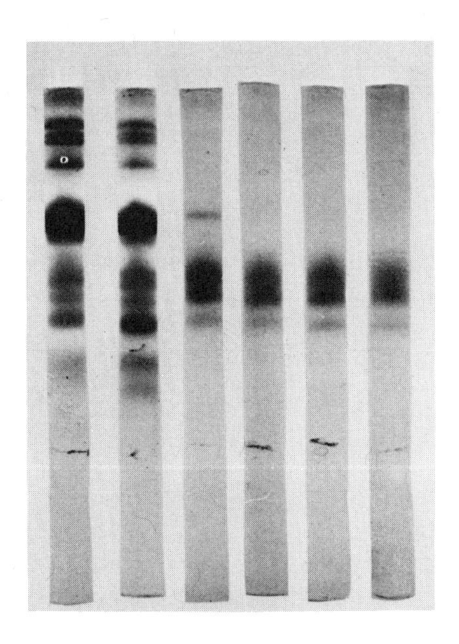

Fig. 7: Effect of external and internal trypsin on membrane proteins of resealed ghosts. Trypsin concentrations (from left to right): 0,500 µg/ml (external); 10,50, 100, 500 µg/ml (internal). Incubation for 45 min. at 37°C. 5% polyacrylamide gels, 0,5%.

for SITS and DNFB which were originally associated with the protein in band 3 are now observed at the location of the two new bands at 58,000 and 48,000 Daltons. No such sites are left at the original position of band 3. Although the resolution of the gels is not perfect,the results in fig. 8 suggest that the common binding sites are about equally distributed among the two bands. Since the label resides in the outer surface of the ghosts, and the enzyme at the inner surface, our results confirm, with a different technique, previous findings which indicated that the protein in band 3 spans the membrane.

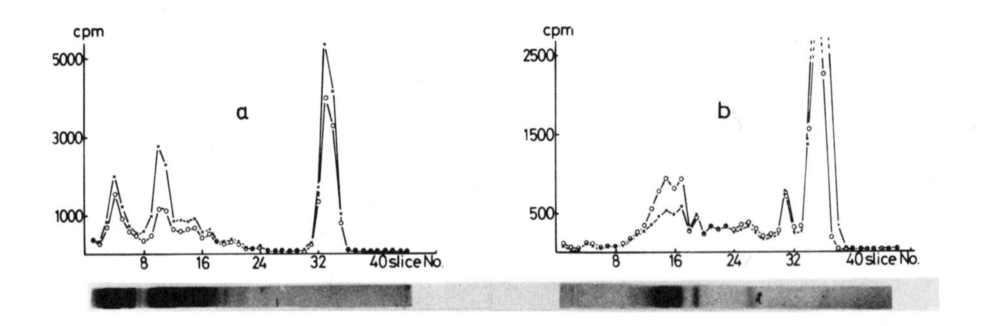

Fig. 8: Location of common binding sites for SITS and DNFB on SDS polyacrylamide gel electropherograms of isolated membranes derived from resealed ghosts after incubation without (a) or with (b) incorporated trypsin. After incubation at 37°C for 45 min., the ghosts were exposed first to SITS (0.5 mM) and subsequently to ^{14}C DNFB (lower tracings) or directly to ^{3}H-DNFB without pretreatment with SITS (upper tracings). After completion of chemical modification, the ^{14}C and ^{3}H labeled ghosts from (a) or (b) were mixed and electrophoresed on the same gels. 5% polyacrylamide, 0.5% SDS. Note that in the SITS treated ghosts less DNFB is bound to the protein in band 3 and to the 55,000 and 48,000 Dalton peptides of tryptic digestion.

In contrast to external chymotrypsin or low concentrations of external pronase, internal trypsin cleaves the protein in band 3 completely. Since the sum of the molecular weights of the two fragments of tryptic digestion is close to the molecular weight of the untreated protein in band 3, it would seem reasonable to assume that these fragments are derived from the same peptide. Any heterogeneity of that peptide would then be confined to segments of the peptide chain which are in contact with the outer membrane surface. Our results would, however, also be compatible with the assumption that trypsin splits two different proteins of 106,000 Daltons, each of which spans the membrane

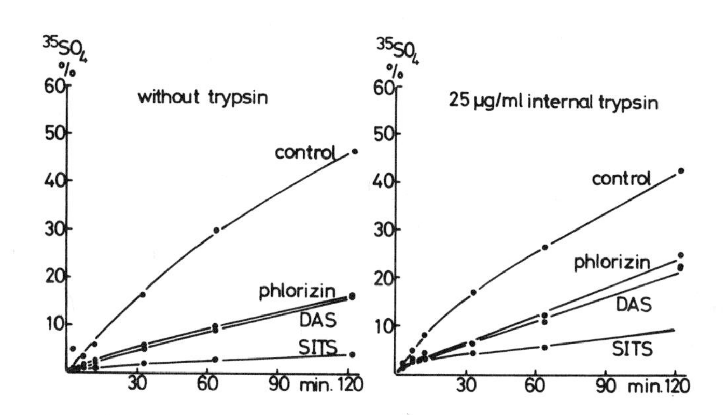

Fig. 9: Effect of various inhibitors of sulfate equilibrium exchange on control ghosts and ghosts exposed to internal trypsin for 45 min. at 37°C prior to the flux measurements. At t = o, the $^{35}SO_4$ containing ghosts were suspended in a medium free of $^{35}SO_4$ and the appearance of the radioactivity in the supernatant was followed : Ordinate: $^{35}SO_4$ in the supernatant of the ghosts suspension (10 Vo.%, pH 7.4, 30°C). Abscissa: time in min. The inhibitors were present in the external medium. Phlorizin: 2mM; DAS: 2.0 mM; SITS: 0.5 mM. Control: no inhibitor present. Figs 8 & 9 are adapted from Passow et al. (23)

and gives rise to one of the H_2DIDS labeled peptides. The other peptides required to bring the sum of the molecular weights of all fragments to 1o6,ooo Daltons may not be labeled by H_2DIDS and be released into the supernatant. Water soluble peptides have indeed been observed after tryptic digestion of membranes which were electrophoresed after treatment with alkaline media and had lost most of their protein except the protein defived from vand 3 (45). Thus, although this would not seem likely it is not impossible that the two labeled products of tryptic digestic are derived from different peptides.

Anion equilibrium exchange in the internally trypsinized ghosts proceded more slowly than in control ghosts which contained no trypsin. The degree of inhibition is, however, difficult to assess. The size of the vesicles formed under the influence of internal trypsin is quite different from the size of the intact ghosts and, more important, varies considerably among the individual members of the population. For this reason it was not yet possible to obtain a measure of the surface/volume ratio which is sufficiently accurate for the calculation of permeability coefficients. Nevertheless, exposure of the trypsin-treated vesicles to a series of the usual inhibitors of anion transport (H_2DIDS, SITS, DNFB, phlorizin, APMB, dipyridamol) produces strong further inhibition (fig. 9). This indicates that in spite of the release or drastic alteration of the major membrane proteins , essential elements of the anion transport system are still intact and strongly suggests that at least one of the peptides derived from the protein in band 3 performs an important function in anion transport.

IV. Conclusions.

The reinvestigation of the relationship between inhibition of anion transport and binding of H_2DIDS or DIDS to the red cell membrane has shown that the protein in band 3 is not the only candidate for some function in anion transport as was originally thought. The assumption of a participation of that protein in anion transport is, however, compatible with the finding that the different effects of three different enzymes on the protein in band 3 are associated with different effects on anion transport. It is supported by the experiments with internally trypsinized ghosts which lost most of their membrane proteins but retained two peptides derived from the protein in band 3 and which continued to show essential features of the normal anion transport system.

If one accepts the described evidence for a participation of the protein in band 3 in anion transport as sufficient proof, it remains to be seen whether or not all of the H_2DIDS binding sites on that protein are involved. Our experiments show that the response of the protein in band 3 to treatment with externally applied chymotrypsin, pronase, and papain is not uniform (see fig. 6). This could be explained in several different ways: (i) All molecules in band 3 are identical. Hydrolysis of some of them leads to the formation of products which protect the remaining molecules against enzymatic hydrolysis. (ii) The conformation or arrangement in the lipid bilayer of protein molecules with the same primary structure may differ such that some of them are susceptible or accessible to enzymatic attack while others are not. (iii) There exist two populations of distinct protein species with different primary structures. So far, we were unable to decide between the various possibilities. However, (iii) and (ii) are more likely than (i). It should also be noted that case (ii) involves the further assumption that molecules in different locations or conformational states do not attein thermodynamical equilibrium within the time of exposure to the enzymes. Finally, it should be pointed out that slight differences in primary structure which manifest themselves in different susceptibilities to enzymatic hydrolysis would not necessarily rule out the possibility that all H_2DIDS binding sites on band 3 are involved in anion transport. H_2DIDS binding sites on slightly different peptide chains could still perform the same function nearly equally well. If so, the protein in band 3 would constitute a mixture of peptides which act like isocymes.

References.

1 BENDER, W.W., GARAN, H. and BERG, H.C.:J.Mol.Biol. 58 (1971) 783.
2 STECK, T.L.: J.Cell.Biol. 62 (1974) 1.
3 TANNER, M.J.A. and BOXER, D.H.: Biochem. J. 129 (1972) 333.
4 JULIANO, R.L. and ROTHSTEIN, A.: Biochim. Biophys. Acta 249 (1971) 227.
5 FAIRBANKS, G., STECK, T.L., WALLACH, D.F.H.: Biochemistry 1o (1971) 26o6.
6 BRETSCHER, M.S.: J.Mol.Biol. 59 (1971) 351.
7 TRIPLETT, R.B. and CARRAWAY, K.L.: Biochemistry 11 (1972) 2897.
8 JULIANO, R.L.: Biochim. Biophys. Acta 3oo (1973)341.
9 BRETSCHER, M.S.: Nature 231 (1971) 229.
1o STECK, R.L.: J. Mol. Biol. 66 (1972) 295.
11 SHIN, B.C. and CARRAWAY, K.L.: Biochim. Biophys. Acta 345 (1974) 141.
12 PINTO DA SILVA, P., MOSS, P.S., FUDENBERG, H.H.: Exptl. Cell. Res. 81 (1973) 127.
13 KIEFER, H. LINDSTROM, J., LENNA, E.S., SINGER, S.J.: Proc. Nat . Acad. Sci. U.S.A. 67 (197o)1688.
14 BELLHORN, M., BLUMENFELD, O.O., GOLLOP, P.M.: Biochem. Biophys. Res. Comm. 39 (197o) 267.
15 NICOLSON, G.L. and PAINTER, R.G.: J. Cell Biol. 59 (1973) 395.
16 PINTO DA SILVA, P. and NICOLSON, G.L.: Biochim. Biophys. Acta 363 (1974) 311.
17 BROWN, P.A., FEINSTEIN, M.B., SHA'AFI, R.I.: Nature New Biol.: 254 (1975) 523.
18 PINTO DA SILVA, P.: Proc. Nat. Acad. Sci. U.S.A. 7o (1973) 1339.
19 TAVERNA, R.D. and LANGDON, R.G.: Biochem. Biophys. Res. Comm. 54 (1973) 593.
2o LIN, S. and SPUDICH, H.A.: Biochem. Biophys. Res. Comm. 61 (1975) 1471.
21 Suggested by A. ROTHSTEIN, personal communication.
22 CABANTCHIK, Z.I. and ROTHSTEIN, A.: J. Membrane Biol. 15 (1974) 2o7.
23 PASSOW, H., FASOLD, H. ZAKI, L., SCHUHMANN, B. and LEPKE, S. In:Biomembranes, Structure and Function (G.Gárdos and I. Szász Edts.) p. 197 - 214 (Proceedings of the 9th FEBS Meeting, Budapest 1974).
24 PASSOW, H.: Progress in Biophys., Mol.Biol. 19 (1969) 424.
25 POENSGEN, J. and PASSOW, H.: J. Membrane Biol. 6 (1971) 21o.
26 SCHNELL, K.F. and PASSOW, H.: Experientia 25 (1969) 46o.

27 ZAKI, L., GITLER, C. and PASSOW, H.: XXV Internat. Congress Physiol. Sci. Munich 1971, p. 616.
28 PASSOW, H.: J. Membrane Biol. 6 (1971) 233.
29 SCHWOCH, G., RUDLOFF, V., WOOD-GUTH, I. and PASSOW, H.: Biochim. Biophys. Acta 339 (1974) 126.
3o KNAUF, P.A. and ROTHSTEIN, A.: J.Gen.Physiol. 58 (1971) 19o.
31 OBAID, A.L., REGA, A.F., GARRAHAN, P.: J.Membrane Biol. 9 (1972) 385.
32 MADDY, A.H.: Biochim. Biophys. Acta 88 (1964) 39o.
33 CABANTCHIK, Z.I. and ROTHSTEIN, A.: J. Membrane Biol. 1o (1972) 311.
34 ZAKI, L., FASOLD, H., SCHUHMANN, B. and PASSOW, H.: J. Cell. Physiol. 86 (1975) 471.
35 ROTHSTEIN, A. et al. J. Biol. Chem. In the press.
36 HO, M.K. and GUIDOTTI, G.: J. Biol. Chem. 25o (1975) 675.
37 ROTHSTEIN, A.,CABANTCHIK, Z.I., BALSHIN, M. and JULIANO, R.: Biochem. Biophys. Res. Comm. 64 (1975) 144.
38 FASOLD, H.: unpublished results.
39 LEPKE, S. and PASSOW, H.: Biochim. Biophys. Acta 298 (1973) 529.
4o GUNN, R.B. In: Oxygen Affinity of Hemoglobin and Red Cell Acid Base Status (Rørth, M. and Astrup Edts.) p. 823 - 827 Copenhagen, Munksgaard (1972).
41 DALMARK, M.: J. Physiol. (Lond.) 25o (1975) 65.
42 LEPKE, S., FASOLD, H., PRING, M. and PASSOW, H.: J. Membrane Biol., in the press.
43 LÄUGER, P.: Science 178 (1972) 24.
44 CABANTCHIK, Z.I. and ROTHSTEIN, A.: J. Membrane Biol. 15 (1974) 227.
45 STECK, T.L., RAMOS, B. and STRAPAZON, E.: Biochemistry 15 (1976) 1154.
46 STECK, T.L. and FOX, C.F.: In: Membrane Molecular Biology (Fox, C.F. and Keith, A.D. Edts.) p. 27 Stamford, Conn. Sinauer Assoc. (1972).
47 SCHWOCH, G. and PASSOW, H.: Mol. Cell. Biochem. 2 (1973) 197.

We thank Drs. V.Rudloff, D.Schubert, Ph.Wood and L.Zaki for reading the manuscript and their comments.

DISCUSSION

PRESSMAN: Your turnover number is two to four orders of magnitude lower than that generally accepted for channels in neural preparations and at the upper edge of what one would expect for a carrier mediated transport. You keep referring to this as a channel even when in the presentation by Dr. Rothstein, the channel is so complicated that the distinction between it and a carrier is somewhat obscure. We victims of ionophoria would much prefer they (ionophoria) be called "pathways." There is no reason to exclude the possibility that these pathways fall closer to the classical definition of carriers than channels. I can't understand the reluctance to concede this.

PASSOW: You have contributed a lot to the development of the concept that anion transport in erythrocytes may be mediated by a carrier and not by channels. I think that the erythrocyte people have accepted this reasoning to a very large extent.

ROTHSTEIN: I was just going to say that I would like it both ways. The model I presented involves a carrier working within a channel. Such a model will fit typical carrier kinetics. The reason I included a channel as part of the model relates to the fact that band 3, which seems to contain the anion transport site, is a dimeric transmembrane protein folded twice through the membrane so that six polypiptide strands may traverse the lipid bilayer. It seems sensible that these strands will be arranged collectively with a hydrophilic core and with hydrophobic groups on the outside in contact with the aliphatic side chains of the lipids. Such an arrangement could provide an aqueous channel through the membrane. We would propose that the carrier for anions operates as a gate within such a channel.

PASSOW: I think it is necessary to define the term "carrier".

PRESSMAN: I am quite satisfied to relegate the whole question to the area of semantics. There is not a black and white distinction between carriers and channels.

PASSOW: A carrier is not necessarily identical with a valinomycin type of an ionophor. A carrier could be a compound which, after combination with a substrate to be transported, makes some sort of a rotational movement which would not necessarily be simple diffusion, as in the case of valinomycin. Thus the movement of the substrate to be transported may be due to the rotation of a side chain as was suggested by Dr. Rothstein or it may be due to the diffusion on an ionophor of the type that you used in your work. So, I think there is no basic discrepancy.

PRESSMAN: Remember, however, that the rotating carrier doesn't have to negotiate a pathway through the lipid side chains.

BRODSKY: If one considers the long string-like polymeric value in the membrane as an oscillating anion carrier, then the frequency of oscillation of the polymer strand would be related directly to the velocity of transport. The oscillation could be a rotation about the long axis of the polymeric structure rather than about the point of its attachment to the membrane. This kind of rotation would occupy a smaller volume of space for any given frequency, and hence might be more economical. Is this kind of guesswork consistent with your working hypothesis?

PASSOW: It was the point I wanted to make. We cannot exclude the possibility that a rotational movement over a shorter distance than the thickness of the bilayer could be so fast that fewer sites may be sufficient to accomplish transport at the observed rate than one would expect for a valinomycin-like transport and still have the kinetics of a carrier system.

LAKOWICZ: What fraction of the total added inhibitor reacts with components other than the proteins possibly with the membrane? Secondly, is that fraction significant to change the surface charge of the membrane?

PASSOW: About 70% of the total binding of H_2DIDS is to the protein in band 3 and 30% to other sites which are partly definable and partly not. There is virtually no binding to the phospholipids. This has been shown in Toronto and we have confirmed this. It has been shown, however, by the Toronto group that when you treat the membrane with trypsin a lot of binding to the lipids occurs. For DIDS the binding to proteins other than those in band 3 is somewhat smaller than for H_2DIDS, about 20%. It is unlikely that the surface charge is substantially altered since neuraminic acid, which sticks its head groups out into the external medium completely dominates the behavior of the surface.

SHAMOO: No matter what the chymotrypsin data are it has been shown just recently that band 3 is more than two proteins (Conrad and Penniston, 1976, J. Biol. Chem. 251:253-255). Are you familiar with it?

ROTHSTEIN: I would agree that band 3 is heterogenious. In addition to the published data, unpublished observation by Passow, by Morrison and by ourselves indicate that the technique of isoelectric focusing can separate band 3 into a major component and three to four minor components.

ON THE NATURE OF THE TRANSPORT PATHWAY USED FOR Ca-DEPENDENT K MOVEMENT IN HUMAN RED BLOOD CELLS.

Joseph F. Hoffman and Ronald M. Blum

Department of Physiology, Yale University
School of Medicine, New Haven, Connecticut 06510

ABSTRACT

This paper is concerned with the mechanism by which energy-depleted human red cells become permeable to K (but not to Na) when they are exposed to Ca. In an attempt to distinguish a diffusion from a mediated process competitive type effects of different ions and their sidedness of action on K transport are considered as well as the action of certain transport inhibitors. While the nature of the interactions implies the involvement of a mediated process (perhaps an altered form of the Na:K pump apparatus) more direct evidence will be needed to make a definitive assessment.

Introduction

--The membrane tenants all
the transport muses.

The topic reviewed in this paper is not new to this conference series (31). The idea that a variety of foreign and apparently unrelated agents selectively alter the permeability of the red cell to K but not to Na represents an intriguing problem still unsolved. The kinds of agents known to markedly stimulate K transport include fluoride (47,12), iodoacetate and adenosine (13), Ca after energy depletion (26), triose reduction (29), lead ions (18), propranolol (9) and the Ca ionophore, A23187 (35). It now seems clear that the feature these various agents have in common (with the possible exception of lead) is a requirement for Ca. And what was once thought to represent only a curious if not capricious feature of the red cell may have relevance in providing a rationale for the systematic removal of cells from the circulation (e.g. ref. 2). Over the past few years new insight has been gained and it is our purpose here to review not all the advances in the field but only those which would appear to define the type of membrane process underlying the increased transport of K. The results are consistent with the view that K moves through the membrane not by a process of simple diffusion but rather by a mediated mechanism (3). It would also appear that the vehicle used for this carriage of K may be an altered form of the Na:K pump apparatus.

Discussion in this paper will draw mainly on studies of the Ca-sensitive K transport system as characterized in energy-depleted human red cells. We do this because the overt features of the altered K permeability displayed by this system appear to be shared by the other Ca-sensitive systems, at least to the extent that comparative experiments have been performed. And also because this is the system upon which our work has mainly been concentrated.

Previous studies have shown that the Na or K permeability of normal red cells is not affected by the presence or absence of Ca in the external medium (19). External Ca (Ca_o) can only affect the permeability of a cell which has been either metabolically manipulated or perturbed directly (26). But the situation is complicated because in different situations the red cell (19,37) can increase its permeability either to K alone or to both Na and K. It seems to be the case that internal Ca is associated with those situations where the change in permeability is exclusive for K (37,46,4). Whether or not the same process is involved or that other membrane sites are recruited in those cases where Na and K permeability are altered by Ca_o and therefore presumably by Ca_i is not clear. But to avoid all ambiguity we will deal exclusively with the circumstances where only the K permeability becomes increased, where cell shrinkage due to net loss of KCl can be demonstrated and that the increase in K presumably can be inhibited say with oligomycin (26,3). These criteria appear to apply exclusively to those situations where the change in permeability is selective for K and not for Na and K.

It is of interest to review briefly the multiple effects with which inside Ca is associated. A number of different methods are now available for trapping Ca inside red cells (33,37,17) or ghosts (19,4,41,8,34,43). One of the earliest approaches used resealed ghosts into which Ca was incorporated (either as the free ion and/or in complexed form) and it was found that the Na:K pump was inhibited by inside Ca (19). In addition, an ATP-dependent pump was identified which operated to extrude Ca from the ghost

(41). While it was attractive to consider that the Na:K pump and the Ca pump were two different aspects of the same membrane mechanism, especially since ATP was identified as the proximate source of energy for both, the evidence now favors the idea that the Na:K pump and the Ca pump are separate processes. This distinction between the two pumps has been made on the basis that the associated phosphoproteins separable by acrylamide-gel electrophoresis behave differently in the sense that the apparent molecular weight of the phosoprotein associated with the Na:K pump is 103,000 compared to 150,000 for the Ca pump (23).

Resealed ghosts were also used early on to show on the one hand that sequestering agents placed inside energy depleted ghosts prevent or decrease the effectiveness of Ca_o from changing the membrane's permeability to K (20,26). On the other hand, incorporation of small concentrations (less than 10 μM) of Ca was found to induce a selective increase in K permeability that showed the same overt characteristics of the process found with intact cells (26). And as little as 10 μM EGTA incorporated inside could be shown to reduce the Ca-induced response by seventy percent (26). More recently ghosts have been prepared where the Ca_i concentration has been controlled by incorporating Ca buffers and aside from confirming the earlier observations set limits on the concentration of ionic Ca_i required to alter maximally the K permeability (e.g. 34,43,10,44). Experiments with intact red cells where Ca entry as well as the concentration Ca_i has been manipulated either by incubation in the presence of Ca_o (37,8,34,10) or by employing the Ca-ionophore, A23187, (35) have also helped to establish the concentration dependence and related parameters of the system.

Results and Discussion

With this brief background we will now consider evidence, seen through the lens of kinetics, which while indicating the complexities of the Ca-dependent system, would appear to begin to define the nature of the transport pathway used by K to cross the membrane. That the transport of K is mediated (gated) in the sense that the flow of K might be distinguished from free diffusion is based on studies concerned with competition and with the effects of certain inhibitors. In addition, to the extent that their mode of action is selective, the effects of inhibitors help to specify the membrane locus for K mediation.

Competition. This section is concerned with the separate effects that internal and external Na as well as TEA (tetraethylammonium ion) have on the Ca-induced efflux of K from energy-depleted human red cells. The conclusion to be reached from these studies is that while the relative effects of these ions are complicated it seems clear that the manner in which they alter the kinetic characteristics of the K transport system is consistent not with simple diffusion but with a mechanism involving interaction within the membrane, that is, mediated transport.

The effect of changing the concentration of internal Na (Na_i) on the K efflux is shown in Fig. 1. The upper curve shows the usual response to changes in the concentration of external K (K_o) that has been described previously (3) and indicates that K efflux is dependent upon and activated by K_o, the maximum effect occurring at about 2 to 3 mM K_o. The inhibition of K efflux by high [K_o] is probably due to K_o decreasing the influx of Ca and preventing activation of the system by Ca_i (18a); this does not seem to be the case at low [K_o] since Ca influx is maximum and essentially unaltered under these circumstances. This is also supported by kinetic analysis of the effect of Ca_o, under the same experimental condition, on the outward rate constant (^ok_K) for K at different [K_o]. Double reciprocal plots of ^ok_K vs Ca_o are straight lines (at least when Ca_o is varied between 0.5 and 20 mM) and show that the apparent K_m of the membrane for [Ca_o] changes from 1.33 to 0.33 to 2.1 mM, when [K_o] is changed from 0.1 to 1.0 to 135 mM, respectively (5). When the concentration of internal Na, [Na_i], is raised, at the expense of internal K, [K_i], then K efflux is decreased and the [K_o] at which the maximum efflux occurs shifts to the right, to about 5 mM K_o. The reduction in K efflux could be due to decreasing [K_i] or to an antagonism exerted by Na_i or both but the shift in the maximum would indicate that Na_i decreases the affinity of the system for K_o. This becomes even clearer in Fig. 2 where cells containing four different ratios of [Na_i] to [K_i], where the sum [$Na_i + K_i$] was kept constant, were prepared and the K efflux was measured at each of three different concentrations of [K_o]. The pattern of K efflux at different [K_o] is different depending upon the ratio [Na_i] to [K_i]. And, in addition, it would also appear that Na_i is effective in altering the pattern even when [K_i] is high (between 100 and 80 mM). Thus Na_i directly antagonizes K_i for K efflux. Perhaps not coincidently, this type of an effect is related to that seen in normal (ATP containing) intact human red cells where Na_i has been shown to antagonize the ouabain-sensitive K efflux that is dependent upon the presence of K_o (25). In the situation (Fig. 2) where [Na_i] and [K_i] are changed reciprocally it is hard to sort out the separate effects of Na compared to K. Therefore experiments were carried out where [Na_i] was varied at constant [K_i], using choline to maintain the total cation concentration the same. The results shown in Table 1 indicate that [Na_i] has a marked effect in activating as well as in inhibiting K efflux when [K_i] is held constant at either high or low values. And further, the effects of Na_i on altering the affinity of the system for K_o are still apparent at

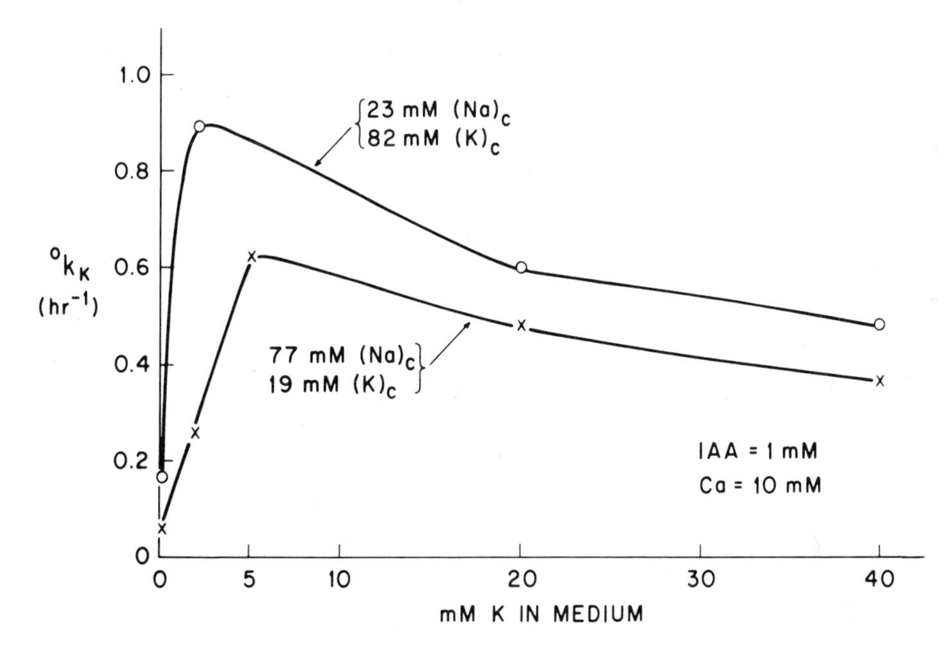

Fig. 1. The effect of varying $[K_o]$ on the Ca-dependent K transport from human red cells prepared to contain two different concentrations of Na and K. $^{o}k_K$ stands for the outward rate constant in units of reciprocal hours. $[Na_i]$ and $[K_i]$, in units of mM/liter cells, were altered using the PCMBS method (see ref. 39) on cells depleted by incubation at 37°C for 21 hours before exposure to PCMBS and being labeled with ^{42}K. The total depletion time was approximately 24 hours. $[K_o]$ was varied by replacing Na with K and the external medium also contained 20 mM glycylglycine at pH 7.15. The flux was measured at 37°C in the presence of 1 mM iodoacetate (IAA) and 10 mM $CaCl_2$ over a 30 min period (after 10 min temperature equilibration) begun by the addition of concentrated $CaCl_2$ to make the final concentration 10 mM. (See reference 3 for details of measurements.)

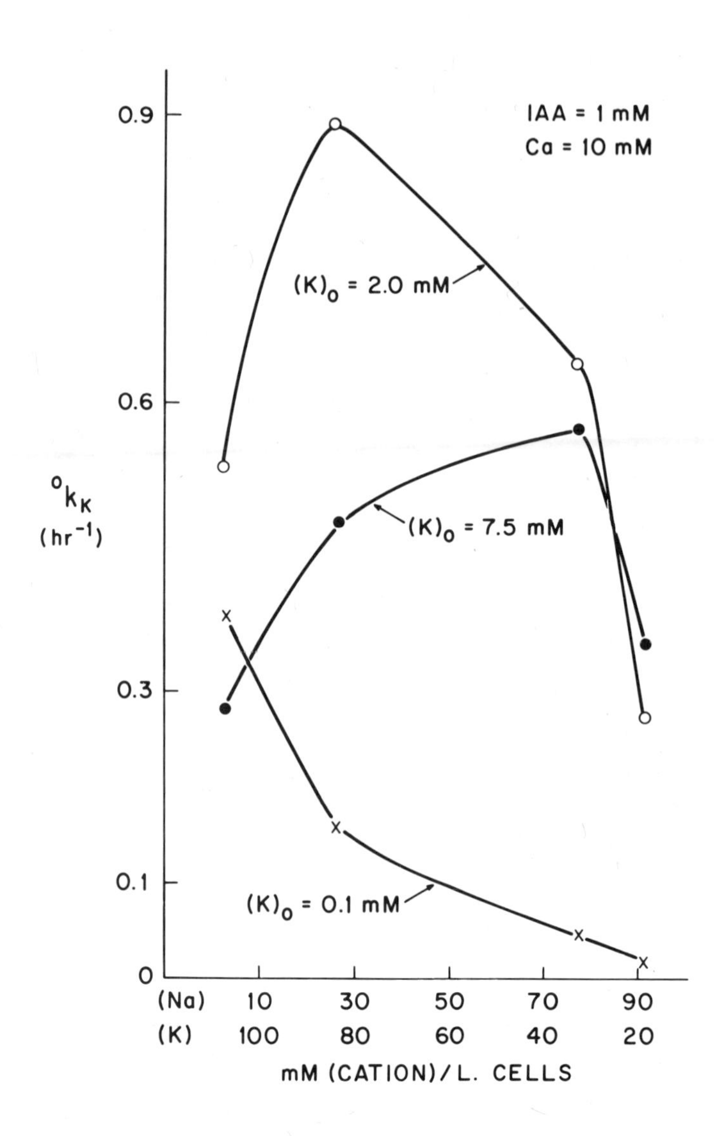

Fig. 2. The effect of altering the internal Na and K composition on the efflux of K from energy-depleted human red cells. Preparation of cells and methods used are the same as described or referred to in the legend to Fig. 1.

low values of K_o. While the meaning of this cross-membrane effect is not clear it has been observed in the Ca-induced system by others (36), under somewhat different circumstances, and in normal intact red cells where the activity of the ouabain-sensitive Na:K pump was studied (e.g. ref. 15).

TABLE I

The effect of varying $[Na_i]$ at constant $[K_i]$ on the Ca-induced K efflux from energy-depleted human red cells, at different $[K_o]$. $[Na_i]$ and $[K_i]$ were changed using the PCMBS method (see ref. 39), using Choline Cl to maintain osmotic balance and the sum of $[Na_i + K_i + Choline_i]$ constant. Otherwise the protocol and flux determination was the same as that described in the legends to Figs. 1 and 2. 1 mM IAA and 10 mM $CaCl_2$ were present in the final incubation medium during the measurement of K efflux.

$[K_i]$ mM/L cells	$[Na_i]$ mM/L cells	${}^{o}k_K$ (hr^{-1}) 0.1 mM K_o	2.0 mM K_o	7.0 mM K_o
66	5	0.31	0.38	0.29
	10	0.24	0.49	0.50
	21	0.14	0.25	0.24
5	5	0.15	0.25	0.24
	14	0.12	0.31	0.35
	26	0.05	0.18	0.16

Turning now from the inside to the outside surface of the membrane, interactions between K_o and other ions in the external medium can also be observed. Fig. 3 shows that the activating and inhibitory actions of K_o on the Ca-induced K effect are influenced by the presence of Na_o. Here, Na_o enhances the effectiveness of K_o at all concentrations of K_o. The effect of varying $[Na_o]$ at one particular and low concentration of K_o (0.1 mM) has also been studied and, as shown in Fig. 4, a linear relationship was found between $[Na_o]$ and K efflux. This type of effect of Na_o depends to some extent on the concentration of K_o and on the

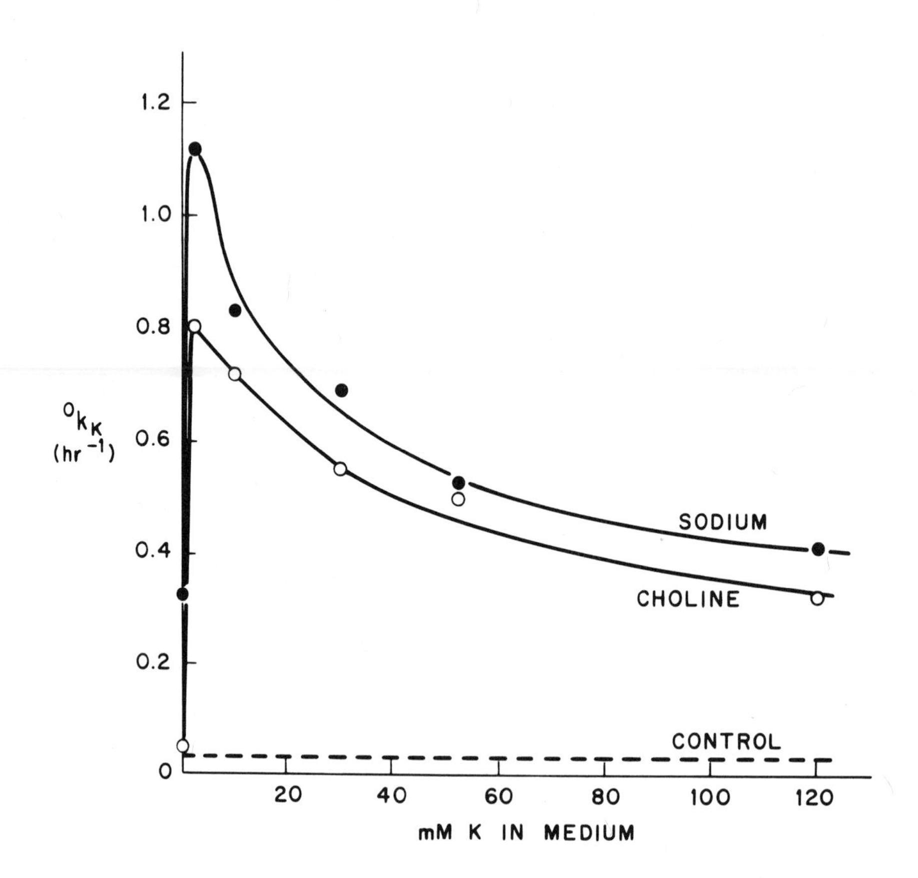

Fig. 3. The effect of external Na on the efflux of K from energy-depleted human red cells as a function of $[K_o]$. $[K_o]$ was varied by substitution with either NaCl or Chloride-Cl. The medium also contained 1 mM IAA and 10 mM $CaCl_2$. The cells used contained approximately 70 mM K and 25 mM Na per liter of packed cells after being energy-depleted following a standard procedure previously described (3). Details of the flux measurements are also given in reference (3).

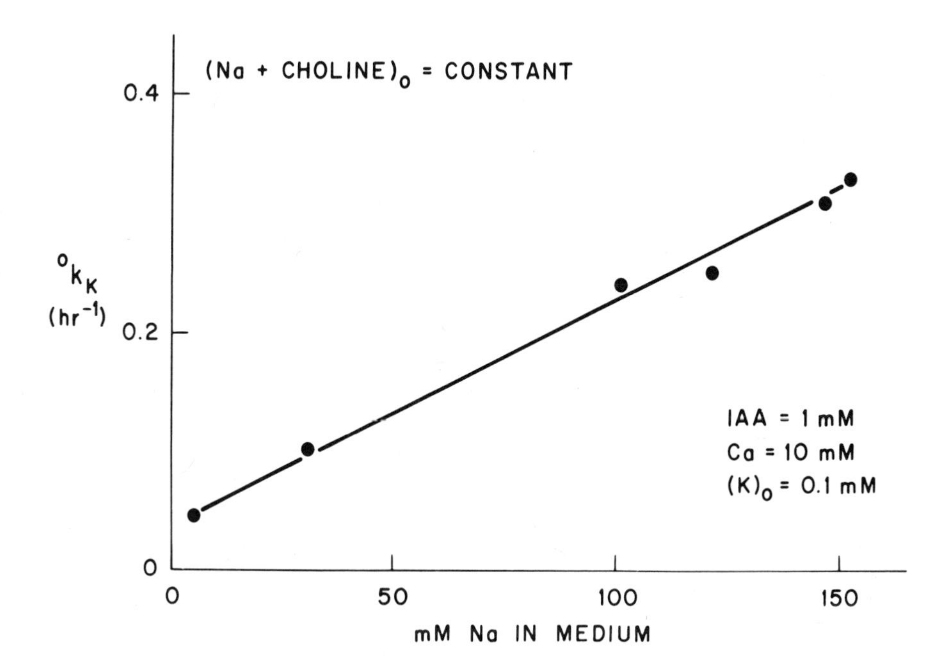

Fig. 4. The activating effect of Na_o on the Ca-induced K efflux from energy-depleted human red cells. Cells were prepared and the flux determined (37°C, pH 7.15) using the procedures previously described (3). [Na_o] was varied by substitution with Choline-Cl.

type of ion used to substitute for Na_o (26). For the type of protocol used in experiments similar to that presented in Fig. 4, no differences were observed when tetramethylammonium Cl was used instead of Choline-Cl to replace Na_o. This would indicate that there is a direct effect of Na_o independent of the nature of the substituting ion. On the other hand, while it is clear that Na_o interacts with the Ca-sensitive K efflux system, it still has to be sorted out whether the effect of Na_o is restricted to the outside or also represents a cross-membrane effect (36,24).

We have assumed so far that because of the complexity of the various effects that Na has on the Ca-induced K transport system, that K must evidently be mediated in its transit through the membrane. The effect of tetraethylammonium ion (TEA) supports this notion as well as implicating the involvement of the Na:K pump apparatus. Fig. 5 shows that TEA inhibits the Ca-dependent efflux of K. That this would appear to be a competitive effect of TEA with K_o is evident both from the form of the inhibition curve and from the fact that TEA becomes less effective in inhibiting K efflux the higher the value of K_o. It has been concluded from studies concerned with the mechanisms of action of TEA on normal human red cells that, although TEA is not transported, TEA behaves like external K with regard to its effect on the pump (38). Thus, TEA was found to inhibit the pump influx of K, the pump efflux of Na, the ouabain-sensitive influx of Na assocciated with Na:Na exchange and the binding of ouabain to the cell (38). In addition, TEA did not appear to have any discernable effect on the passive ouabain-insensitive permeabilities of Na or K. To the extent that TEA interacts with the same membrane components in the Ca-induced system as in the normal cell, the locus would appear to be the Na:K pump complex.

<u>Inhibitors</u>. In this section we will be dealing with the effects of ouabain, furosemide and oligomycin on the same system as discussed above, namely, the Ca-induced stimulation of K transport as seen in energy-depleted human red cells. Figure 6 shows that the Ca-dependent efflux of K is inhibited by ouabain and by furosemide and that the magnitude of the inhibition by either drug is affected by $[K_o]$. As reported previously (3) the maximum inhibition of K efflux observed with ouabain averages 26% when $[K_o]$ is less than 0.4 mM. But when $[K_o]$ is more than about 2 mM, as indicated in Fig. 6, the K efflux becomes essentially insensitive to the presence of ouabain. This type of behavior is not paralleled by furosemide although the fraction of the total flux that is inhibited by furosemide is influenced by $[K_o]$. K efflux can also be activated by using Cs_o in place of K_o but the maximum efflux rate is shifted to the right to about 6 to 10 mM Cs_o, in line with the cells (that is the pump's) lower affinity for Cs (see ref. 19). Ouabain inhibits the activation of K

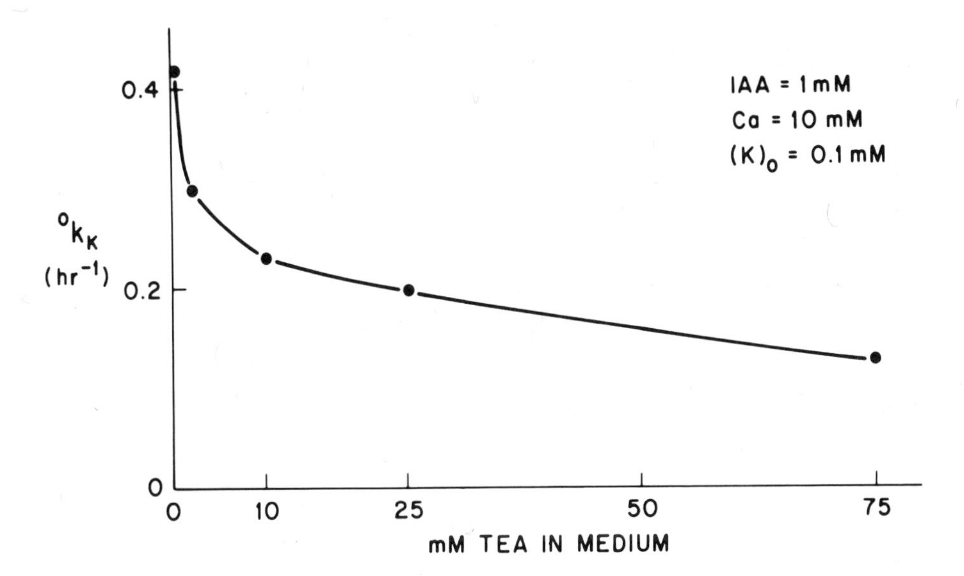

Fig. 5. The inhibition by tetraethylammonium (TEA) of the Ca-dependent K efflux from energy-depleted human red cell. Cells were prepared and the flux determined using the procedures previously described (3). [TEA_o] was varied by substitution with NaCl.

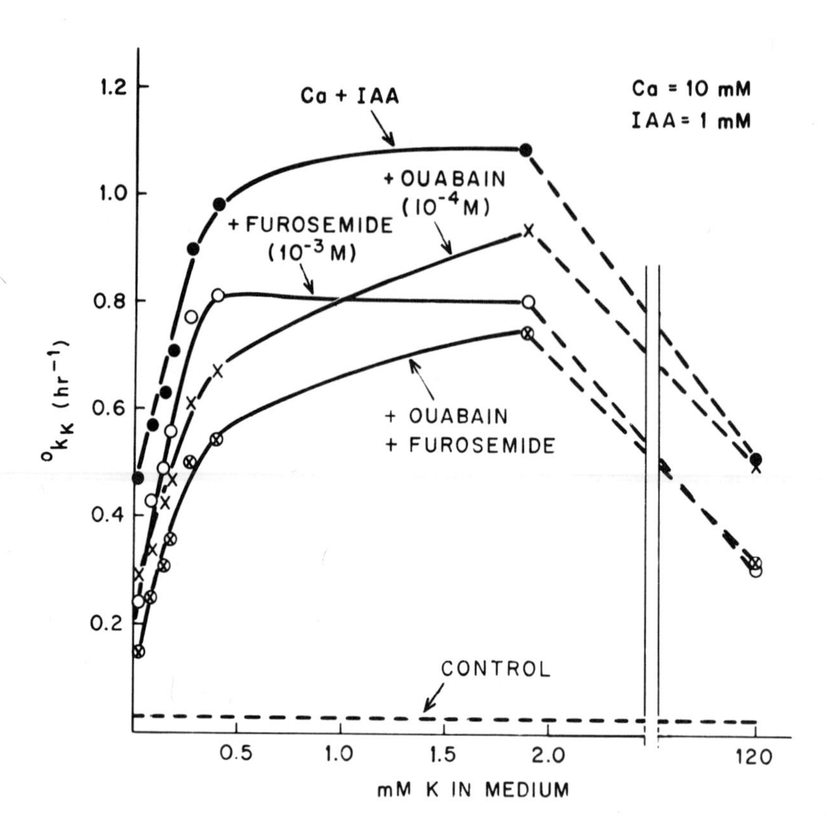

Fig. 6. The inhibitory effects of ouabain and furosemide on the rate of Ca-induced K efflux from energy-depleted human red cells at varying [K_o]. After the cells had been depleted for 24 hours they were washed and resuspended in a Na containing K-free medium and incubated another hour with part of the cells being exposed to 1 x 10^{-4}M ouabain. [Na_i] and [K_i] in control and ouabain exposed cells were the same and did not change after this one hour incubation. The cells were washed again and resuspended in media containing 1 mM IAA and varying [K_o]. When present Ca was 10 mM, ouabain was 1 x 10^{-4}M and furosemide was 1 x 10^{-3}M. Exposure to furosemide only took place in the final incubation medium whereas with ouabain the cells were preexposed as well. [K_o] was varied by substitution with NaCl. Cell preparation and flux determination carried out as previously described (3).

efflux by Cs_o but the system does not become insensitive to ouabain until $[Cs_o]$ approaches 50 mM.

It is interesting that the effects of ouabain and furosemide, as seen in Fig. 6, tend to be additive at least in the range that ouabain is effective. But it is also important to note that the effects of ouabain and furosemide acting together were found to be consistently smaller than the sum of each acting alone, implying that furosemide inhibits a ouabain-sensitive as well as a ouabain-insensitive component of the K efflux. For instance, at 0.1 mM K_o the efflux of K is inhibited approximately 40% by ouabain (10^{-4}M), 25% by furosemide (10^{-3}M) and 55% when both are present together; at 1.9 mM K_o, ouabain exerts 13% inhibition; furosemide, 26%, and the combination, ouabain + furosemide, 32%. In another experiment, when the furosemide concentration was increased to 2.5 mM, the inhibition by 10^{-4}M ouabain + furosemide accounted for 84% of the Ca-activated K efflux. The results presented in Fig. 7 show that ouabain and furosemide alter the net efflux (loss) of K from the cells in an analogous way to the efflux measured with ^{42}K (Fig. 6). The inhibitory effects of furosemide but not of ouabain were found to be reversible when the cells were washed prior to exposure to Ca.

Of the three transport inhibitors studied, oligomycin was found to have the most profound effects on the Ca-dependent increase in K efflux. The dose response curve is shown in Fig. 8 in which it is apparent that the induced K efflux is essentially completely inhibited by oligomycin. For the experiment presented in Fig. 8, the control K efflux rate constant measured in the absence of Ca (with or without oligomycin) is between 0.02 to 0.03 hr^{-1}, compared to the value, 0.025 hr^{-1}, obtained with 5 μg/ml oligomycin in the presence of Ca. However, as shown in Fig. 9, the magnitude of the inhibition produced by oligomycin depends upon $[K_o]$. Increasing $[K_o]$ decreases the effectiveness of oligomycin in inhibiting the K efflux but to a much lesser extent than that seen with ouabain (compare Fig. 6). There is a type of competition displayed between oligomycin and K_o in the sense that the higher the $[K_o]$ the more oligomycin needed to reach the same level of inhibition, at least up to approximately 25 mM K_o. When $[K_o]$ is higher than 25 mM increasing the concentration of oligomycin above 10 μg/ml does not increase the level of inhibition observed. The sequence of addition is also important since prebound oligomycin is not easy to displace by increasing $[K_o]$ and cells preexposed to oligomycin at varying $[K_o]$ retain less oligomycin (determined by its subsequent inhibitory effect) the higher the $[K_o]$.

Since, as stated before, we are mainly concerned with the character of the transport mechanism involved in the Ca-induced

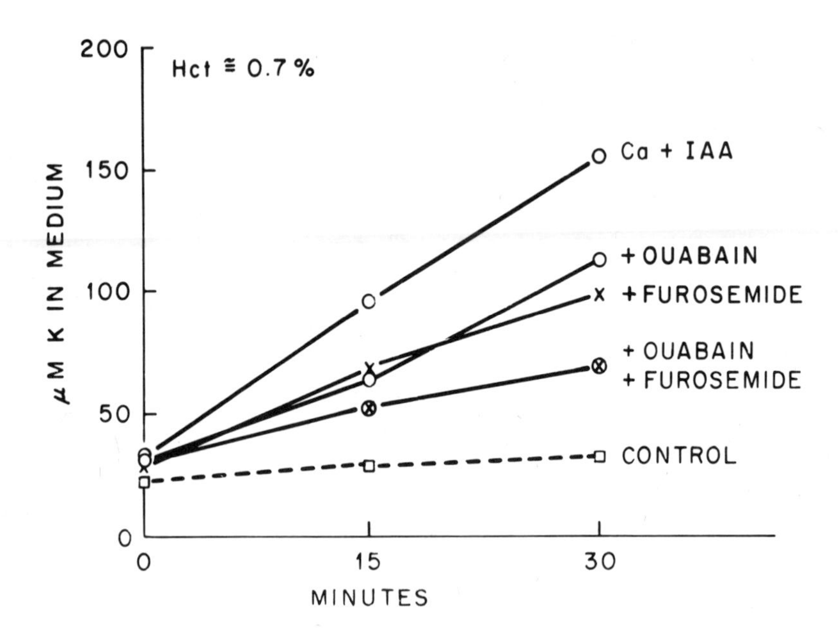

Fig. 7. The inhibiting effects of ouabain and furosemide on the rate of net loss of K induced by Ca from energy-depleted human red cells. The protocol used for this experiment was the same as for the one described in Fig. 6 except that the cells were incubated at a low hematocrit (0.7%) in order to follow the change in $[K_o]$ that occurred with time. The cells in each circumstance were exposed to 1 mM IAA and all except the control were exposed to Ca and either ouabain (1×10^{-4}M) or furosemide (1×10^{-3}M) or both.

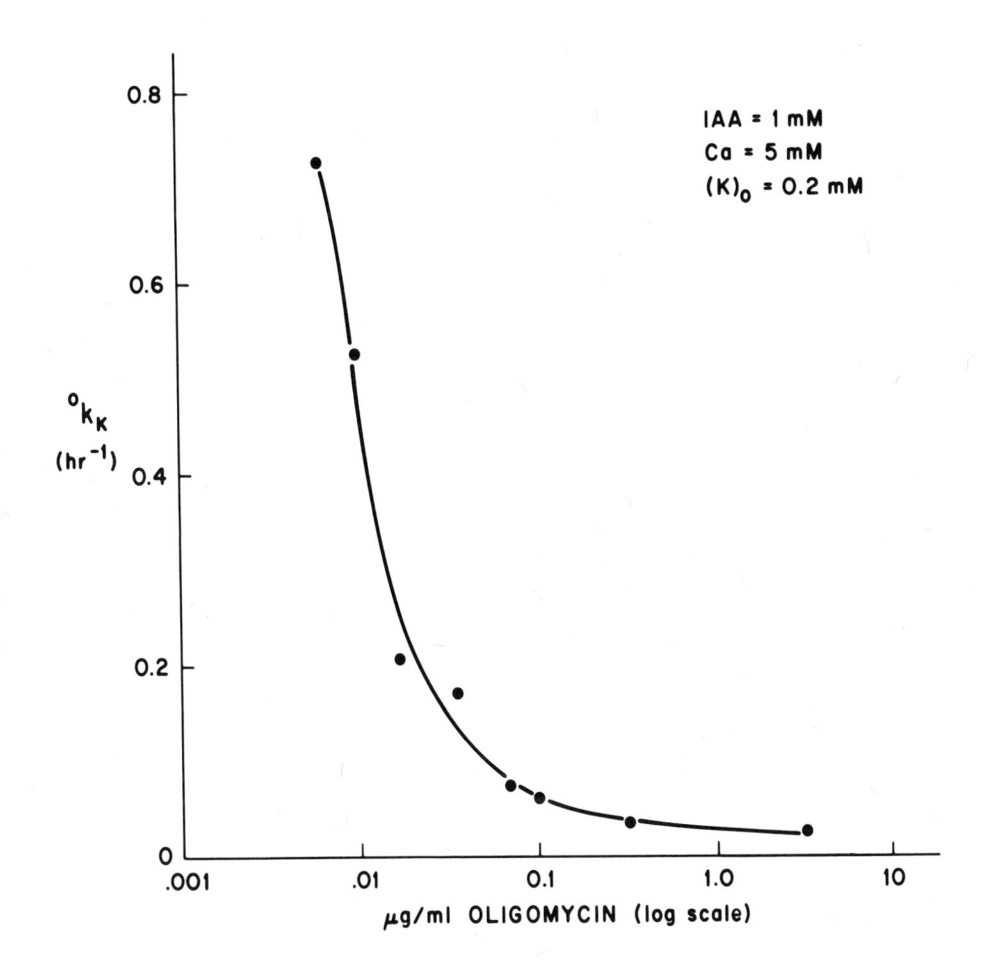

Fig. 8. The dose-response curve of oligomycin on tne Ca-induced K efflux from energy-depleted human red cells. Cells were prepared and the flux was determined using the procedures previously described (3). Oligomycin at the indicated concentrations was present in the final incubation medium, where $[IAA_o]$ was 1 mM, $[K_o]$ was 0.2 mM and the final concentration of Ca_o was 5 mM.

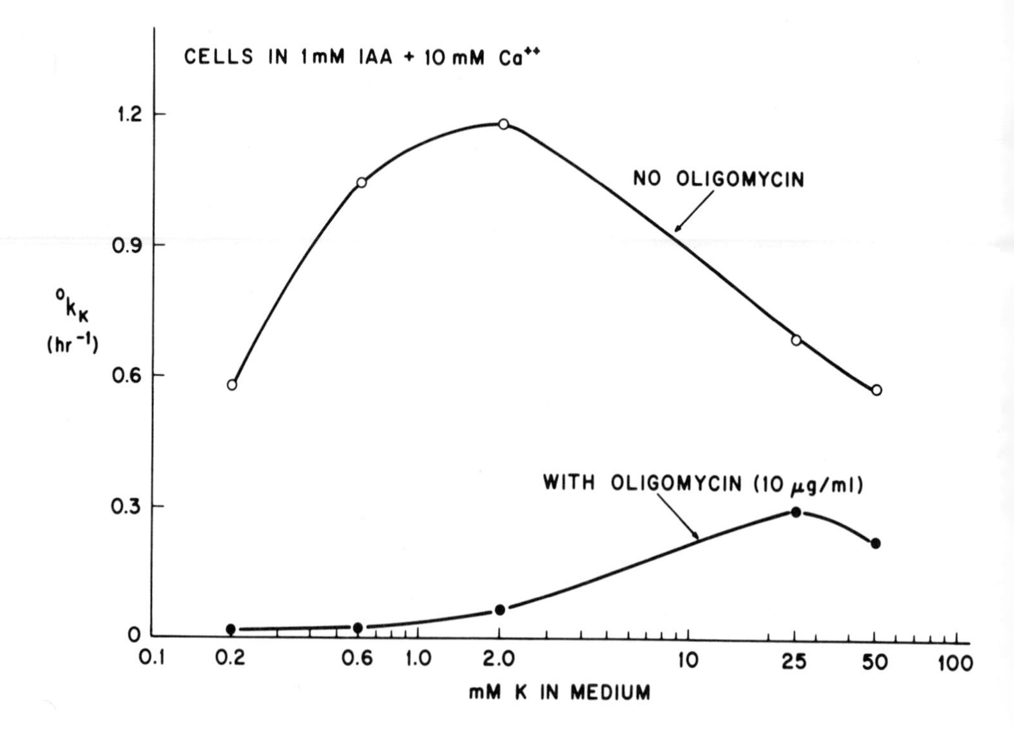

Fig. 9. The inhibition by oligomycin (10 μg/ml) of Ca-induced K efflux from energy-depleted human red cells as a function $[K_o]$. Oligomycin, 1 mM IAA and 10 mM $CaCl_2$ were present in the final incubation medium. $[K_o]$ was varied by replacement with Na. Cell preparation and the flux determination was carried out as previously described (3).

process, we can ask to what extent can the action of these different inhibitors help define the nature of the K transport pathway. Part of the answer to this question depends on whether or not a diffusion or a mediated process is involved and this in turn is tied to the specificity of action of these drugs in terms of whether or not the same membrane site is involved when they inhibit the Ca-dependent system (in altered cells) compared to when they inhibit the Na:K pump in normal cells. The utility of furosemide and oligomycin is limited in this regard since it is not known how many types of sites are involved in their action. But the fact that their inhibitory characteristics on the Ca-induced system parallel their action (40,45) on the Na:K pump provides inferential evidence that perhaps only one type of site is involved. These various inhibitors could, of course, act in ways independent of any action they might exert on the Na:K pump. They might act indirectly to inhibit the K efflux by decreasing the influx of Ca into the cell. This type of an effect is unlikely, since the increase in the K efflux rate initiated by the addition of Ca can be inhibited by the subsequent addition of furosemide or oligomycin, but proper evaluation would require direct measurements of Ca uptake rates. Another indirect way these inhibitions might exert their effects is by altering the cell's membrane potential. If the membrane potential takes on the characteristics of a K diffusion potential and becomes hyperpolarized (inside negative with respect to outside) upon the addition of Ca, as appears to be the case (16,21), then furosemide or oligomycin could act by altering the cell's chloride permeability (P_{Cl}) rather than its K permeability (P_K). Thus any change in the ratio, P_K/P_{Cl}, could reflect a change in the membrane potential. Since there is evidence (6) that furosemide markedly inhibits the self-exchange of Cl (and presumably, therefore, P_{Cl}) the K efflux could be retarded or inhibited because of the increase in the hyperpolarization of the membrane. This type of an effect would also predict that the influx of K would be increased by the inhibitor. In previous tests of this notion using the anion inhibitors, dipyridamole and SITS (21), the Ca-induced influx of K was found to be enhanced and K efflux inhibited according to the change in magnitude of the membrane potential. But oligomycin, which is without effect on anion permeability, was found to inhibit both K influx as well as efflux, a result which indicates that oligomycin acts directly on the cell's permeability to K (21). And we have also observed that furosemide, contrary to expectation based on furosemide's effect on P_{Cl}, also inhibits K influx by about 40% (5). This indicates that furosemide affects P_K as well as P_{Cl}, consistent with the idea that furosemide can act at the level of the Na:K pump apparatus as referred to before.

In contrast, ouabain is more suitable for our purpose of discriminating mechanism because it is thought to act with great specificity in inhibiting the Na:K pump on red blood cells as well as in other cell types. In addition to the evidence already presented, we have previously reviewed the action of ouabain on the pump in relation to its action on the Ca-induced K transport system as studied in energy-depleted cells (3). However, other workers (e.g. 36,14,26) have either experienced difficulty in reproducing the inhibitory effect of ouabain, but not the oligomycin effect (30), on K efflux or have offered other explanations for the observed behavior (27). Part of the problem would appear to be related to the conditions under which the effect of ouabain was tested, e.g. $[K_o]$ too high, less than optimal Ca concentrations, perhaps pH, depletion time, intracellular concentrations of Na, K, P_i, iodoacetate, or different species of red blood cells. On the other hand, in an instance where our observations have been claimed to have been repeated (28), under somewhat different conditions, it is stated that ouabain partially inhibited the Ca-induced K efflux when $[ATP_i]$ was above but not below 10^{-6}M. And it was also suggested that the consequent reduction in K efflux seen in the presence of ouabain was due to an indirect effect. This conclusion is based on the idea that ATP_i, in being spared by ouabain inhibiting its breakdown by the Na:K pump, results in a lower $[Ca_i]$ and a diminished K efflux, by the operation of the ATP-dependent Ca extrusion mechanism (42). While the logic of the argument is admirable we are attracted by a more interesting explanation. But first let us consider the relation between $[ATP_i]$ and the inhibition of the Ca-dependent K efflux by ouabain as shown in Table II. Here the change in $[ATP_i]$ is given as a function of the depletion time over a period of from 18 to 36 hours; and the effectiveness of ouabain in inhibiting the K efflux is shown in two different experiments (A and B) to which the $[ATP_i]$ analyses apply. It is apparent that the fractional inhibition produced by ouabain is not only maintained by lengthening the depletion time but actually increased with decreasing values of $[ATP_i]$. Since ouabain was presented to the cells only at the moment when Ca was added to initiate the efflux measurement it is unlikely that $[ATP_i]$ would be different in the cells exposed to ouabain compared to the control (no ouabain), especially since the pump was already inhibited if not inoperative at the low value of $[K_o]$ used. If $[ATP_i]$ is reduced by adding adenosine prior to measurement of K efflux (iodoacetate was already present) the inhibitory effect of ouabain disappeared consistent with the observations referred to above (28). On the other hand, when cells are energy-depleted for 16 hours and 10^{-4}M ouabain is then added the fractional inhibition of K efflux by ouabain is maintained over the next 22 hours. While $[ATP_i]$ was not measured in this type experiment, we might expect that early on, $[Ca_i]$ might be kept lower at the expense of spared $[ATP_i]$, but even this type of effect was missing. Obviously more detailed and systematic studies are required but the foregoing information might serve to reduce the web of sophistry that has evidently sprung up around our results. The explanation we favor envisages ouabain, binding to its usual site on the Na:K pump, inhibiting directly the subsequent activity or turnover of an altered form of the pump apparatus. Now, consistent with the results described before, $[ATP_i]$, above a concentration of 10^{-6}M, would be necessary in order for ouabain to display its retarding effect on the operation of the pump complex. In terms of mechanism, it might be that since the pump appears to have two binding sites for ATP_i, occupation of one of the sites by ATP_i is necessary in order for ouabain to slow a K-transporting conformational change of the apparatus when it is stimulated by Ca_i (22). Preliminary results in which UTP was incorporated into ATP-free ghosts, made from energy-depleted red cells (4), indicate that ouabain could be observed to inhibit the Ca-dependent K efflux when the ghosts contained rather low values of $[UTP_i]$. UTP does not appear to be a substrate for the Na:K pump (19) but it is not known whether or not the Ca pump is operative under these circumstances (but see ref. 7). However instructive this type of experiment might appear it is still necessary and desirable to develop a more direct label in order to identify the membrane component(s) involved in the transfer of K.

TABLE II

Data taken from reference (3). Ouabain (1×10^{-4}M) was present only during the K efflux measurement. $[K_o]$ was 0.05 mM and the final incubation medium also contained 1 mM IAA. Percent ouabain inhibition refers to the inhibition of K efflux when ouabain was present compared to when it was absent. $[ATP_i]$ was determined fluorometrically similar to methods previously described (17) on aliquots of cells taken just prior to estimating K efflux.

HOURS DEPLETED	18	24	30	36
ATP (μM/L cells)	62	24	13	9

EXPT.	HOURS DEPLETED	% OUABAIN INHIBITION
A	19	34
	24	42
	27	42
B	18	16
	33	26
	39	39

Before leaving ouabain as a topic, it is of interest to comment on a recent observation that ouabain inhibits Ca uptake into human red cells, suspended in hypertonic media, provided K_o is present (32). It is not clear what relation this effect of ouabain has to that discussed above, but if ouabain alters Ca uptake, and the magnitude of K efflux is dependent upon [Ca_i], then the characteristics of Ca entry are rather different in the two systems. In hypertonic media, Ca uptake and K movement was found to be retarded by increasing [Na_o] but with cells suspended in isotonic media (Figs. 3 and 4) increasing Na_o increased K efflux. Further, in our studies, ouabain applied after Ca exposure (when Ca entry has presumably already occurred) results in approximately the same level of inhibition as when it is added before or together with Ca. We also failed to observe any competition between Ca and ouabain (3). And finally, there is a lack of parallel between changes in K efflux and sensitivity to [K_o] and inhibition by ouabain from cells suspended in isotonic media (see Fig. 6) compared to cells suspended in hypertonic media.

By way of summary, this paper has reviewed evidence which is aimed at defining the membrane process responsible for the transfer of K under the unusual conditions of energy-depletion and dependence on Ca. While the K transfer process is susceptible to a range of influences, from sided and cross membrane effects of different ions to inhibition by various drugs, it is only possible to speculate on the nature of the actual underlying mechanism. The information available is hardly definitive in distinguishing a diffusion from a mediated process and therefore even less helpful in specifying the membrane locus. Either type of process is consistent with a flux ratio type analysis (26) and models can be constructed which show many of the properties of the Ca-induced system (26,1). In considering a more direct if not new approach, it may be possible to study the properties of red cell membranes fused say with a thin lipid membrane. In this circumstance it might be possible to obtain evidence distinguishing channels from carriers for the Ca-induced transport of K (48).

REFERENCES

1 ADRIAN, R.H.: Rectification in muscle membrane, Ch. 8, Prog. in Biophys. and Molec. Biol. (Butler, J.A.V., Noble, D. Eds.). Pergamon Press. New York 19 (1961) 341.

2 ASTRUP, J.: Na and K in human red cells. Variations among centrifuged cells. Scand. J. Lab. and Clin. Invest. 33 (1974) 231.

3 BLUM, R.M. and HOFFMAN, J.F.: The membrane locus of Ca-stimulated K transport in energy depleted human red blood cells. J. Memb. Biol. 6 (1971) 315.

4 BLUM, R.M. and HOFFMAN, J.F.: Ca-induced K transport in human red cells: Localization of the Ca-sensitive site to the inside of the membrane. Biochem. Biophys. Res. Comm. 25 (1972) 1146.

5 BLUM, R.M. and HOFFMAN, J.F.: Unpublished results.

6 BRAZY, P.C. and GUNN, R.B.: Furosemide inhibition of chloride transport in human red blood cells. Physiologist 18 (1975) 151.

7 CHA, Y.N., SHIN, B.C., and LEE, K.S.: Active uptake of Ca and Ca-activated MgATPase in red cell membrane fragments. J. Gen. Physiol. 57 (1971) 202.

8 COLOMBE, B.W. and MACEY, R.I.: Effects of calicum on potassium and water transport in human erythrocyte ghosts. Biochim. Biophys. Acta. 363 (1974) 226.

9 EKMAN, A., MANNINEN, V. and SALMINEN, S.: Ion movements in red cells treated with propranolol. Acta Physiol. Scand. 75 (1969) 333.

10 FERRIRA, H.G. and LEW, V.L.: Use of ionophore A23187 to measure cytoplasmic Ca buffering and activation of the Ca pump by internal Ca. Nature. 259 (1976) 47.

12 GARDOS, G.: The role of calcium in the potassium permeability of human erythrocytes. Physiol. Sci. Acad. Hung. 15 (1959) 121.

13 GARDOS, G.: The permeability of human erythrocytes to potassium. Acta. Physiol. Sci. Acad. Hung. 10 (1956) 185.

14 GARDOS, G., SZASZ, I. and SARKADI, B.: Mechanism of Ca-dependent K-transport in human red cells. Biomembranes: Structure and Function (Gardos, G. and SZASK, I. Eds). North Holland, Amsterdam. FEBS Proc. 35 (1975) 167.

15 GARRAHAN, P.J. and GARAY, R.P.: A kinetic study of the Na pump in red cells. Its relevance to the mechanism of active transport. Annals New York Acad. Sci. 242 (1974) 445.

16 GLYNN, I.M. and WARNER, A.E.: Nature of the calcium dependent potassium leak induced by (+)-propranolol, and its possible relevance to the drugs antiarrhythmic effect. Brit. J. Pharmac. 44 (1972) 271.

17 GLYNN, I.M. and HOFFMAN, J.F.: Nucleotide requirements for sodium-sodium exchange catalyzed by the sodium pump in human red cells. J. Physiol. 218 (1971) 239.

18 GRIGARZIK, H. and PASSOW, H.: Versuche zum mechanismus der bleiwirkung aul die kalium-permeabilitat roter blutkorperchen. Pflugers Archiv. 267 (1958) 73.

18a HARDY, JR., M.A. and LEW, V.L.: Captacion de calcio y flujo de potasio en eritrocitos humanos depletados. Proc. XVI Meeting Soc. Argentina Invest. Clin. Cordoba, Argentina (1971) 199.

19 HOFFMAN, J.F.: Cation transport and structure of the red cell plasma membrane. Circulation. 26 (1962) 1201.

20 HOFFMAN, J.F.: The red cell membrane and the transport of sodium and potassium. Am. J. Med. 41 (1966) 666.

21 HOFFMAN, J.F. and KNAUF, P.A.: The mechanism of the increased K transport induced by Ca in human red blood cells. Erythrocytes, thrombocytes and leukocytes (Gerlach, E., Moser, K., Deutsch, E., and Wilmanns, W. Eds.). Georg Thieme. Stuttgart (1973) 66.

22 HOFFMAN, J.F. AND PROVERBIO, F.: Membrane ATP and the functional organization of the red cell Na:K pump. Annals New York Acad. Sci. 242 (1974) 459.

23 KNAUF, P.A., PROVERBIO, F. AND HOFFMAN, J.F.: Electrophoretic separation of different phosphoproteins associated with Ca-ATPase and Na, K-ATPase in human red cell ghosts. J. Gen. Physiol. 63 (1974) 324.

24 KNAUF, P.A., RIORDAN, J.R., SCHUMANN, B., WOOD-GUTH, I., and PASSOW, H.: Calcium-potassium-stimulated net potassium efflux from human erythrocyte ghosts. J. Memb. Biol. 25 (1975) 1.

25 KNIGHT, A.B. and WELT, L.G.: Intracellular potassium. A determinant of the sodium-potassium pump rate. J. Gen. Physiol. 63 (1974) 351.

26 KREGENOW, F. and HOFFMAN, J.F.: Some kinetic and metabolic characteristics of calcium-induced potassium transport in human red cells. J. Gen. Physiol. 60 (1972) 406.

27 LEW, V.L.: Effect of ouabain on the Ca-dependent increase in K permeability in depleted guinea pig red cells. Biochim. Biophys. Acta 249 (1971) 236.

28 LEW, V.L.: On the mechanism of the Ca-induced increase in K permeability observed in human red cell membranes. Comparative Biochemistry and Physiology of Transport (Bolis, L., Block, K., Luria, S.E., and Lynen, F. Eds.). North-Holland, Amsterdam. (1974) 310.

29 PASSOW, H.: Zusammenwirken von Membranstruktur und Zellstrffwechsel bei Regulierung der Ionenpermeabilität roten Blutkörperchen. Biochemie Des Aktiven Transports. Colloq. Ges. Physiol. Chem. Mosbach/Baden, Springer-Verlag, Berlin (1961) 54.

30 PASSOW, H.: Metabolic control of passive cation permeability in human red cells. Cell Surface Interactions (Brown, H.D. Ed.). Scholaris Library, New York (1963) 57.

31 PASSOW, H.: The red blood cell: Penetration, distribution and toxic action of heavy metals. Effects of Metals on Cells, Subcellular Elements and Macromolecules (Maniloff, J., Coleman, J.R. and Miller, M. Eds.). pp. 291-340. Charles C. Thomas, Springfield, Ill. (1970).

32 PLISHKER, G. and GITELMAN, H.J.: Calcium transport in intact human erythrocytes. J. Gen. Physiol. 68 (1976) 29.

33 PONDER, E.: Volume changes, ion exchanges, and fragilities of human red cells in solutions of the chlorides of the alkaline earths. J. Gen. Physiol. 36 (1953) 767.

34 PORZIG, H.: Comparative study of the effects of propranolol and tetracaine on cation movements in resealed human red cell ghosts. J. Physiol. 249 (1975) 27.

35 REED, P.W.: Effects of the divalent cation ionophore A23187 on K permeability of rat erythrocytes. J. Biol. Chem. 25 (1976) 3489.

36 RIORDAN, J.R. and PASSOW, H.: The effects of calcium and lead on the potassium permeability of human erythrocytes and erythrocyte ghosts. Comparative Physiology (Bolis, L., Schmidt-Nielsen, K., and Mandrell, S.H.P. Eds.). North-Holland, Amsterdam (1973) 543.

37 ROMERO, P.J. and WHITTAM, R.: The control by internal calcium of membrane permeability to sodium and potassium. J. Physiol. 214 (1971) 481.

38 SACHS, J.R. and CONRAD, M.E.: Effect of tetraethylammonium on the active cation transport system of the red blood cell. Am. J. Physiol. 215 (1968) 795.

39 SACHS, J.R.: Sodium movements in the human red blood cell. J. Gen. Physiol. 56 (1970) 322.

40 SACHS, J.R.: Ouabain-insensitive sodium movements in the human red blood cell. J. Gen. Physiol. 57 (1971) 259.

41 SCHATZMANN, H.J.: ATP dependent Ca extrusion from human red cells. Experientia. 22 (1966) 364.

42 SCHATZMANN, H.J. and VINCENZI, F.F.: Calcium movements across the membrane of human red cells. J. Physiol. 201 (1969) 369.

43 SIMONS, T.J.B.: The preparation of human red cell ghosts containing calcium buffers. J. Physiol. 256 (1976) 209.

44 SIMONS, T.J.B.: Calcium-dependent potassium exchange in human red cell ghosts. J. Physiol. 256 (1976) 227.

45 WHITTAM, R., WHEELER, K.P. and BLAKE, A.: Oligomycin and active transport reactions in cell membranes. Nature. 203 (1964) 720.

46 WHITTAM, R.: Control of membrane permeability to potassium in red blood cells. Nature. 219 (1968) 610.

47 WILBRANDT, W.: A relation between the permeability of the red cell and its metabolism. Trans. Farad. Soc. 33 (1937) 956.

48 This work was supported by USPHS grants HL 09906 and AM 17433.

DISCUSSION

LAKOWICZ: Demetrios Papahadjopoulos has shown that synthetic vesicles are most permeable at their base transition (Membrane Fusion-Molecular Aspects and Biological Implications, The Molecular Biology of Membrane, U. of Minnesota, 5/19/76). With synthetic phospholipids, especially phosphytitalserine, calcium will induce a phase transition, or change the transition temperature of the phospholipid, and change the permeability greatly.

Now, there is an asymetric distribution of phospholipids in erythrocyte. Can this permeability play any role?

HOFFMAN: I do not know whether or not calcium alters the physical state of the lipid in the membrane for the circumstance I've described. It is possible that lipids or phase transition are involved but the evidence indicates that the red cell has no obvious phase transition due to its cholesterol content. And the changes in the apparent liquidity of the membrane which are thought to occur would be difficult to associate with this kind of a process. If the site of action of calcium is the Na:K pump apparatus then calcium could of course alter the state of the associated lipids. I'm not aware of any evidence indicating asymmetry in the distribution of lipids associated with the pump but this would represent an interesting finding in itself.

PASSOW: Could you tell us the percent of the cases in which you had an oubaine effect?

HOFFMAN: We observe ovabain effects in more than 95% of the instances tested.

SHA'AFI: Could this mechanism account for the difference in high K and low K sheep red cells' permeability to Na and K?

HOFFMAN: I assume you are referring to the dimorphism seen in sheep and goat red blood cells. If this is so then the answer is no since it has not been possible to induce the specific change in K permeability in these cells by agents which do induce the change in human red cells.

TOXIC CHEMICALS AS PROBES OF NERVE MEMBRANE FUNCTION

Toshio Narahashi

Duke University Medical Center
Department of Physiology and Pharmacology
Durham, North Carolina 27710

ABSTRACT

Certain toxins and chemicals have been used as probes to characterize nerve membrane ionic channels. These neuroactive agents exert highly specific actions on sodium or potassium channel. Tetrodotoxin and saxitoxin block the sodium channel at very low concentrations thereby impairing nervous conduction. They are being used to separate the membrane current into a sodium and potassium component, to eliminate the sodium current, to study the mechanism of synaptic and neuromuscular transmission, and to characterize, isolate and identify the sodium channel. A new toxin isolated from *Gonyaulax tamarensis* exerts a very similar sodium channel blocking action. Tetraethylammonium is known to block the potassium channel. Aminopyridines inhibit the potassium current in a manner dependent upon the membrane potential, time and stimulus frequency, and are effective by external application. These potassium blocking agents are useful tools to eliminate the potassium current, to study synaptic transmission, and to characterize the potassium channel. Grayanotoxins selectively increase the resting sodium permeability thereby causing a large depolarization without much effect on the sodium permeability increase responsible for the action potential generation. The resting sodium channel exhibits a very low permselectivity to various organic and inorganic cations, whereas it shows a high selectivity during activity. Grayanotoxins do not affect the selectivity ratio either at rest or during activity. Batrachotoxin and veratridine also increase the resting sodium permeability. These depolarizing agents are useful probes to characterize the resting sodium channel.

I. CHEMICAL DISSECTION OF NERVE MEMBRANES

The use of toxins and chemicals as tools to characterize excitable membranes has become increasingly important in recent years. In most cases, the usefulness of these neuroactive agents relies upon their specific affinities for certain parameters associated with excitation. The history of such tools, however, goes back to the last century. Langley made use of the ganglionic blocking action of nicotine to map the autonomic nervous system (118-120). Curare has been used by a number of investigators to selectively block neuromuscular junctions. Similarly many other alkaloids have been utilized including atropine to block muscarinic receptors, physostigmine to inhibit cholinesterase, and reserpine to deplete nerve terminals of norepinephrine.

The present paper limits its scope to the "pharmacological dissection" of nerve membranes. This is one of the areas in which truly remarkable progresses have been made during the past decade or so as a result of discovery of unique features of several toxins and chemicals. Unlike physiology of synapses and neuromuscular junctions, use of chemicals as tools is relatively new in the field of axon physiology. Prior to such development, local anesthetics such as procaine and cocaine and some general anesthetics such as chloroform and ethyl ether were commonly used for the purpose of blocking nervous conduction in various experiments. However, these anesthetics have a limited usefulness owing to their relatively nonspecific mechanisms of action as compared to highly specific chemicals that have been discovered recently. The anesthetics were used only when nervous conduction was to be blocked for certain purposes. No drugs had been used until some ten years ago to elucidate the function of specific nerve membrane sites such as ionic channels.

II. SPECIFIC INHIBITORS OF SODIUM CHANNELS

A. Tetrodotoxin

One of the most remarkable pharmacological agents utilized as tools in axon physiology is tetrodotoxin (TTX). It is contained in the ovary and liver of th puffer fish, the toxicity of which has been recognized in China and Egypt for a long time. Since early this century, the puffer fish poison has been a subject of chemical and pharmacological investigations especially in Japan where the puffer fish is regarded as one of the most delicious. Tahara (190) extracted a toxic component from the puffer fish and named it tetrodotoxin. Since then a number of physiological and pharmacological investigations have been carried out at the systemic level and with isolated nerve muscle preparations (84). It was well established that TTX blocks excitation and conduction of peripheral nerves and skeletal muscles, but the mechanism underlying this action remained obscure.

It is of interest that TTX has been found not only in the puffer fish but also in a few other animals such as the California newt Tarica. Several review articles have been published dealing with the history, biology, chemistry and mode of action of TTX (37, 42,45,50,69,84,85,132,137,142-146,167,197). The chemical structure of TTX is illustrated in Fig. 1.

The first experiment aimed at elucidating the cellular mechanism of action of TTX was carried out using frog sartorius muscle (155). Intracellular microelectrode experiments revealed that the muscle action potential was blocked by low concentrations of TTX without any change in resting membrane potential, resting membrane

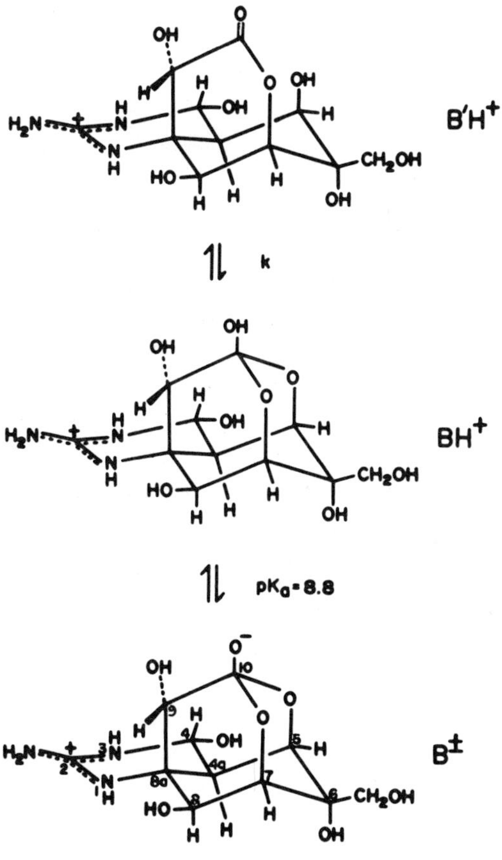

Figure 1. Structure of tetrodotoxin (158).

resistance and delayed rectification. It was suggested that sodium conductance increase normally occurring during activity is selectively blocked by TTX. This hypothesis was indeed demonstrated by voltage clamp experiments with lobster giant axons (160). Peak transient sodium current associated with step depolarization was blocked by TTX, while steady-state potassium current remained unimpaired (Fig. 2). TTX was effective at extremely low concentrations in the order of 1×10^{-8} M. The effective concentration of procaine is estimated to be about 1 mM. Thus TTX is much more potent than procaine in blocking the nerve, the difference being 100,000-fold. Another feature of TTX action which is even more

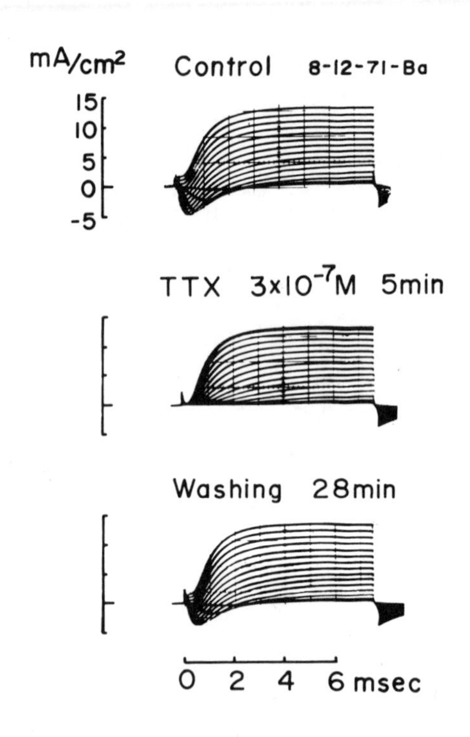

Figure 2. Families of membrane currents associated with step depolarizations (10 mV steps) in a squid giant axon before and during external application of 3×10^{-7} M tetrodotoxin (TTX) and after washing with toxin free medium. TTX blocks peak transient sodium current without any effect on steady-state potassium currents (145).

striking is its highly selective affinity for the sodium mechanism. Procaine and other local anesthetics suppress both sodium and potassium conductances. These two unique features of TTX made it possible to use as a tool in a variety of experiments.

More detailed characteristics of TTX action were further elucidated. TTX had no specific affinity for sodium ions per se, and was capable of blocking sodium channels regardless of the species of ions that carried the current. It caused no rectification blocking the sodium current flowing in either inward or outward direction across the nerve membrane (133). TTX did not change the kinetics of sodium current (191). This is important not only because TTX is different from certain local anesthetics which slow the sodium kinetics (159,195), but also because it indicates the selective action on the term $\bar{g}_{Na}$ in the Hodgkin-Huxley formulation (77) without effect on m.

Despite its highly potent action when applied outside, TTX had no effect on the sodium conductance if directly applied to the internal surface of the nerve membrane. No change was observed in either action potential or sodium current during internal perfusion of TTX at a concentration of 1×10^{-5} M or 1×10^{-6} M (151,152). Since TTX is not soluble in most organic solvents, it is presumably difficult to penetrate the nerve membrane which is composed of lipids. Thus the site of action of TTX is located on or near the external surface of the nerve membrane. TTX has a pK_a of 8.8, and therefore exists in both cationic and zwitterionic forms at physiological pH values (Fig. 1). The potency of TTX in blocking action potential and sodium current increased as the external pH was decreased (28,68,158,168). This result indicates that TTX acts on the sodium channels in the cationic form.

It should be noted that TTX is also effective in decreasing resting sodium conductance. A hyperpolarization of a few millivolts was observed by external application of TTX or 1 mM Na solution. However, TTX did not cause a hyperpolarization when applied in 1 mM Na solution (43,44,150). Resting sodium flux was partially inhibited by TTX (15,63). These observations can be interpreted as being due to the TTX-induced decrease in resting sodium permeability.

The effects of TTX on excitable membranes other than those of giant axons have been extensively examined. In short, the nerve membranes that produce action potentials by sodium conductance increase were susceptible to TTX action with a few exceptions of the nerves from some clams and shellfish (198) and from the newt and puffer fish (86). Giant neurons of *Aplysia* produce action potentials by both sodium and calcium, and only the sodium dependent component was blocked by TTX (51,52).

The skeletal muscles that produce action potentials by sodium current flow were sensitive to TTX as in the case of frog muscle (155), whereas those that produce action potentials by calcium current flow were insensitive as in the case of crustacean muscles (57,189). However, the muscles of the puffer fish and the newt, which are sodium dependent, were insensitive to TTX (59,189). The mammalian muscles, which are sensitive to TTX, became less sensitive after denervation (61,173,174). In this connection, it is interesting to see that the developing rat myoblast and embryonic chick skeletal muscle are insensitive to TTX (83,106).

The smooth muscles are generally not blocked by TTX (7,53,54, 62,112,114). However, it is of interest that TTX has a direct relaxant action on the vascular smooth muscle (88,121,140). This action is thought to be mainly responsible for severe hypotension observed with systemic administration of TTX.

The transient inward current recorded from the voltage clamped cardiac muscle was blocked by TTX or by elimination of sodium from the bathing medium (17,18,38,39,126,175,176,193). This current is responsible for the upstroke of the action potential. The secondary, slow transient inward current, which is related to the plateau of the cardiac action potential, was unaffected by TTX but blocked by manganese, a calcium current blocking agent (166,175,193,203). It was dependent on both external calcium and sodium concentrations (17,18,126,166,175).

The generator potentials from various sensory receptors are relatively insensitive to TTX, while spike discharges are blocked. Thus TTX affords a convenient tool to study the kinetics of the generator potentials in the stretch receptor of the crayfish (125) and in the Pacinian corpuscles (125,164,170). The membrane current and potential change produced in the barnacle photoreceptors by light stimulation were unaffected by TTX (25,26). Thus these generator potentials are thought to be produced by a mechanism different from that in nerve membranes.

Another unique feature of TTX action has been found in synapses and neuromuscular junctions. TTX blocked conduction of presynaptic and postsynaptic nerve and muscle, but had no effect on the transmitter release and the sensitivity of the postsynaptic membrane to the transmitter (31,40,41,48,89,90,92,131). Thus it was possible to release the transmitter by depolarizing the nerve terminal without producing an action potential in the presence of TTX, and TTX has become an extremely useful tool to study the excitation-secretion coupling. Delayed rectification associated with a sustained depolarization of the presynaptic nerve membrane can be eliminated by iontophoretic injection of tetraethylammonium in the nerve terminal. These techniques have been widely used for frog neuromuscular junctions (81,91,93,94,97), for squid giant synapses (24,49,92,95, 96,98,99,115-117,122,123), and spinal neurons (31,34). Release of norepinephrine in the cat spleen was also unaffected by TTX (109). It should be noted that the end-plate membrane undergoes conductance increases to sodium and potassium upon transmitter action which are not blocked by TTX. This constitutes one of the major differences between the sodium system in axonal membranes and that in end-plate membranes.

Another way of utilizing TTX as a tool is to separate membrane current into sodium and potassium components. In the voltage clamped nerve membrane, a transient sodium current is followed by a steady-state potassium current during step depolarization. The sodium component can be completely eliminated by application of TTX, thereby permitting observation of the kinetics of pure potassium current. This method is particularly useful when the sodium current is prolonged. For example, in the lobster axon poisoned with the insecticide DDT, the sodium inactivation is greatly prolonged (147,148). Therefore, the steady-state current flows inward at certain membrane potentials as a result of a large, inward, steady-state sodium current overlapping with an outward, steady-state potassium current. The kinetics of both sodium and potassium currents in such conditions can be analyzed separately by the use of TTX or saxitoxin (Fig. 3). The same technique applies to any other chemicals and drugs which prolong the sodium inactivation. Such examples, among many, include *Condylactis* toxin (139,161), the insecticide allethrin (139,205), and scorpion venom (162).

The fact that TTX has no effect on gating currents affords an excellent opportunity to analyze them. The gating current is generated as a result of movement of charges associated with rearrangements of membrane macromolecules during opening and closing of the ionic channels (11,12,22,101,102,128,129,165,182). Thus it can be recorded in the axon in which sodium current has been eliminated by exposure to TTX.

Perhaps the most remarkable way of utilizing TTX as a tool is an attempt to characterize, isolate and identify sodium channels. The first study has been performed to count the density of sodium channels in the lobster walking leg nerve (135). The amount of TTX absorbed on the nerve as measured by bioassay, and the upper limit to the TTX binding site density was estimated to be 13 per square microns of nerve membrane. Since TTX interacts with sodium channel on a one-to-one stoichiometric basis (33), the density of TTX binding sites represents that of sodium channels. Similar measurements of sodium channel density have since been made with various nerves using either bioassay or tritiated TTX. The density thus estimated ranges from $3/\mu^2$ in the garfish olfactory nerve to $75/\mu^2$ in the rabbit vagus nerve (16,32,56,104). The sodium channel density of squid axon membranes has recently been estimated to be

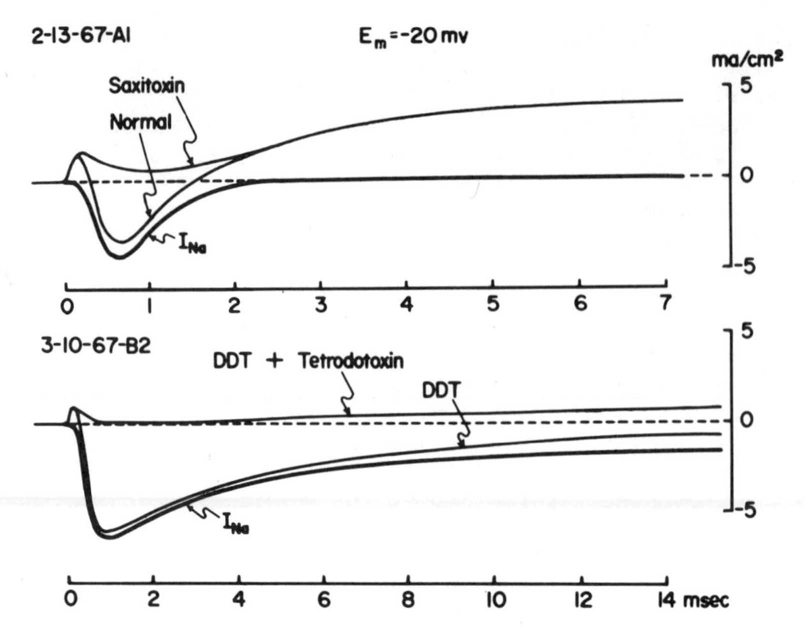

Figure 3. Separation of membrane current into sodium and potassium currents by use of saxitoxin or tetrodotoxin (3×10^{-7} M) in a normal and a DDT-treated lobster giant axon. The membrane current in saxitoxin and that in DDT (5×10^{-4} M) plus tetrodotoxin represent the potassium current. The sodium current (I_{Na}) is obtained by subtraction of the potassium current from the total membrane current (148).

$522/\mu^2$ from the rate of action of TTX (103) and $483/\mu^2$ from the gating current (102). The channel densities of the surface membrane and the transverse tubular membrane of the frog sartorius muscle have been measured using the method by which the tubular system was disrupted with glycerol (82). The density in the surface membrane is about $175/\mu^2$ and that in the transverse tubular membrane is $41\text{-}52/\mu^2$.

The TTX-receptor binding has been studied using homogenates and solubilized preparations of various nerves (65,66). The homogenized garfish olfactory nerve bound TTX with a dissociation constant of 8.3×10^{-9} M, a value which agrees reasonably well with the apparent dissociation constant to block the action potential (20,64). Proteases such as trypsin, α-chymotrypsin and pronase inhibited TTX binding in support of the notion that the TTX binding sites are proteins (21).

The mechanism whereby TTX binds to the sodium channel remains largely obscure. Kao and Nishiyama (87) suggested that the guanidinium group of TTX molecules plugs in the sodium channel thereby preventing the passage of sodium ions. The rest of the TTX molecule is too large to penetrate the channel. This hypothesis was further elaborated by Hille (72-74) who proposed that the guanidinium group of TTX molecule sticks the narrow selectivity filter of the sodium channel. The selectivity filter is assumed to be composed of a 3 by 5Å construction of the pore formed by a ring of six oxygen atoms, one of which is an ionized carboxylic acid. TTX molecule presumably binds to the selectivity filter of the sodium channel through the coulombic force and hydrogen bonds by virtue of the guanidinium group and five hydroxy group.

B. Saxitoxin

Saxitoxin (STX) is contained in toxic clams and shellfish such as the Alaska butter clam _Saxidomas giganteus_, but it is originally derived from the dinoflagellate _Gonyaulax catenella_ (84,179). The chemical structure has recently been established as shown in Fig. 4 (178). Unlike TTX, STX contains two guanidinium groups.

The mechanism of action of STX on nerve membranes is essentially the same as that of TTX. STX blocks the sodium channel without any effect on the potassium channel (68,157). The only small difference so far found is that the recovery after STX block is slightly faster than that after TTX (157,204). Aside from this STX shares many features with TTX (see 142,144). Thus STX has been used as a useful tool in a variety of neurophysiological experiments to characterize sodium channels (144).

Several monovalent, divalent and trivalent cations have been found to compete with STX for the binding site (65). The dissociation constant was smallest for trivalent cations (10^{-3} M), intermediate for divalent cations (10^{-2} M), and largest for monovalent cations (0.5 M), except for thallous ion (Tl^+) which has an unusually low dissociation constant of 2×10^{-2} M. These results are in keeping with the view that STX directly blocks the sodium channel, since the apparent dissociation constants of these cations in blocking the sodium channel are of the same order of magnitude. It is also interesting that STX competes with TTX for the binding site (16,32,64,65,204).

C. Toxin from _Gonyaulax tamarensis_

Gonyaulax tamarensis, a dinoflagellate which causes red tide as in _G. catenella_, contains toxins including STX and other new components (187). One of the unidentified toxins, tentatively called _Gonyaulax tamarensis_ toxin (GTTX), has recently been found

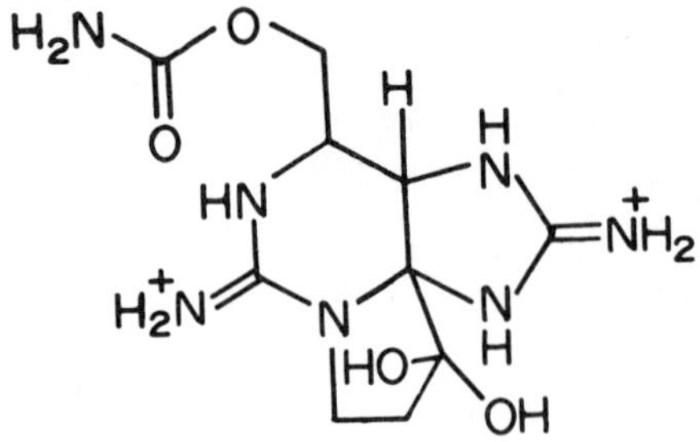

Figure 4. Structure of saxitoxin.

to exert a nerve blocking action in a manner identical with that of TTX and STX (145,153). GTTX inhibited the sodium current without any effect on the potassium current (Fig. 5), and was effective only from outside of the nerve membrane.

Whereas the chemical structure of GTTX has not completely been identified yet, there is evidence that it is different from STX or TTX (187). Thus we have now at least three toxins which are chemically different but physiologically almost identical. The common denominator(s) in chemical structure shared by TTX, STX and GTTX would be of great value in identifying and characterizing the binding site and sodium channel in the nerve membrane.

III. SPECIFIC INHIBITORS OF POTASSIUM CHANNELS

A. Tetraethylammonium

In the strict sense of the word, no specific inhibitor has so far been found for potassium channels which is comparable to TTX or STX for sodium channels. No natural toxin has been known to specifically block the potassium channels. This is not so surprising from the biological point of view, because the most effective ways of blocking or disturbing nervous conduction would be to modify the sodium system function, by blocking the opening of the sodium channels as in the case of TTX, STX and GTTX, by inhibiting the sodium inactivation as in the case of Condylactis toxin (161), or by opening up the sodium channels without stimulation thereby causing a large depolarization as in the case of grayanotoxins and batrachotoxin which are to be described later in

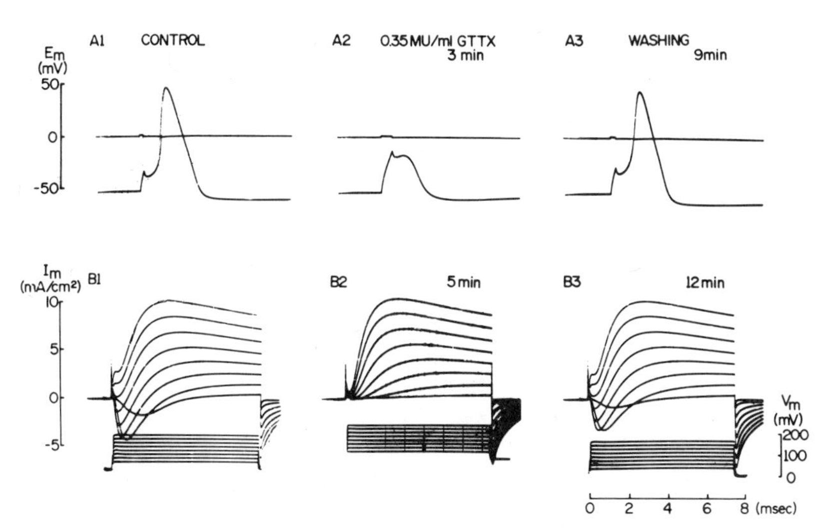

Figure 5. Effects of external application of 0.35 mouse unit (MU)/ml Gonyaulax tamarensis toxin (GTTX) on the action potential (A1-A3) and the membrane currents (B1-B3) associated with step depolarizations (20 mV steps) from the holding membrane potential of -70 mV in a squid giant axon (153).

this paper. In any of these cases, the ultimate effect can be exerted with a minimum amount of energy. In contrast to these modifications of sodium channel function, changes in potassium channels do not cause effects which are large enough to seriously impair normal functioning of the nerve fiber. For example, an increase or decrease in resting potassium permeability does not cause a sizable change in membrane potential comparable to that produced by sodium permeability change. Inhibition of potassium channel opening would cause a prolongation of the action potential, yet the degree of disturbance in nervous function as a whole would be less serious compared with that caused by sodium channel block.

A few chemicals have been found to inhibit potassium channels with potencies much less than those of TTX and STX in blocking sodium channels. The most classical and well-known chemical is tetraethylammonium (TEA). The discovery of the TEA blockage of potassium channel was made by Tasaki and Hagiwara (194) with squid giant axons. They found that TEA, when injected inside of the squid axon, caused a great prolongation of the falling phase of the action potential forming plateau resembling a cardiac action potential. Under voltage clamp conditions, TEA suppressed the steady-state current without much effect on the peak transient current. However, TEA had no such effect when applied outside of the squid axon. More detailed analyses have been performed with squid axons (8,9,13) nodes of Ranvier of frog and *Xenopus* (14,67,107,108,138,181,202), cockroach giant axons (172), snail neurons (58,163), nerve cells of puffer fish (141), and frog skeletal muscle fibers (188). Many nerves other than squid axons were affected by external application of TEA.

It should be borne in mind that the TEA concentrations required to block potassium current are very high (10-50 mM) and that the affinity of TEA is not absolutely specific for potassium channels. For example, in squid giant axons, a TEA concentration of 15-30 mM is needed to block potassium current, the concentration which also suppresses sodium current by 10-30 percent. This is not so surprising because many other quaternary ammonium compounds suppress both sodium and potassium currents (23,156,159).

Kinetic analyses have revealed that TEA occludes the potassium channel only in its open configuration (9,10). The potassium channel appears to have two distinct parts, one being a wide inner mouth that can accept a hydrated potassium ion or a TEA-like ion and the other being a narrower portion that can accept a dehydrated or partially dehydrated potassium ion but not TEA. Certain derivatives of TEA exhibited a time- and voltage-dependent block of the potassium current when injected or internally applied to the squid axon or the node of Ranvier (9,10,14). With increasing a side chain of TEA molecule, the potency in blocking the potassium current increased and the block became time independent, that is to say, increased with prolonging pulse duration. Thus these derivatives caused a potassium inactivation to appear. However, when applied to the external surface of the node of Ranvier, TEA and its derivatives blocked the potassium channel in a manner independent of time and voltage. Thus the mechanisms of the TEA block of potassium channel in the node of Raniver are different in external and internal applications.

B. Aminopyridines

4-Aminopyridine (4-AP) (Fig. 6) has been found to selectively block potassium current in cockroach giant axons (171). It was effective by external application, and it was suggested that 4-AP would become a useful tool to eliminate potassium current. We have embarked on more detailed kinetic analysis using squid giant axons (208-210).

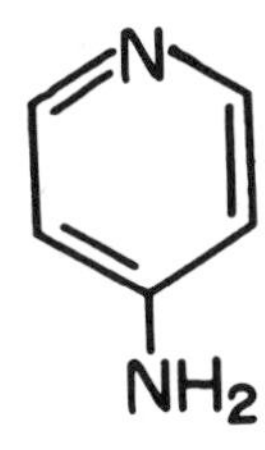

Figure 6. Structure of 4-aminopyridine.

4-AP was effective in blocking the potassium current from either outside or inside of the nerve membrane. The current-voltage relations for peak transient sodium current and steady-state potassium current are illustrated in Fig. 7. The potassium current was effectively suppressed by 2 mM 4-AP, while the sodium current was totally unaffected. One very striking feature became apparent from the potassium current-voltage curve in Fig. 7. The potassium current was completely inhibited at small and moderate depolarizing levels, whereas it was much less affected at large depolarizing levels. Despite such voltage dependency of 4-AP block of potassium current, there was no rectification in the 4-AP block as revealed by the instantaneous current measurements in the isotonic potassium solution. This is in contrast to TEA which exhibits rectification (13).

The kinetics of potassium current underwent drastic changes after exposure to 4-AP. The potassium current in the axon treated with 4-AP rose very slowly, reaching a steady state in several tens of milliseconds as against the control potassium current which attained a steady state in a few milliseconds at $10^{\circ}C$ (Fig. 8). The frequency of stimulations has been found to drastically modify the potassium current in the presence of 4-AP. The onset of potassium current was greatly accelerated by the second stimulus. These observations are compatible with the notion that depolarization progressively removes 4-AP molecules from the binding site. Thus 4-AP causes a voltage, time and frequency dependent block of potassium channels. The use of 4-AP as a tool is naturally limited by these factors, and one must be fully aware of this nature of the 4-AP block before using it as a tool. Similar observations with squid axons were made by Meves and Pichon (130). A kinetic model has

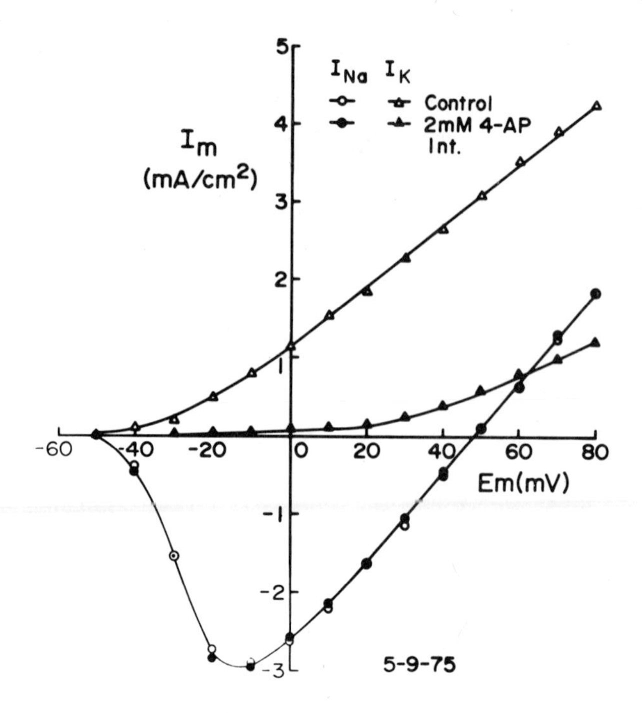

Figure 7. Current-voltage relations for peak transient sodium current (I_{Na}) and steady-state potassium current (I_K) before and during internal perfusion of 2 mM 4-aminopyridine (4-AP) in a squid giant axon. Holding membrane potential was -80 mV, and step depolarizations were applied at 30-second intervals to avoid frequency-dependent effects (208).

been proposed in which aminopyridine molecules bind to the closed configuration of potassium channels and are released from the open channel configuration (209,210). Computer simulations based on this model were able to reproduce membrane currents under a variety of experimental conditions.

3-Aminopyridine (3-AP) and 2-aminopyridine (2-AP) also suppressed the potassium current in the same way as 4-AP (208). Although 3-AP was almost equipotent with 4-AP, 2-AP was less effective than 4-AP.

Aminopyridines have recently been demonstrated to be useful tools to study synaptic transmission (124). When the squid giant synapse was externally exposed to TTX and 3-AP or 4-AP, the sodium and potassium conductance increases associated with presynaptic

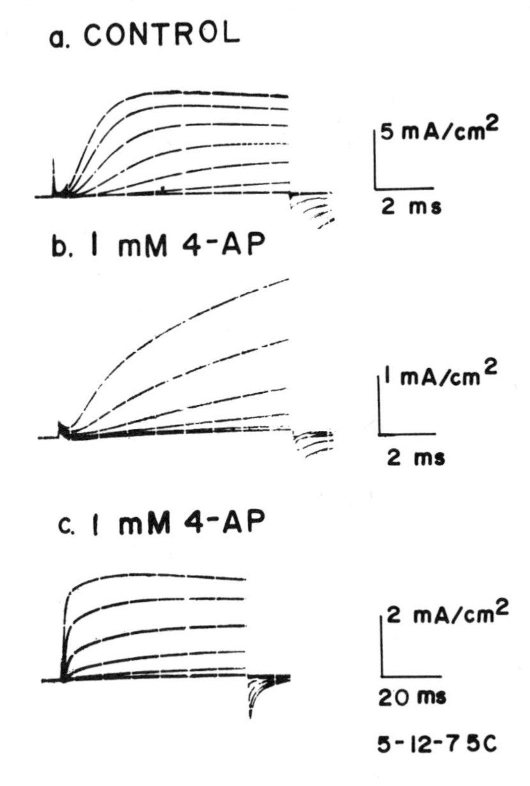

Figure 8. Time dependency of block of potassium currents by 4-aminopyridine (4-AP) applied externally to the squid giant axon in the continuous presence of 3 x 10^{-7} M tetrodotoxin outside to eliminate sodium currents. Holding membrane potential was -70 mV, and the membrane was step depolarized to levels ranging from -20 mV to +100 mV at 30-second intervals to avoid frequency-dependent effects. Note different current densities and time scales in each set of records (208).

depolarization were completely blocked, yet the postsynaptic potentials were evoked by presynaptic depolarizations. Thus aminopyridines do not seem to impair the calcium conductance increase which is thought to be responsible for transmitter release. Since the external application of aminopyridines blocks the potassium current relatively quickly (5-10 min), they are much more convenient than TEA which must be injected in the presynaptic nerve terminal by iontophoresis taking a long period of time. However, caution must be exercised in using aminopyridines for this purpose, because the aminopyridine block of potassium conductance is voltage and time dependent.

IV. AGENTS THAT INCREASE RESTING SODIUM CONDUCTANCE

During the past several years, studies of the agents that increase the resting sodium conductance have become increasingly popular. Most of them are natural toxins, and have been found to cause a large depolarization of various excitable membranes through a specific increase in resting sodium permeability. As is discussed before, the increase in resting sodium permeability is an effective means by which the membrane undergoes a sizable depolarization thereby blocking the excitability, since the sodium permeability is much smaller than the potassium permeability under normal resting conditions.

The agents in question include grayanotoxins (GTXs), batrachotoxin (BTX) and veratradine. The present discussion is primarily concerned with the effects of GTXs with which detailed analyses have been performed. Some references will also be made to the studies of BTX and veratridine.

A. Grayanotoxins

Grayanotoxins are contained in the leaves and flowers of various plants such as *Leucothoe*, *Rhododendron*, *Andromeda* and *Kalmia*. There are several toxic components, grayanotoxin I (GTX I), grayanotoxin II (GTX II) and grayanotoxin III (GTX III) being the major compounds (Fig. 9). GTX I used to be called andromedotoxin. Some studies were made for the pharmacology of GTXs (46,47,60,111, 136,192,207). In general, GTXs had potent actions on the cardiovascular and respiratory systems. Bradycardia, hypotension and respiratory depression were among the symptoms of poisoning. Recent finding of a positive inotropic action of GTXs is worth mentioning, because this action may be related to an overloading of Na-pump as a consequence of increased sodium influx (111).

The first study of the effect of GTXs on cell membranes was made by Deguchi and Sakai (36). They found that GTX I caused membrane depolarization and sustained after-depolarization following action potentials in frog muscle. Seyama (183) also made similar observations using rat skeletal muscle. It was found that decreasing sodium concentration in the external medium restored the GTX-induced depolarization. These observations were taken as indicating that the depolarization by GTX I was due to a specific increase in resting sodium permeability. However, we cannot exclude the possibility that permeabilities to other ions such as potassium and chloride are involved as well.

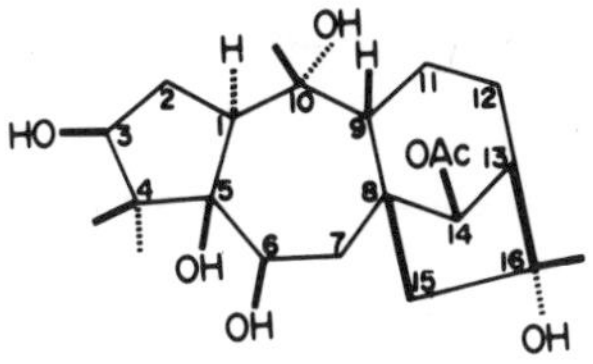

Grayanotoxin I

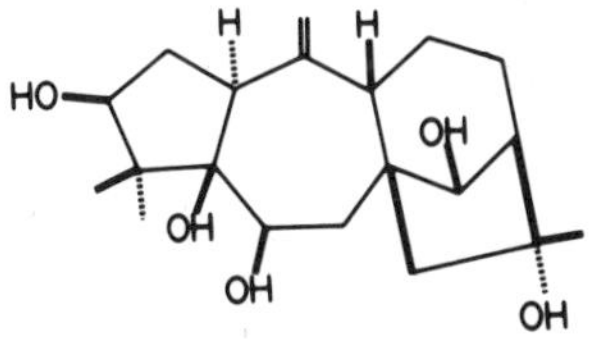

Grayanotoxin II

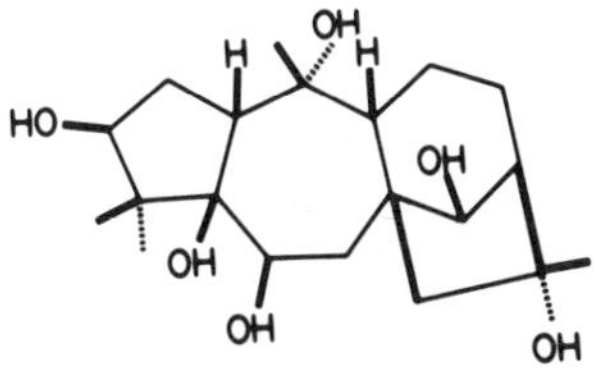

Grayanotoxin III

Figure 9. Structures of grayanotoxins I, II and III (144).

More detailed analyses have been performed with squid giant axons (146,149,184) with which not only external but also internal ionic composition can be modified and voltage clamp techniques are applicable with a high degree of accuracy. GTX I and α-dihydrograyanotoxin II (α-2H-GTX II) were used, and they exerted essentially the same effects except that the latter was somewhat more effective than the former. GTXs were effective by either external or internal application, although the effect appeared more quickly with internal application than with external application.

Fig. 10 illustrates the changes in membrane potential recorded from a squid giant axon as GTX I was applied externally in the presence of various sodium concentrations and TTX outside. The membrane was gradually depolarized after introduction of GTX I (5×10^{-5} M), and the membrane potential attained a level of about -15 mV. A drastic decrease in external sodium concentration from the normal value of 449 mM to 1 mM caused recovery of the membrane

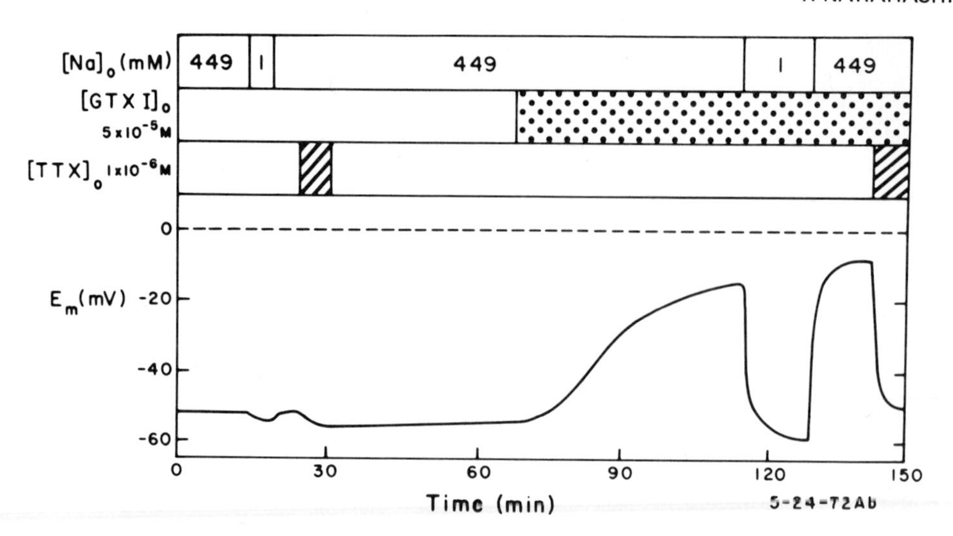

Figure 10. Changes in resting membrane potential (E_m) of an intact squid axon caused by external application of 5 x 10^{-5} M grayanotoxin I (GTX I). Decrease of external sodium concentration ($[Na]_o$) to 1 mM hyperpolarized the membrane in GTX I. Externally applied tetrodotoxin (TTX) also reversed depolarization (149).

potential which attained a level more negative than the value achieved in 1 mM Na in the absence of GTX I. External application of TTX (1 x 10^{-6} M) antagonized the GTX-induced depolarization in the presence of the normal sodium concentration outside. These observations are in support of the notion that GTX I increases the resting sodium permeability thereby causing a depolarization.

However, other possibilities such as the decrease in potassium permeability and the increase in chloride permeability must be excluded. The most straightforward approach would be to see whether GTX I causes depolarization in the absence of sodium in both external and internal media. This idea becomes apparent from the constant field equation (55,78):

$$E_m = \frac{RT}{F} \ln \frac{P_K[K]_o + P_{Na}[Na]_o + P_{Cl}[Cl]_i}{P_K[K]_i + P_{Na}[Na]_i + P_{Cl}[Cl]_o} \qquad (1)$$

where E_m represents the membrane potential, P_K, P_{Na} and P_{Cl} are permeability coefficients for K, Na and Cl, respectively, [K], [Na] and [Cl] are the concentrations (more strictly the activities) of K, Na and Cl, respectively, with subscripts o and i referring to outside and inside phases of the axon, and R, T and F are the gas constant, the absolute temperature and the Faraday constant, respectively. If P_{Na} is the only parameter that is affected by GTX I, then in the absence of sodium inside and outside of the axon GTX I should have no effect on the membrane potential. This was indeed demonstrated to be the case (149,184).

The aforementioned experiments do not give quantitative information concerning the magnitude of sodium permeability increase. Voltage clamp experiments have proved very straightforward and powerful for this purpose. However, the method used here is somewhat different from the conventional voltage clamp which is used to measure ionic conductance changes associated with step potential changes. In the present study, we are concerned with measurements of <u>resting</u> ionic conductances.

The external and internal ionic compositions were so adjusted as to make the equilibrium potentials for potassium (E_K) and chloride (E_{Cl}) equal to -70 mV. By definition, no potassium and no chloride current flows at this membrane potential. The membrane potential was then clamped at -70 mV, and the holding membrane current was measured. This current should have been carried mostly by sodium ions, because currents carried by other ions such as calcium, glutamate and fluoride were much smaller in magnitude. Thus the resting sodium conductance (g_{Na}) was calculated from the equation

$$g_{Na} = \frac{I_{Na}}{E_m - E_{Na}} \qquad (2)$$

where I_{Na} and E_{Na} refer to sodium current and sodium equilibrium potential, respectively. Figure 11 depicts an example of such an experiment. The membrane was unclamped and clamped every several seconds by means of a microswitch so that both the membrane potential and membrane current could be recorded from the same axon. As the membrane started depolarizing following internal perfusion of GTX I (1×10^{-5} M), an inward sodium current started flowing. When the external sodium concentration was decreased, the GTX-induced depolarization and sodium current decreased. The average increase in resting sodium conductance was estimated to be 10-20 fold depending on the concentration of GTX I. The resting sodium permeability was calculated by the constant field equation

$$P_{Na} = -\frac{I_{Na}RT}{F^2E_m} \frac{1-e^{-E_mF/RT}}{[Na]_o - [Na]_i\, e^{-E_mF/RT}} \qquad (3)$$

The mean resting sodium permeability increased to a value of 1.31×10^{-6} cm/sec, which represents an increase by a factor of 20.

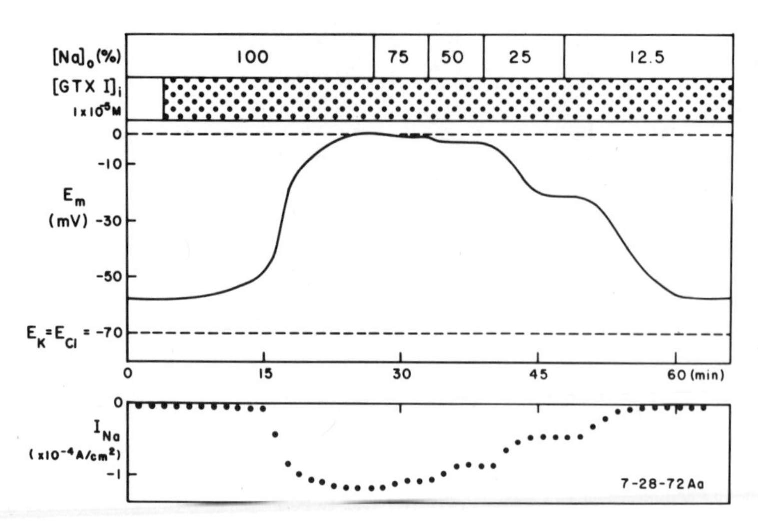

Figure 11. Changes in membrane potential (E_m) and holding membrane current carried by sodium (I_{Na}) by internal perfusion of 1 x 10^{-5} M grayanotoxin I (GTX I). Current with a minus sign denotes an inward current. External sodium concentration ($[Na]_o$) is 449 mM in normal artificial sea water, and reduced $[Na]_o$ is expressed as a percentage of that solution. The membrane potential was voltage clamped for 7 seconds at 90 second intervals at -70 mV which was made equal to the equilibrium potentials for potassium (E_K) and chloride (E_{Cl}) by alteration of the ionic composition of internal perfusate (149).

The kinetics of the TTX antagonism has been studied in detail (149). The magnitude of the depolarization induced by GTX is not an acceptable parameter for this purpose, since the depolarization is the final product as a result of a series of events. The increase in sodium permeability by GTX would cause a depolarization which would in turn increase the potassium permeability. The increased potassium permeability would then lead to a hyperpolarization as expected from the constant field equation (1) and a depolarization as a result of the accumulation of potassium in the periaxonal space (35). Thus the observed depolarization is a sum of all of these changes in membrane potential. Therefore, in order to perform a reliable kinetic analysis, the increase in resting sodium permeability which is the primary action of GTX must be used as a measure of activity. The Lineweaver-Burk plots of resting sodium current are illustrated in Fig. 12. Statistical test indicates that the two curves, one with GTX I and the other with GTX I plus TTX, meet on the abscissa not on the ordinate. Thus TTX antagonizes the GTX action in a noncompetitive manner. The inhibition constant of TTX was estimated to be 40 nM.

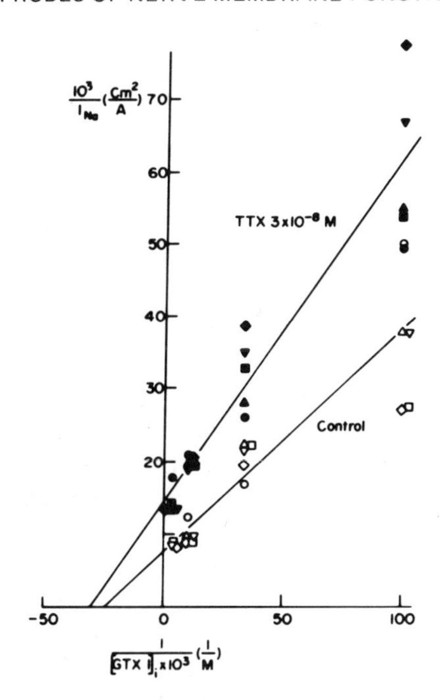

Figure 12. Lineweaver-Burk plots of increase in sodium current (I_{Na}) of squid axons caused by internally applied grayanotoxin I (GTX I) and the antagonistic action of tetrodotoxin (TTX) applied externally. Each symbol represents separate series of experiments. External sodium concentration was reduced to 25% of normal to avoid electrode polarization (149).

In spite of a large increase in resting sodium permeability by GTXs, the mechanism whereby the action potential is generated was almost unimpaired. The action potential or peak transient sodium current of normal amplitude could be produced as long as the membrane potential was brought back to the original level. This raised a question as to whether the resting sodium permeability was operationally different from the sodium permeability change normally occurring upon depolarization. In order to give a clue to this problem, cation permeability ratios have been studied at rest and during activity before and during application of GTX I (75).

The external and internal ionic compositions were adjusted to make E_K equal to E_{Cl} at -70 mV as described before. The inward holding current associated with step hyperpolarization to -70 mV, which was carried mostly by sodium ions, was measured. Then a test cation was substituted for external sodium, and the same measurement of the holding current was repeated. The inward holding current in the test cation present outside was composed of an inward current carried by the test cation and an outward current carrried by sodium. Therefore, the permeability ratio to test cation (X) is given by

$$\frac{P_X}{P_{Na}} = \frac{1}{[X]_o} \{ \frac{I'_X}{I_{Na}} ([Na]_o - [Na]_i \, e^{E_m F/RT}) + [Na]_i \, e^{E_m F/RT} \} \quad (4)$$

where I'_X refers to the inward holding current in the presence of test cation outside.

The resting permeability ratios were estimated to be Na:Li: formamidine:guanidine:Cs:methylguanidine:methylamine = 1:0.83:1.34: 1.49:0.87:0.86:0.78 and 1:0.95:1.27:1.16:0.47:0.72:0.46 before and during application of GTX I, respectively. It should be noted that the ratio of the most permeabile ion guanindine to the least permeable ion methylamine is about 2, indicating a low permselectivity. It is also worth emphasizing that GTX I, though greatly increases the resting permeabilities to sodium and other test cations, does not change the permeability ratio. This makes it rather unlikely that GTX I creates a new channel. Perhaps GTX I increases the number of open sodium channels or increases the permeability of individual sodium channels. The solution for this problem awaits further experimentation.

The poor selectivity for resting cation permeability is in sharp contrast with a high selectivity during peak transient current. With the nodes of Ranvier of the frog, Hille (70,71) obtained a selectivity ratio Na:Li:formamidine:guanidine:Cs: methylguandine:methylamine = 1:0.93:0.14:0.13:<0.013:<0.01:<0.007. Our recent measurements with squid axon membranes gave a ratio 1:1.12:0.19:0.19:0.085:0.061:0.036 (76). GTX I did not appreciably change this ratio.

The important features of GTX action may be summarized as follows: GTXs reversibly increase the resting sodium permeability by a factor of 10-20 thereby causing a sizable depolarization. TTX antagonizes this GTX action in a noncompetitive manner with an inhibition constant of 40 nM, a value 13-fold greater than the TTX dissocation constant to inhibit the peak sodium current (33). The action potential mechanism or the sodium conductance increase upon depolarization is not greatly impaired by GTXs. The resting sodium channel exhibits a poor selectivity for cations, the permeability ratio of the most permeable cation guanidine to the least permeable cation methylamine being about 2. The permeability ratio is not appreciably influenced by GTX I. The sodium channel responsible for peak sodium current shows a high selectivity, the permeability ratio of the most permeable cation lithium to the least permeable cation methylamine being estimated to be 31. The permeability ratio during activity is not much affected by GTX I. These observations are compatible with the notion that the resting sodium channel is operationally different from the sodium channel responsible for peak transient current.

B. Batrachotoxin

Batrachotoxin is a toxic principle contained in the skin secretion of the Colombian arrow poison frog Phyllobates aurotaenia. It is a steroidal compound with a unique 7-member ring containing an oxygen and nitrogen (Fig. 13) (2,144,146,196). BTX is a potent poison acting on nerves and muscles. It caused arrhythmias, A-V block, multifocal ventricular ectopic beats, ventricular tachycardia and fibrillation in the cardiac tissues, and blocked ganglionic transmission through impairment of preganglionic nervous conduction (100).

BTX caused a large depolarization in skeletal muscle fibers, cardiac muscle fibers and squid axons, and the effect was antagonized by low external sodium or TTX (5,6,79,150). However, it had no effect on the resting membrane potential of lobster and crayfish muscles (4). As in GTXs, BTX failed to depolarize the squid axon membrane if the external and internal media were devoid of sodium (154). This demonstrates that BTX selectively increases the resting sodium permeability thereby causing a depolarization. In support of these electrophysiological measurements, the efflux of ^{22}Na from the garfish olfactory nerves treated with ouabain was increased by BTX (63).

BTX resembles GTXs in that it is effective from either side of the nerve membrane and that the depolarization is due primarily to a selective increase in resting sodium permeability, but differs in that the effect is not reversed by washing with drug-free media. BTX is more potent than GTXs. Comparison of an irreversible and

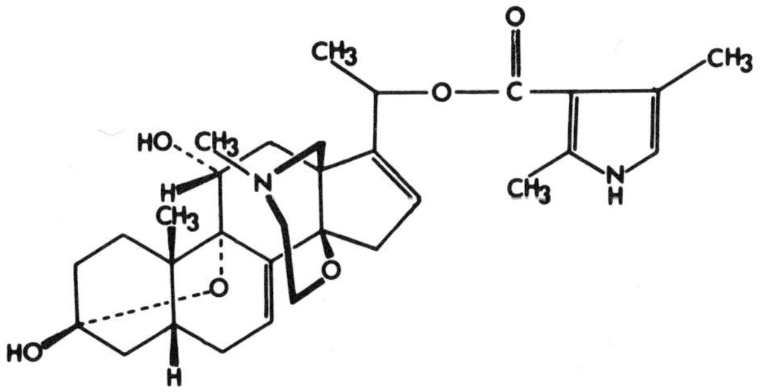

Figure 13. Structure of batrachotoxin (143).

reversible chemical is difficult to make; BTX is effective in causing a moderate depolarization at a concentration of 25 nM, whereas GTX I causes a 50% maximum depolarization at a concentration of 40 μM.

Procaine has been found to antagonize the BTX-induced depolarization in a very unique manner different from TTX (5). When BTX and TTX were applied together to the squid axon, no depolarization occurred. After washing out both BTX and TTX, the membrane was gradually depolarized as the effect of TTX was removed. This suggests that BTX has bound to the receptor in the presence of TTX, although no effect is visible in the form of depolarization. Procaine stopped BTX-induced depolarization when applied in the course of the depolarization. When procaine and BTX were applied together, no change in membrane potential was produced. After washing out both procaine and BTX, however, nothing happened to the membrane potential (Fig. 14). This suggests that procaine prevents BTX from binding to its receptor.

In the frog's node of Ranvier exposed to BTX, the peak transient sodium current was followed by an inward steady-state current during step depolarization (105). The slow inward current which was not inactivated, exhibited a negative conductance at membrane potentials more negative than those where the peak current exhibited a negative conductance. Procaine had discriminating actions on these two component, blocking the peak current more effectively than the slow current. It was suggested that BTX converts a portion of sodium channels to a non-inactivating, procaine-insensitive group. Similar non-inactivating sodium currents were also observed in the node treated with aconitine (180).

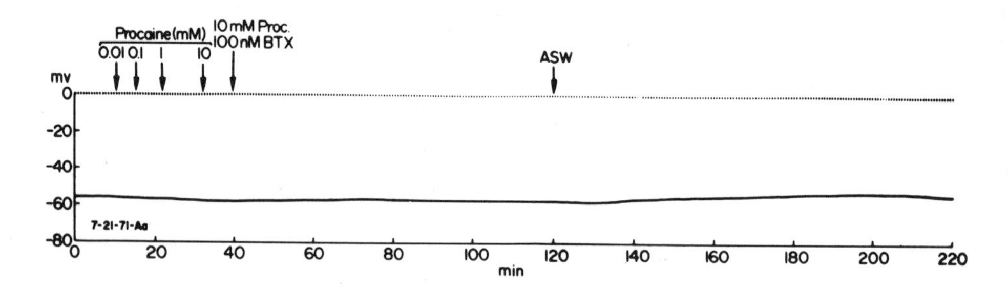

Figure 14. Effects of various concentrations of procaine and simultaneous application of 10 mM procaine and 100 nM batrachotoxin (BTX) on the resting membrane potential of a squid giant axon (5).

Sulfhydryl groups of the membrane proteins appear to play an important role in BTX action (3). After treatment of lobster giant axons with p-chloromercuribenzenesulfonate, dithiothreitol or N-ethylmaleimide, BTX no longer exerted depolarizing action, while TTX could still block action potentials. Apparently TTX and BTX act on different sites of macromolecules. Another evidence for the different TTX and BTX receptors was obtained with the denervated skeletal muscle of the rat (1). TTX became less effective after denervation, whereas BTX potency remained unchanged.

Sodium uptake by electrically excitable neuroblastoma cells was greatly enhanced by BTX, veratridine, aconitine and venom from scorpion *Leiurus quinquestriatus* (29,30). These effects were antagonized by TTX in a noncompetitive manner. It is of interest that the sodium uptake increased by BTX and veratridine was competitively antagonized by divalent cations which showed specificity in the order of $Mn^{++} > Co^{++} > Ni^{++} > Ca^{++} > Mg^{++} > Sr^{++}$. There is evidence to indicate that BTX and veratridine compete with each other for the binding site.

C. Veratrum Alkaloids

Veratrum alkaloids are contained in the plants that belong to the tribe Veratreae, including *Veratrum*, *Schoenocaulon* and *Zygadenus*. The hypotensive action of certain veratrum alkaloids has been known for a long time, and several reviews have been published dealing with the sources, chemistry, pharmacology and therapeutic uses of the veratrum alkaloids (19,110,113,185,186,200).

The action of veratrum alkaloids on excitable membranes differs depending on the structure. The effect may be divided into three groups: 1) Increase in depolarizing (negative) after-potential leading to repetitive after discharges; 2) Depolarization of the membrane without stimulation; 3) Blockage of impulse conduction without change in membrane potential (169). The veratrum alkaloids that exert actions (1) and (2) above are represented by veratridine cevadine and protoveratrines A and B (Fig. 15), and those that exert action (3) above by veratramine, muldamine, 5-veratranine-3β, 11α-diol, and isorubijervine. We are here mainly concerned with actions (1) and (2).

In the voltage clamped node of Ranvier or squid axon exposed to veratridine, the peak transient sodium current associated with step depolarization was followed by a secondary, slow, non-inactivating sodium current (80,134,199,200). The membrane depolarization caused by veratridine is presumably due to this slow, non-inactivating sodium current (200,201,206). Similar secondary

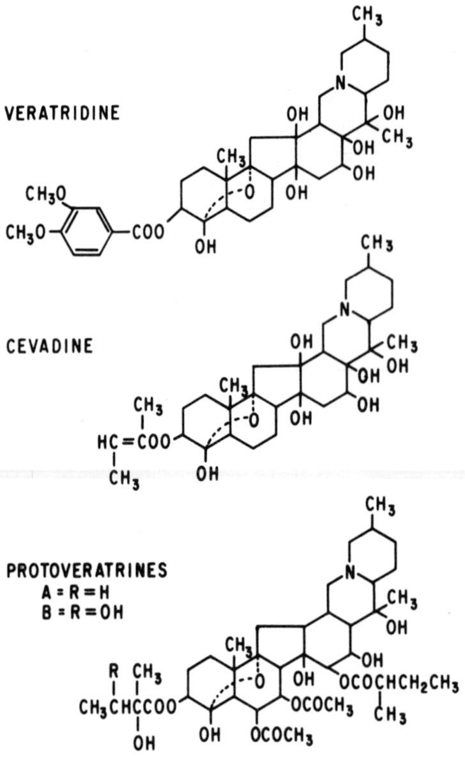

Figure 15. Structures of veratridine, cevadine and protoveratrines A and B (169).

slow currents were also observed in the node exposed to venom of scorpion *Centruroides* (27). TTX can antagonize the slow sodium current and depolarization. Squid axon experiments similar to those for GTXs and BTX have been performed (169). Veratridine did not induce depolarization if the external and internal perfusates were devoid of sodium, indicating that the resting sodium permeability was selectively increased. TTX effectively antagonized the depolarization induced by veratridine. It appears that veratridine acts from inside of the membrane, because the onset and offset were much faster with the internal perfusion of veratridine than with the external perfusion (127,169). The ^{22}Na efflux from the garfish olfactory nerves treated with ouabain was increased by veratrine, and the effect was antagonized by TTX or STX (63).

The rat L6 myotubes have provided a unique opportunity to study the pharmacological characteristics of sodium channels (177). The action potential was highly resistant to high concentrations of TTX (10^{-5} M) and STX (10^{-6} M), but was prolonged by veratridine.

After prolonged application of veratridine, the membrane was depolarized. Most intriguing is the finding that TTX and STX antagonized the veratridine-induced depolarization without any effect on the veratridine-induced prolongation of the action potential. α-Dihydrograyanotoxin II had no effect on this preparation. The differential actions of TTX and STX are in support of the notion that the resting sodium channel is operationally different from the sodium channel responsible for the action potential generation.

ACKNOWLEDGMENTS

The author's studies in this paper were supported by NIH Grant (NS 10823). Unfailing secretarial assistance by Virginia Arnold, Arlene McClenny and Delilah Munday is greatly appreciated.

REFERENCES

1. ALBUQUERQUE, E.X. and WARNICK, J.E.: J. Pharmacol. Exp. Ther. 180 (1972) 683.

2. ALBUQUERQUE, E.X., DALY, J.W. and WITKOP, B: Science 172 (1971) 995.

3. ALBUQUERQUE, E.X., SASA, M., AVNER, B.P. and DALY, J.W.: Nature-New Biol. 234 (1971) 93.

4. ALBUQUERQUE, E.X., SASA, M. and SARVEY, J.M.: Life Sci. 11 (1972) 357.

5. ALBUQUERQUE, E.X., SEYAMA, I. and NARAHASHI, T.: J. Pharmacol. Exp. Ther. 184 (1973) 308.

6. ALBUQUERQUE, E.X., WARNICK, J.E. and SANSONE, F.M.: J. Pharmacol. Exp. Ther. 176 (1971) 511.

7. ANDERSON, N.C., RAMON, F. and SNYDER, A: J. Gen. Physiol. 58 (1971) 322.

8. ARMSTRONG, C.M.: J. Gen. Physiol. 50 (1966) 491.

9. ARMSTRONG, C.M.: J. Gen. Physiol. 54 (1969) 553.

10. ARMSTRONG, C.M.: J. Gen. Physiol. 58 (1971) 413.

11. ARMSTRONG, C.M.: Quart. Rev. Biophys 7 (1975) 179.

12. ARMSTRONG, C.M. and BEZANILLA, F.: Nature 242 (1973) 459.

13. ARMSTRONG, C.M. and BINSTOCK, L.: J. Gen. Physiol. 48 (1965) 859.

14. ARMSTRONG, C.M. and HILLE, B.: J. Gen. Physiol. 59 (1972) 388.

15. BAKER, P.F., BLAUSTEIN, M.P., KEYNES, R.D., MANIL, J., SHAW, T.I. and STEINHARDT, R.A.: J. Physiol. (London) 200 (1969) 459.

16. BARNOLA, F.V., VILLEGAS, R. and CAMEJO, G.: Biochim. Biophys. Acta 298 (1973) 84.

17. BEELER, G.W., Jr. and REUTER, H.: J. Physiol. 207 (1970) 165.

18. BEELER, G.W., Jr. and REUTER, H.: J. Physiol. 207 (1970) 191.

19. BENFORADO, J.M.: Part D, Physiological Pharmacology, IV (Root, W.S., Hofmann, F.G. Eds.). Academic Press (1968) 331.

20. BENZER, T.I. and RAFTERY, M.A.: Proc. Nat. Acad. Sci. 69 (1972) 3634.

21. BENZER, T.I. and RAFTERY, M.A.: Biochem. Biophys. Res. Comm. 51 (1973) 939.

22. BEZANILLA, F. and ARMSTRONG, C.M.: Science 183 (1974) 753.

23. BLAUSTEIN, M.P.: J. Gen. Physiol. 51 (1968) 309.

24. BLOEDEL, J., GAGE, P.W., LLINÁS, R. and QUASTEL, D.M.J.: Nature 212 (1966) 49.

25. BROWN, H.M., HAGIWARA, S., KOIKE, H. and MEECH, R.M.: J. Physiol. 208 (1970) 385.

26. BROWN, H.M., HAGIWARA, S., KOIKE, H. and MEECH, R.W.: Fed. Proc. 30 (1971) 69.

27. CAHALAN, M.D.: J. Physiol. London 244 (1975) 511.

28. CAMOUGIS, G., TAKMAN, B.H. and TASSE, J.R.P.: Science 156 (1967) 1625.

29. CATTERALL, W.A.: Proc. Nat Acad. Sci. 72 (1975) 1782.

30. CATTERALL, W.A.: J. Biol. Chem. 250 (1975) 4053.

31. COLOMO, F. and ERULKAR, S.D.: J. Physiol. 199 (1968) 205.

32. COLQUHOUN, D., HENDERSON, R. and RITCHIE, J.M.: J. Physiol. 227 (1972) 95.

33. CUERVO, L.A. and ADELMAN, W.J., Jr.: J. Gen. Physiol. 55 (1970) 309.

34. DAMBACH, G.E. and ERULKAR, S.D.: J. Physiol. 228 (1973) 799.

35. deGROOF, R.C. and NARAHASHI, T.: Europ. J. Pharmacol. (1976) in press.

36. DEGUCHI, T. and SAKAI, Y.: J. Physiol. Soc. Japan 29 (1967) 172.

37. DETTBARN, W.D.: Vol. 1, Neuropoisons, Their Pathophysiological Actions (Simpson, L.L. Ed.). Plenum Press. New York (1971) 169.

38. DUDEL, J., PEPER, K., RÜDEL, R. and TRAUTWEIN, W.: Pflug. Arch. ges. Physiol. 292 (1966) 255.

39. DUDEL, J., PEPER, K., RÜDEL, R. and TRAUTWAIN, W.: Pflügers Archiv. 295 (1967) 213.

40. ELMQVIST, D. and FELDMAN, D.S.: Acta Physiol. Scand. 64 (1965) 475.

41. EVANS, M.H.: Brit. J. Pharmacol. 43 (1971) 681.

42. EVANS, M.H.: Internat. Rev. Neurobiol. 15 (1972) 83.

43. FREEMAN, A.R.: Fed. Proc. 28 (1969) 333.

44. FREEMAN, A.R.: Comp. Biochem. Physiol. 40A (1971) 71.

45. FUHRMAN, F.A.: Sci. American 217 (1967) 60.

46. FUKUDA, H., KUDO, Y. and ONO, H.: Europ. J. Pharmacol. 26 (1974) 136.

47. FUKUDA, H., KUDO, Y., ONO, H. YASUE, M., SAKAKIBARA, J. and KATO, T.: Chem. Pharmacent. Bull. 22 (1974) 884.

48. FURUKAWA, T., SASAOKA, T. and HOSOYA, Y.: Japan.J. Physiol. 9 (1959) 143.

49. GAGE, P.W.: Fed. Proc. 26 (1967) 1627.

50. GAGE, P.W.: Vol. 1, Neuropoisons, Their Pathophysiological Actions (Simpson, L.L., Ed.). Plenum. New York (1971) 187.

51. GEDULDIG, D. and GRUENER, R.: J. Physiol. 211 (1970) 217.

52. GEDULDIG, D. and JUNGE, D.: J. Physiol. 199 (1968) 347.

53. GERSHON, M.D.: Brit. J. Pharm. and Chemother. 29 (1967) 259.

54. GOLDENBERG, M.M.: J. Pharm. Pharmacol. 23 (1971) 621.

55. GOLDMAN, D.E.: J. Gen. Physiol. 27 (1943) 37.

56. HAFEMANN, D.R.: Biochim. Biophys. Acta 266 (1972) 548.

57. HAGIWARA. S. and NAKAJIMA, S.: J. Gen. Physiol. 49 (1966) 793.

58. HAGIWARA, S. and SAITO, N.: J. Physiol. 148 (1959) 161.

59. HAGIWARA, S. and TAKAHASHI, K.: J. Physiol. 190 (1967) 499.

60. HARDIKAR, S.W.: J. Pharmacol. Exp. Ther. 20 (1923) 17.

61. HARRIS, J.B. and THESLEFF, S.: Acta Physiol. Scand. 83 (1971) 382.

62. HASHIMOTO, Y., HOLMAN, M.E. and MCLEAN, A.J.: Nature 215 (1967) 430.

63. HENDERSON, R. and STRICHARTZ, G.: J. Physiol. 238 (1974) 329.

64. HENDERSON, R. and WANG, J.H.: Biochemistry 11 (1972) 4565.

65. HENDERSON, R., RITCHIE, J.M. and STRICHARTZ, G.R.: J. Physiol. 235 (1973) 783.

66. HENDERSON, R., RITCHIE, J.M. and STRICHARTZ, G.R.: Proc. Nat. Acad. Sci. 71 (1974) 3936.

67. HILLE, B.: J. Gen. Physiol. 50 (1967) 1287.

68. HILLE, B.: J. Gen. Physiol. 51 (1968) 199.

69. HILLE, B.: Progr. Biophys. Mol. Biol. 21 (1970) 1.

70. HILLE, B.: J. Gen. Physiol. 58 (1971) 599.

71. HILLE, B.: J. Gen. Physiol. 59 (1972) 637.

72. HILLE, B.: Biophys. J. 15 (1975) 615.

73. HILLE, B.: J. Gen. Physiol. 66 (1975) 535.

74. HILLE, B.: Fed. Proc. 34 (1975) 1318.

75. HIRONAKA, T. and NARAHASHI, T.: Fed. Proc. 34 (1975) 360.

76. HIRONAKA, T. and NARAHASHI, T.: Biophys. J. 16 (1976) 187a.

77. HODGKIN, A.L. and HUXLEY, A.F.: J. Physiol. 117 (1952) 500.

78. HODGKIN, A.L. and KATZ, B.: J. Physiol. 108 (1949) 37.

79. HOGAN, P.M. and ALBUQUERQUE, E.X.: J. Pharmacol. Exp. Ther. 176 (1971) 529.

80. HONERJAGER, P.: Abstr. 15th. Ann. Meet. Biophys. Soc. (1971) 54a.

81. HUBBARD, J.I., JONES, S.F and LANDAU, E.M.: J. Physiol. 197 (1968) 639.

82. JAIMOVICH, E., VENOSA, R.A., SHRAGER, P. and HOROWICZ, P.: J. Gen. Physiol. 67 (1976) 399.

83. KANO, M. and SHIMADA, Y.: J. Cell. Physiol. 81 (1973) 85.

84. KAO, C.Y.: Pharmacol. Rev. 18 (1966) 997.

85. KAO, C.Y.: Fed. Proc. 31 (1972) 1117.

86. KAO, C.Y. and FUHRMAN, F.A.: Toxicon 5 (1967) 25.

87. KAO, C.Y. and NISHIYAMA, A.: J. Physiol. 180 (1965) 50.

88. KAO, C.Y., NAGASAWA, J., SPIEGELSTEIN, M.Y. and CHA, Y.N.: J. Pharmacol. Exp. Ther. 178 (1971) 110.

89. KATZ, B. and MILEDI, R.: J. Physiol. 185 (1966) 5P.

90. KAT7, B. and MILEDI, R.: Proc. Roy. Soc. B167 (1967) 8.

91. KATZ, B. and MILEDI, R.: J. Physiol. 189 (1967) 535.

92. KATZ, B. and MILEDI, R.: J. Physiol. 192 (1967) 407.

93. KATZ, B. and MILEDI, R.: J. Physiol. 195 (1968) 481.

94. KATZ, B. and MILEDI, R.: J. Physiol. 199 (1968) 729.

95. KATZ, B. and MILEDI, R.: J. Physiol. 203 (1969) 459.

96. KATZ, B. and MILEDI, R.: Pubbl. Staz. Zool. Napoli 37 (1969) 303.

97. KATZ, B. and MILEDI, R.: J. Physiol. 203 (1969) 689.

98. KATZ, B. and MILEDI, R.: J. Physiol. 207 (1970) 789.

99. KATZ, B. and MILEDI, R.: J. Physiol. 216 (1971) 503.

100. KAYAALP, S.O., ALBUQUERQUE, E.X. and WARNICK, J.E.: Eur. J. Pharmacol. 12 (1970) 10.

101. KEYNES, R.D. and ROJAS, E.: J. Physiol. 233 (1973) 28P.

102. KEYNES, R.D. and ROJAS, E.: J. Physiol. 239 (1974) 393.

103. KEYNES, R.D., BEZANILLA, F., ROJAS, E. and TAYLOR, R.E.: Phil. Trans. Roy. Soc. London B 270 (1975) 365.

104. KEYNES, R.D., RITCHIE, J.M. and ROJAS, E.: J. Physiol. 213 (1971) 235.

105. KHODOROV, B.I., PEGANOV, E.M., REVENKO, S.V. and SHISHKOVA, L.D. Brain Res. 84 (1975) 541.

106. KIDOKORO, Y.: Nature New Biol. 241 (1973) 158.

107. KOPPENHÖFER, E.: Pflüg. Arch. ges. Physiol. 293 (1967) 34.

108. KOPPENHÖFER, E. and VOGEL, W.: Pflügers Arch. 313 (1969) 361.

109. KRAUSS, K.R., CARPENTER, D.O. and KOPIN, I.J.: J. Pharmacol. Exp. Ther. 173 (1970) 416.

110. KRAYER, O.: Pharmacology in Medicine (Drill, V. A., Ed.). McGraw-Hill. New York (1958) 515.

111. KU, D., AKERA, T., FRANK, M. and BRODY, T.M.: Pharmacologist 17 (1975) 218.

112. KUMAMOTO, M. and HORN, L.: Microvascular Res. 2 (1970) 188.

113. KUPCHAN, S.M. and FLACKE, W.E.: Antihypertensive Agents (Schlittler, E., Ed.). Academic. New York (1967) 429.

114. KURIYAMA, H., OSA, T. and TOIDA, N.: British J. Pharm. and Chemother. 27 (1966) 366.

115. KUSANO, K.: J. Neurobiol. 1 (1970) 435.

116. KUSANO, K.: J. Neurobiol. 1 (1970) 459.

117. KUSANO, K., LIVENGOOD, D.R. and WERMAN, R.: J. Gen. Physiol. 50 (1967) 2579.

118. LANGLEY, J.N.: J. Physiol. 20 (1896) 223.

119. LANGLEY, J.N.: J. Physiol. 27 (1901) 224.

120. LANGLEY, J.N. and DICKINSON, W.L.: Proc. Roy. Soc. (London) 46 (1889) 423.

121. LIPSIUS, M.R., SIEGMAN, M.J. and KAO, C.Y.: J. Pharm. Exp. Ther. 164 (1968) 60.

122. LLINÁS, R. and NICHOLSON, C.: Proc. Nat. Acad. Sci. 72 (1975) 187.

123. LLINÁS, R., JOYNER, R.W. and NICHOLSON, C.: J. Gen. Physiol. 64 (1974) 519.

124. LLINÁS, R., WALTON, K. and BOHR, V.: Biophys. J. 16 (1976) 83.

125. LOEWENSTEIN, W.R., TERZUOLO, C.A. and WASHIZU, Y.: Science 142 (1963) 1180.

126. MASCHER, D. and PEPER, K.: Pflüg. Arch. Europ. J. Physiol. 307 (1969) 190.

127. MEVES, H.: Pflüg. Arch. ges. Physiol. 290 (1966) 211.

128. MEVES, H.: J. Physiol. 243 (1974) 847.

129. MEVES, H.: Phil. Trans. Roy. Soc. London B 270 (1975) 493.

130. MEVES, H. and PICHON, Y.: J. Physiol. 251 (1975) 60P.

131. MILEDI, R.: J. Physiol. 192 (1967) 379.

132. MOORE, J.W. and NARAHASHI, T.: Fed. Proc. 26 (1967) 1655.

133. MOORE, J.W., BLAUSTEIN, M.P., ANDERSON, N.C. and NARAHASHI, T.: J. Gen. Physiol. 50 (1967) 1401.

134. MOORE, J.W., HAAS, H.G. and TARR, M.: Proc. Internat. Union Physiol. Sci. 7 (1968) 303.

135. MOORE, J.W., NARAHASHI, T. and SHAW, T.I.: J. Physiol. 188 (1967) 99.

136. MORAN, N.C., DRESEL, P.E., PERKINS, M.E. and RICHARDSON, A.P.: J. Pharmacol. Exp. Ther. 110 (1954) 415.

137. MOSHER, H.S., FUHRMAN, F.A., BUCHWALD, H.D. and FISCHER, H.G.: Science 144 (1964) 1100.

138. MOZHAYEVA, G.N. and NAUMOV, A.P.: Biochim. Biophys. Acta 290 (1972) 248.

139. MURAYAMA, K., ABBOTT, N.J., NARAHASHI, T. and SHAPIRO, B.I.: Comp. Gen. Pharmacol. 3 (1972) 391.

140. NAGASAWA, J., SPIEGELSTEIN, M.Y. and KAO, C.Y.: J. Pharmacol. Exp. Ther. 178 (1971) 103.

141. NAKAJIMA, S.: J. Gen. Physiol. 49 (1966) 629.

142. NARAHASHI, T.: Fed. Proc. 31 (1972) 1124.

143. NARAHASHI, T.: Fundamentals of Cell Pharmacology (Dikstein, S., Ed.). C. C. Thomas. Springfield (1973) 395.

144. NARAHASHI, T.: Physiol. Res. 54 (1974) 813.

145. NARAHASHI, T.: Proc. First Internat. Conf. Toxic Dinoflagellate Blooms (LiCicero, V.R., Ed.). Mass. Sci. and Technol. Fund. Massachusetts (1975) 395.

146. NARAHASHI, T.: The Nervous System (Tower, D.B., Ed.). Raven Press. New York (1975) 101.

147. NARAHASHI, T. and HAAS, H.G.: Science 157 (1967) 1438.

148. NARAHASHI, T. and HAAS, H.F.: J. Gen. Physiol. 51 (1968) 177.

149. NARAHASHI, T. and SEYAMA, I.: J. Physiol. 242 (1974) 471.

150. NARAHASHI, T., ALBUQUERQUE, E.X. and DEGUCHI, T.: J. Gen. Physiol. 58 (1971) 54.

151. NARAHASHI, T., ANDERSON, N.C. and MOORE, J.W.: Science 153 (1966) 765.

152. NARAHASHI, T., ANDERSON, N.C. and MOORE, J.W.: J. Gen. Physiol. 50 (1967) 1413.

153. NARAHASHI, T., BRODWICK, M.S. and SCHANTZ, E.J.: Environ. Letters 9 (1975) 239.

154. NARAHASHI, T. DEGUCHI, T. and ALBUQUERQUE, E.X.: Nature New Biol. 229 (1971) 221.

155. NARAHASHI, T., DEGUCHI, T., URAKAWA, N. and OHKUBO, Y.: Am. J. Physiol. 198 (1960) 934.

156. NARAHASHI, T., FRAZIER, D.T. and MOORE, J.W.: J. Neurobiol. 3 (1972) 267.

157. NARAHASHI, T., HAAS, H.G. and THERRIEN, E.F.: Science 157 (1967) 1441.

158. NARAHASHI, T., MOORE, J.W. and FRAZIER, D.T.: J. Pharmacol. Exp. Ther. 169 (1969) 224.

159. NARAHASHI, T., MOORE, J.W. and POSTON, R.N.: J. Neurobiol. 1 (1969) 3.

160. NARAHASHI, T., MOORE, J.W. and SCOTT, W.R.: J. Gen. Physiol. 47 (1964) 965.

161. NARAHASHI, T., MOORE, J.W. and SHAPIRO, B.I.: Science 163 (1969) 680.

162. NARAHASHI, T., SHAPIRO, B.I., DEGUCHI, T., SCUKA, M. and WANG, C.M.: Am. J. Physiol. 222 (1972) 850.

163. NEHER, E. and LUX, H.D.: Pflüg. Arch. 336 (1972) 87.

164. NISHI, K. and SATO, M.: J. Physiol. 184 (1966) 376.

165. NONNER, W., ROJAS, E. and STÄMPFLI, R.: Pflüg. Arch. Eur. J. Physiol. 354 (1975) 1.

166. OCHI, R.: Pflüg. Arch. Eur. J. Physiol. 316 (1970) 81.

167. OGURA, Y.: Vol. 1, Neuropoisons, Their Pathophysiological Actions (Simpson, L.L., Ed.). Plenum Press. New York (1971) 139.

168. OGURA, Y. and MORI, Y.: Europ. J. Pharmacol. 3 (1968) 58.

169. OHTA, M., NARAHASHI, T. and KEELER, R.F.: J. Pharmacol. Exp. Ther. 184 (1973) 143.

170. OZEKI, M. and SATO, M.: J. Physiol. 180 (1965) 186.

171. PELHATE, M. and PICHON, Y.: J. Physiol. 242 (1974) 90P.

172. PICHON, Y.: C. R. Soc. Biol. 163 (1969) 952.

173. REDFERN, P. and THESLEFF, S.: Acta Physiol. Scand. 82 (1971) 70.

174. REDFERN, P., LUNDH, H. and THESLEFF, S.: Eur. J. Pharmacol. 11 (1970) 263.

175. ROUGIER, O., VASSORT, G., GARNIER, D., GARGOUIL, Y.M. and CORABOEUF, E.: Pflüg. Arch. 308 (1969) 91.

176. ROUGIER, O., VASSORT, G. and STÄMPFLI, R.: Pflüg. Arch. ges. Physiol. 301 (1968) 91.

177. SASTRE, A. and PODLESKI, T.R.: Proc. Nat. Acad. Sci. 73 (1976) 1355.

178. SCHANTZ, E.J., GHAZAROSSIAN, V.E., SCHNOES, H.K., STRONG, F.M., SPRINGER, J.P., PEZZANITE, J.O. and CLARDY, J.: J. Am. Chem. Soc. 97 (1975) 1238.

179. SCHANTZ, E.J., LYNCH, J.M., VAYVADA, G., MATSUMOTO, K. and RAPOPORT, H.: Biochemistry 5 (1966) 1191.

180. SCHMIDT, H. and SCHMITT, O.: Pflüg. Arch. 349 (1974) 133.

181. SCHMIDT, H. and STÄMPFLI, R.: Pflüg. Arch. ges. Physiol. 287 (1966) 311.

182. SCHNEIDER, M.F. and CHANDLER, W.K.: Nature 242 (1973) 244.

183. SEYAMA, I.: Japan J. Physiol. 20 (1970) 381.

184. SEYAMA, I. and NARAHASHI, T.: J. Pharmacol. Exp. Ther. 184 (1973) 299.

185. SHANES, A.M.: Pharmacol. Rev. 10 (1958) 59.

186. SHANES, A.M.: Pharmacol. Rev. 10 (1958) 165.

187. SHIMIZU, Y., ALAM, M., OSHIMA, Y. and FALLON, W.E.: Biochem. Biophys. Res. Comm. 66 (1975) 731.

188. STANFIELD, P.R.: J. Physiol. London 209 (1970) 209.

189. STARKUS, J.G.: Effects of α-dihydrograyanotoxin II on excitable and inexcitable membranes, and divalent cation antagonism (Thesis). Duke University. North Carolina (1976).

190. TAHARA, Y.: Biochem. Zeitsch. 10 (1910) 255.

191. TAKATA, M., MOORE, J.W., KAO, C.Y. and FUHRMAN, F.A.: J. Gen. Physiol. 49 (1966) 977.

192. TAKEMOTO, T., NISHIMOTO, Y., GEGURI, H. and KATAYAMA, K.: J. Pharm. Soc. Japan 75 (1955) 1441.

193. TARR, M.: J. Gen. Physiol. 58 (1971) 523.

194. TASAKI, I. and HAGIWARA, S.: J. Gen. Physiol. 40 (1957) 859.

195. TAYLOR, R.E.: Am. J. Physiol. 196 (1959) 1071.

196. TOKUYAMA, T., DALY, J. and WITKOP, B.: J. Am. Chem. Soc. 91 (1969) 3931.

197. TSUDA, K.: Naturwissenschaften 53 (1966) 171.

198. TWAROG, B.M., HIDAKA, T. and YAMAGUCHI, H.: Toxicon 10 (1972) 273.

199. ULBRICHT, W.: J. Cell. Comp. Physiol. 66 (1965) 91.

200. ULBRICHT, W.: Ergeb Physiol. Biol. Chem. Exp. Pharmacol. 61 (1969) 18.

201. ULBRICHT, W. and FLACKE, W.: J. Gen. Physiol. 48 (1965) 1035.

202. VIERHAUS, J. and ULBRICHT, W.: Pflüg. Arch. Eur. J. Physiol. 326 (1971) 88.

203. VITEK, M. and TRAUTWEIN, W.: Pflüg. Arch. 323 (1971) 204.

204. WAGNER, H.H. and ULBRICHT, W.: Pflüg. Arch. Eur. J. Physiol. 359 (1975) 297.

205. WANG, C.M., NARAHASHI, T. and SCUKA, M.: J. Pharmacol. Exp. Ther. 182 (1972) 442.

206. WELLHÖNER, H.H.: Naunyn-Schmied. Arch. Pharmakol. 267 (1970) 185.

207. WOOD, H.B., Jr., STROMBERG, V.L., KERESZTESY, J.C. and HORNING, E.C.: J. Amer. Chem. Soc. 76 (1954) 5689.

208. YEH, J.Z., OXFORD, G.S., WU, C.H. and NARAHASHI, T.: Biophys. J. 16 (1976) 77.

209. YEH, J.Z., OXFORD, G.S., WU, C.H. and NARAHASHI, T.: Biophys. J. 16 (1976) 188a.

210. YEH, J.Z., OXFORD, G.S., WU, C.H. and NARAHASHI, T.: J. Gen. Physiol. (1976) in press.

DISCUSSION

ELDEFRAWI: You did mention that the tetrodotoxin had its site of action on the external part of the membrane? Would you care to comment about the sites for action of some of the toxins that affect this same sodium translocation?

NARAHASHI: Data indicate that grayanotoxins are presumably acting from inside. There isn't any concrete evidence for this notion, but they act more quickly and more effectively when applied inside. Some of the grayanotoxin derivatives are very effective for the inside but not so effective from the outside.

NARAHASHI: Cesium permeability is also increased by grayanotoxin; but the ratio of cesium to sodium permeability remains unchanged. To my knowledge, no toxins have been known to affect the ratio.

ELDEFRAWI: Has anybody attempted to immobilize any of these toxins on inert carriers and then test the biological activity to see if it is still effective?

NARAHASHI: Several people are working on trying to isolate the sodium channel using affinity chromotography, but with slight changes in structure the whole activity is easily lost.

BRODSKY: Are you saying that it binds and doesen't penetrate?

NARAHASHI: That is correct.

BRODSKY: So, that is the one from the rhododendron.

NARAHASHI: No, that is from puffer fish. But that site of action is different from the site of action of grayanotoxin.

ELDEFRAWI: How about scorpion toxin?

NARAHASHI: That has a different mode of action. It acts on the sodium inactivation mechanism.

GATZY: Has anyone checked the sensitivity of the Columbia arrow poison frog to grayanotoxin or the veratrum alkaloids?

NARAHASHI: Yes, that frog was insensitive to batrachotoxin (Albuquerque, et al., J. Pharmacal. Exp. Ther. 184: 315, 1973). Puffer fish and California newt contain tetrodotoxin, and is insensitive to tetrodotoxin.

LAKOWICZ: Is it true that these agents may be acting at the same receptor site? Certainly their actions appear to be the same.

NARAHASHI: To my knowledge no such study has been done. However, one of my graduate students, John Starkus, studied the effect of grayanotoxin on the sartorius muscle of the California newt. This muscle is resistant to tetrodotoxin but is sensitive to grayanotoxin. However, some other muscles such as crayfish muscle which produce action potential by calcium are not sensitive to grayanotoxin.

BRODSKY: Are these tissues ever sensitive to amiloride?

NARAHASHI: We don't yet know.

LAKOWICZ: Have you used methylmercury and chlorinated hydrocarbons in these tests?

NARAHASHI: Yes, methylmercury, depending on the concentration, can affect the sodium mechanism.

PRESSMAN: Has anyone checked the effect of TTX on puffer fish that don't produce TTX?

NARAHASHI: That is an interesting question. I understand that at least some species in the Atlantic Coast are not very toxic. Dr. C. Y. Kao of Downstate Medical Center in New York, has shown that it is not effective on puffer fish: (Kao & Fuhrman, Toxicon 5: 25, 1967).

SHAMOO: With respect to ameloride, it has been done at NIH. When I was at NIH (1972-1973) I gave them a sample to test. They tested ameloride directly, and it was found to have no effect on axon action potential.

NARAHASHI: Is that from the outside or the inside?

SHAMOO: I don't know. It was on myxicola giant axon and the experiment was done by Leonard Binstock and found that it had no effect. (Private Communication).

Effects of Membranes and Receptors

INTERACTIONS OF ACETYLCHOLINE RECEPTORS WITH ORGANIC MERCURY COMPOUNDS*

M. E. Eldefrawi , N. A. Mansour**, and A. T. Eldefrawi

Dept. of Pharmacology & Experimental Therapeutics

Univ. of Maryland School of Medicine, Baltimore, MD 21201

ABSTRACT

Micromolar concentrations of methylmercury and several organic mercury fungicides were found to block binding of [^{3}H]acetylcholine (ACh) to the ACh-receptor of the electric organ of the electric ray, Torpedo ocellata. The same compounds had little or no effect on the catalytic activity of ACh-esterase of the same tissue. [^{14}C]Methylmercury bound to the purified ACh-receptor with high affinity (K_d = 7μM) and there were 6.5 ± 0.5 binding sites for each ACh-binding site. Binding of methylmercury was highly cooperative with a Hill coefficient of 2.6. This binding was irreversible by redialysis in methylmercury - free medium, however, the bound [^{14}C]methylmercury was easily displaced from the receptor protein with μM concentrations of BAL or penicillamine. Methylmercury also blocked binding of [^{3}H] nicotine and [^{3}H]pilocarpine to the nicotinic and muscarinic ACh-receptors of the rat brain, respectively. The data suggest that the ACh-receptor may be a target for methylmercury and other organic mercury compounds.

---

*This research was supported in part by National Science Foundation grant BMS 75-06760 and National Institutes of Health grant AI 12543 and NS 11280.

**Permanent address: Department of Plant Protection, Faculty of Agriculture, University of Alexandria, Alexandria, Egypt.

Introduction

The acetylcholine (ACh) receptor is an intrinsic regulatory protein of cholinergic postsynaptic membranes. In excitatory synapses, binding of ACh to its receptor is suggested to cause conformational changes in the receptor which lead to increases in the flux of Na^+ and K^+ across the membrane resulting in localized depolarization.

The ACh-receptor is now purified to a high degree in several laboratories (1, 2, 10, 24, 27, 28, 31, 32, 37). This was achieved through four major advances. First was the discovery that the electric organ of the electric ray, *Torpedo* sp. (and to a lesser degree that of the electric eel, *Electrophorus electricus*), was the richest known tissue in nicotinic cholinergic synapses (for reviews see (3,16,31,34,36,39,40). Along with this was the success in the *in vitro* detection of the ACh-receptor, made possible by its binding of a variety of labeled cholinergic drugs, its transmitter ACh and α-neurotoxins from snake venoms (4, 11, 14, 15, 33). This was followed by solubilization of the receptor, which was accomplished only by detergents (18, 33). Finally, the development of the method of affinity chromatography (5) rapidly advanced the purification of the receptor.

Several of the molecular properties of this ACh-receptor are now known. Its molecular weight is estimated to be around 330,000 (8). The molecule is suggested to be made up of two major subunits (plus possibly 2 minor ones), only one of which (MW ≃ 40,000) carries the sites that bind activators, inhibitors and Ca^{2+} (42, 43). Using fluorescent probes, the data suggest that binding of ACh to the receptor molecule causes it to release Ca^{2+} (for which it has a very high affinity (19) and which may have physiologic significance (43). The purified receptor can be reconstituted in synthetic lipid membranes and exhibit selective monovalent cation flux upon its binding of activators (16, 48). However, this flux is still different in some characteristics from what occurs *in vivo*, and it is possible that part of the ion conductance modulator may have been removed during purification or that the perfect conditions for reconstitution are still to be discovered.

During the course of our studies on this receptor molecule, we discovered that methylmercury, but not mercuric chloride, bound with high affinity to the ACh-receptor of *Torpedo* electroplax (49). This finding raised the following questions: Is this high affinity restricted only to methylmercury or does it also extend to other organic mercurials? Do ACh-receptors of the central nervous system also bind methylmercury with a high affinity? How specific is this effect to ACh-receptors? What is the nature of the binding formed between methylmercury and the receptor?

The data presented to this conference attempt to answer these questions, and as usual they raise additional ones.

INHIBITION OF ACH BINDING TO ITS RECEPTOR BY ORGANIC MERCURY COMPOUNDS

We used an ACh-receptor enriched membrane fraction isolated from the electric organ of the electric ray, Torpedo ocellata to study the binding of [^{3}H]ACh to its receptor and the effect of organic mercurials. Binding was measured by equilibrium dialysis at 4°C for 16 h, as previously described (13, 14). Each 1 ml of the membrane fraction (containing 1.5 mg protein and 1.4 nmoles of receptor sites) was incubated with 100 μM of the organic mercury compound for 1 h at 23°C, followed by incubation for 1 h with 100 μM DFP (diisopropyl fluorophosphate), so as to inhibit the ACh-esterase present. This concentration of DFP has no effect on the binding of [^{3}H]Ach to its receptors (12). Then the receptor preparation was transferred to a dialysis bag tied at both ends and placed in a bath of 100 ml Krebs original phosphate (pH 7.4) containing 1 μM [^{3}H] ACh (sp. act. 49.5 Ci/mole from New England Nuclear), 10 μM DFP and 100 μM of the organic mercurial. After equilibration, triplicate samples were taken from bath as well as bag contents, and their radioactivity counted in a Packard liquid scintillation spectrometer Model 3380. Excess radioactivity in bag samples represented the amount of [^{3}H] bound. As Fig. 1 demonstrates, methylmercury inhibits ACh binding and this inhibition is concentration dependent. It is significant that micromolar concentrations of methylmercury are inhibitory of ACh binding to its receptor, since concentrations of methylmercury detected in the blood of mammals exposed to this drug range from 4-80 μg/g (21, 22, 51) which

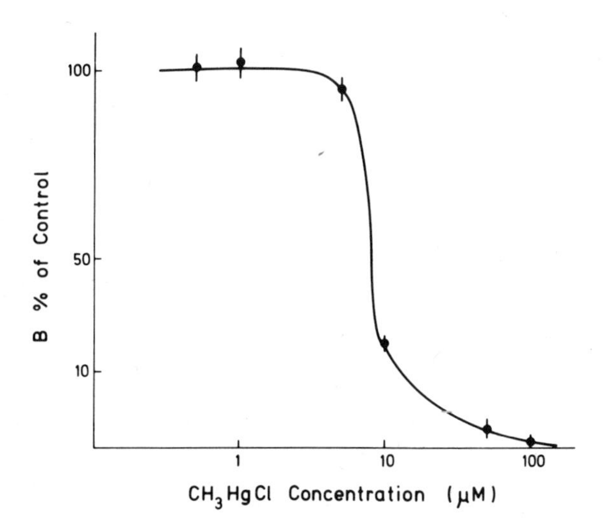

Fig. 1 The effect of increasing concentrations of methylmercury on the binding of [^{3}H]Ach (1 μM) to the membrane bound ACh-receptor from T. ocellata B, amount bound.

is equivalent to 16-319 μM. About 90% of this concentration is found bound to red blood cells and lipoproteins.

Organic mercurials are used commercially for seed protection against fungal diseases (50), we obtained a few from the Environmental Protection Agency (EPA) purity 97-100%, and tested their effect on [^{3}H]ACh binding (at 1 μM) to its membrane-bound receptor. The organic mercurials were used at 100 μM except for the borate and hyroxide salts of phenylmercury which were used at 10 μM because of their lower solubility. As shown in Table 1, it is quite evident that several of the organic mercurials are as strong inhibitors of ACh-binding to the receptor as, or more so than, _d_-tubocurarine which is a specific inhibitor of neuromuscular nicotinic ACh-receptors (29). It should be emphasized here that in these binding studies both activators and inhibitors of ACh-receptors would act similarly, by blocking binding of ACh to its receptor. Therefore, we cannot conclude as to whether these mercurials would be inhibitors or activators of ACh-receptors _in vivo_. For this, electrophysiological measurements of postsynaptic potentials are required. It is also evident that the organic mercurials differ in their effect on the ACh-receptor, with diphenylmercury not inhibiting [^{3}H]ACh-binding and methylmercury and MEMMI inhibiting the binding totally. This suggests that binding of methylmercury to the receptor is quite specific and inhibition is due neither to the mercury atom alone nor to the molecule being organic. A certain steric structure is required for maximum inhibition of the binding of ACh to the receptor.

INHIBITION OF ACH-ESTERASE BY ORGANIC MERCURIALS

In cholinergic synapses, choliesterases play an important role in transmission by hydrolysing ACh, thus terminating its action. There are many similarities between acetylcholinesterases and the ACh-receptor. Many of the drugs known as activators or inhibitors of the ACh-receptor, such as decamethonium and _d_-tubocurarine, are known to react with ACh-esterase as well (41). In addition ACh-esterase inhibitors, such as DFP, also bind to the ACh-receptor, though at much higher concentrations (12). Therefore, we tested the effect of organic mercurials on the catalytic activity of ACh-esterase (by the Ellman _et al_. method (20)) present in the same membrane fraction of electric organ containing the ACh-receptors. The mercurial compound was incubated at 23°C for 1 h with the membranes, then the enzyme activity measured. We discovered that, contrary to their effect on the ACh-recptor, organic mercurials (at 100 μM) had little or no effect on ACh-esterase activity while the organophosphates, DFP and Tetram gave >99% inhibition (Table 1).

This finding is most interesting particularly in view of the fact that the ACh-receptor and ACh-esterase have similar amino acid composition and both have free SH groups (9, 10, 17). This leads us to conclude that the action of the organic mercury compounds at the cholinergic synapse is directed selectively against the ACh-receptor.

TABLE I. Effects of Organic Mercury Fungicides on Binding of [^{3}H]ACh (at 1μM) to ACh-receptors and on ACh-esterase Activity in Electric Organ Membranes of *Torpedo ocellata*.

DRUG (100 μM)		% Inhibition	
Chemical Name	EPA Code	ACh-receptor Mean ± SD	ACh-esterase Mean ± SD
Methylmercury chloride	4560	102.2 ± 8.3	4.8 ± 2.4
Ethylmercury Chloride	3400	61.8 ± 6.2	0.8 ± 2.8
MEMMI*	4440	99.7 ± 6.3	17.5 ± 0.7
Cyano (Methylmercury Granidine	1560	92.2 ± 3.7	11.9 ± 2.4
Phenylmercury acetate	5680	55.5 ± 2.2	11.9 ± 2.4
Phenylmercury chloride	5480	50.3 ± 3.7	6.3 ± 7.6
Diphenylmercury	2640	1.5 ± 2.4	16.3 ± 1.8
Phenylmercury borate**	5460	4.5 ± 4.2	12.7 ± 3.4
Phenylmercury hydroxide**	5485	1.3 ± 3.1	3.2 ± 2.4
d-Tubocurarine*		58.6 ± 1.5	4.8 ± 2.1
DFP*†		0.5 ± 2.0	99.7 ± 0.2
Tetram*		1.8 ± 1.5	99.5 ± 0.1

*MEMMI is *N*-methylmercuri-1,2,3,6-tetrahydro-3,6-3ndomethano-3,4,5,6,7,7-hexachlorophtalimide; DFP is disophroply fluorophosphate; Tetram is 0, 0-diethyl *S*(β-diethylamino)ethyl phosphorothiolate.

**The borate and hydroxide salts of phenylmercury were used at 10μM, while all other drugs were used at 100μM.

†DFP is added to the ACh-receptor preparation at 100μM, and to the dialysis bath at 10μM when binding of [^{3}H]ACh is determined, so as to inhibit the ACh-esterase present, the value of 99.7 ± 0.2% inhibition of the enzyme of DPF is after 1 h incubation, with higher inhibition obtained with longer incubation.

NATURE OF METHYLMERCURY BINDING TO THE ACETYLCHOLINE RECEPTOR

To investigate the nature of the binding between the organic mercurials and the ACh-receptor, we used [^{14}C]methylmercury (sp. act. 3.36 Ci/mole, from New England Nuclear). Since methylmercury may bind to many structural proteins such as Ca^{2+}, Mg^{2+} - ATPase (49) and presynaptic components (26), we used the pure ACh-receptor for this study. We purified it from the electric organ of T. ocellata by affinity adsorption, using as the affinity ligand the α-neurotoxin from the venom of the cobra, Naja naja siamensis. As previously described (10, 17), the solutions used contained 0.1 mM DFP and 1mM EDTA to reduce oxidation or proteolytic and esteratic breakdown of the receptor molecules. The ACh-receptor bound 12 moles of ACh or α-bungarotoxin/mg protein.

Binding of [^{14}C]methylmercury to the pure ACh-receptor of T. ocellata, was measured by equilibrium dialysis. The upward curvature in the double reciprocal plot of the data (Fig. 2) is indicative of positive cooperativity. The Hill coefficient is 2.6. At saturation, the ACh-receptor binds 80 ± 5 nmoles of methylmercury per mg protein. By comparison, each mg of receptor protein in this preparation binds 12 nmoles ACh, also with positive cooperativity, but a Hill coeffieient of 1.6. Accordingly, the ACh-receptor monomer (carrying one ACh binding site), which has an estimated molecular weight of 83,000, carries about 6.5 ± 0.5 methylmercury binding sites. A similar value was reported for the ACh-receptor purified from T. californica (49).

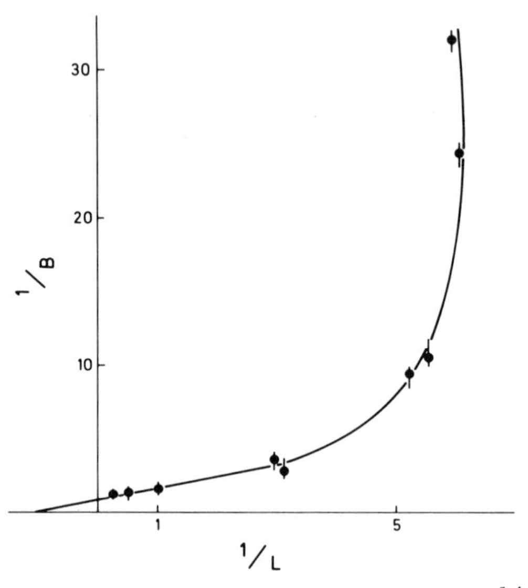

Fig. 2. Double reciprocal plot of the binding of [^{14}C]methylmercury to the ACh-receptors purified from T. ocellata. B, amount bound in 0.1 µmoles per mg protein; L, concentration of [^{14}C]methylmercury in µM.

Our analysis of the free SH groups of the pure ACh-receptor from Torpedo gave 2 SH groups per ACh binding site (17). This means that the binding of methylmercury is not restricted to free SH groups. In addition, we have shown that alkylation of all free SH groups on the pure ACh-receptor with p-chloromercuribenzoate reduced [^{3}H]ACh binding to the receptor by 25% (17), whereas methylmercury at the same concentration inhibited [^{3}H]ACh binding totally (Table I). Thus, groups other than SH, which bind methylmercury, may be closely associated with the active site that binds ACh. An added evidence is our finding that methylmercury inhibits [^{3}H]ACh binding competitively, as judged by the common intercept of the ordinate in Fig. 3.

To determine the reversibility of [^{14}C]methylmercury binding to the ACh-receptor purified from T. ocellata, the receptor was first dialyzed at 4°C for 16 h in Ringer original phosphate solution containing [^{14}C]methylmercury free solution redialyzed in [^{14}C]methylmercury free solution for 24 hr and the radioactivity in 0.1ml samples of receptor solution compared. As shown in Table II, the [^{14}C]methymercury bound was not removed from the receptor, even with a second redialysis in Ringer or in Ringer containing 10 μM ACh. We selected this concentration of ACh because it is the highest concentration of ACh that may occur at a cholinergic synapse (35, 44). What is interesting is that the two antidotes for heavy metal poisoning, BAL (British Anti Lewisite) and penicillamine, could remove [^{14}C]methyl-

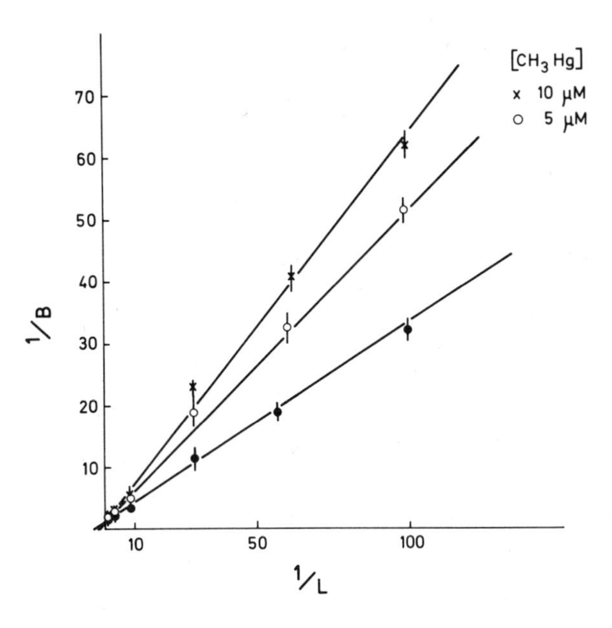

Fig. 3. Double reciprocal plot of the binding of [^{3}H]ACh to the ACh-receptors purified from T. ocellata. B, amount bound in nmoles per mg protein; L, concentration of [^{3}H]ACh in μM.

Table II. Displacement of [^{14}C]Methylmercury Bound to *T. ocellata* Purified ACh-receptors.

Treatment	cpm in 0.1 ml of sample	% Displacement of [^{14}C]methylmercury
Equilibrium dialysis in 2 μM [^{14}C]methylmercury	3763.6 ± 75	
Redialysis in Ringer free of methylmercury	3828.5 ± 45	0
Second redialysis in Ringer	3905.7 ± 35	0
Second redialysis in Ringer containing 10 μM ACh	3877.7 ± 92	0
Second redialysis in Ringer containing 10 μM BAL	1178.8 ± 34	69.8
Second redialysis in Ringer containing 10 μM penicillamine	1168.3 ± 13	70.1

mercury bound to the pure ACh-receptor. When series of concentrations of either of the two reagents were tested as to their effectiveness in displacing [^{14}C]methylmercury, they were equipotent (Fig. 4). However, the displacement gave a nonlinear plot, which suggests the involvement of multiple bond energies. We have no evidence at present whether or not the bonds formed between methylmercury and the ACh-receptor are covalent.

BINDING OF METHYLMERCURY TO BRAIN ACH RECEPTORS

We were interested in establishing whether the high affinity binding of methylmercury to ACh-receptors of electric organs (which pharmacologically are of the neuromuscular nicotinic type) also occurs ot central ACh-recptors. The first step was to establish the subcellular localization of [^{14}C]methylmercury binding in rat brains. This was achieved by incubating the brain homogenate with [^{14}C]methylmercury for 1 h at 23°C prior to separating the subcellular fractions according to the method of Gray and Whittaker (23). The crude mitochondrial membrane fraction was separated on discontinuous sucrose gradient into three membrane fractions (myelin, synaptosomal

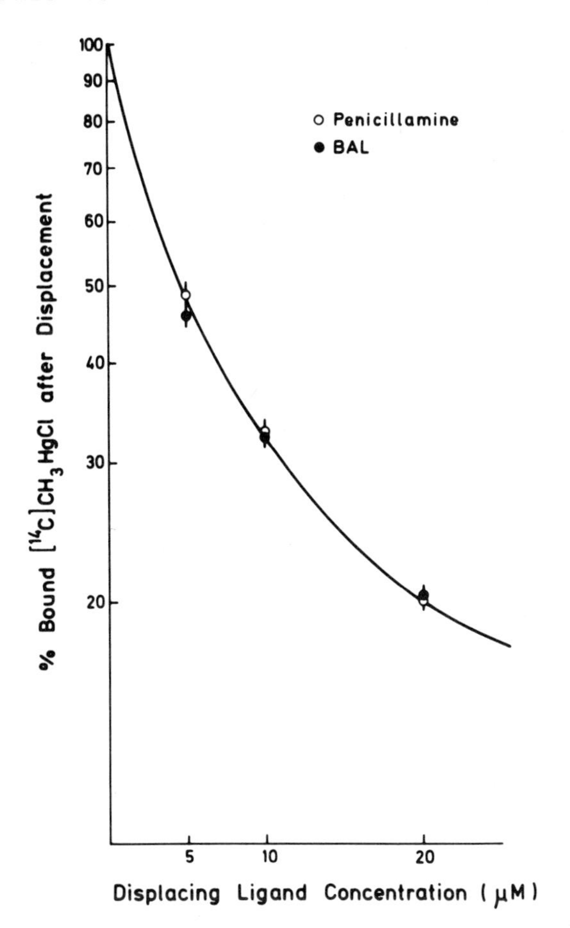

Fig. 4. Displacement by penicillamine and BAL of the [^{14}C]methyl-mercury bound at 2 μM ACh-receptors purified from *T. ocellata*.

and mitochondrial) and their radioactivity and protein contents assayed. As shown in Table III, synaptosomal fraction contained the highest concentration of [^{14}C]methylmercury per mg protein.

We had previously shown that this synaptosomal fraction, compared to other sulcellular fractions, contained the highest concentration of nicotinic ACh-receptors as well as muscarinic ACh-receptors (46). The nicotinic and muscarinic receptors of mouse brain were identified *in vitro* by their specific binding of [^{3}H] nicotine and [^{3}H] pilocarpine, respectively (46). Therefore, the synaptosomal fraction was incubated at 23°C with methylmercury (at 10μM) for 1 h, then its binding of 0.01 mM [^{3}H]nicotine (sp. act. 2750 Ci/mol; from Amersham Searle and 0.01 mM [^{3}H]pilocarpine (sp. act. 18,8000 Ci/Mol; from New England Nuclear) determined. Equilibrium dialysis

Table III. Subcellular Distribution of [^{14}C]methylmercury bound to Rat Brain

Fraction	cpm/100μl	cpm/mg protein
Myelin	1140	1156 ± 54
Synaptosomal	630	2040 ± 88
Mitochondrial	2400	1137 ± 134

was performed at 4°C for 16 h, with methylmercury (10μM) also present in the bath, and the binding data compared to synaptosomal membranes not exposed to any methylmercury. The results (Table IV) clearly demonstrate that methylmercury inhibits the binding of both [^{3}H]nicotine and [^{3}H]pilocarpine to brain cholinergic receptors.

Table IV. Inhibition by Methylmercury of the Binding of [^{3}H]Nicotine and [^{3}H]Pilocarpine to the Synaptosomal Fraction of Rat Brain. Binding of Cholinergic Ligands is Measured by Equilibrium Dialysis.

Treatment	Amount bound (pmoles/g brain)			
	[^{3}H]nicotine (.01 μM)	%Inhibition	[^{3}H]pilocarpine (.01 μM)	%Inhibition
Control	6.43 ± 0.38		3.99 ± 1.12	
After exposure to methylmercury (10 μM)	2.64 [0.35	59	0.91 [0.2	77

CONCLUDING REMARKS

Poisoning by organic mercurials is usually characterized by neurotoxic substances. The early symptoms are often described as non-specific, consisting of fatigue, headache, and impairment of vision, memory, and concentration. Advanced toxicosis is usually associated with ataxia and muscular incoordination. In animal tests, when high doses of organic mercurials are administered, the clinical symptoms of acute toxicity usually include tremors, abnormal gait, labored respiration and sometimes total paralysis (6, 7, 25, 38, 45). In addition to the delayed neurotoxic effects of organic mercurials, such as expressed in the minamata disease (30), the diversity of neural disturbances caused by organic mercurial poisoning suggest that multiple targets are involved. Possible targets are Ca^{2+} + Mg^{2+}-ATPase Ca^{2+} transport (49), plasmalogens (47), and the ACh releasing mechanism (26). The high affinity presently observed of organic mercurials for ACh-receptors, whether peripheral or central, make these receptors excellent targets for such drugs.

REFERENCES

1. Biesecker, G. Biochemistry 12 (1973) 4403.

2. Chang, H. W. Proc. Nat. Acad. Sci. USA 71 (1974) 2113.

3. Changeux, J. P. Ch. 7, Handbook of Psychopharmacology (Iverson, L. L., Iverson, S. D. and Synder, S. H. Eds) Plenum, NY (1975).

4. Changeux, J. P., Kasai, M. and Lee, C. Y. Proc. Nat. Acad. Sci. USA 67 (1970) 1241.

5. Charlton, K. M., Can. J. Comp. Med. 38 (1974) 75

6. Cuatrecasas, P. and Anfinsen, C. B. Annu. Rev. Biochem 40 (1971) 259.

7. Diamond, S. S. and Sleight, S. D. Toxicol. Appl. Pharmacol. 23 (1972) 197.

8. Edelstein, S. J. Beyer, W. B., Eldefrawi, A. T. and Eldefrawi, M. E. J. Biol. Chem. 250 (1975) 6101.

9. Eldefrawi, M. E. Ch. 7 The Peripheral Nervous System (Hubbard, J. I., Ed) Plenum, New York (1974) 181.

10. Eldefrawi, M. E. and Eldefrawi, A. T. Arch. Biochem. Biophys. 159 (1973) 362.

11. Eldefrawi, M. E., Britten, A. G. and Eldefrawi, A. T. Science 173 (1971) 338.

12. Eldefrawi, M. E., Britten, A. G. and O'Brien, R. D. Pestic. Biochem. Physiol 1 (1971) 101.

13. Eldefrawi, M. E., Eldefrawi, A. T. and O'Brien, R. D. Ag Food Sci. 18 (1970) 1113.

14. Eldefrawi, M. E., Eldefrawi, A. T. and O'Brien, R. D. Mol. Pharmacol. 7 (1971) 104.

15. Eldefrawi, M. E., Eldefrawi, A. T. and O'Brien, R. D. Annu. Rev. Pharmacol 12 (1972) 19.

16. Eldefrawi, M. E., Eldefrawi, A. T. and Shamoo, A. E. Ann. N. Y. Acad. Sci. 264 (1975) 183.

17. Eldefrawi, M. E., Eldefrawi, A. T. and Wilson, D. B. Biochemistry 14 (1975) 4304

18. Eldefrawi, M. E., Eldefrawi, A. T. Seifert, S. and O'Brien, R. D. Arch. Biochem. Biophys. 150 (1972) 210.

19. Eldefrawi, M. E., Eldefrawi, A. T., Penfield, L. A. O'Brien, R. D. and Van Campen, D., Life Sciences 16 (1975) 925.

20. Ellman, G. L. Courtney, K. D., Andres, V. Jr. and Featherstone, R. M. Biochem. Pharmacol. 7 (1961) 88.

21. Friberg, L. A. M. A. Arch. Industr. Health 20 (1959) 42.

22. Gage, J. C. Brit. J. Industr. Med 21 (1964) 197.

23. Gray, E. G. and Whittaker, V. P. J. Anat. 96 (1962) 79.

24. Heilbronn, E. and Mattson, C. J. Neurochem. 22 (1974) 315.

25. Iverson, F., Downe, R. H., Paul, C. and Treholm, H. L. Toxicol. Appl. Pharmacol. 24 (1973) 545.

26. Juang, M. S. and Yonemura, K. Nature 256 (1975) 211.

27. Karlin, A. and Cowburn, D. Proc. Nat. Acad. Sci. U. S. A. 70 (1973) 3636.

28. Klett, R. P., Fulpius, B. W., Cooper, D., Smith, M., Reich, E. and Possani, L. D. J. Biol. Chem. 248 (1973) 6841.

29. Koelle, G. B. Ch. 28, The Pharmacological Basis of Therapeutics (Goodman, L. S. and Gilman, A., Eds) Macmillan, New York (1975) 575.

30. Kurland, L. T., Faro, S. N. and Siedler, H. World Neurol 1 (1960) 370.

31. Lindstrom, J. and Patrick, J. Syneptic Transmission and Neuronal Interaction (Bennett, M. V. L. ed.) Raven, New York (1974) 191.

32. Michaelson, D., Vandlen, R., Bode, J. Moody, T. Schmidt, J. and Raftery, M. A. Arch. Biochem. Biophys 165 (1974) 796.

33. Mideli, R., Molinoff, P. and Potter, L. T. Nature 229 (1971) 554.

34. Nachmansohn, D. Ch. 2, Principles of Receptor Physiology (Loewenstein, W. R., Ed) Springer, Berlin (1971) 18.

35. Negrette, J., Del Castillo, J. Escobar, I. and Yankelevich, G. Nature (New Biol.) 235 (1972) 158.

36. O'Brien, R. D., Eldefrawi, M. E. and Eldefrawi, A. T. Ann Rev. Pharmacol. 12 (1972) 19.

37. Olsen, R. W. Meunier, J. C. and Changeux, J. P. FEBS Lett. 28 (1972) 96.

38. Piper, R. C., Miller, V. L. and Dickinson, E. O. Am. J. Vet. Res. 32 (1971) 263.

39. Potter, L. and Molinoff, P. Ch. 2, Perspectives in Neuropharmacology (Snyder, S. H. Ed.) Oxford University Press, Oxford (1972) 9.

40. Raftery, M. A. Ch. 10, Functional Linkage in Biomolecular Systems (Schmitt, F. O. Schneider, D. M., and Crothers, D. M.) Raven New York (1975) 215.

41. Robaire, B. and Kato, G. Mol. Pharmacol 11 (1975) 722.

42. Rubsamen, H., Hess, G. P., Eldefrawi, A. T. and Eldefrawi, M. E. Biochem. Biphys. Res. Comm 68 (1976) 56.

43. Rubsamen, H., Montgomery, M., Hess, G. P., Eldefrawi, A. T. and Eldefrawi, M. E. Biochem Biophys. Res. Comm (in press).

44. Salpeter, M. M. and Eldefrawi, M. E. J. Histochem. Cytochem 21 (1973) 769.

45. Salvaterra, P., Lown, B. Morganti, J. and Massaro, E. J. Acta Pharmacol. et toxicol 33 (1973) 177

46. Schleifer, L. S. and Eldefrawi, M. E. Neuropharm. 13 (1974) 53.

47. Segall, H. J. and Wood, J. M. Nature 248 (1974) 456.

48. Shamoo, A. E. and Eldefrawi, M. E. J. Membr. Biol. 25 (1975) 47.

49. Shamoo, A. E., MacLennan, D. H. and Eldefrawi, M. E. Chem-Biol. Interactions 12 (1976) 41.

50. Thompson, J. F. Pesticide Reference Standards and Supplemental Data U. S. Government Printing Office (1974) pp. 52.

51. Ulfvarson, U. Int. Arch. Gewerbepathol. Gewerbehyg. 19 (1962) 412.

DISCUSSION

GATZY: Can the calcium on the receptors be removed? And, if it can, what is the effect on acetylcholine binding and on the ion fluxes across the vesicle?

ELDEFRAWI: Sixty percent of all the calcium can be removed in response to cholinergic drugs. If you make a receptor deficit in calcium, so that it has only 10% of the normal calcium content and then look at the binding of acetylcholin, you discover that it does not affect the binding very much. So, acetylcholine will still bind very efficiently with the same kinetics to a calcium-deficit receptor. But high calcium concentration will block acetylcholine binding. By high I mean about 5 millimolar of calcium; 10 millimolar of calcium would efficiently start blocking acetylcholine binding.

BRODSKY: The effect of acetylcholine seemed to be blocked by the methylmercury and a couple of other mercurials. When you discussed snake poisons, you said nothing about atropine which is supposed to be a well-known blocker of cholinergic effects. Why?

ELDEFRAWI: I thought the slide showed a lot because on this particular type of receptor, the nicotine receptor of electric organ which has been characterized to be a pure nicotinic receptor, similar to neuromuscular receptors, should not respond to atropine except at very high concentrations. And, at the concentrations that were used here it did not have an effect. We considered this as a very strong point in favor that the binding assay that we are using is very sensitive and that can distinguish between drugs.

BRODSKY: What kind of stimulus does atropine block? What is it supposed to do?

ELDEFRAWI: It is supposed to block acetylcholine binding. Atropine would block acetylcholine binding and the physiological parasympathetic response at musculinic receptors, at junctions, at central musculinic receptors, but not at nicotinic junctions.

WEINER: Some of the organic molecules were effective and some were not. Does the difference in molecular structure give you any information about the type of binding?

ELDEFRAWI: We are asking this question right now. We are collecting as many of the organic materials as we possibly can. For instance, we are starting to look at the organic material diuretics and so forth, to see if the structure really has a meaning there. But one of the things I didn't have time to say, was that a couple of the compounds that had zero effect had a solubility problem. We could not use them at the same concentration. They were 5 to 10-fold lower concentrations so there may be a concentration problem right there.

LAKOWICZ: Methylmercury can also exist as methylmercury chloride, which is less polar. Is there any effect of the chloride concentration inhibition?

ELDEFRAWI: No, not at all because the chloride ion concentrations in our dialysis media are much higher than would be by the methylmercury chloride.

LAKOWICZ: But would it affect the balance of each one available?

ELDEFRAWI: Yes, but the phenylmercury compounds, the chloride, and the hydroxyl salts were equally effective while in the acetate and the borate were much less. So whether the accompanying ion here is playing any role at all in the toxicity of the organomercurials is hard to say, but we don't know yet for certain. Studies are still too premature to make such a judgement.

LEAD ACTIONS ON SODIUM-PLUS-POTASSIUM-ACTIVATED ADENOSINETRIPHOSPHATASE FROM ELECTROPLAX, RAT BRAIN, AND RAT KIDNEY

George J. Siegel, Suzanne K. Fogt, and Mary Jane Hurley

Neurology Research Laboratory, Neurology Department
University of Michigan Medical Center
Ann Arbor, Michigan 48109

ABSTRACT

Inorganic lead ion, in micromolar concentrations, reversibly inhibits the sodium-plus-potassium-activated adenosinetriphosphatase (ATPase) and potassium-activated p-nitrophenylphosphatase (NPPase) activities of microsomal fractions from electric organ, rat kidney, and rat brain. In the presence of 3 mM $MgCl_2$ and 3 mM ATP, the concentrations of $PbCl_2$ producing half-maximal inhibition of the ATPase from these tissues are 4 X 10^{-6} M, 20 X 10^{-6} M, and 55 X 10^{-6} M, respectively. The corresponding values for inhibition of the NPPase are 10^{-6} M, 53 X 10^{-6} M, and 22 X 10^{-6} M. $PbCl_2$ also stimulates the phosphorylation by [γ-^{32}P]ATP of a microsomal protein from all three tissues in the absence of added sodium ion. This reaction was extensively studied with electroplax microsomes. In common with the well-known Na^+-dependent phosphorylation of (Na^+ + K^+)-ATPase, the Pb^2 -dependent reaction is inhibited by ouabain, specific for ATP, dependent on Mg^{2+}, and yields an acid-stable phosphoprotein with a molecular weight of 98,000 in sodium dodecylsulfate. The Pb^{2+}-dependent phosphoprotein, however, is not sensitive to K^+. These observations are pertinent to the biochemistry and toxicity of inorganic lead in tissues and to the molecular mechanism of the cation transport enzyme.

INTRODUCTION

Inorganic lead is widely encountered as a trace metal in plant and animal tissues but there is no evidence that might indicate a physiologic function for this metal ion. On the other hand, the toxicity of excessive lead has been amply documented (4, 7, 15, 31, 37). Inorganic lead poisoning is an important health problem among children of urban slums (19, 28) and has been incriminated as a factor associated with learning and behavioral disorders in children (11). The major toxic effects of inorganic lead involve the central and peripheral nervous systems, the kidney, and hematopoiesis.

Lead ion is complexed readily by sulfhydryl and, under appropriate conditions, by carboxyl and phosphate groups, and by certain amino acid side chains (2, 53, 54). It is known that Pb^{2+} inhibits a number of sulfhydryl-containing enzymes, interferes with oxidative phosphorylation, and inhibits the biosynthesis of heme and of certain other proteins (54). Recent reports show that Pb^{2+} inhibits the activities in vitro of NADH: cytochrome c reductase (24), NADPH: cytochrome P450 reductase (6), and adenyl cyclase (35). The most potent specific enzyme effects of Pb^{2+} at less than 10^{-5} M appear to be on delta-aminolevulinic acid dehydratase and on adenyl cyclase. In vivo lead has been reported to increase the activity of certain enzymes but there is no evidence for Pb^{2+} functioning in metalloenzymes or as a metal activator of enzymes (53, 54), although substitution of Pb^{2+} in carboxypeptidase can alter substrate specificity (8). Recent evidence indicates that Pb^{2+} produces mitogenic activity and stimulation of protein and nucleic acid biosynthesis in kidney tissue (5). Thus, lead toxicity in tissues probably is caused by a number of biochemical effects.

One of the more severe toxic manifestations of lead in brain is the appearance of brain swelling with increased intracranial pressure (37). It is possible that alteration of cation transport by lead might contribute to cell swelling and accumulation of edema fluid in brain as well as in other organs. If present, altered cation transport can also lead to abnormal regulation of special tissue functions dependent on transport, such as neuronal excitability and renal tubular secretion. $(Na^+ + K^+)$-ATPase[1] represents the membrane molecular machinery that subserves active sodium-potassium transport (17, 21, 49). The available information concerning Pb^{2+} effects on $(Na^+ + K^+)$-ATPase is sparse. It is reported that Pb^{2+} can inhibit $(Na^+ + K^+)$-ATPase activity in

[1] Abbreviations: $(Na^+ + K^+)$-ATPase, $(Na^+ + K^+)$-activated adenosinetriphosphatase; K^+-NPPase, K^+-activated p-nitrophenylphosphatase; SDS, sodium dodecylsulfate; BAL, 2,3-dimercaptopropanol; EDTA, ethylenediaminetetraacetic acid; EGTA, ethyleneglycol-bis (β-aminoethyl ether) N,N^1-tetraacetic acid; DOC, deoxycholate.

preparations from red cells (20, 30, 41) and kidney (25, 42, 52). This paper provides a description of some properties of Pb^{2+} actions on $(Na^+ + K^+)$-ATPase in microsomal preparations from the electric organ of E. electricus and from rat brain and kidney.

Purified $(Na^+ + K^+)$-ATPase extracted from membranes consists of two dissimilar polypeptides as found in sodium dodecylsulfate solutions. The smaller polypeptide (M_r of about 50,000) is a sialoglycoprotein whose function in the intact enzyme is not yet known. The larger protein (M_r of about 100,000) contains the catalytic center which, in the intact enzyme, is phosphorylated on an aspartyl residue in a $(Mg^{2+} + Na^+)$-dependent reaction to form an acid stable phosphoenzyme. This reaction involves transfer of the terminal phosphate of ATP to the enzyme and has been believed to have absolute dependence on sodium ion (10). In the native enzyme, the phosphorylation by ATP is reversible and Na^+-dependent ADP-ATP transphosphorylation can be measured under conditions of low Mg^{2+} concentrations (13). The extent of Na^+-dependent phosphorylation is known to be reduced by K^+ in parallel with K^+-activation of ATP hydrolysis (43). In addition, preparations of $(Na^+ + K^+)$-ATPase also catalyze K^+-stimulated hydrolysis of p-nitrophenylphosphate in the presence of Mg^{2+}. The K^+-NPPase is believed to represent the phosphatase moiety of the enzyme (10).

Thus, the overall activity of the enzyme appears to involve kinase, phosphoryl acceptor, and phosphatase functions (1b). Several kinetic models (1a, 16) for the enzyme activity and physical models (38, 50) for the coupled cation translocation have been discussed.

METHODS

Tris salts of nucleotides were products of Sigma or Calbiochem. [γ-^{32}P]ATP was obtained from ICN or New England Nuclear.

Microsomal $(Na^+ + K^+)$-ATPase was prepared from electric organ (46) and from rat brain (1a) as described. Microsomal preparations were obtained from rat kidneys as follows. Kidneys were removed from 200 g Sprague-Dawley rats immediately after decapitation of the animals. The outer capsules were removed and the kidneys were stored in liquid nitrogen for periods up to a week. The kidneys were thawed and the inner white medullas were dissected away. The remainders were first minced with scissors and then dispersed in glass homogenizers by hand in 20 volumes of an ice cold solution consisting of 30 mM imidazole, 0.25 M sucrose, 1 mM EDTA, and 1 mM

EGTA, adjusted to pH 6.8. The homogenate was filtered through two layers of gauze and sodium deoxycholate was added to a concentration of 2.4 mM with stirring. After two additional strokes of the pestle, the entire suspension was centrifuged at 10,800 x g for 30 minutes at 0°. The supernatant portions were then centrifuged at 50,000 x g for one hour. The resulting pellets were resuspended to a concentration of about 10 mg protein/ml in 30 mM imidazole (pH 7.0), distributed into aliquots, and stored in liquid nitrogen. Prior to assay of enzyme activity, the microsomes were diluted to 1 mg/ml and exposed to 0.2% deoxycholate in 30 mM imidazole (pH 7.0) at room temperature for 30 minutes. This mixture was further diluted 10 times in 0.15 M imidazole (pH 7.8), chilled in an ice bath, and used immediately.

Assays of $(Na^+ + K^+)$-ATPase activity in the kidney microsomes were made with 2-5 μg of microsomal protein in 40 μl of media containing 3 mM $MgCl_2$, 3 mM Tris [γ-^{32}P]ATP, 80 mM NaCl, 10 mM KCl, and 40 mM imidazole (pH 7.8) for 15 minutes at 37°. Release of ^{32}Pi was measured as described (43). NPPase assays were performed with the same DOC-treated microsomes, 5 μg protein, in 40 μl of media containing 5 mM $MgCl_2$, 5 mM p-nitrophenylphosphate, 20 mM KCl, and 25 mM imidazole (pH 7.8) at 37° for 15 minutes. Assays were terminated by the addition of 200 μl of 0.1 N NaOH. p-Nitrophenol was determined from the absorbancy at 420 nm. Media excluding Na^+ and K^+ were used to obtain Mg^{+}-baseline activities. When ouabain inhibition was measured, the DOC-treated kidney microsomes were diluted 20 fold in imidazole (pH 7.8) containing 2 mM $MgCl_2$ with or without 4 mM ouabain and let stand for 30 minutes at 2° prior to assay. The final concentration of ouabain in the assay was 2 mM.

Standard conditions for assays of $(Na^+ + K^+)$-ATPase and K^+-NPPase activities of electroplax and brain microsomes were the same as for kidney except that the microsomes were stored as supensions in water, there was no exposure to deoxycholate, the media buffer was 75 mM imidazole (pH 7.4), and the electroplax enzyme was assayed at 27° rather than at 37°.

Incorporation of ^{32}P into microsomal protein from [γ-^{32}P]ATP was measured using the same conditions and procedures described earlier (46) for each of the tissue sources. Standard incubations were at 2° for 45 seconds. Media contained 100 μg microsomal protein, 3 mM $MgCl_2$, 1 mM Tris [γ-^{32}P]ATP, 75 mM imidazole (pH 7.4), and either 100 mM NaCl or an indicated concentration of $PbCl_2$. In some experiments the imidazole was replaced by Tris HCl. Appropriate controls showed no difference in results.

Protein measurements (29) and polyacrylamide gel electrophoresis in sodium dodecylsulfate (23) were carried out according to published methods.

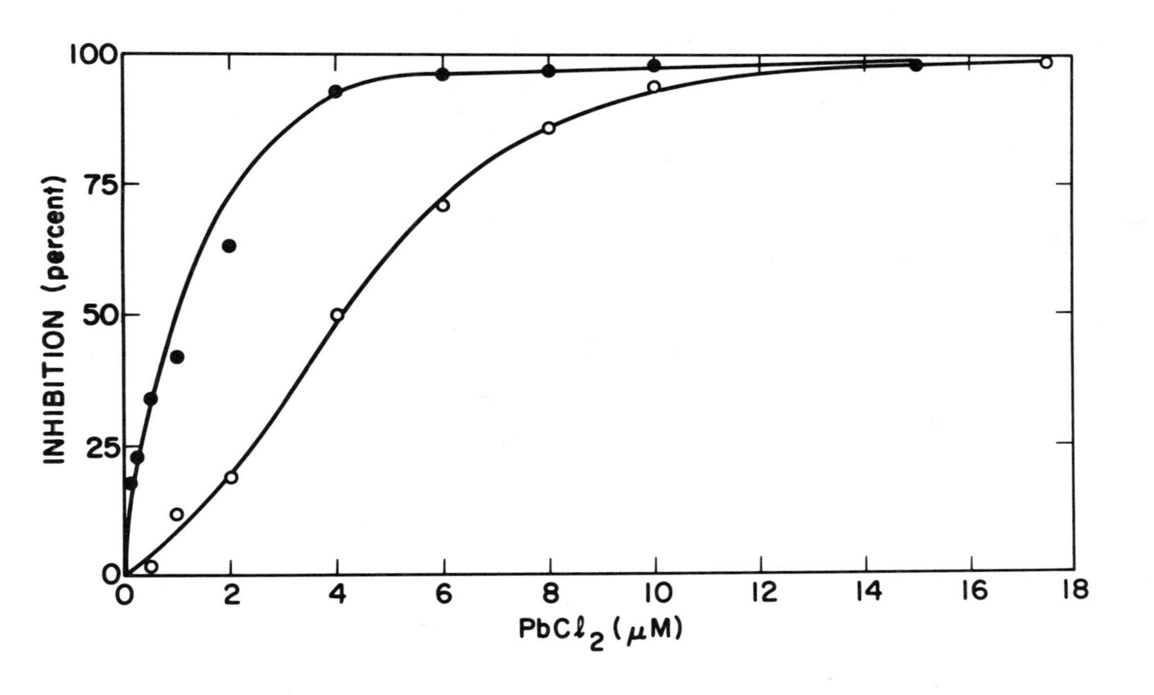

Figure 1. Inhibition of electroplax $(Na^{+} + K^{+})$-ATPase and K^{+}-NPPase activities by $PbCl_2$. The Mg^{2+}-baseline activities are subtracted. o-o, $(Na^{+} + K^{+})$-ATPase; ●-●, K^{+}-NPPase.

RESULTS

Inhibition of $(Na^+ + K^+)$-ATPase and of K^+-NPPase

Figure 1 shows the effects of $PbCl_2$ on electroplax enzyme activity. The values of $[PbCl_2]_{0.5}$ for inhibition of $(Na^+ + K^+)$-ATPase and K^+-NPPase activities are 4 μM and 1 μM, respectively. Since ATP binds Pb^{2+}, an accurate comparison of the kinetics of Pb^{2+} inhibition of ATPase and NPPase activities depends on measurements of free lead ion activities. Without considering other species of bound lead, an approximate comparison may be made by calculating apparent $[Pb^{2+}]$ based on the dissociation constants of $MgATP^{2-}$ and $PbATP^{2-}$ at approximately neutral pH. Assuming $K_D = 10^{-4}$ for $MgATP^{2-}$ (3) and $K_D = 0.74 \times 10^{-4}$ for $PbATP^{2-}$ (52), the calculated apparent $[Pb^{2+}]_{0.5}$ is 0.5 μM for inhibition of $(Na^+ + K^+)$-ATPase. This value is sufficiently close to that for inhibition of K^+-NPPase that the inhibitions of both activities may be related to the same lead binding site on the enzyme. Table I shows that various dithiols and chelating agents are able to prevent the lead ion inhibition of ATPase activity.

Figure 2 shows the effects of $PbCl_2$ on rat brain enzyme activity. The values of $[PbCl_2]_{0.5}$ for inhibition of the $(Na^+ + K^+)$-ATPase and K^+-NPPase activities are 55 μM and 22 μM, respectively. The calculated apparent $[Pb^{2+}]_{0.5}$ for inhibition of ATP hydrolysis is 7.4 μM. This value is at least three-fold less than that for inhibition of K^+-NPPase. The brain and electroplax enzyme

TABLE I. Reversal of Pb^{2+}-inhibition of electroplax ATPase.

Microsomes were pretreated at 2°C for 10 minutes with 27 μM $PbCl_2$ in 75 mM imidazole HCl (pH 7.4) and then assayed for ATP hydrolysis employing the complete medium as described in the text. Final $PbCl_2$ concentrations were 10 μM during incubations. Results are expressed as inhibition relative to activities obtained with native microsomes in the presence of each agent without $PbCl_2$. BAL plus $PbCl_2$ produced apparent activation.

Addition	Pb^{2+}-Inhibition
0.1 mM	percent
none	91
Dithiothreitol	22
DL-Penicillamine	14
EDTA	10
EGTA	4
BAL	-37

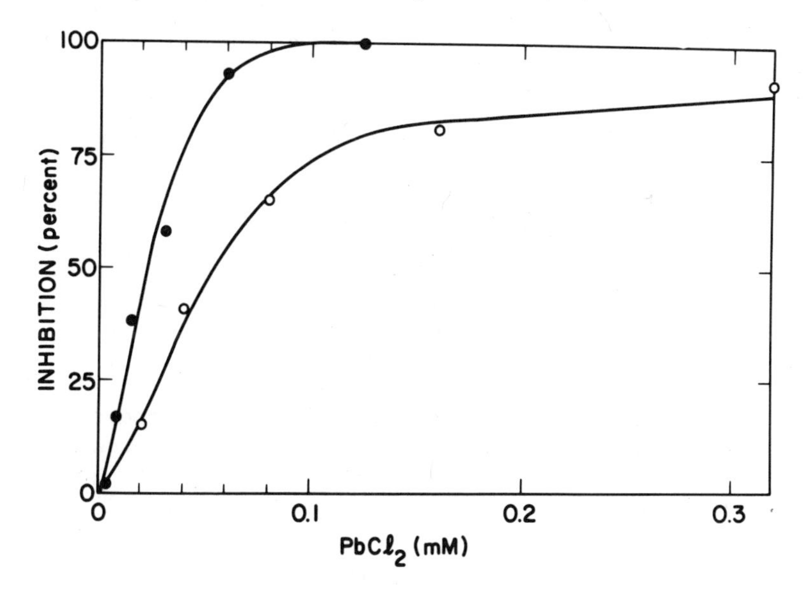

Figure 2. Inhibition of rat brain $(Na^+ + K^+)$-ATPase and K^+-NPPase activities by $PbCl_2$. The Mg^{2+} baseline activities are subtracted. o-o, $(Na^+ + K^+)$-ATPase; ●-●, K^+-NPPase.

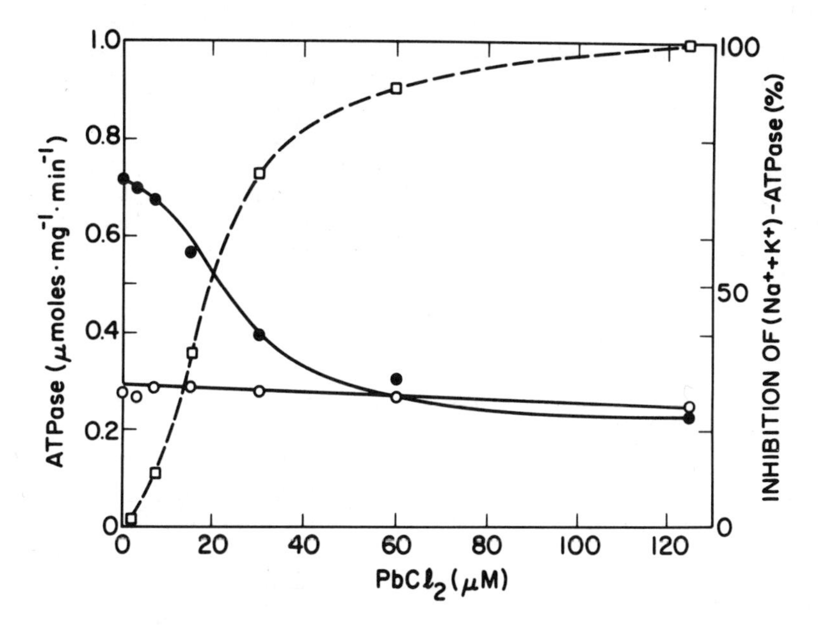

Figure 3. Inhibition of rat kidney $(Na^+ + K^+)$-ATPase by $PbCl_2$. Left ordinate: ●-●, $(Mg^{2+} + Na^+ + K^+)$-ATPase; o-o, Mg^{2+}-ATPase. Right ordinate: □- -□, percent inhibition of increments due to Na^+ plus K^+.

preparations are dissimilar in this respect and also with regard to their sensitivities to lead ion.

The Pb^{2+} inhibition of $(Na^+ + K^+)$-ATPase is readily reversed by dilution (Table II). In this experiment, the brain enzyme was first exposed to an inhibitory concentration of $PbCl_2$ (0.2 mM) for 15 minutes at room temperature and then diluted 40 times in the assay media so that the lead concentration fell below that required for inhibition. Under these conditions, there was no inhibition. The reversibility of inhibition implies that one may not necessarily expect to find inhibited enzyme activity in homogenates of tissues taken from lead intoxicated animals.

Responses of a rat kidney microsomal enzyme preparation to $PbCl_2$ are shown in Figure 3. The unwashed kidney microsomes contain substantial portions of ouabain-resistant Mg^{2+}-ATPase; however, the Mg^{2+} -ATPase activity is not significantly affected by $PbCl_2$. The $(Na^+ + K^+)$-increments in ATP hydrolysis, on the other hand, are completely inhibited; $[PbCl_2]_{0.5}$ is 20 µM. The calculated apparent $[Pb^{2+}]_{0.5}$ is 3 µM. K^+-NPPase activity of the same preparation is also completely inhibited; $[PbCl_2]_{0.5}$ is 53 µM (Fig. 4). The Mg^{2+}-NPPase activity, however, is not significantly inhibited by as much as 0.2 mM $PbCl_2$.

Table III shows that the Pb^{2+} inhibition is completely reversible by dilution of the $PbCl_2$. In this experiment, kidney microsomes, 5 µg protein/µl, were first exposed to 0.2% deoxycholate in 30 mM imidazole (pH 7.0) at room temperature for 30 minutes after which a portion of this suspension was exposed to 125 µM $PbCl_2$. The suspensions with and without added $PbCl_2$ stood another 10 minutes at room temperature and were then diluted 40 times in the standard assay media. The $(Na^+ + K^+)$-ATPase activities of samples containing 125 µM $PbCl_2$ during both the preincubation and assay

TABLE II. Rat brain ATPase: reversibility of Pb^{2+}-inhibition by dilution.

$PbCl_2$ in Preincubation	$PbCl_2$ in Assay	ATPase Inhibition ΔMg^{2+}	ATPase Inhibition $\Delta(Na^+ + K^+)$
µM	µM	percent	percent
200	5	0	0
200	200	3	82
0	200	10	88
0	5	0	0

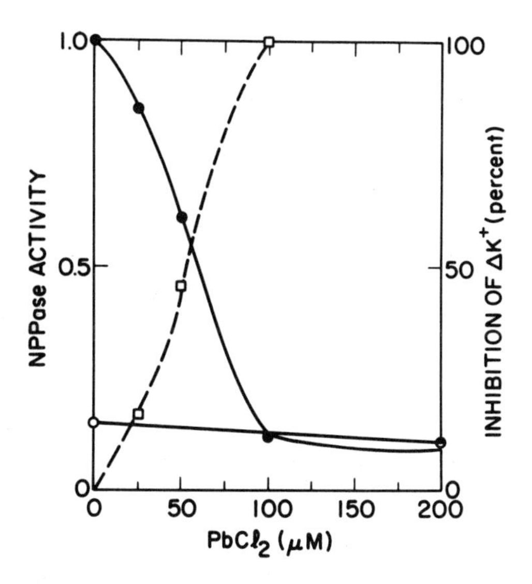

Figure 4. Inhibition of rat kidney K^+-NPPase by $PbCl_2$. Left ordinate: ●-●, $(Mg^{2+}+K^+)$-NPPase; o-o, Mg^{2+}-NPPase. Right ordinate: □ - -□, percent inhibition of increments due to K^+.

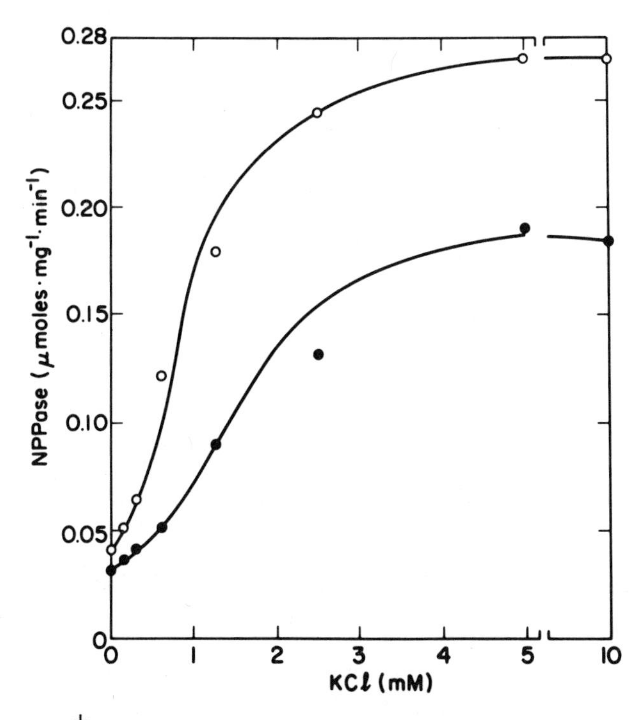

Figure 5. K^+-activation of rat kidney NPPase. o-o, no $PbCl_2$; ●-●, in the presence of 38 μM $PbCl_2$.

TABLE III. Rat kidney ATPase: reversibility of Pb^{2+}-inhibition by dilution.

$PbCl_2$ in Preincubation	$PbCl_2$ in Assay	ATPase Inhibition	
		ΔMg^{2+}	$\Delta(Na^+ + K^+)$
μM	μM	percent	percent
125	3	19	0
125	125	37	100
0	125	37	100
0	3	4	4

periods or only during the assay period were completely inhibited while samples in which the $PbCl_2$ was diluted to 3 μM during the assay showed full activity.

The $(Na^+ + K^+)$-ATPase activity of kidney appears more sensitive than that of brain to Pb^{2+} while the K^+-NPPase activity of kidney appears less sensitive than that of brain to lead ion. These differences in the apparent $[PbCl_2]_{0.5}$ values are not explained only by differences in non-specific binding of Pb^{2+} to various microsomal components since the sensitivities of the two enzyme activities in kidney and brain change in opposite directions. Another possibility is a differential effect of deoxycholate on the ATPase and NPPase activities in the kidney microsomes. Further studies of the interactions of Pb^{2+} with the various ligands are required. Preliminary data failed to show evidence for an interaction of Pb^{2+} with K^+ as measured by NPPase activity of the kidney microsomes (Fig. 5). In this experiment, $[K^+]_{0.5}$ for activation of NPPase is 0.85 mM and 0.95 mM in the absence and presence of 38 μM $PbCl_2$, respectively.

Pb^{2+}-stimulated enzyme phosphorylation

The studies of Pb^{2+} inhibition of ATP and NPP hydrolysis indicate that the turnover of the enzyme is slowed but do not point out the partial reaction(s) on which the rate-limiting effect of Pb^{2+} is exerted. Investigations of the effects of Na^+, K^+, Mg^{2+} and ATP on the Pb^{2+}-inhibition of both ATPase and NPPase activities are necessary to help elucidate this probelm. Another approach is to determine the effects of Pb^{2+} on rates of enzyme phosphorylation and dephosphorylation and on steady-state levels of phosphoenzyme.

In testing the effects of $PbCl_2$ on the steady-state levels of phosphoenzyme using electroplax microsomes, the surprising observation was made that $PbCl_2$ stimulates phosphate incorporation into protein in the absence of added NaCl (Fig. 6). In this and in other experiments, the extents of Pb^{2+}-dependent phosphorylation are almost equal to the Na^+-dependent levels and the total phosphorylation is not additive in the presence of saturating concentrations of both Pb^{2+} and Na^+ (Fig. 6). In addition, the Na^+-dependent phosphoprotein level is not reduced by $PbCl_2$ except at the highest $PbCl_2$ concentration. Therefore, either the same phosphate acceptor site is involved in both Pb^{2+}- and Na^+-stimulated reactions, or one site is stimulated exactly in parallel with inhibition of another phosphate acceptor site.

Table IV shows that ouabain, a specific inhibitor of (Na^+ + K^+)-ATPase (1a, 10), also inhibits both the Pb^{2+}- and the Na^+-stimulated phosphorylation of electroplax microsomes. Similar experiments with brain microsomes showed the same results (Table V). Kidney microsomes also exhibit Pb^{2+}-dependent phosphorylation (Table VI) but further tests have not yet been performed with this tissue.

The ouabain-sensitive stimulation of phosphate incorporation appears to be unique for Pb^{2+} among other divalent cations tested. Table VIIA shows that, in the presence of Mg^{2+}, only Ba^{2+} and Fe^{2+} at 0.1 mM produced more than 20% of the increment in phosphorylation produced by 0.1 mM Pb^{2+}. The effects of Ba^{2+} and Fe^{2+} were therefore tested further for ouabain sensitivity. Table VIIB shows that the ouabain-sensitive increment due to Ba^{2+} is 10% of that due to Pb^{2+} while Fe^{2+} and Zn^{2+} produce little or no ouabain-sensitive increments.

TABLE IV. Ouabain inhibition of phosphorylation of electroplax microsomes.

Microsomes were pre-treated with 75 mM imidazole HCl (pH 7.4), 3 mM $MgCl_2$, with or without 0.5 mM ouabain at 2°C for 1 hour. Ouabain, 0.1 mM, was included in phosphorylating media with ouabain-treated microsomes. Data are from (46).

NaCl	$PbCl_2$	^{32}P-phosphoprotein	
		Native	Ouabain
mM	μM	pmoles ^{32}P/mg protein	
0	0	267	9
100	0	951	78
0	80	761	74
100	80	814	195

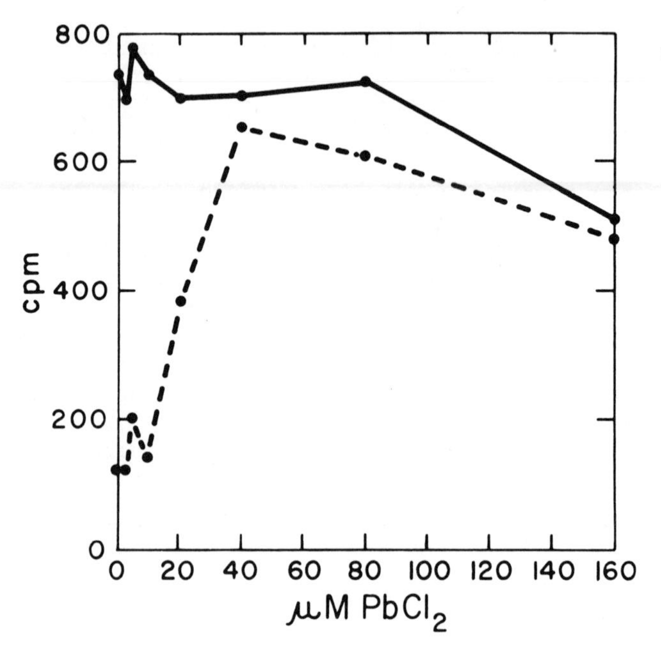

Figure 6. Pb^{2+}-stimulated phosphorylation of electroplax microsomes. Ordinate shows cpm of ^{32}P incorporated into acid washed pellets after subtraction of values for blanks. ●-●, 0.1 M NaCl present; ●- -●, no NaCl present. From (46) with permission.

TABLE V. Ouabain inhibition of phosphorylation of rat brain microsomes.

Microsomes were first exposed to 50 mM imidazole (pH 7.4) and 2 mM $MgCl_2$ plus or minus 1 mM ouabain for 3 hours at 2°C. These samples were then diluted by half in respective phosphorylating media.

NaCl	$PbCl_2$	^{32}P-phosphoprotein	
		Native	Ouabain
mM	mM	pmoles ^{32}P/mg protein	
0	0	56	51
100	0	301	124
0	0.2	200	67
100	0.2	193	89

In order to determine whether the Pb^{2+}-dependent phosphorylation is related to the catalytic unit of ($Na^+ + K^+$)-ATPase, further characterization was undertaken utilizing the electroplax enzyme preparation in the following experiments.

It was found that the Pb^{2+}-stimulation of phosphate incorporation is at least partially dependent on added Mg^{2+} (Table VIII). Although there is a small amount of Pb^{2+}-stimulated phosphorylation in the absence of added $MgCl_2$, the fact that EDTA reduces the baseline value suggests the presence of endogenous Mg^{2+} in the microsomes.

TABLE VI. Na^+- and Pb^{2+}-stimulated phosphorylation of rat kidney microsomes.

NaCl	$PbCl_2$	^{32}P-phosphoprotein
mM	mM	pmoles ^{32}P/mg protein
0	0	53
100	0	162
0	0.25	144
100	0.25	116

TABLE VII. Effects of divalent cations on phosphorylation of electroplax microsomes.

Electroplax microsomes, 100 μg protein, were incubated at 2° for 45 seconds in 3 mM $MgCl_2$, 1 mM [γ-^{32}P]ATP, 75 mM Tris HCl (pH 7.4), and 0.01 mM or 0.1 mM of various additional divalent cations as chloride salts except MoO_3 (Part A). Incorporation of ^{32}P into protein was measured as stated in the text. In part B, microsomes were first exposed to 3 mM $MgCl_2$, 50 mM Tris HCl (pH 7.4) with or without 5 mM ouabain for 3 hours at 0° prior to phosphorylation in media as in part A except that 0.5 mM ouabain was present in the indicated samples. Added divalent cations were in concentrations of 0.1 mM. MoO_3 solutions contained .4 mM NaOH. Data are from (46).

A. Cation	0.01 mM	0.1 mM
	pmoles ^{32}P/mg protein	
none	117	117
Ca	131	138
Sr	149	150
Ba	230	314
Mo	124	188
Mn	122	142
Fe	132	239
Co	149	133
Ni	129	134
Cu	123	122
Zn	152	102
Cd	149	184
Hg	135	115
Pb	135	508

B. Cation	+ouabain	-ouabain	ouab-sensitive
	pmoles ^{32}P/mg protein		
none	56	94	38
Ba	108	193	85
Fe	77	109	32
Zn	63	95	32
Pb	114	602	488

TABLE VIII. Partial requirement of Mg^{2+} for Pb^{2+}-stimulated phosphorylation of electroplax microsomes.

Media contained 1 mM Tris [γ-^{32}P]ATP, 75 mM Tris HCl (pH 7.4) with or without 80 µM $PbCl_2$ and various concentrations of $MgCl_2$. Data are from (46).

Addition	$-PbCl_2$	80 µM $PbCl_2$
	pmoles ^{32}P/mg protein	
2 mM EDTA	4	
none	16	205
mM $MgCl_2$		
0.01	65	278
0.05	118	439
0.10	193	515
0.20	188	671
1.00	234	597
3.00	125	564
6.00	34	513

Table IX shows that the extent of phosphorylation by [γ-^{32}P] ATP is not significantly inhibited by 10-fold excesses of other unlabeled nucleotides which indicates the relative specificity for

TABLE IX. Effects of other nucleosides on Pb^{2+}-dependent phosphorylation of electroplax microsomes by ATP.

Media contained 75 mM Tris HCl (pH 7.4), 0.1 mM Tris [γ-^{32}P] ATP, 3 mM $MgCl_2$, and 80 µM $PbCl_2$. Other additions to the media were made as shown. Results are compared to the controls containing 0.1 mM ATP. Data are from (46).

Addition, 1 mM	^{32}P incorporation
	percent of control
none	100
ATP	172
adenosine	117
AMP	112
ADP	84
GTP	143
UTP	131
CTP	142
Tris PO_4#	31

Precipitate forms

for ATP. The 31% to 42% stimulation produced by GTP, UTP, and CTP might represent a non-specific nucleotide activating effect but this assumption would require further tests. The addition of Tris phosphate produces a lead phosphate precipitate. In other experiments it was found that Tris ^{32}Pi and [^{14}C]-ATP yielded no significant isotope incorporation into microsomes in the presence of either 80 μM $PbCl_2$ or 100 mM NaCl under the same conditions in which phosphorylation by [γ-^{32}P]ATP was observed. Thus, both the Pb^{2+}- and Na^+-stimulated phosphorylation reactions involve the transfer of the gamma phosphate of ATP to a microsomal component.

The effects of K^+ on Na^+- and Pb^{2+}-dependent reactions are compared in Tables X and XI. It is known that Na^+-stimulated phosphorylation may be reduced by KCl through activation by K^+ of dephosphorylation and, under certain conditions, through inhibition by K^+ of the initial phosphate incorporation (48). Table X shows that KCl added to Na^+-phosphorylating media after phosphorylation is achieved has the effect of reducing the phosphorylation level whether or not 80 μM $PbCl_2$ is also present. However, this effect of K^+ is not seen in Pb^{2+}-media without Na^+ added. If K^+ were able to inhibit Pb^{2+}-stimulated phosphorylation, it might be possible to observe a reduction in levels by including K^+ at the onset of incubations. Table XI shows that KCl present from the start of incubations lasting 15 to 180 seconds produces no reduction in the Pb^{2+}-dependent product. However, these data, while excluding a K^+ effect on the steady-state level of phosphoprotein in the absence of Na^+, do not eliminate the possibility of a decreased rate of Pb^{2+}-dependent phosphorylation due to K^+.

TABLE X. K^+-dependent reduction in electroplax microsomal phosphorylation in the presence of Na^+.

Phosphorylation was carried out as described in the text in the presence of either Na^+ or Pb^{2+} for 45 seconds (t_o). In additional samples, at the end of 45 seconds, 10 μl of either H_2O or 250 mM KCl were added and 15 seconds later protein was precipitated with trichloroacetic acid. Data are from (46).

Phosphorylation conditions		^{32}P-protein at t_o	^{32}P-protein at 15 sec after	
			H_2O	50 mM KCl
mM Na^+	mM Pb^{2+}	pmoles/mg	pmoles/mg	
0	0	144	172	4
100	0	771	779	393
0	0.08	694	648	632
100	0.08	691	649	394

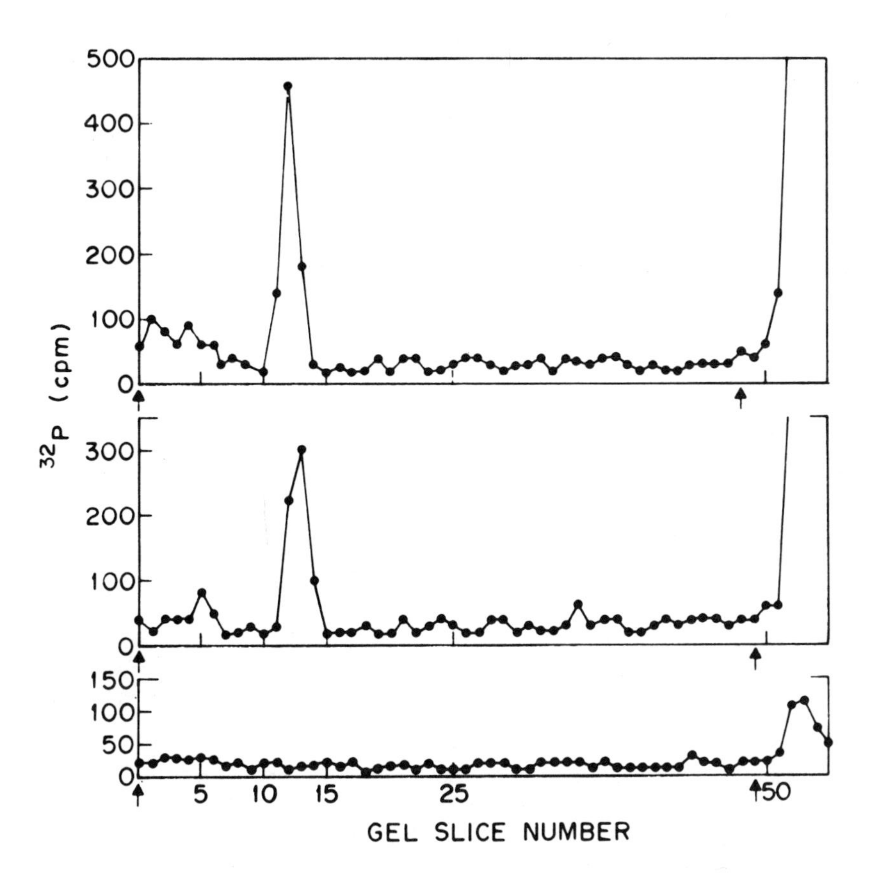

Figure 7. Polyacrylamide gel electrophoresis of electroplax microsomal phosphoproteins. Upper: Pb^{2+} medium; middle: Na^+ medium; lower: 50 mM KCl present, no Pb^{2+} or Na^+. From (46) with permission.

TABLE XI. Steady-state levels of electroplax Pb^{2+}-dependent and Na^+-dependent ^{32}P-Protein: effects of K^+.

Phosphorylation was carried out as described in the text in the presence of either Pb^{2+} or Na^+. Each set was performed with and without 50 mM KCl present throughout the incubations. Data are from (46).

Incubation time	80 μM $PbCl_2$		0.1 M NaCl	
	$-K^+$	$+K^+$	$-K^+$	$+K^+$
seconds	pmoles ^{32}P/mg protein			
15	581	574	670	333
30	573	545	802	340
60	528	530	772	463
120	480	486	729	321
180	429	509	767	354

The stabilities of the Na^+- and Pb^{2+}-dependent phosphoproteins are compared in Table XII. Both acid-precipitated phosphoproteins exhibit the same time course for release of ^{32}P at pH 2 and pH 7.4

Finally, the molecular sizes of the acid-denatured phosphorylated products dissolved in sodium dodecylsulfate were estimated by the method of polyacrylamide gel electrophoresis. Figure 7 shows that a single [^{32}P]-phosphoprotein band is found in solubilized samples of both Na^+- and Pb^{2+}-dependent phosphorylated microsomes. The R_f of the peak corresponds to $\underline{M}_r$ of 98,000 which value agrees with those obtained for the catalytic unit of (Na^+ + K^+)-ATPase purified from several different tissues (10, 22). Experiments with rat brain microsomes have yielded similar results (47).

DISCUSSION

Pb^{2+} inhibition of (Na^+ + K^+)-ATPase and of K^+-NPPase

$PbCl_2$, in micromolar concentrations, is a reversible inhibitor of the (Na^+ + K^+)-ATPase in microsomal preparations from electroplax, rat kidney, and rat brain, in order of decreasing sensitivity. It appears that the sensitivity to Pb^{2+} depends on the species and

TABLE XII. Stabilities of Na^+-dependent and Pb^{2+}-dependent ^{32}P-proteins.

Electroplax microsomes, 1 mg protein, were first phosphorylated in 0.5 ml of media containing ATP and $MgCl_2$ plus either 100 mM NaCl or 80 μM $PbCl_2$. The washed, acid-precipitated ^{32}P-protein pellets were finally resuspended in either 50 mM glycine HCl (pH 2) in part A or 75 mM imidazole HCl (pH 7.4) in part B. The resuspended pellets were then incubated at 40°C while duplicate 20 μl samples were removed from each at the intervals shown and transferred to 20 μl of ice-cold 20% trichloroacetic acid. These were sedimented and portions of the supernatant fractions were transferred to scintillation solvent for measurement of released ^{32}P. The data are expressed as the percent liberated of radioactivity in the original washed pellets. Data are from (46).

Condition	Na^+-dependent	Pb^{2+}-dependent
A. pH 2.0, 40°C		
minutes	percent of ^{32}P released	
0	22	13
15	30	29
30	45	39
60	53	53
120	81	74
B. pH 7.4, 40°C		
minutes	percent of ^{32}P released	
0	28	16
15	60	57
30	88	80

tissue. The method of preparation, particularly the presence of traces of chelators, thiols, and, possibly, detergents, are additional factors. The fact that ATP chelates Pb^{2+} and Mg^{2+} almost equally indicates that the enzyme kinetics will depend on ratios of $[Pb^{2+}]:[Mg^{2+}]:[ATP^{4-}]$ as well as on their individual concentrations (52).

Inhibition of electroplax (Na^+ + K^+)-ATPase appears to ensue from the binding of Pb^{2+} to a site on the enzyme rather than from formation of $PbATP^{2-}$ or from competition of Pb^{2+} with Mg^{2+} for ATP.

This is indicated by the facts that the concentrations of $MgCl_2$ and ATP are more than 100 times greater than inhibitory concentrations of $PbCl_2$, that Pb^{2+} also inhibits K^+-NPPase (see Results), and that increasing concentrations of ATP decrease the inhibition due to Pb^{2+} while increasing concentrations of $MgCl_2$ increase the inhibition (45).

The fact that the K^+-NPPase and $(Na^+ + K^+)$-ATPase activities of electroplax are inhibited under almost similar concentrations of apparent free $[Pb^{2+}]$ is consistent with but does not prove the assumption of a single inhibitory site for Pb^{2+}. The principal effect of Pb^{2+} appears to be a reduction in the turnover rate of the enzyme without obviously altering the steady-state levels of the phosphoenzyme intermediate. However, the reactivity of the residual phosphoprotein found in complete media containing Pb^{2+}, Na^+, and K^+ is not yet known.

The enzyme preparations from rat tissues behave somewhat differently with respect to Pb^{2+}. Both brain and kidney enzymes show lower affinities for Pb^{2+} as measured by K^+-NPPase than those affinities as measured by $(Na^+ + K^+)$-ATPase activities. These different sensitivities indicate either multiple Pb^{2+}-binding sites or interactions of Pb^{2+} with other ligands. The absence of ATP and Na^+ in the NPPase assay media might be pertinent and further studies are required.

Pb^{2+}-stimulated phosphorylation

Pb^{2+}-stimulated phosphate incorporation into microsomes is encountered in the three tissue extracts so far tested. This reaction was studied extensively utilizing electroplax preparations. The concentrations of $PbCl_2$ and the apparent $[Pb^{2+}]_{0.5}$ required for phosphorylation are substantially higher than those necessary for inhibition of ATPase. These two effects cannot, therefore, be assigned to the same Pb^{2+} binding site. It seems more likely from present data to consider a high affinity site related to inhibition of turnover and a low affinity site related to stimulation of phosphorylation. *(see Addendum)

The Pb^{2+}-dependent and Na^+-dependent products are, in common, partially dependent on Mg^{2+}, specific for ATP, inhibited by ouabain, stable in acid, of similar extents, and of the same molecular size in SDS-PAGE. Thus, it appears that the same protein is involved in both phosphorylated products. However, this does not mean necessarily that the identical amino acid residue is phosphorylated.

The difference so far observed between the Na^+- and Pb^{2+}-dependent products is the apparent insensitivity of the latter to K^+. The fact that K^+ does not stimulate the rate of ATP hydrolysis in the presence of Pb^{2+} and absence of Na^+ (45) taken together with the lack of K^+ effect on steady-state levels of the Pb^{2+}-dependent phosphoprotein (Results) indicates that K^+ does not increase the rate of dephosphorylation, in contrast to the K^+ effect on the Na^+-dependent phosphoprotein. Whether this failure of response to K^+ is related to binding of Pb^{2+} at either of its presumed sites or the absence of Na^+ is problematic.

The fact that K^+ does reduce the steady-state levels of phosphorylation in the presence of Na^+ and 80 μM $PbCl_2$ together despite almost complete inhibition of hydrolysis suggests that the reduction is not solely a consequence of accelerated turnover. This is supported by other studies which indicate that K^+ is able to inhibit enzyme phosphorylation under certain conditions (48). That K^+ does reduce the phosphorylation in the presence of Na^+ or Na^+ plus Pb^{2+} but not Pb^{2+} alone raises the intriguing possibility that the effect of K^+ depends on Na^+ binding.

The molecular mechanisms involved in Na^+ activation of enzyme phosphorylation and in the complex Na^+-K^+ interactions are not understood, although they are supposed to involve conformational interconversions of the enzyme (1a). The K^+ actions are shared by other monovalent cations but it has been believed heretofore that the Na^+ action is absolutely specific (10, 49). It now appears that Pb^{2+} mimics an action that leads to enzyme phosphorylation but not one that permits certain effects of K^+. There is no obvious clue presently to the unique feature that Pb^{2+}, of all other metal ions tested, holds in common with Na^+ to account for its action on phosphorylation. The binding of Pb^{2+} is presumed to modify enzyme conformational restraints otherwise specifically sensitive to Na^+ (46). This action is dissimilar from that of N-ethylmaleimide and of ouabain in that the former does not produce phosphorylation in the absence of Na^+(14) and the latter stimulates incorporation of Pi (44). It is also difficult to find an analogy among other enzymes. Although other divalent cations function in phosphoryl transfers (34), this is the first known instance of Pb^{2+}-stimulated kinase activity.

Further studies may help in elucidating actions of the physiologic ligands on this enzyme. Furthermore, this information may contribute to understanding other biochemical effects of Pb^{2+} in tissues. For example, certain stimulatory effects of Pb^{2+} on biosynthesis (5) might involve analagous molecular actions of Pb^{2+} on other enzymes.

Implications for Pb^{2+}-toxicity in tissues

In experimental lead intoxication of suckling rats which developed paraplegia, the brain lead concentrations[2] were in the order of 30 μM to 58 μM (27, 32, 51). At 55 μM $PbCl_2$, $(Na^+ + K^+)$-ATPase of rat brain in vitro is 50% inhibited in the present study. At lower levels of lead ingestion, the lead concentrations in brains from suckling rats made hyperactive by the intoxication were 2.5μ M to 6 μM in various studies (33, 39). These concentrations produce less than 10% inhibition of the $(Na^+ + K^+)$-ATPase in vitro. Therefore, inhibition of cation transport could be involved in the examples of severe lead intoxication but probably not in the mild cases unless there is substantial sequestration of Pb^{2+} adjacent to pump sites in vivo. It is noteworthy that lead concentrations ranging from 6 μM to 70 μM have been found in brains from children and adults dying from lead toxicity (9, 36). These concentrations are in the range that significantly inhibit the rat brain enzyme in vitro and the same implications could be extended to toxicity in humans.

Kidney tissue generally accumulates more lead than does brain, and the present study shows that the kidney enzyme is more sensitive than the brain enzyme is to lead ion. Concentrations of lead in kidneys from rats exposed for one week to diets containing 0.5% lead acetate were found to be 100 μM (40), a concentration which is twice that necessary for complete inhibition of the rat kidney $(Na^+ + K^+)$-ATPase in vitro in the present study. Rat kidneys containing even lower quantities of lead exhibit a number of morphologic changes including edema (18) to which inhibition of cation transport could be a contributory factor. In addition, if sodium reabsorption (26) and potassium secretion (12) in kidney tubules depend partly on $(Na^+ + K^+)$-ATPase, as suggested, then this amount of lead in the kidney might be expected to interfere with renal tubular function. This remains to be established. It is of significance that lead concentrations in kidneys of lead intoxicated humans range from 10 μM to 100 μM (9), and the same possibilities regarding renal function could have implications for lead toxicity in humans. Since the Pb effects on $(Na^+ + K^+)$-ATPase in vitro are reversible, adequate tests of these suggestions require comparisons of Pb^{2+} actions on enzyme activity and cation fluxes in whole tissue slices or intact cells.

[2]Assuming complete solution and uniform distribution of lead in the tissue for the purpose of comparison to in vitro data; this assumption is obviously limited since all the loci of sequestration and extent of binding of lead are unknown.

Acknowledgment: This work was supported by National Science Foundation Grant #PCM 75-05979 and grants from General Research Support and Institute for Environmental Quality of the University of Michigan.

*Addendum: Subsequent experiments with electroplax enzyme have shown that high concentrations of microsomes, such as used in phosphorylation assays, raise the $(PbCl_2)_{0.5}$ to 15 μM in the presence of 3 mM $MgCl_2$, 1 mM ATP, and 0.1 mg microsomal protein/ 40 μl at 2°. Therefore, both effects of Pb^{2+}, inhibition of hydrolysis and stimulation of phosphorylation, may actually be assigned to a single Pb^{2+} binding site.

REFERENCES

1a ALBERS, R.W., KOVAL, G.J. and SIEGEL, G.J.: Studies on the interaction of ouabain and other cardioactive steroids with sodium-potassium-activated adenosine triphosphatase. Mol. Pharmacol. 4 (1968) 324-336.

1b ALBERS, R.W. and KOVAL, G.J.: Sodium-potassium-activated adenosine triphosphatase VII. Concurrent inhibition of $(Na^+ + K^+)$-ATPase and activation of K^+-nitrophenylphosphatase activities. J. Biol. Chem. 247 (1972) 3088.

2 BJERRUM, J., SCHWARZENBACH, G. and SILLEN, L.G.: Stability Constants of Metal-Ion Complexes, with Solubility Products of Inorganic Substances. The Chemical Society, Burlington House. London (1957).

3 BOCK, R.M.: Adenine nucleotides and properties of pyrophosphate compounds, Ch. 1, The Enzymes (Boyer, P.E., Lardy, H. and Myrback, K. Eds.). 2nd Edition, Academic Press. New York (1960) vol 2, pp. 16.

4 CHISOLM, J.J.Jr.: Treatment of lead poisoning. Mod. Treat. 8 (1971) 593-611.

5 CHOIE, D.D. and RICHTER, G.W.: Cell proliferation in mouse kidney induced by lead. Lab. Investig. 30 (1974) 652.

6 CHOW, C.C. and CORNISH, H.H.: personal communication.

7 COFFIN, R., PHILLIPS, J.L., STAPLES, W.I. and SPECTOR, S.: Treatment of lead encephalopathy in children. J. Pediat. 69 (1966) 198-206.

8 COLEMAN, J.E. and VALLEE, B.L.: Metallocarboxypeptidases: stability constants and enzymatic characteristics. J. Biol. Chem. 236 (1961) 2244-2249.

9 CUMINGS, J.N.: Heavy Metals and the Brain pp. 113-120. C.C. Thomas. Springfield (1959).

10 DAHL, J.L. and HOKIN, L.E.: Sodium-potassium-adenosine triphosphatase. Annu. Rev. Biochem. 43 (1974) 327.

11 DAVID, O., CLARK, J. and VOELLER, K.: Lead and hyperactivity. Lancet 2 (1972) 900.

12 EPSTEIN, F.H. et al: Metabolic adjustments of the kidney involved in the adaptation to potassium loading. Med. Clin. North. Am. 59 (1975) 763.

13 FAHN,S., KOVAL, G.J. and ALBERS, R.W.: Sodium-potassium-activated adenosine triphosphatase of Electrophorus electric organ. I. An associated sodium-activated transphosphorylation. J. Biol. Chem. 241(8) (1966) 1882-1889.

14 FAHN, S., KOVAL, G.J. and ALBERS, R.W.: Sodium-potassium-activated adenosine triphosphatase of Electrophorus electric organ. Phosphorylation by adenosine triphosphate-^{32}P. J. Biol. Chem. 243 (1968) 1993.

15 FELTON, J.S.: Moderator, Heavy metal poisoning: mercury and lead. Annals of Int. Med. 76 (1972) 779-792.

16 FUKUSHIMA, Y. and TONOMURA, Y.: Properties of conversion of an enzyme-ATP complex to a phosphorylated intermediate in reaction of Na^+ - K^+ - dependent ATPase. J. Biochem. 77 (1975) 533.

17 GOLDIN, S.M. and SWEADNER, K.J.: Reconstitution of active transport by kidney and brain (Na^+ + K^+)-ATPase. Ann. N.Y. Acad. Sci. 264 (1975) 387.

18 GOYER, R.A.: Lead and the kidney. Curr. Topics Pathol. 55 (1971) 147.

19 GUINEE, V.F.: Lead poisoning in New York City. Trans. N.Y. Acad. Sci. 33 (1971) 539.

20 HASAN, J., VIHKO, V. and HERNBERG, S.: Deficient red cell membrane ($Na^+ + K^+$)-ATPase in lead poisoning. Arch. Environ. Health 14 (1967) 313-318.

21 HILDEN, S. and HOKIN, L.E.: Active potassium transport coupled to active sodium transport in vesicles reconstituted from purified sodium and potassium ion-activated adenosine-triphosphatase from rectal gland of Squalus acanthias. J. Biol. Chem. 250 (1975) 6296.

22 HOKIN, L.E.: Purification and properties of the (sodium + potassium)-activated adenosinetriphosphatase and reconstitution of sodium transport. Ann. N.Y. Acad. Sci. 242 (1974) 12.

23 HOKIN, L.E. et al: Studies on the characterization of the sodium-potassium transport adenosine triphosphatase X. Purification of the enzyme from the rectal gland of Squalus acanthias. J. Biol. Chem. 248 (1973) 2593.

24 IANNACCONE, A. et al: *In vitro* effects of lead on enzymatic activities of rabbit kidney mitochondria. Experientia 30 (1974) 467.

25 JACOBSEN, N.O. and JORGENSEN, P.L.: A quantitative biochemical and histochemical study of the lead method for localization of adenosine triphosphatehydrolyzing enzymes. J. Histochem. Cytochem. 17 (1969) 443-453.

26 KATZ, A.I. and EPSTEIN, F.H.: Physiologic role of sodium-potassium-activated adenosinetriphosphatase in the transport of cations across biologic membranes. New Engl. J. Med. 278 (1968) 253.

27 KRIGMAN, M.R. et al: Lead encephalopathy in the developing rat: effect upon myelination. J. Neuropath. Exptl. Neurol. 33 (1974) 58.

28 LIN-FU, J.S.: Undue absorption of lead among children - a new look at an old problem. New Eng. J. Med. 286 (1972) 702.

29 LOWRY, O.H., ROSEBROUGH, N.J.,FARR, A.L. and RANDALL, R.J.: Protein measurement with the Folin phenol-reagent. J. Biol. Chem. 193 (1951) 265.

30 MARCHESI, V.T. and PALADE, G.E.: The localization of Mg-Na-K-activated adenosine triphosphate on red cell ghost membranes. J. Cell Biol. 35 (1967) 385-404.

31 MARSDEN, H.B. and WILSON, V.K.: Lead poisoning in children: correlation of clinical and pathological findings. Brit. Med. J. (i) (1955) 324-326.

32 MICHAELSON, I.A.: Effects of inorganic lead on RNA, DNA and protein content in the developing neonatal rat brain. Toxicol. and Appl. Pharmacol. 26 (1973) 539-548.

33 MICHAELSON, I.A. and SAUERHOFF, M.W.: An improved model of lead-induced brain dysfunction in the suckling rat. Toxic. and Appl. Pharmacol. 28 (1974) 88-96.

34 MORRISON, J.F. and HEYDE, E.: Enzymic phosphoryl group transfer. Annu. Rev. Biochem. 41 (1972) 29-54.

35 NATHANSON, J.A. and BLOOM, F.E.: Lead-induced inhibition of brain adenyl cyclase. Nature 255 (1975) 419.

36 PENTSCHEW, A.: Morphology and morphogenesis of lead encephalopathy. Acta Neuropathol. 5 (1965) 133-160.

37 RAIMONDI, A.J., BECKMAN, F. and EVANS, J.P.: Fine structural changes in human lead encephalopathy. Trans. Am. Neurol. Ass. 91 (1966) 322-323.

38 REPKE, K.R.H. and SCHON, R.: Flip-flop model of (NaK)-ATP-ase function. Acta Biol. Med. Germ. 31 (1973) K 19-K 30.

39 SAUERHOFF, M.W. and MICHAELSON, I.A.: Hyperactivity and brain catecholamines in lead-exposed developing rats. Science 182 (1973) 1022.

40 SCHIBECI, A. and MOUW, D.: unpublished data.

41 SECCHI, G.C. and ALESSIO, L.: Ricerche Sul meccanismo d'inibizione della ($Na^+ + K^+$)-ATPasi eritocitaria ad opera del plombo. Med. Lavaro. 60 (1969) 670-673.

42 SECCHI, G.C., ALESSION, L. and GERVASINI, N.: Ricerche sulla ($Na^+ + K^+$)-ATPasi renale nella intossicazione saturnina sperimentale. Med. Lavaro. 60 (1969) 674-677.

43 SIEGEL, G.J. and ALBERS, R.W.: Sodium-potassium-activated adenosine triphosphatase of Electrophorus electric organ. IV. Modification of response to sodium and potassium by arsenite plus 2,3-dimercaptopropanol. J. Biol. Chem. 242 (1967) 4972.

44 SIEGEL, G.J., KOVAL, G.J. and ALBERS, R.W.: Sodium-potassium-activated adenosine triphosphatase VI. Characterization of the phosphoprotein formed from orthophosphate in the presence of ouabain. J. Biol. Chem. 244 (1969) 3264-3269.

45 SIEGEL, G.J. and FOGT, S.K.: Inhibition of electroplax (Na^+ + K^+)-adenosine-triphosphatase by lead ion. Pharmacol. 16 (1974) 294.

46 SIEGEL, G.J. and FOGT, S.K.: Lead ion activates phosphorylation of electroplax (Na,K)-ATPase in the absence of sodium ion. Arch. Biochem. Biophys. 174 (1976) 744-746.

47 SIEGEL, G.J. and FOGT, S.K.: Effects of lead ion on brain microsomes: inhibition of cation transport ATPase and stimulation of phosphorylation. Trans. Am. Neurol. Ass. in press.

48 SIEGEL, G.J. and GOODWIN, B.B.: Sodium-potassium-activated adenosine-triphosphatase. Potassium-regulation of enzyme phosphorylation. Sodium-stimulated, potassium-inhibited uridine triphosphate hydrolysis. J. Biol. Chem. 247 (1972) 3630-3637.

49 SKOU, J.C.: The (Na^++ K^+)-activated enzyme system and its relationship to transport of sodium and potassium. Quart. Rev. Biophys. 7 (1975) 401.

50 STEIN, W.D. et al: A model for active transport of sodium and potassium ions as mediated by a tetrameric enzyme. Proc. Nat. Acad. Sci. USA 70(1) (1973) 275-278.

51 THOMAS, J.A. and THOMAS, I.M.: The pathogenesis of lead encephalopathy. Indian J. Med. Res. 62 (1974) 36-41.

52 TICE, L.W.: Lead-adenosine triphosphate complexes in adenosine triphosphatase histochemistry. J. Histochem. Cytochem. 17 (1969) 85-94.

53 VALLEE, B.L. and COLEMAN, J.E.: Metal coordination and enzyme action, vol 12, pp. 165-235, Comprehensive Biochemistry (Florkin, M. and Stotz, E.H. Eds.). Elsevier. Amsterdam (1964).

54 VALLEE, B.L. and ULMER, D.D.: Biochemical effects of mercury, cadmium and lead. Ann. Rev. Biochem. 41 (1974) 91-128.

DISCUSSION

GATZY: If you are looking for a brain region that sequesters lead you might check the choroid plexus. There is recent evidence that shows short-term accumulation of pulse injections to a concentration many times greater than those found in other brain regions (O'Tuama, L. A., Kim, C. S., Gatzy, J. T., Krigman, M. R. and Mushak, P., Tox. Appl. Pharmacol. 36: 1-9, 1976).

SIEGEL: When speaking about sequestration I was really referring to the cellular level, i.e., sequestration within the cell.

GATZY: That is true, but whole brain analyses probably provide even less information on localization in cells than regional analyses.

PRESSMAN: I would like to raise a question concerning the extension of *in vitro* studies of toxicity of lead in *in vivo* conditions. Lead has such a tremendously high affinity for both sulfhydryls and phosphates. You would expect when it is ingested, rather than equilibrating with the organism, it would get stuck on those affinity sites that it sees first, possibly in the GI tract. So, it may never get to the sites on the isolated enzymes that you find are sensitive to it.

SIEGEL: Of course, that is possible. I don't really wish to make any conclusion regarding *in vivo* effects of the lead, except to say that if it gets there, then these are the kinetics that will permit assessing whether lead ion possibly produces inhibition. However, it is true that lead, when ingested, does get into the brain, kidney, and many other tissues. So, all of it is not trapped by the intestine and/or by the bloodstream.

PRESSMAN: But its high affinity indicates that there are geometric factors, that is, that those groups the lead is exposed to first are strongly determinative of its ultimate effects, as well as its equilibrium affinities for the various organs and the sensitivity of various organs to the direct toxic effects of lead.

BERG: I want to make sure that I understood the concentrations in your system. Had you much more ATP than lead?

SIEGEL: Correct. The concentrations were usually 3mM ATP compared to around 10^{-5}M Pb Cl_2.

BERG: So, these remarkably low concentrations are the actual total lead?

SIEGEL: Yes.

BERG: On your first slide, I question the reaction where you have ATP as the substrate and magnesium as a co-factor. I suggest that the substrate is magnesium ATP chelate, and you are looking at a competition between magnesium-ATP chelate and lead-ATP chelate for an enzyme site. Not an effect of lead, but an effect of lead-ATP chelate.

SIEGEL: I think that Pb ATP^{2-} or Mg ATP^{2-} is the substrate for phosphorylation. Lead ion, however, inhibits the hydrolysis.

BERG: If this is the case, then magnesium ATP plus sodium ion causes phosphorylation and a conformational change; lead ATP, in the absence of sodium, causes phosphorylation without a conformational change. Without conformational change there would be no potassium effect.

SIEGEL: With regard to the lack of K^+ effect in the absence of Na^+ and presence of Pb^{2+}, I believe that something of that sort is occurring.

BERG: An important factor to consider is the ratio at the target cell of magnesium ATP chelate and lead ATP chelate. Can that be measured?

SIEGEL: In the cell?

BERG: Can you get some sort of *in vitro* and *in vivo* comparison?

SIEGEL: I think that that is very difficult and represents one of the problems here in trying to draw a good comparison to an *in vivo* action from an *in vitro* one. My suspicion is that varieties of substances in the cell will bind lead; but there is great difficulty in doing ionic activity measurements from intact cells. Do you have a suggestion as to how to approach this problem?

KNAUF: Have you compared the phosphopeptides by paper electrophoresis after partial proteolysis (with pepsin or pronase) to see if the site of phosphorylation is the same in the presence of lead as it is in the presence of sodium?

SIEGEL: We are working on that now.

LIPID MODEL MEMBRANE STUDIES ON IMMUNE CYTOTOXIC MECHANISMS

Robert Blumenthal*, John N. Weinstein* & Pierre Henkart**

*Laboratory of Theoretical Biology and **Immunology Branch

National Institutes of Health, National Cancer Institute

Bethesda, Maryland 20014

ABSTRACT

The immune lysis of cells is thought to involve an initial breakdown of the membrane permeability barrier to small ions, with eventual colloid osmotic swelling and disruption of the cell. Lipid model membranes have been used by several workers to study complement-mediated cytotoxicity. We have introduced a system based on the planar lipid bilayer in which mechanisms of lymphocyte-mediated cytotoxicity can be investigated. We have shown that human lymphocytes induce membrane conductance increases (i.e., ion permeability increases) of several orders of magnitude in bilayers containing a hapten (dinitrophenyl) if specific antibody (IgG anti-trinitrophenyl) is added. The conductance increase occurs only when the membrane voltage is positive on the lymphocyte side, as would be the case with a target cell membrane. A variety of controls (in which one or another component of the system is altered or omitted) are all negative, indicating that this effect has essentially the same immunospecificity as lymphocyte-mediated killing of antibody-coated target cells. The results suggest that killer lymphocytes can act on membranes directly to cause permeability increases without participation of other target cell components. Moreover, target cell membrane protein appears unnecessary for the effect. Measurements of membrane potential in the presence of salt concentration differences suggest that the conductance-inducing material has a mild (3:1) selectivity for anions over cations.

Introduction

The lysis of foreign cells by the immune system can be accomplished by two classes of mechanisms: those that are humoral, involving substances found in the serum, and those that are cell-mediated, involving the interaction of immunocompetent lymphocytes with specific sites on the target cell membrane. In the humoral system a series of nine serum protein components known collectively as complement is triggered by the interaction of specific antibodies with antigens in the target cell membrane (1). With attachment to the antigenic site on the membrane the antibody molecule undergoes a change which promotes its interaction with C_1, the first of the nine components of complement.

The most intensively studied type of cell-mediated lytic mechanism is that involving T-lymphocytes from animals previously immunized with foreign cells. We will concentrate, however, on antibody-dependent cell-mediated cytotoxicity, in which the Fab portions of an antibody molecule bind to a target cell surface antigen, and the Fc portion is then recognized by the Fc receptor on the lymphocyte surface (2). Interaction of the antigen-antibody complex with C_1 in the humoral case and with the Fc receptor in antibody-dependent lymphocyte-mediated cytoxicity triggers a cascade of events ending in lysis of target cells. In each case immunolysis is thought to be mediated by an initial breakdown of the permeability barrier to small ions, leading to subsequent cell swelling and colloid osmotic lysis (3,4).

Thermodynamic Analysis

The events leading to colloid osmotic lysis can be formulated in terms of irreversible thermodynamics, which gives an expression for volume flow (dV/dt) out of a cell as a function of osmotic gradients (5)

$$dV/dt = L_p(\sum \Delta\pi_{macromol.} - \sum \sigma\Delta\pi_{small}) \quad (1)$$

In equation (1) $\sum \Delta\pi_{macromol.}$ is the colloid osmotic pressure difference (between cell interior and the external medium) of all the macromolecules and $\Delta\pi_{small}$ is the osmotic pressure difference due to a small ion or nonelectrolyte. The summations are taken over all chemical species of a given type. L_p is the hydraulic coefficient and σ the reflection coefficient (6), which is the ratio of the osmotic flow caused by a gradient of a test molecule to the flow caused by the same gradient of a molecule known to be impermeant. This ratio is 1 for an impermeant molecule and decreases progressively for increasingly permeant molecules until $\sigma=0$ is reached for a solute as permeant as water (7). Cells generally respond as osmometers to changes in extracellular salt concentration (3), so, in the normal state, the osmotic pressure differences in eq (1) are balanced ($\sum \Delta\pi_{macromol} = \sum \sigma \Delta\pi_{small}$) and, there is no volume flow (dV/dt=0). If the permeability of the target cell membrane to small molecules is increased by interaction with complement or lymphocytes the reflection coefficients decrease towards zero and the cell swells (perhaps to lysis) according to eq (2):

$$dV/dt \rightarrow L_p \sum \Delta\pi_{macromol.} \tag{2}$$

With even larger disruptions of the membrane, macromolecules leak out as well.

Bilayer Immunology

Lipid model membranes of two radically different types have been used to characterize permeability changes induced by various "ionophorous" antibiotics on target cell membranes: The liposome (8) and the planar bilayer lipid membrane (BLM) (9). The former are generally either in the form of 200 - 500 A diameter single-lamellar vesicles or in the form of large multilamellar "onion-skins". The usual assay technique for liposome permeability depends on release into the medium of a marker trapped inside the liposome. In contrast, permeability of the planar BLM (to small ions) is usually determined from electrical measurements. Liposomes are advantageous in that the large aggregate surface area allows the permeation rates of a large number of both ionic and non-ionic species to be determined (generally by radioactive or enzymatic methods). Planar BLM's on the other hand are advantageous in that electrical measurements are extremely sensitive (detecting the opening of single conductive channels (10)) and in that concentrations on both sides of the membrane, as well as the membrane potential, can easily be controlled.

Kinsky (11) and his associates have used liposomes in an extensive series of experiments on complement-mediated lysis. They have found that complement will release trapped glucose from antigenic liposomes if antibody is present. Their results suggest that non-membrane target cell components are not necessarily involved in lysis, though caution must, as always, be exercised in imputing characteristics of model systems to biological processes. Other experiments also show that complement does not involve any detectable enzymatic reaction (i.e., one that breaks down target cell components). Although liposomes can be used to assay the lipid, antigen and antibody specificity in complement lysis, planar bilayers have some advantages, especially in studies of the nature of the lesion incurred.

Use of the planar BLM as a tool in immunology has not approached Kinsky's success story with liposomes. Del Castillo and his col-

Table I: Planar BLM studies of antibody and complement action

Lipid(antigen)	Antibody(complement)	Effect	Reference
mixed brain lipids (BSA)	antiserum to BSA	transient conductance increase	(12)
Sphingomyelin α-tocopherol (insulin) (lysozyme) (ribonuclease)	antiserum to insulin, lysozyme, ribonuclease (GPC)	10-1200x conductance increase	(13)
Sphingomyelin α-tocopherol cholesterol (BSA serum)	anti-BSA Serum (GPC)	10-1000x conductance increase, rupture	(14)
oxidized cholesterol (DNP - PE)	purified anti-TNP IgG (GPC)	rupture	(Blumenthal & Henkart, unpublished observations)
lecithin	(purified C5b-9 components)	100-1000x stable conductance increase	(16)

laborators (12) reported ten years ago that large transient conductance changes occur when bovine serum albumin (BSA) and anti-BSA antiserum are added to the bathing solution on one side of a BLM. Those experiments were followed up by Barfort et al. (13), using lysozyme, insulin and ribonuclease as antigens, and rabbit antibodies against them. For reasons still not known, they observed conductance changes only when the antigen and antibody were added to opposite sides of the BLM. Heat-treated antiserum did not produce conductance changes, but addition of freshly-thawed quinea pig complement (GPC) to the heat-treated antiserum initiated a conductance change similar to the one noted with fresh antiserum. (It is not clear whether Del Castillo's antiserum contained complement. Complement is initially in antisera but its activity is very dependent upon the way the serum is handled. To prevent inactivation complement is normally stored at - 80°C and only used immediately after thawing.) Wobschall and McKeon (14) as well as Blumenthal and Henkart (unpublished observations) found that the presence of antigen, antibody and complement at concentrations used by Barfort et al. led to rapid rupture of BLM's. Wobschall and

McKeon however carried out experiments at pre-rupture concentrations of complement in an attempt to find discrete changes in conductance indicative of channel-like ionophorous behavior. They concluded that complement channels form 2.2 nm diameter holes in the target membrane.

An important advantage of the planar BLM is its exquisite sensitivity to mediators of ion conductance, but the blessing is not unmixed. In many cases the BLM will change its conductance when soluble proteins are added to the bathing solution at higher concentrations (due to some ill-defined detergent-like mechanism). Such non-specific conductance changes presumably do not correspond to the physiological role of those proteins, which are not designed to be mediators of ion conductance (15). Because of this "detergency" effect purified complement components are required for the characterization of complement-induced conductance pathways. Michaels et al. (16), have recently shown that purified components C5b-9, added in sequence, can give rise to stable and reliable conductance changes in BLM's. Attempts to study immune mechanisms using BLM's are summarized in Table I.

Antibody-Dependent Cell-Mediated Cytotoxicity

We now turn to the second major class of immune lytic mechanisms, those mediated by lymphocytes. To test the hypothesis that lymphocytes bearing Fc receptors interact with antigen-antibody complexes on target cells to break down their permeability barriers, we constructed a planar BLM model of the interaction (17). We decided to use the planar BLM rather than liposomes for a variety of reasons, among them the notion that released marker might be reabsorbed by the lymphocytes. (We have actually shown that liposomes containing a fluorescent dye are taken up by lymphocytes and that the contents of the liposomes are released into the cytoplasm of the lymphocyte (18)). In our experiments we used the dinitrophenyl (DNP) and anti-trinitrophenyl (anti-TNP) antibody system developed in Kinsky's laboratory (19) for complement-mediated lysis in liposomes. The lipid bilayers were formed from a solution of oxidized cholesterol and DNP - phosphatidyl ethanolamine (PE) or DNP - caproyl (Cap) PE in octane or decane painted across a 1 mm diameter hole in a thin-bottomed teflon cup (inside) immersed in a Lucite container (outside). All experiments were carried out in Hank's balanced salt solution (B.S.S.) without serum protein, that is, without extraneous materials which could cause nonspecific detergent-like effects such as plagued some of the early attempts at BLM immunology. Affinity purified anti-TNP antibody was added to a final concentration of approximately 1 μg/ml and was allowed a 20 minute period of incubation to form membrane antigen-antibody complexes on the bilayer. Human peripheral blood lymphocytes (PBL)

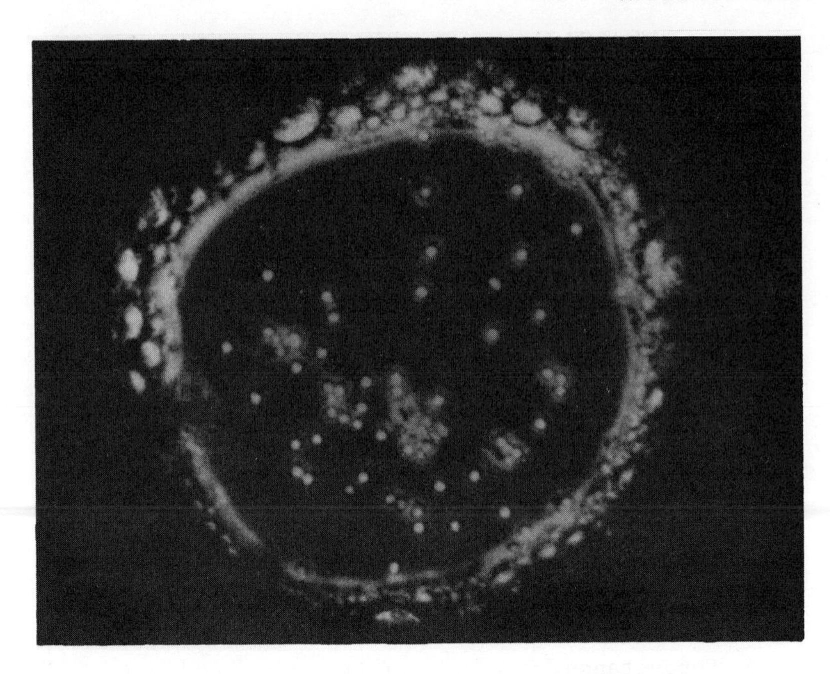

Figure 1. Lymphocytes resting on an antibody-coated lipid bilayer membrane. (17)

were then injected in a 5 μl volume (of a suspension of 10^7 cells/ml) into the solution in the cup, directly toward the bilayer.

A portion of the lymphocytes settled on the bilayer within about five minutes, as monitored with a phase microscope (see Figure 1). The torus is a thick layer of lipid and bubbles which surrounds the bilayer; it appears dark under phase optics. Approximately 50 lymphocytes can be seen on the bilayer, which has an area of approximately 1 mm^2. The lipid bilayer could support dozens of lymphocytes for periods of more than an hour before membrane-breakage occurred. The conductance was measured at all stages of incubation with antibody and with lymphocytes. As shown in Figure 2 the "antigenic" oxidized cholesterol$_2$- DNP-PE bilayer had an initial conductance of 10^{-7} - 10^{-8} mho/cm^2. Exposure to antibody had no effect on the bilayer (in contrast to Del Castillo's experiments (12)). Figures A and B represent typical experiments in which we added anti-TNP to the medium over the lipid bilayer shortly after formation of the BLM. Twenty minutes later the lymphocytes (1c) were injected into the medium over the bilayer, with numbers as indicated in Figure 2 settling on the membrane. Figure 2C is a control experiment with no antibody (the cells were added shortly after BLM formation). In each experiment the

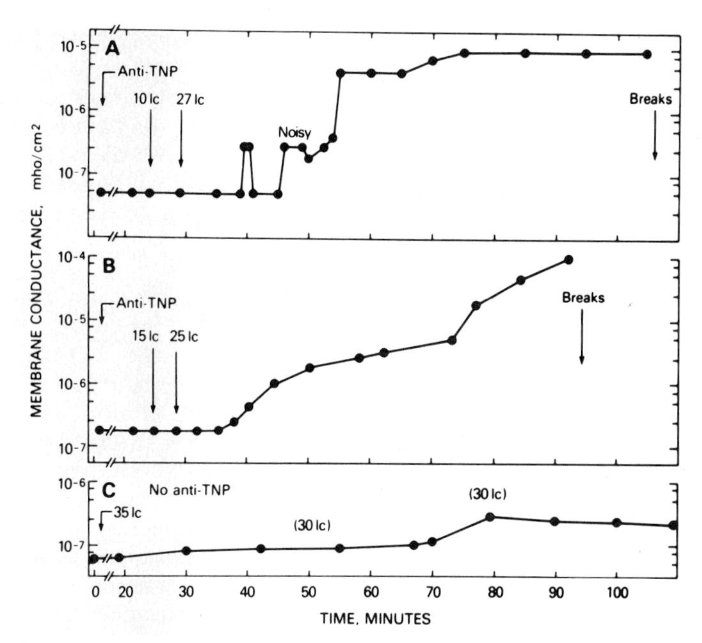

Figure 2: Conductance changes caused by lymphocytes acting on antibody-coated lipid bilayers. (17)

membrane voltage was held constant at +20 to +60 mV positive (lymphocyte side) except for a brief period in the experiment of Figure 2A between 41 and 45 minutes during which the membrane potential was held at - 60 mV. The conductance increase occurred only when the potential on the lymphocyte side was positive; when a negative membrane potential was imposed at the start of the experiment, the conductance did not increase in spite of the presence of antibody and many lymphocytes. In Figure 2A an initial conductance increase occurred while the BLM was held at +60 mV (at 38 minutes). At 41 minutes the voltage was reversed to -60 mV, and the conductance returned to its baseline value. When the membrane voltage was returned to +60 mV at 45 min., the conductance again increased. At higher conductance levels negative voltage pulses only reversed part of the conductance increase induced by a positive voltage. The polarity of the voltage required for a conductance increase is the same as that pertaining in the physiological situation, since cells have negative inside (i.e., positive outside) membrane potentials.

With positive imposed potentials a marked conductance change was generally observed beginning 2 - 20 minutes after cells arrived on the membrane. The rate and extent of conductance change varied from experiment to experiment. In experiments such as that shown in Figure 2A the initial change was a conductance increase of 2 - 10x accompanied by an increase in current noise level. After a period of 1 - 10 minutes a more substantial and often rapid conductance increase of between one and three orders of magnitude developed. Figure 2B shows a less frequently observed pattern of increase, a slow, steady rise. We observed conductance increases with as few as five cells and commonly with fewer than 20 cells. In Table II the BLM responses are compared with antibody-dependent lymphocyte-mediated radioactive chromium (^{51}Cr) release by tinitrophenylated (TNP) red blood cells (RBC).

Table II shows that the conductance change in the BLM follows the cellular and antibody requirements of the lymphocyte-mediated lysis of target cells: lymphocytes resting on the BLM in the absence of antibody do not induce significant conductance changes, as shown in Figure 2C. Antibody-dependence is also a characteristic of lymphocyte-mediated cytotoxicity as is shown in Table II. Anti - TNP $F(ab')_2$ antibodies did not sensitize the DNP bilayer to conductance increases although the antigen-binding ability of the antibodies was intact. Antibody-coated BLM's in the absence of lymphocytes have stable low conductance. In each experiment the BLM was incubated with antibody alone for 20 minutes; conductance increases greater than 5x were never seen. In two experiments the BLM's were incubated with antibody alone for one hour. A lymphocyte cell population depleted of cells bearing Fc receptors was inactive both in killing antibody-coated target cells and in increasing the conductance of antibody-coated BLM's.

In order to test the notion that the lymphocytes could produce a conductance change by crosslinking antigen-antibody complexes on the BLM, we prepared an affinity purified IgG fraction of goat antibody to rabbit anti-TNP IgG, which presumably crosslinks the antigen-antibody complexes on the BLM. We allowed the rabbit anti-TNP to incubate with the antigenic BLM for 20 minutes, washed off the excess rabbit IgG and added the goat antibody at a final con-

Table II. Antibody-dependent lymphocyte-mediated cytotoxicity

Antibody	Lymphocytes	$\%^{51}Cr$ release by TNP-RBC (no. of Exps)	BLM conductance increase (no. of Exps)	
Rabbit anti-TNP	Human PBL	100	250x 20,000x	(22)
None	Human PBL	0-2(4)	0 - 16x	(6)
Anti-TNP-$F(ab')_2$	Human PBL	4-5(2)	0	(1)
Anti-TNP	None	1-8(4)	0 - 5x	(2)
Anti-TNP	Fc receptor neg.	3-15(4)	0	(1)
Anti-TNP	None + Goat anti Rabbit IgG	-	0	(3)

centration of 1 μgm/ml. Under those conditions no BLM conductance increases were observed, as indicated in Table II.

When BLM's were made from membrane-forming solutions of low molar ratio (less than 1/100) of antigen to oxidized cholesterol no responses were observed, indicating that antigen had to be present in sufficient concentration if a conductance change was to occur. Pure oxidized cholesterol BLM's tended to rupture more frequently in the presence of antibody and lymphocytes than did BLM's containing antigen. We observed a similar effect with antibody and complement. Possible crosslinking of antigen-antibody complexes on the BLM could render the BLM more resistent to nonspecific detergency effects.

Using fluorescent antibody David Wolff (unpublished observations) has observed that antibody indeed resides on the bilayer, but he has also noted that some antibody adsorbs to the BLM nonspecifically in the absence of antigen. This adsorption is not sufficient to cause a response, consistent with the hypothesis that triggering of the cytotoxic response requires specific binding of the antibodies to antigen in order to induce a change in the Fc portion of the antibody.

Nature of the Conductance-inducing Material

As in the case of damage by complement (11), the BLM experiments on lymphocyte-mediated cytotoxicity indicate that immunospecific permeability changes can occur without the mediation of target cell cytoplasm or membrane proteins. The experiments do not distinguish yet whether the conductance-inducing material is released from the lymphocyte and subsequently inserted into the bilayer or whether the conductance change is induced while the material is still associated with the lymphocyte. In some cases we have observed that conductance, once increased, reverted spontaneously to baseline. In those cases we observed that lymphocytes had moved away from the center of the bilayer onto the torus. The observation that conductance decrease correlated with dissappearance of the lymphocyte from the black (thin) area, suggests that the cytotoxic material is still attached to the lymphocyte surface. The spontaneous reversals occurred both with a positive potential maintained on the lymphocyte side and, more frequently, when the potential was zero or negative. In a number of cases we re-formed a BLM which had ruptured after producing a conductance increase in the presence of antigen, antibody and lymphocytes. The conductance of the re-formed BLM did not increase over its baseline level, indicating either that the increase in conductance is not mediated by release of material or that such material cannot easily be re-incorporated into the BLM.

In order to study cation versus anion selectivity of the lymphocyte-mediated conductance pathway we created ion concentration differences between the outer and inner compartments by adding small aliquots of concentrated salt solution to the outer compartment. The voltage source was then switched to open circuit, and membrane potentials (E), defined as the potential of the outer compartment with respect to the inner compartment, were measured with silver-silver chloride electrodes. The results of one experiment are shown in Figure 3.

Small aliquots of a solution of 8M $NaNO_3$ in B.S.S. were added ($NaNO_3$, rather than NaCl, because the silver-silver chloride electrodes respond to changes in chloride concentration as well as to membrane potential. NO_3^- and CL^- have essentially identical free-solution mobilities.) In the case of lymphocyte-mediated conductance the diffusion potential changed by 29 mV/decade as the outside-to-inside concentration ratio of sodium ($[Na_o]/[Na_i]$) was altered (circles in Figure 3). From this observation we calculated a selectivity of 3.0/1 for anions over cations, assuming chloride and nitrate ion to be equivalent in terms of their ion selectivity. A selectivity ratio of 3.3/1 is obtained when we fit the data in Figure 3 to the Goldman (20) equation. The dashed line in Figure 3 indicates the potentials expected if the membrane were completely anion-selective. The line of "no selectivity" in Figure 3 was calculated from the mobility ratio of nitrate and sodium in water. It represents the result to be expected if the conductance pathways were very large water-filled channels. DNP-PE-oxidized cholesterol bilayers themselves did not show selectivity between anions and cations (triangles in Figure 3).

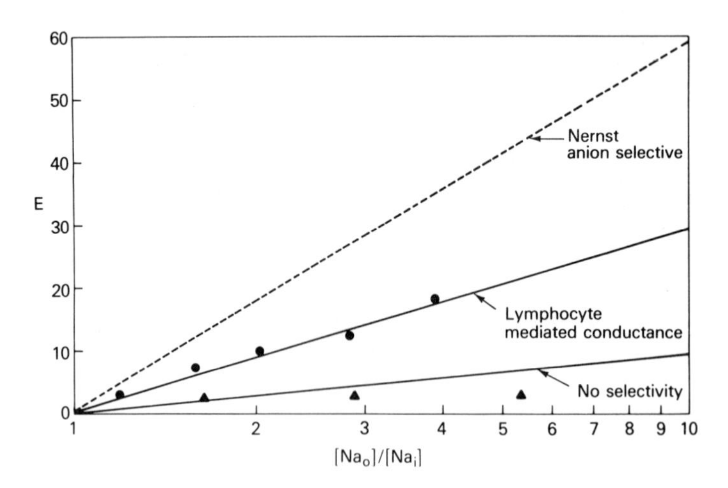

Figure 3. Diffusion Potential (E, in mV) as a function of sodium concentration ratio.

The moderate anion selectivity of the lymphocyte-mediated conductance change suggests that the ionophorous material has some positive charge. This observation fits with the data on voltage-dependence, which could result from the necessity of driving a positive charge into the membrane to produce a conductance change. The non-selectivity of the unmodified DNP-PE oxidized cholesterol bilayer indicates that the positive charge does not come from the lipid. We have also found that using the larger monovalent cation tetraethylammonium do not increase the anion versus cation selectivity. This observation indicates that the lymphocyte produces large holes in the bilayer (with some positive charge).

Perspective

Lipid model membranes have been applied with considerable success in studying the mechanisms of antibiotic toxicity (21). Such models have not so easily been adapted to studies of immune toxicity, in large part because of non-specific effects (detergency and adsorption) such as those often encountered in reconstitution of transport systems (15). We expect, however, that the problems will gradually be overcome with increasing availability of purified molecular components of each type of cytotoxic interaction. The several experimental systems discussed in this paper represent the early stages of that process.

But why produce artificial versions of an immune cytotoxic event when the real thing is available in such abundance? Three general aims can be distinguished: (i) to ascertain whether intracellular components of the target cell need be involved in the effect, (ii) to determine which (if any) components of the target cell membrane are involved, and (iii) to isolate components of the phenomenon in a stripped-down system amenable to controlled study. The model studies of complement- and lymphocyte-mediated cytoxicity discussed here suggest, in each case, that the mechanism need only involve the membrane of the target cell, and only its lipid constituents, at that. But, in a sense, that conclusion is implicit in the fact that this paper was written in the first place; if it were not possible to find lipid model systems suggesting such answers, our results, and those of the other workers mentioned, would have been uninterestingly negative and hardly worth the mention. Even though the specific requirements in each model system correspond in striking ways to those of its physiological counterpart, one must always be wary of assuming that the two systems actually do work in the same way. That is the ever present caveat in work with biological analogues.

References

1 Müller-Eberhard, H.J.: Harvey Lectures 66 (1972) 75.

2 Cerottini, J.C. and Brunner, K.T.: Adv. Immunol. 18 (1974) 67.

3 Robinson, J.R.: Pathobiology of Cell Membranes, Vol. I, Ch. 4, (Trump, B.F. and Arstila, A.U. eds.) Academic, New York (1975).

4 Seeman, P.: Fed. Proc. 33 (1974) 2116.

5 Kedem, O. and Katchalsky, A.: Biochim. Biophys. Acta 27 (1958) 229.

6 Staverman, A.S.: Rec. Trav. Chim. 70 (1951) 344.

7 Diamond, J. and Wright, E.M.: Annu. Rev. Physiol. 31 (1969) 581.

8 Bangham, A.D., Hill, M.W. and Miller, N.G.A.: Methods in Membrane Biology, Vol. I, Ch. 1, (Korn, E.D., ed.) Plenum, New York (1974).

9 Mueller, P., Rudin, D.O., Tien, H.T. and Westcott, W.C.: 1 (1964) 379.

10 Bean, R.C., Shepherd, W.C., Chan, H. and Eichner, J.: J. Gen. Physiol. 53 (1969) 741.

11 Kinsky, S.C.: Biochim. Biophys. Acta. 265 (1972) 1.

12 Del Castillo, J., Rodriquez, A., Romero, C.A., and Sanchez, V.: Science 153 (1966) 185.

13 Barfort, P., Arquilla, E.R. and Vogelhut, P.O.: Science 160 (1968) 119.

14 Wobschall, E. and McKeon, C.: Biochim. Biophys. Acta. 413 (1975) 317.

15 Blumenthal, R. and Shamoo, A.E.: Ann. N.Y. Acad. Sci. 264 (1975) 483.

16 Michaels, D.W., Abramovitz, A.S., Hammer, C.H., and Mayer, M.M.: Proc. Nat. Acad. Sci. USA (1976) In press.

17 Henkart, P. and Blumenthal, R.: Proc. Nat. Acad. Sci. USA 72 (1975) 2789.

18 Weinstein, J.N., Yoshikami, S., Henkart, P., Blumenthal, R. and Hagins, W.A.: Biophys. J. 16 (1976) 104a.

19 Uemura, K., and Kinsky, S.C.: Biochemistry 11 (1972) 4085.

20 Goldman, D.E.: J. Gen. Physiol. 27 (1943) 37.

21 McLaughlin, S. and Eisenberg, M.: Annu. Rev. Biophys. & Bioeng. 4 (1975) 335.

DISCUSSION

SZABO: What is the nature of the conductance of the lymphocyte? What happens when you take it away?

BLUMENTHAL: In a few experiments we squirted the lymphocyte away from the black lipid area and observed a decrease in conductance.

ELDEFRAWI: Did you mention that one of the possible effects probably for complement would be a protease-like effect?

BLUMENTHAL: Not the complement, because Kinsky has very nicely demonstrated that one can obtain complement mediated lysis on pure lipid membranes(S.C. Kinsky, Biochim. Biophys. Acta 265, 1972). My speculation came more from Dr. Allison's work (this volume), that the lymphocyte mediated change might be mediated by some protease.

ELDEFRAWI: Well, this is essentially the point I wanted to refer to, that if you have especially a complement effect, in terms of an antibody-antigen complement-fixing reaction, the major catalytic activity as changed with complement is usually a phosphylipase-like effect.

BLUMENTHAL: I think Kinsky has very nicely shown that there is no phospholipids involved in complement. He has started with pure phospholipid membranes and analyzed the lipids after complement mediated lysis to see if he finds any breakdown products of the lipids. He has done an enormous number of controls to study possible phospholipase activity in the complement system which has been claimed by a number of workers (Inoue, K. and S. C. Kinsky, Biochemistry 9: 47-67, 1970).

DIFFUSIONAL TRANSPORT OF TOXIC MATERIALS IN MEMBRANES STUDIED BY FLUORESCENCE SPECTROSCOPY

Joseph R. Lakowicz, Delman Hogan

Freshwater Biological Institute
University of Minnesota
P. O. Box 100
Navarre, MN 55392

ABSTRACT

As a nation we are currently concerned with the effects of a multitude of synthetic chemicals on life processes. The diversity of opinions on these issues reflects, in part, a lack of understanding of the molecular aspects of toxicity and bioaccumulation. Integral to these concerns is the effect of toxic materials on cell membranes and the permeability barriers which these membranes impose to xenobiotics.

Fluorescence spectroscopy provides a powerful tool for investigating many aspects of membrane sensitivity to toxic materials. Chlorinated hydrocarbons, olefins, and amines act as diffusional quenchers of fluorescence. Measurement of the fluorescence lifetimes of probes embedded in biological membranes can reveal the probe-quencher collisional frequency, and hence the xenobiotic's diffusion coefficient in the membrane. Such information, coupled with the rates of exchange of foreign materials between serum proteins and membranes, may possibly allow predictions of the bioaccumulation potential of toxic materials.

Fluorescence quenching studies can also be used to determine the xenobiotic's membrane-water partition coefficient. A range of values from 10 to 10^8 appear to be experimentally accessible. Localization of foreign materials in either the glycerol or acyl side chain region of a membrane may be revealed by investigations using localized fluorescent probes. In favorable circumstances it appears likely that one can measure both the xenobiotic's lateral diffusion rate across the membrane's surface and the transport rate through the bilayer.

INTRODUCTION AND RATIONALE

Fluorescence is the emission of photons from the excited singlet state of aromatic molecules. Typically 10^{-8} seconds, or 10 nanoseconds, is the time between absorption and emission of the photon. Small molecules may diffuse over considerable distances in 10 nsec. For example, using the diffusion coefficient of oxygen in aqueous solution ($2.5 \times 10^{-5} cm^2/sec$) and the Einstein equation

$$\Delta x^2 = 2Dt,$$

one finds an oxygen molecule can diffuse 70A°, or the thickness of a biological membrane in 10 nsec. Oxygen acts as a collisional quencher of fluorescence. If the excited fluorophore collides with an oxygen molecule the energy is dissipated without the emission of a photon. The fluorophore can thus be used to indicate the rate of collision between itself and oxygen, and hence yield the oxygen diffusion coefficient.

A large variety of toxic compounds which probably interact with membranes are also collisional quenchers of fluorescence. These include chlorinated hydrocarbons such as $CHCl_3$, DDT and mirex; amines, such as triethylamine and procaine; and olefins such as 1,2-dimethylcyclohexene. We shall describe techniques whereby one may use the quenching of fluorescent probes embedded in membranes to determine both the diffusion coefficient of the foreign molecule in the membrane and its lipid-water partition coefficient. Additionally, we will describe methods by which one might distinguish lateral diffusion of a toxic molecule on one side of a bilayer from transport across the bilayer. These techniques promise to yield much fundamental information about the dynamics of foreign molecules in membranes and their interactions with membrane components.

Passive Diffusional Transport in Membranes

The transport properties of biological membranes are of great interest since it is these properties which maintain a suitable internal environment for the metabolic activities of all living cells. We presently view a membrane as a phospholipid bilayer, of complex chemical composition, in which a variety of proteins are embedded. Concentration gradients are maintained by energy requiring active transport systems, such as the sodium-potassium ATPase or the sodium electrogenic pump. Cell membranes often provide permeability barriers to small essential molecules (e.g., sugars and amino acids). In such cases transport systems have evolved which either simply assist transport across membranes (facilitated or exchange diffusion), or permit concentration of a desired species by coupling its uptake with the transport of a second molecule down its concentration gradient (co-diffusion). An example of exchange diffusion is a carrier in mitochondrial membranes which exchanges internal ATP with external ADP + Pi. Glucose uptake by epithelial cells in the intestine is driven by co-transport of sodium into the cells.

The above types of transport require the synthesis of protein carriers and are thus only used for molecules of great metabolic importance. For numerous small molecules, examplified by oxygen, and low molecular weight alcohols, amides, esters, etc., passive diffusional transport through the lipid or protein portions of the membrane most likely accounts for the observed permeation rates. Additionally, it appears unlikely that transport systems would have evolved for the many man-made drugs and pesticides which do permeate cell membranes and, in some cases, bioaccumulate. Clearly, knowledge of the dynamic and equilibrium behavior of small molecules in membranes is important for understanding the permeability barrier encountered by small molecules for which specific transport systems do not exist.

Bioaccumulation of Toxic Materials. Uptake and transport of small foreign molecules such as pesticides, polychlorinated biphenyls (PCBs) and toxic organometallic compounds such as methyl mercury, is presently of great interest as a result of the need to explain their environmental transport and bioaccumulation in higher organisms. Such transport and accumulation has been observed on many occasions. For example, high levels of PCBs are found in both freshwater and marine organisms (7), and DDT has been detected in birds and fish at concentrations much higher than in their environment or in the food they consume (21).

The process of bioaccumulation involves a number of more fundamental events such as:

1) partitioning of the foreign molecule under consideration between the environment and some surface of the organism,

2) diffusional transport of these molecules across cell membranes,

3) transport mediated by body fluids, such as exchange between blood vessels and serum lipoproteins,

4) concentration of the foreign molecule in various tissues depending upon its affinity for certain biomolecules, such as nerve lipids,

5) biodegradation of the foreign material.

The bioaccumulation process is thus seen to be a result of both kinetic (diffusional transport and biodegradation) and equilibrium (partitioning) processes. The relative importance of these processes is at present undecided. Intuition dictates that a molecule will not bioaccumulate in an organism if its degradation rate is greater than its accumulation rate. Our experience with DDT may be considered a massive experiment in which we eventually concluded that degradation occurred too slowly compared to the transport and partitioning of DDT into the higher levels of the food chain; thus permitting toxic levels to result.

Bioavailability of Drugs. There is no doubt that drugs play a crucial role in maintaining and improving our health. Equally important to the mechanism of drug action is the bioavailability (quantity) at the site of drug action. In spite of much research the search continues for improved ways to predict movements of drugs throughout an organism and across cell membranes. Cultures of mammalian cells are an often used model system (24). A major difficulty in using whole cells is the inability to differentiate among several mechanisms (eight in the previous reference), so that the rate limiting steps are not understood. Although passive diffusion is not the mode of transport for all drugs, it is certainly of importance for many. Studies of diffusional transport rates in membranes can define reasonable limits for passive diffusion, and possibly provide predictive powers for newly synthesized, untested drugs.

Mechanisms of Pesticide and Drug Action. General and local anaesthetics probably function as a result of their interactions with membranes and their effects on membrane function. The symptoms of exposure to chlorinated hydrocarbon pesticides indicate some interference with both peripheral and central nervous system function. It appears likely that these pharmacological activities result from interactions between the foreign compounds and biological membranes, especially the lipids involved in the transmission of nerve impulses. To date, no mechanism of action has been elucidated for either anaesthesia or the pesticidal activity of chlorinated hydrocarbons. The specificity of pesticidal activity leads us to search for interactions between these molecules and membrane components. Stability and low water solubility are not sufficient criteria for explaining the observed insecticidal activities or the bioaccumulation potentials. Figure 1 compares these properties for three similar molecules. DDE bioaccumulates as readily as DDT, but is not an active insecticide (18). The trifluoromethyl derivative of DDT neither bioaccumulates (10) nor shows insecticidal activity. Of the eight possible steroisomers of lindane (1,2,3,4,5,6-hexachlorocyclohexane) the γ isomer (Figure 5) is the most active by factors ranging from 50 to 10,000. As a result of these considerations it appears valuable to develop techniques which can potentially reveal the relevant interactions between foreign molecules and membrane components. By the use of model membrane systems, and modern instrumental techniques, one can hope to elucidate these specific interactions. We shall

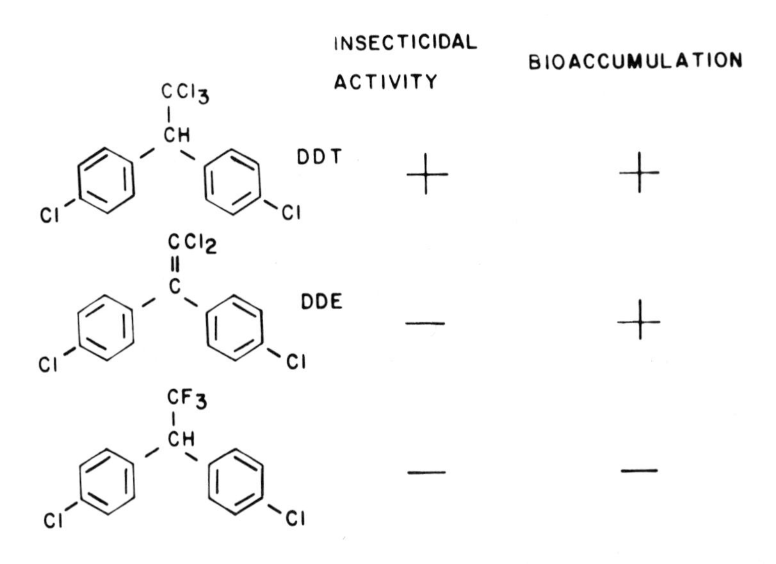

Figure 1. Comparison of the bioaccumulation potential and insecticidal activity of DDT and two analogues.

describe in detail the utilization of the phenomenon of collisional quenching of fluorescence to investigate the interactions of foreign molecules with biological membranes.

Importance of the Synthetic Chemicals for Mankind

In the United States we have banned the use of DDT, chlordane, heptachlor, aldrin, dieldrin, and several similar pesticides as a result of their adverse ecological effects and their action as suspect carcinogens. Thus one might conclude that such compounds have outlived their usefulness and will disappear from the marketplace, making fundamental research into their mode of action an academic exercise. However, I feel such an attitude is the result of a lack in understanding of present health conditions in underdeveloped nations and the great suffering which can result from insect-born diseases. Between 1347-1350 about 25% of the population of Europe (about 25,000,000 persons) died as a result of bubonic plague (30), a disaster which could have been prevented by dusting the populase with DDT since transmission is primarily by flea bites. Between 1917 and 1921 2 to 3 million Russians died from typhus, which is transmitted by body lice (29). During World War II a typhus epidemic broke out in Naples as the Allies landed, but this was brought quickly under control by dusting the populace with DDT. In spite of the fact that practically no toxic effects in humans have resulted from exposure to DDT, and that millions of lives have been saved by its use, widespread public outcry and doomsday predictions would most likely result from any proposed DDT dusting of the populace. As a result of its low cost, effectiveness, and low human toxicity, DDT is still a widely utilized pesticide on a worldwide basis. The pressures of increased worldwide population on available food supplies and resources are likely to continually result in increases in the distribution of synthetic molecules in the environment. The major social benefits to be gained by the use of synthetic molecules will not allow other nations the luxury of years of testing prior to the widespread dispersion of these compounds in the environment. The ecological effects of synthetic molecules dispersed by other nations are likely to be felt on a global basis, hence within our boundaries. This nation presently has the scientific resources and opportunity to identify useful synthetic chemicals which do not pose threats to human health or to other forms of life upon which we are dependent. Towards this end we must develop a fundamental understanding of the mechanisms of toxicity and bioaccumulation so as to allow predictions of ecological side effects based on known mechanisms of action.

DIFFUSIONAL PROCESSES STUDIED BY COLLISIONAL QUENCHING OF FLUORESCENCE

Quenching of Fluorescence by Halogenated Molecules

For our purposes fluorescence is considered to be the emission of previously absorbed photons from aromatic molecules which are in the lowest excited singlet state. An energy rate diagram which describes the total process is shown in Figure 2. Typical structures of fluorescence molecules (fluorophores) used in this study are shown in Figure 3. Absorption of a photon is very rapid ($\sim 10^{-15}$sec) and does not allow any change in the nuclear displacements of either the fluorophore or the solvent (Franck-Condon Principle). However, the electronic charge distribution is changed, typically resulting in an increased dipole moment in the excited state. After absorption the solvent molecules reorientate around the new dipole moment at a rate dependent upon the solvent fluidity. A rate of 10^{12}sec^{-1} is typical for fluid solvents such as water or ethanol at room temperature. This relaxation results in a lowering of the excited state energy level and is responsible for the longer wavelength of the fluorescence relative to the absorbed light (Stokes shift). After solvent relaxation the fluorophore may either emit a photon with a rate k_f, or return to the ground state by a variety of deactivation mechanisms, Σk_i. The quantum yield of fluorescence is defined as the fraction of absorbed photons which

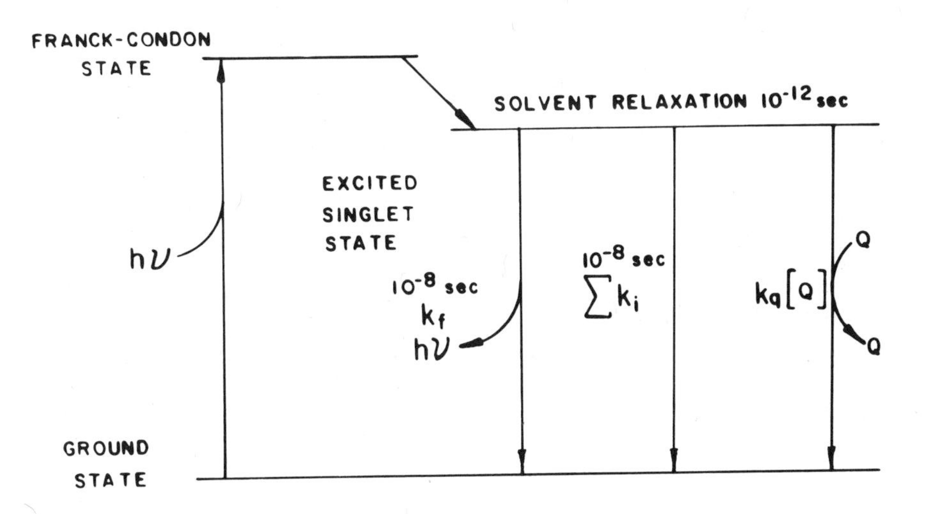

Figure 2. Energy-rate diagram describing the phenomenon of fluorescence.

PYRENE

N-DANSYL ETHANOLAMINE
(DNS-EA)

N-DANSYL PHOSPHATIDYL ETHANOLAMINE (DNS-PE)

3-DIMETHYLAMINO-9-ETHYL
CARBAZOLE
(3-DMA-9-EC)

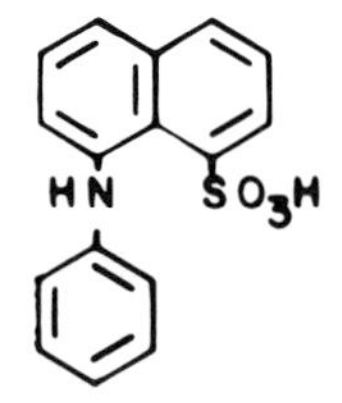

8-ANILINO-1-NAPHTHALENE
SULFONIC ACID
(ANS)

Figure 3. Structures of the fluorescent probes used in this study.

are reemitted. This fraction is determined by a competition between the emissive (k_f) and non-emissive (Σk_i) processes.

$$\text{Quantum yield} = F_o = \frac{k_f}{k_f + \Sigma k_i} \quad (1)$$

The fluorescence lifetime is defined as the time required for the excited population to decrease to 1/e of its original value after illumination has ceased, and is given by

$$\tau_o = \frac{1}{k_f + \Sigma k_i} \quad (2)$$

Fluorescence lifetimes are typically in the range of 10^{-8}sec or 10 nanoseconds (nsec), but may vary from 0.1 to 400 nsec depending upon the fluorophore and the solvent in which it is dissolved.

The quenching of fluorescence refers to the decrease in the observed fluorescence intensity, or photon flux emanating from an irradiated fluorophore. This phenomenon has been recognized for many years (22,28), and a variety of mechanisms are possible. From our standpoint of trying to observe diffusional transport in membranes, dynamic or collisional quenching is of greatest importance. Assume a molecular species Q has the property of inducing the excited fluorophore to return to the ground state via a non-radiative route upon molecular contact. Such contact results from the mutual diffusion of these species while in solution. This dynamic or collisional process serves as another route by which the fluorophore may return to the ground state (Figure 2). The quantum yield and fluorescence lifetime under quenching conditions are given by

$$F = \frac{k_f}{k_f + \Sigma k_i + k_Q [Q]} \quad (3)$$

$$\tau = \frac{1}{k_f + \Sigma k_i + k_Q [Q]} \quad . \quad (4)$$

The bimolecular quenching constant, k_Q, is of greatest interest because it contains the diffusion coefficients of the probe (fluorophore) and quencher.

$$k_Q = \gamma \, 4\pi \, a \, N'(D_P + D_Q) \quad . \quad (5)$$

This expression may be understood by reference to Figure 4, which portrays a probe (pyrene) and a chlorinated hydrocarbon (mirex) in a phospholipid bilayer. D_P and D_Q are the diffusion coefficients of the probe and quencher respectively, N' is avogadros' number divided by 10^3 (needed to have correct units in k_Q), a is the sum of

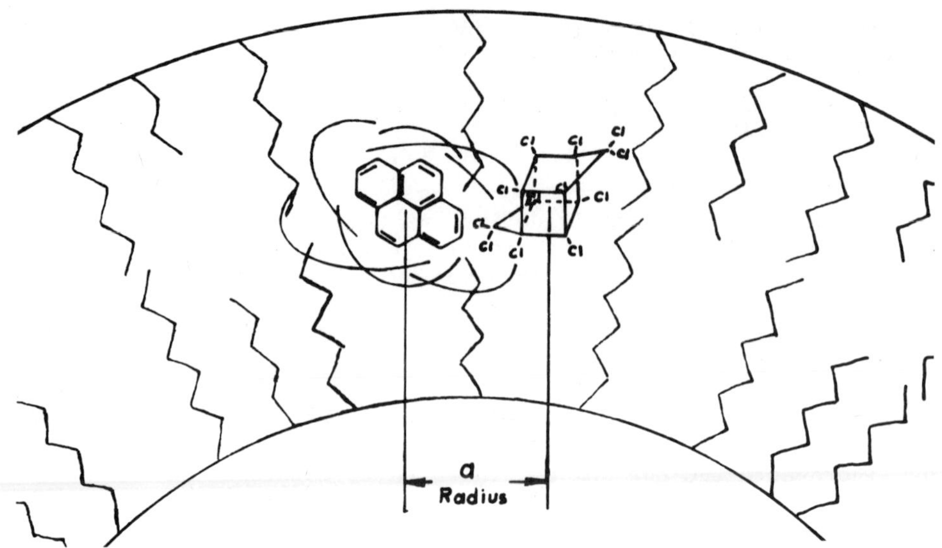

Figure 4. Quenching of the fluorescence probe pyrene, by the pesticide mirex, in a biological membrane.

the molecular radii of the probe and quencher, and γ is the probability that a single collision results in deactivation of the excited state. The diffusion coefficient of a molecule represents the net flux of this molecule through a unit area due to a unit concentration gradient. The collisional rate is thus given by the product of diffusion coefficients and the area of the collision sphere ($4\pi a$). Thus, if one knows the quenching efficiency for a particular probe-quencher pair and the quencher concentration, $D_P + D_Q$ may be calculated from a measured value of k_Q.

Quenching data are generally presented and analyzed according to the Stern-Volmer equation,

$$\frac{F_o}{F} = 1 + k_Q \tau_o [Q] = 1 + K [Q], \quad (6)$$

which provides a convenient method to determine k_Q. F_o and F are the fluorescence quantum yields or intensities in the absence and presence of quencher respectively, and $K = k_Q \tau_o$ is the Stern-Volmer quenching constant. This equation may be easily derived from equations 1, 2 and 3. Measurement of F_o/F as a function of quencher concentration, combined with an independent determination of τ_o, allows k_Q to be calculated. K is the reciprocal of the concentration of Q needed to quench the fluorescence to one-half of its original value. Thus, given equivalent quenching efficiencies, a longer lived probe is sensitive to lower concentrations of quencher.

Quenching may occur by static as well as by dynamic processes. In general, quenching may be divided into these two classes, regardless of mechanism. Consider the formation of a non-fluorescent ground state complex between probe and quencher.

$$P + Q \rightleftharpoons P \cdot Q \tag{7}$$

$$K_c = [P \cdot Q] \;/\; [P][Q] \tag{8}$$

The fractional fluorescence intensity in the presence of Q will be equal to the fraction of uncomplexed probe,

$$\frac{F}{F_o} = \frac{[P]}{[P] + [P \cdot Q]} \quad \frac{1}{1 + K_c[Q]} \;. \tag{9}$$

Thus we see that complex formation results in an identical dependence on quencher concentration as is found for dynamic quenching. Static and dynamic quenching are best distinguished by measurements of fluorescence lifetimes. Dynamic quenching provides an additional rate process which depopulates the excited state, thus decreasing the lifetime in proportion to the yield. Manipulation of equations 1 to 4 show

$$\tau_o/\tau = F_o/F \tag{10}$$

for collisional quenching. When static quenching occurs the complexed species are non-fluorescent, and hence non-observable. The observed lifetime remains equal to τ_o, that of the uncomplexed fluorophore. Agreement of observed data with equation 10 may be considered proof of dynamic quenching. Alternately, lifetime measurements may be considered to ignore complex formation and to report solely on collisional processes.

A wide variety of molecules and atoms act as collisional quenchers. If the fluorescent molecule is located within a biological structure, such as a membrane, the quenching phenomenon may be used to investigate the passive diffusion of quenchers in the membrane. Known diffusional quenchers of fluorescence include the paramagnetic gases such as oxygen (9) and nitric oxide (1), halogens and halogenated compounds such as Br^-, I^- (19), bromobenzene (17), and CBr_4 (26), SO_2 (3), xenon (6) and nitroxides (2). Quenching by amines such as triethylamine (4,15) or N,N-dimethylaniline (16) is dependent upon their ionization potential. Recently, quenching of α-cyanonaphthylene was observed by a variety of non-aromatic olefins (27). The mechanism of quenching is probably different for each

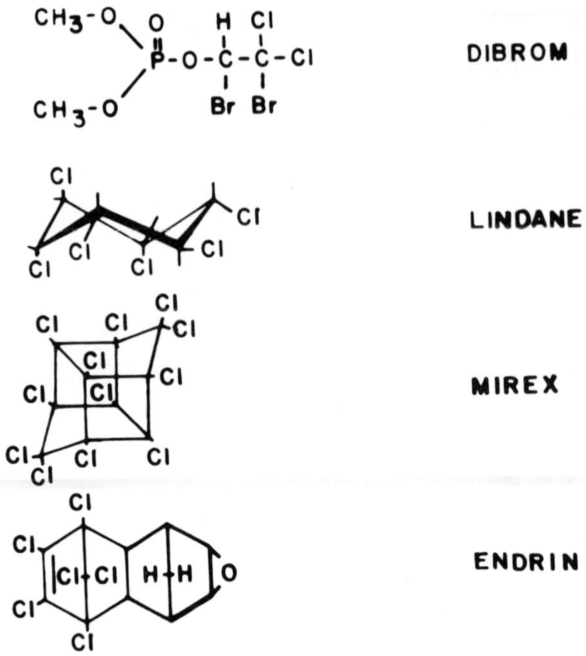

DIBROM

LINDANE

MIREX

ENDRIN

Figure 5. Structures of some commonly used pesticides which act as quenchers of fluorescence.

fluorophore-quencher pair. For example, amines and olefins probably quench by formation of excited state charge transfer complexes, and paramatic species probably catalyze formation of an excited triplet state from the singlet state. Regardless of the mechanism involved, lifetime measurements can yield the probe-quencher collisional frequency.

Quenching in Homogeneous Solutions-Efficiency and Mode of Quenching

Mirex (Figure 5) is a pesticide often used to combat fire ants and their infestations. Figure 6 shows the quenching of pyrene by mirex as observed by both fluorescence yield and lifetime measurements . The equivalence of τ_0/τ with F_0/F demonstrates the dynamic nature of the observed quenching. No alteration in the emission spectrum of

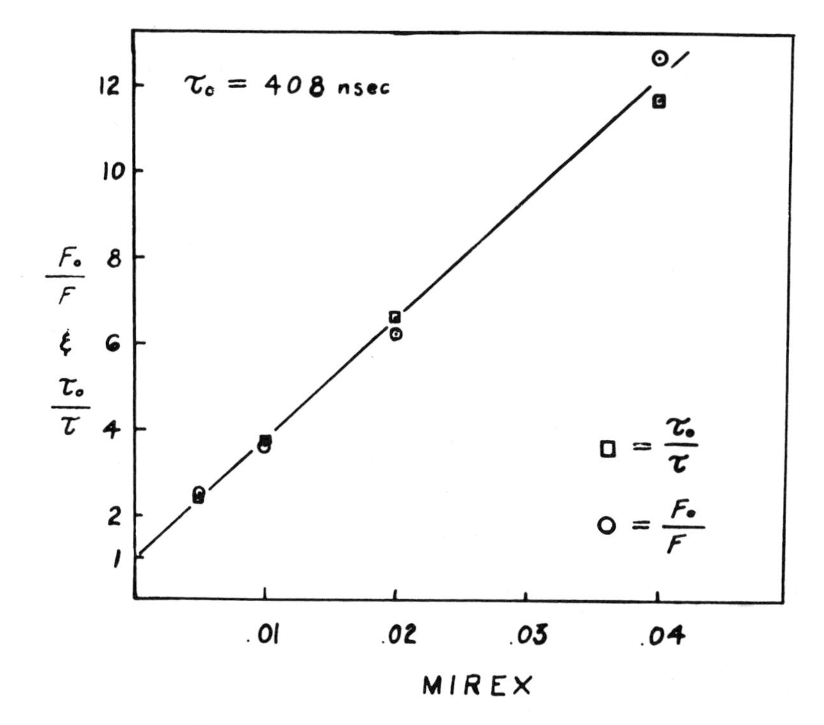

Figure 6. Quenching of Pyrene by Mirex as observed by both fluorescence lifetimes and yields: Solvent = dodecane, T = 25°C. Lifetimes by the demodulation method at 10 MHz. The fluorescence emission was observed at 396 nm.

pyrene was observed in the presence of mirex. One should check that no drastic change in the emission spectrum occurs since excited state charge-complex formation can result in the appearance of a new fluorescent species (16). While this phenomenon may also be used to investigate diffusion in membranes, an alternate mode of data acquisition and analysis would be required.

The efficiency of quenching may be estimated from a comparison of the observed k_Q with that calculated for $\gamma = 1$.

$$\gamma = \frac{k_Q}{4\pi a\ (D_Q + D_P) N'} \qquad (11)$$

Unfortunately, diffusion coefficients for the various probes and quenchers used in this study are not generally available. However, a fair estimation of D may be obtained from the Stokes-Einstein equation,

$$D = kT/6\pi\eta r \tag{12}$$

where η is the viscosity of the medium in poise and r is the molecular radius. r may be calculated from the known density and molecular weight, or from molecular models. Quenching of pyrene by mirex is found to be about 13% efficient, thus about 1 collision in 8 results in quenching. Table I lists pyrene bimolecular quenching constants for a variety of quenchers, and the estimated quenching efficiency. Pyrene is quenched with maximal efficiency by dibrom, a commonly used organophosphorus pesticide, but is not quenched efficiently by DDT, endrin or lindane. This is unfortunate because the long lifetime of pyrene could make it sensitive to diffusion over larger distances, or lower quencher concentrations. Tribromoethanol and iodobenzene are reasonably efficient. Thus one could obtain DDT analogues with high quenching efficiency by substituting the trichloromethyl group by a tribromomethyl, or by replacing the p-chlorophenyl ring by a p-iodophenyl moiety. Although this is a reasonable approach we were concerned that the use of analogues could prevent us from observing possible specific interactions between chlorinated hydrocarbons and membrane components (see Figure 1 and related discussion). Thus we felt it necessary to identify fluorescent probes which are quenched efficiently by chlorinated hydrocarbons in widespread use. After examining numerous fluorophores we found that indole, carbazole and two of its derivatives are efficiently quenched by a wide spectrum of chlorinated hydrocarbons including commonly used pesticides (Figure 5). It is apparent from the data in Table II that lipid probes based on carbazole may be used to investigate diffusional transport of a wide variety of molecules in membranes. We feel confident that even longer lived probes of equal sensitivity can be identified through further investigations.

Quenching of Probes Located in Membranes

The quenching of fluorophores localized in membranes presents a number of difficulties not encountered in homogeneous solution. We shall attempt to describe these difficulties and suggest methods to resolve them. The amount of quenching observed for a membrane bound probe depends on three factors: 1) the quenching efficiency for the probe-quencher pair *in the membrane*, 2) the local concentration of quencher in the membrane, and 3) the sum of D_Q and D_P. For simplicity we shall assume that the quenchers are small relative to the phospholipids, so that their diffusion

TABLE I

Quenching of Pyrene by Halogen Containing Hydrocarbons

Quencher	$k_Q x10^{-8} (M^{-1} sec^{-1})$	Approximate Quenching Efficiency (γ)
Dibrom	27.1	.45
Mirex	6.6	.13
DDT	<.01	<.0002
Endrin	<.03	<.0005
Lindane	<.03	<.0005
Iodobenzene	2.0	.03
Bromobenzene	0.07	<.001
p-dichlorobenzene	<.01	<.0002
CBr_3CH_2OH	16.6	.19
CCl_4	3.5	.06
$CHCl_3$	<.01	<.0002
CH_2Cl_2	<.003	<.00005

All lifetimes were measured by the demodulation method, using a modulation frequency of 10 MHz. Absolute ethanol was the solvent in all cases except mirex, where dodecane was used. T = 25°C. Excitation wavelength = 336 nm. Emission was observed through a Corning 0-52 filter, or a 378 interference filter. All solutions were purged with N_2 to remove dissolved oxygen.

τ_o (dodecane) = 408 nsec

τ_o (ethanol) = 310 nsec

TABLE II

Quenching of Carbazoles and Indole by Halogen Containing Hydrocarbons

PROBE Quencher	[Q](M)	τ(nsec)	$k_Q \times 10^{-8}(M^{-1}sec^{-1})$*
INDOLE			
None	-	4.3	-
Lindane	.097	1.7	36.6
CARBAZOLE			
None	-	12.7	-
CCl_4	.5	0.4	48.4
CBr_3CH_2OH	.5	0.6	31.8
o-dibromobenzene	.5	1.5	11.8
$CHCl_3$	.5	8.3	0.8
N-ETHYL CARBAZOLE			
None	-	13.3	-
Lindane	.086	6.4	9.4
Endrin	.052	11.3	2.6
p,p'-DDT	.012	10.6	2.2
3-AMINO-9-ETHYL CARBAZOLE			
None	-	27.4	-
CCl_4	.25	0.6	65.
$CHCl_3$	.25	3.4	10.
CH_2Cl_2	.25	24.8	.15
o-dichlorobenzene	.25	23.5	.24
p-dichlorobenzene	.25	24.7	.16
bromobenzene	.25	24.2	.19
o-dibromobenzene	.25	0.8	48.
iodobenzene	.25	0.8	48.
acrylamide	.25	0.7	56.
lindane	.086	3.7	27.
endrin	.010	13.0	40.
p,p'-DDT	.012	11.6	41.

All lifetimes were measured using the phase shift method with a 10 MHz modulation frequency, except for indole where 30 MHz was used. The solvent and experimental conditions are given in the legend to Table I and in Table IV.

*These bimolecular quenching constants should only be regarded as approximate values. They are based only on two data points; the unquenched lifetime and the lifetime in the presence of the quencher concentration indicated.

and concentration may be adequately described as if the quenchers were located in an isotropic oil-like phase. Typically the probe concentration is in the range of one probe per 100 to 1000 phospholipids, so as to minimize its perturbation on the membrane. Some typical lipid probes are shown in Figure 3.

Quenching Efficiency (γ). As we described earlier the efficiency of quenching may be estimated from a comparison of the calculated collisional rate in the solvent of interest ($\gamma = 1$) with the observed bimolecular quenching constant. A priori one cannot assume an identical efficiency in all solvents. For very efficient quenchers, such as O_2, the efficiency appears to be 100% regardless of solvent. However quenching by amines is very solvent dependent (15). This process involves the formation of charge-transfer complex, a process which is favored by polar solvents. Although quenching by bromides and iodides is probably a result of catalyzed formation of the excited triplet state (which is in turn quenched to non-observable yields), quenching by chlorine containing molecules may involve charge-transfer complex formation or photochemical reaction. Quenching of indole by $CHCl_3$ proceeds with formation of products which fluoresce the blue (12). Indole fluoresces in the ultraviolet. Photochemical degradation was observed in many of our studies, but since the products were nonfluorescent they did not interfere with the lifetime measurements (see Materials and Methods).

In Table III we list data describing the quenching of DNS-EA (Figure 3) by three brominated quenchers, and the solvent dependence of CCl_4 quenching. DNS-EA is used as a model of the phospholipid analogue DNS-PE (Figure 3) which we used as a fluorescent probe in lipid bilayers. The CCl_4 quenching efficiency appears to vary about six-fold, depending upon the solvent polarity. Only minor changes were observed in the emission spectra of DNS-EA at 0.3M CCl_4 in the three alcohols, while blue shifts were observed in benzene and cyclohexane. These changes may be related to the differences in the apparent k_Q values. Alternately, associations of solvent dipoles with the dimethylamino group of the fluorophore may decrease the collisional rate.

Since the emission spectrum of DNS-PE when localized in vesicles resembles that of DNS-EA in ethanol, and the efficiency varies only slightly between the three alcohols, it seems reasonable to use the quenching efficiency in ethanol as an estimate of the quenching efficiency in membranes.

TABLE III

Quenching of DNS-EA by Halogen Containing Hydrocarbons

Solvent = 100% Ethanol

Quenchers	τ_o(nsec)	$k_Q x10^{-8}(M^{-1}sec^{-1})$
Dibrom	19.5	16.8
CBr_3CH_2OH	19.2	16.5
$CHBr_3$	19.4	19.2
Quencher = CCl_4		
Methanol	16.9	1.7
Ethanol	19.4	2.3
Butanol	19.0	1.6
Benzene	18.4	7.1
Cyclohexane	14.5	9.6

Lifetimes were measured by the phase shift technique using a modulation frequency of 10 MHz. T = 25°C. Other instrumental conditions are listed in Table IV.

Partitioning of Quencher Between the Lipid and Water Phases

Low molecular weight quenchers, such as CCl_4, possess sufficient water solubility so that a fraction of the added quencher will be located in both the lipid and water phases. Calculation of the quencher's membrane diffusion coefficient from the observed quenching data requires knowledge of the actual quencher concentration in the lipid $[Q]_L$. We will examine the simple case where $[Q]_L$ is determined by a simple partitioning between the lipid and aqueous phases.

One may visualize the membrane sample as consisting of an oil-like and water phase with volumes V_L and V_w, respectively. The partition coefficient defines the ratio of the quencher concentration in each phase.

$$P = \frac{[Q]_L}{[Q]_w} \tag{13}$$

The total concentration of Q added, which is determined by the experimental procedure, distributes among the water and lipid phases according to

$$[Q]_T V_T = [Q]_L V_L + [Q]_W V_W \tag{14}$$

From these two equations, and by defining

$$\alpha_L = V_L/V_T \tag{15}$$

to be the volume fraction of the lipid,one may obtain

$$[Q]_W = \frac{[Q]_T}{P\alpha_L + (1 - \alpha_L)} \quad . \tag{16}$$

Phospholipid vesicles have a density near 1g/ml, so that a 1mg/ml lipid sample has $\alpha_L = 10^{-3}$. If $P = 10^3$, 50% of the quencher would be found in each phase. The observed quenching may now be described by

$$\frac{1}{\tau} = \frac{1}{\tau_o} + k_Q P[Q]_W \tag{17}$$

$$\frac{1}{\tau} = \frac{1}{\tau_o} + \frac{k_Q P\ [Q]_T}{P\alpha_L + (1 - \alpha_L)} \quad . \tag{18}$$

It is convenient to describe the observed quenching in lipid systems by an apparent quenching constant (k_{app}) where

$$k_{app} = \frac{k_Q P}{P\alpha_L + (1-\alpha_L)} \quad . \tag{19}$$

Since the volume fraction of lipid is generally small (i.e., $1-\alpha_L \simeq 1$), one may write

$$\frac{1}{k_{app}} = \frac{\alpha_L}{k_Q} + \frac{1}{k_Q \cdot P} \quad . \tag{20}$$

A plot of k_{app}^{-1} versus α_L allows the diffusion and partition dependence of the observed quenching to be separated.

Figure 7 illustrates the variability of quenching which can be observed as a result of variation in the lipid concentration. This data describes the quenching of DNS-PE (Figure 3) labeled DML (dimyristoyl-L-α-lecithin) vesicles by CCl_4. This dramatic dependence indicates the need to carefully control membrane concentrations in all quenching studies. Figure 8 demonstrates how we can turn this dependence to our advantage for the determination of the lipid-water CCl_4 partition coefficient. These data indicate that CCl_4 concentrates in DML vesicles about 500 fold (3.7 kcal), and a bimolecular quenching constant of 1.6×10^8. Note from the slope of the 1 mg/ml sample in Figure 7 that k_{app} is 5×10^{10}, which is about 10 fold greater than the diffusion controlled limit. Such data indicate that the quencher is concentrated in the membrane. We should note that the quenching becomes non-linear above 2mM CCl_4, which is when our fluorescent probe indicates alterations in the bilayer structure (see Figure 13 and related discussion).

Figure 9 shows similar data for the diffusion and partitioning of tribromoethanol in DML vesicles. In this case $P \simeq 35$ and $k_Q = 4 \times 10^8$. The dependence of k_{app} on α_L is less than was observed for CCl_4. This is a result of the smaller partition coefficient. Accurate determination of P requires that α_L be varied over the range where the fraction of the quencher in the lipid phase (f_L) varies from 0.1 to 0.9. From equations 13 to 15 one sees that

$$f_L = \frac{[Q]_L}{[Q]_T} = \frac{P\alpha_L}{P\alpha_L + (1-\alpha_L)} \tag{21}$$

For tribromoethanol the maximum fraction which was in the lipid phase was about 0.3 at $\alpha_L = 0.012$, whereas for CCl_4 this fraction was 0.84. Since phospholipid suspensions as concentrated as 100mg/ml ($\alpha_L = 0.1$) can be prepared, partition coefficients as

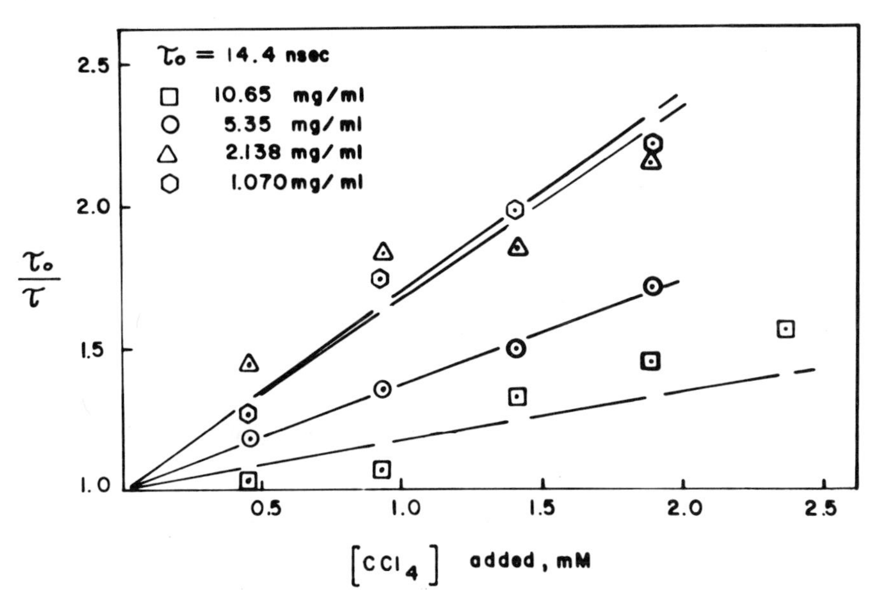

Figure 7. Quenching of DNS-PE labeled DML by CCl_4, at different lipid concentrations. T = 31°C. Lifetimes were measured by the phase shift technique at a 10 MHz modulation frequency. Concentrations of phospholipid, in mg/ml, are indicated.

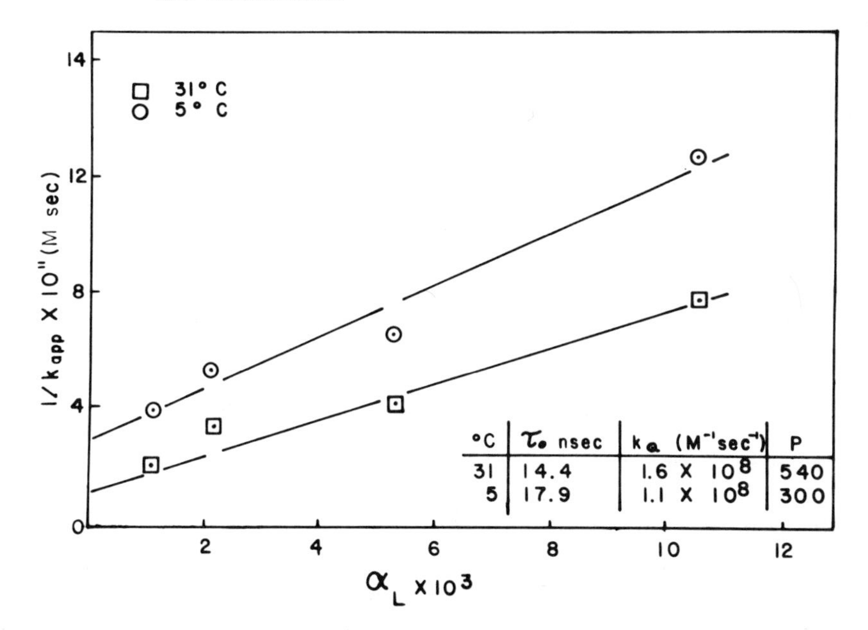

Figure 8. Separation of Diffusion and Partition of CCl_4 in DNS-PE labeled DML vesicles. See legend of Figure 7. α_L = mg lipid/ml x 10^{-3}.

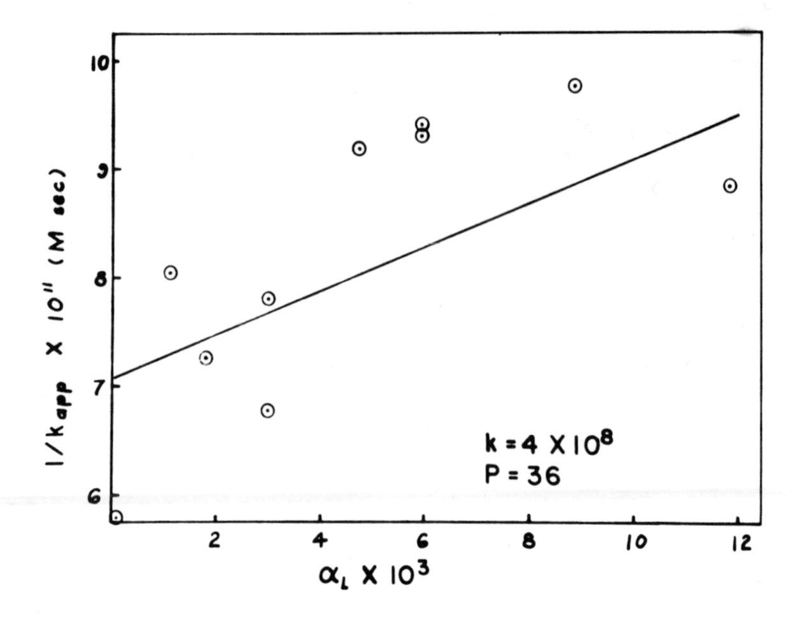

Figure 9. Separation of diffusion and partition of tribromoethanol in DNS-PE labeled DML vesicles. T = 31°C. See legend of Figure 7.

weak as 10 can be determined. Assuming a detection limit of 10^{-10}M for the probe, and thus requiring a 10^{-8}M lipid suspension, partition coefficients as strong as 10^8 could be measured. Thus this technique can be used to determine partition coefficients over a wide range of values.

In order to ascertain the possibility of the quencher causing disruption of the lipid bilayer, we can calculate the amount of quencher present in the membranes under our experimental conditions. In the case of tribromoethanol quenching of DNS-PE labeled DML (Figure 10a) $\tau_o/\tau = 2$ at $[Q]_T = 6$mM. Using equation 21 with $\alpha_L = 0.86 \times 10^{-3}$ (1.2mM Pi) we find $f_L = 0.03$, or $[Q]_L = 0.18$mM. This is equivalent to 1 tribromoethanol for each 6.8 DML molecules, or a volume fraction of tribromoethanol in the membrane of less than 3%. Such a small fraction of quencher is unlikely to perturb the membrane to the extent of making the data invalid.

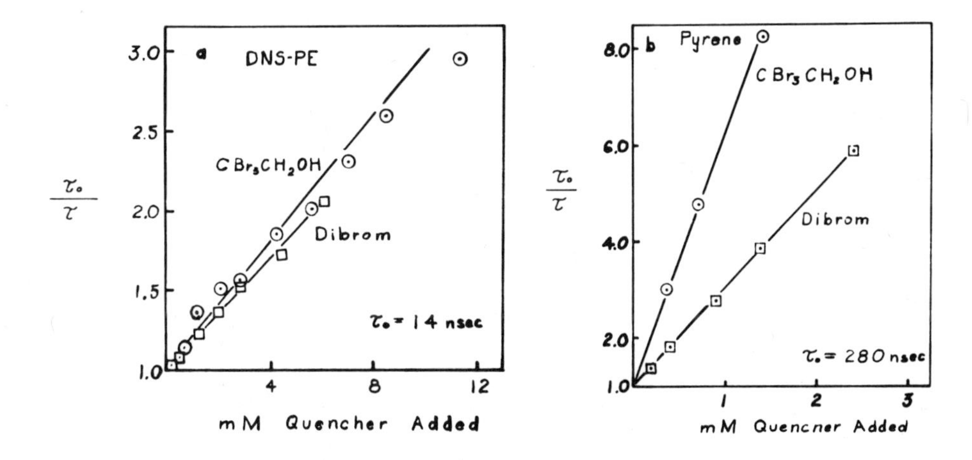

Figure 10. Quenching of DNS-PE labeled (a) and pyrene labeled (b) DML vesicles by CBr_3CH_2OH and dibrom. T = 31°C. DNS-PE lifetimes by the phase shift technique, and pyrene lifetimes by the demodulation method, both at 10 MHz. [Pi] = 1.23mM for (a) and [Pi] = 1.4 and 1.1mM for (b), CBr_3CH_2OH and dibrom, respectively.

Calculation of the Quencher's Diffusion Coefficient in a Membrane

k_Q is proportional to $D_P + D_Q$. The emission spectrum of DNS-PE in DML vesicles is similar to that of DNS-EA in ethanol. If we assume the quenching efficiency and radius for the DNS-PE CCl_4 pair in lipid to be the same as DNS-EA CCl_4 in ethanol, then

$$\frac{(k_Q)_L}{(k_Q)_{EtOH}} = \frac{(D_P + D_Q)_L}{(D_P + D_Q)_{EtOH}} \quad . \tag{22}$$

The diffusion coefficients of DNS-EA and CCl_4 in ethanol may be estimated from equation 12. Assuming DNS-EA has a density near $1.2 g/cm^3$ then these diffusion coefficients are 2.7×10^{-6} and 3.7×10^{-6} for DNS-EA and CCl_4, respectively. By using the bimolecular quenching constants found in ethanol (Table III) and in DML vesicles at 31°C (Figure 8), one finds $(D_P + D_Q)_L = 4.5 \times 10^{-6} cm^2/sec$. Since the probe is covalently attached to the phospholipid D_P is about $10^{-8} cm^2/sec$ (20,13), which is small compared to $(D_P + D_Q)_L$. Hence, the diffusion coefficient of CCl_4 in DML vesicles at 31°C is about 120% of its diffusion rate in ethanol. A similar calculation for the DML vesicles at 5°C, which is below the phase transition of the membrane, indicates a diffusion rate of about 80% of the ethanol rate. In these calculations we made no attempt to correct for temperature differences between the membrane and ethanol solutions because the temperature dependence of CCl_4 diffusion in ethanol is slight (less than a 50% change over the temperature range in these studies), and the errors resulting from the use of the Stokes-Einstein equation are probably dominant. Nonetheless this data indicates a very rapid diffusion of CCl_4 in the membrane where the fluorescent probe is located.

Intuitively, these numbers appear to be too large. As we saw in Table III, the CCl_4 quenching efficiency of DNS-EA can vary by a factor of 6 depending upon the solvent. Additionally, CCl_4 may diffuse rapidly across the surface of the bilayer, but not as rapidly in the bilayer. Such two-dimensional diffusion can increase the distance covered by a diffusing molecule by a factor of $\sqrt{2}$. At this time we cannot rule out the possibility that CCl_4 concentrates preferentially in the glycerol backbone region of the lipid, which is where the probe is located. Assuming this concentration effect is not important, CCl_4 diffuses in DML vesicles at least 1/10 of its diffusion rate in ethanol.

A similar calculation for CBr_3CH_2OH indicates a diffusion coefficient in DML vesicles of 1.5×10^{-6} cm^2/sec, or about 40% fo the ethanol diffusion rate. Since the quenching efficiency for CBr_3CH_2OH in ethanol is close to unity, it is unlikely to increase in the membrane. Thus we conclude that the diffusion rates of CBr_3CH_2OH are very rapid in the glycerol region of the membrane. By localization of our probes at various depths in the bilayer, we hope to determine the permeability profile of these membranes.

Localization of Foreign Molecules in Membranes

Lipid-like probes may be designed to probe any desired region of the phospholipid bilayer. The dansyl moiety of DNS-PE is probably found in the glycerol region (25), whereas pyrene most likely localizes in the aliphatic side chain region of the bilayer. Thus one may determine the accessibility of each of these regions to foreign molecules. Figure 10 compares the quenching of the above mentioned probes in DML vesicles by tribromoethanol and dibrom. At present we do not have adequate partition information on dibrom to separate diffusion and partition contributions to quenching. Nonetheless, these data indicate that pyrene is less accessible to dibrom than to tribromoethanol, possibly indicating a localization of the dibrom on the membrane surface, away from the pyrene. Pyrene diffusion to the surface could account for the observed quenching. We need to localize the probe at defined depths in the bilayer to arrive at any firm conclusions regarding the surface localization of dibrom.

Effect of Membrane Composition on Quenching

Cholesterol is a major component of biological membranes. NMR studies have shown that cholesterol decreases side chain mobility. Fluorescence polarization has indicated an increased microviscosity in membranes containing cholesterol (5). We examined quenching in vesicles of DPL containing cholesterol. Tribromoethanol quenching of DNS-PE and pyrene labeled vesicles is shown in Figure 11. As one would expect in a more viscous membrane, quencher diffusion is slowed by the presence of cholesterol. Note that the initial slopes, which are proportional to k_Q assuming equal partitioning for vesicles, differ by a factor of 2 for DNS-PE, but by a factor of 10 for pyrene. These data are consistent with the notion that cholesterol decreases membrane fluidity primarily in the side chain region of the bilayer. Note from Figure 8 that cooling DML below its phase transition did not greatly affect CCl_4 diffusion in the glycerol backbone region of the membrane.

Figure 12 compares the dibrom quenching of the same vesicles. Again, the cholesterol effect is much more pronounced on the pyrene labeled vesicles. The distinct biphasic nature of these Stern-Volmer plots may indicate that the membrane becomes saturated with dibrom. Thus, simple partitioning will probably not be adequate to describe the concentration of all quenchers in membranes, but then quenching could be used to reveal the binding isotherms.

Effects of Foreign Materials on Membranes

Over the past 10 years a great deal of effort has gone into the development and understanding of the phospholipid vesicles as models for biological membranes. It now seems appropriate to use this system for experiments aimed towards understanding the molecular aspects of toxicity.

The emission spectra of many fluorescent molecules are sensitive to the environment in which they are located. This sensitivity is a result of the relaxation of the adjacent environment around the excited state dipole (Figure 2). Naphthylamine sulfonamides such as DNS-EA and DNS-PE are especially sensitive. Figure 13 shows the emission spectrum of DNS-EA in several solvents of varying polarity. This figure also shows the emission spectra of DNS-PE labeled DML as CCl_4 is added. Concentrations of CCl_4 up to 2mM cause very little shift in the emission spectra. Thus our diffusion and partition information for CCl_4 was obtained with minimum perturbation of the vesicles. Continued addition of CCl_4 results in a progressive blue shift in the probe's emission, indicating a change in the average environment around the probe. Such spectral changes can be useful indicators of the effects of toxic materials on membrane structure.

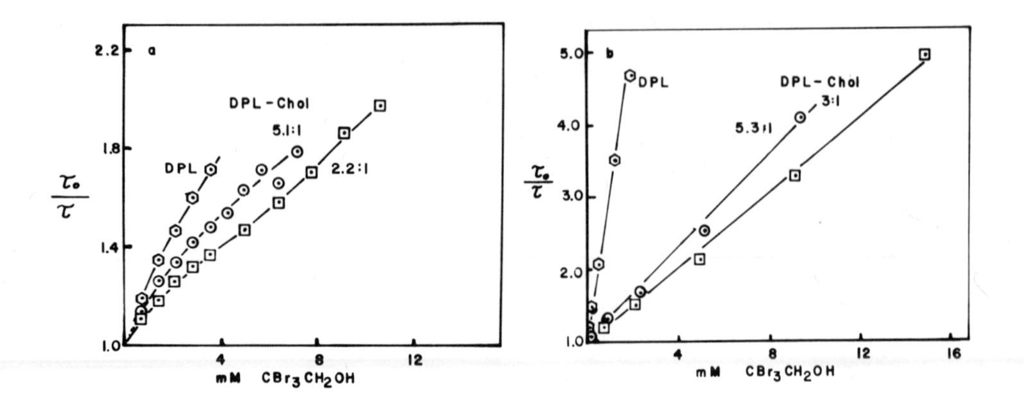

Figure 11. Quenching of DNS-PE (a) and pyrene (b) labeled DPL vesicles containing cholesterol by CBr_3CH_2OH. T = 31°C. Lifetimes by the demodulation method at 10 MHz.

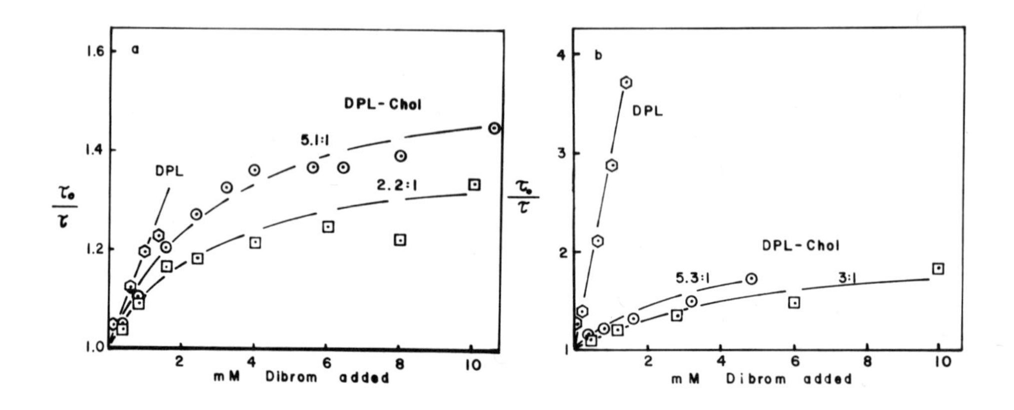

Figure 12. Quenching of DNS-PE (a) and pyrene (b) labeled DPL vesicles containing cholesterol by dibrom. T = 31°C. Lifetimes were measured by the demodulation method at 10 MHz.

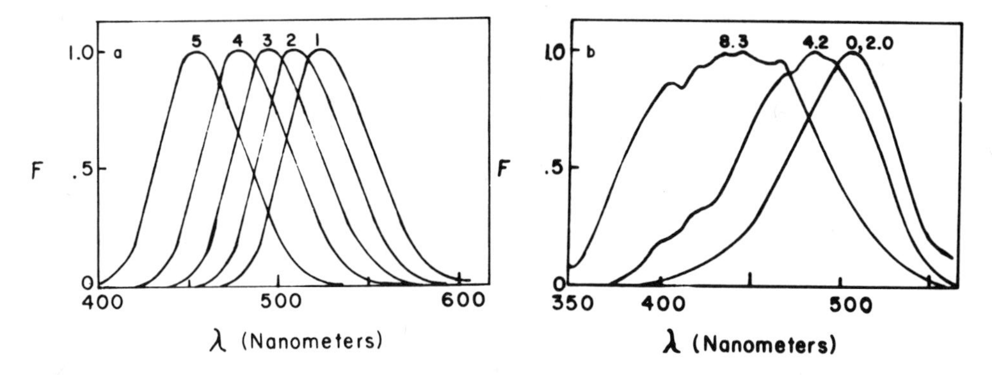

Figure 13. Normalized fluorescence emission spectra (a) DNS-EA in solvents of varying polarity: and b) DNS-PE labeled DML vesicles containing the indicated concentration of CCl_4 (mM), [Pi] = 1.6mM. The solvents were, 1 through 5 respectively, methanol-H_2O 1:1; methanol, n-butanol, benzene and cyclohexane.

Recently Tsong (23) described a probe-lipid system which could be very useful in determining the effects of foreign materials on membrane permeability and integrity. The fluorescent probe ANS (Figure 3) is practically non-fluorescent in water, but highly fluorescent in organic solvents or when bound to lipids. If an ANS solution is added to a DML vesicle suspension at 5°C (below the phase transition) the probe only binds to the outer layer of the vesicle. Upon heating the vesicles above the transition temperature (~40°C) the ANS permeates the vesicle and binds to both surfaces, yielding a 60% increase in fluorescence upon recooling to 5°C. Thus, disruption of vesicles by foreign molecules or an increase in anion permeability, can be detected by an increased rate of ANS binding to the inner bilayer detected via the increased fluorescence. Figure 14 shows the time dependence of the fluorescence of this ANS-DML system in the presence and absence of dicryl, a commonly used herbicide. Whether the increase in intensity is caused by a change in anion permeability or is due to disruption of the vesicles is not known at this time. Nonetheless, this data demonstrates the possibility of using this system to study perturbations caused by foreign molecules in membranes.

As we mentioned earlier, many fluorophores are insensitive to quenching by chlorinated hydrocarbons. Additionally, many compounds which exert toxic effects on membrane will not act as quenchers. Non-quenchable probes can be used to report on the response of other membrane parameters, such as microviscosity, without complications due to quenching.

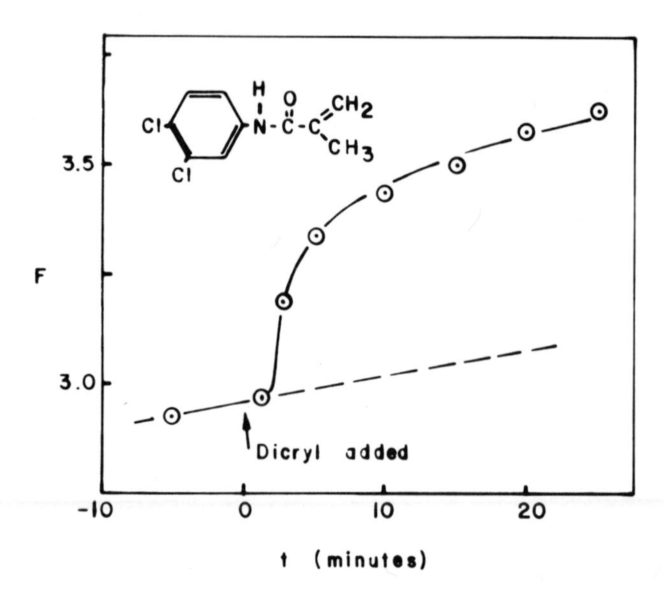

Figure 14. Fluorescence intensity of ANS labeled DML upon addition of the herbicide dicryl. [DML] = 0.3mM, $[DICRYL]_{added}$ = 0.1mM, [ANS] = 2.7 x 10^{-5}M.

Separation of Lateral and Transverse Diffusion in Membranes

The dynamics of phospholipid mobility in membranes are not adequately described by a single isotropic diffusion coefficient. The lateral diffusion coefficients of lipids across the membrane surface are about $10^{-8}cm^2/sec$ (20,13). Using the Einstein equation for two dimensions

$$\Delta x^2 = 4Dt \tag{23}$$

and a distance of 2 microns to represent the size of a bacterium, one finds that a phospholipid may passively diffuse the length of a bacterium in about one second. The rate at which phospholipids or cholesterol exchange between the inner and outer bilayers of a vesicle is much slower. The half-life for this exchange is at least as long as hours (11) and may be as long as weeks (14). We thus have no reason to expect that the diffusional transport of polar quenchers such as CBr_3CH_2OH or dibrom, or that of large molecules such as DDT, can be described in simple isotropic terms. Fluorescence quenching techniques provide a means to determine these separate diffusion rates.

Probes such as DNS-PE or ANS, if incorporated into vesicles at the time of sonication, are distributed between the inner and outer layers. These probes probably localize in the glycerol backbone region of the bilayer. Due to the slow rate of probe transport between inner and outer surfaces (23), this rate cannot provide a significant contribution to quenching which requires collisions on the nanosecond timescale. Thus, the inner and outer probe populations report independently on the concentration and diffusion of quenchers on its surface.

We shall describe one simple case to illustrate this potential measurement. With reference to Figure 15, assume that k_P, the rate of partitioning into the outer bilayer, is rapid compared to k_T, the transmembrane transport rate. Additionally, assume sufficient $[Q]_W$ so that $[Q]_{out}$, the quencher concentration in the outer layer is not depleted by k_T. The amount of quenching on each surface is assumed to be described by

$$F_{in} = \frac{F_o}{1 + K[Q]_{in}} \qquad (24)$$

$$F_{out} = \frac{F_o}{1 + K[Q]_{out}} \qquad (25)$$

where K describes the quenching due to lateral diffusion and thus contains the lateral diffusion coefficients of quencher and probe.

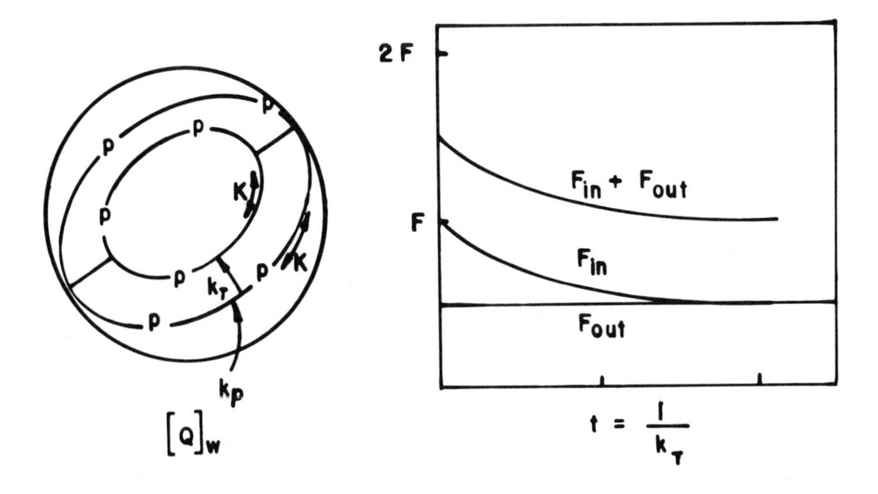

Figure 15. Measurement of lateral and transverse diffusion rates of foreign molecules in membranes by fluorescence quenching.

For simplicity we have assumed an equivalent amount of probe on both sides of the bilayer. If labeled vesicles are rapidly mixed with quencher the concentration of Q on the inner surface will increase according to

$$[Q]_{in} = [Q]_{out} \, (1 - e^{-k_T t}). \tag{26}$$

Under conditions where sufficient quencher is present to give $F_o/F = 2$ we find

$$F_{in}(t) = \frac{F_o}{2 - e^{-k_T t}} . \tag{27}$$

The fluorescence yield from the inner surface decays exponentially to one-half of its original value.

For CCl_4, CBr_3CH_2OH and dibrom quenching of DNS-PE or pyrene labeled vesicles no time dependent decrease in lifetime could be observed after quencher addition. Thus transport of these species is complete within a few minutes. Rapid mixing techniques will be needed to determine these rates. It should be mentioned that the photochemical degradation caused by these quenchers makes the interpretation of fluorescent yields very difficult, but does not affect the lifetime measurements. The phase-modulation lifetime instrument in our laboratory (Materials and Methods) allows determination of a lifetime by both the phase shift and the demodulation method, precise to ±0.5 nanoseconds, in one second. This will allow transport rates on the minute timescale to be easily measured.

DISCUSSION

In this paper we have described the use of collisional quenching of fluorescence to investigate the diffusional rates of quenchers in membranes and the partitioning of these quenchers between water and lipid phases. Since a large variety of molecules act as quenchers, and because fluorescent probes may be located in any desired region of the lipid bilayer, we shall be able to investigate the permeability barriers which cell membranes present to molecules for which specific transport systems do not exist. The techniques which we describe are not limited to phospholipid vesicles, but should prove equally applicable to whole cells, or sections of blood vessels or fish gills. Thus, these techniques should allow one to determine the rates at which blood proteins may exchange foreign molecules into cell membranes, leading to a better understanding of the transport of xenobiotics throughout an organism.

MATERIALS AND METHODS

Fluorescence Instrumentation

All fluorescence lifetimes were determined by an SLM Instruments, Inc., Phase-Modulation Spectrofluorometer (Champaign, Illinois) using a modulation frequency of 10 or 30 MHz. The exciting source consisted of a 450 watt xenon lamp and a double concave holographic grating monochrometer. Phase shift and demodulation information was collected simultaneously and analyzed by an interfaced Hewlett-Packard 9810 calculator.

Emission spectra were determined on an SLM Instruments, Inc., Spectrofluorometer. Where necessary, oxygen was removed from all solutions by nitrogen purging.

In the presence of quencher the emission intensity sometimes decayed upon illumination, presumably due to photochemical degradation. This process did not affect the lifetime measurements because the products did not have significant fluorescence, and the illumination intensity in our lifetime instrument is five to ten fold less than in our spectrofluorometer. We are able to determine the emission spectrum without complications resulting from degradation by using a second channel of the spectrofluorometer to observe the total emission, and normalizing each wavelength intensity reading by this value.

Phospholipid Vesicles

Synthetic β-γ-dimyristoyl-L-α-lecithin was purchased from Sigma, β-γ-dipalmitoyl-L-α-lecithin from P. L. Biochemicals, Inc., and Cholesterol from J. T. Baker, Ultrex grade, were used without further purification. All vesicles were prepared by sonication of lipid suspensions in 0.01M tris·HCl + 0.05M KCl, pH = 7.5 for about 10 minutes at about 45°C. Suspensions were then centrifuged at 48,000 G for 1.5 hours. Phospholipid concentrations were determined after centrifugation by the Bartlett assay for inorganic phosphate, as described by Kates (8), cholesterol was also determined as described by Kates.

Reproducible quenching data from volatile compounds like CCl_4 is difficult to obtain as a result of material being lost to the gas phase, especially if nitrogen purging is used to remove oxygen. For the CCl_4 quenching studies described here, vesicle samples were loaded anaerobically into small screw cap vials so that no air space remained. Microliter quantities of CCl_4 in ethanol were added with Hamilton syringes through a polyethylene lined cap. Mixing was effected by sonication of the vial in a bath type sonicator.

Non-volatile materials such as tribromoethanol and dibrom were added in μl quantities as concentrated solutions in ethanol. In no instance did the volume added exceed 1% of the sample volume.

Fluorescence Probes

N-dansyl phosphatidylethanolamine (DNS-PE) was synthesized according to Waggoner and Stryer (25), except as noted. Egg phosphotidylethanolamine was obtained from P.L. Biochemicals, Inc. The product was isolated on a silica preparative thin layer plate, developed with $CHCl_3/MeOH/H_2O$,V/V, 65/35/5, at $R_f = 0.8$. The elemental analysis agreed with that expected for saturated C_{18} side chains.

N-dansylethanolamine (DNS-EA) was prepared by the reaction of equimolar quantities dansyl chloride with ethanolamine in $CHCl_3$. The mixture was washed with 0.02M HCl and 0.01M NaOH, then dried over $MgSO_4$. Purified product was isolated on preparative silica thin layer plates using ethyl ether/benzene/ethanol/acetic acid, 40/50/2/0.2,V/V, at $R_f \simeq 0.25$.

The remaining fluorophores were obtained from commercial sources, and purified by crystallization.

TABLE IV

Excitation Wavelengths and Filters for each Probe

Probe	λ_{EX}	F_{EX}	F_{EM}
DNS-EA	340[a]	7-54	3-70+$NaNO_2$
DNS-PE	340[a]	7-54	3-70+$NaNO_2$
Indole	280	7-54	0-54
Carbazole	324	7-54	0-52
N-Ethyl Carbazole	324	7-54	0-52
3-Amino-N-Ethyl Carbazole	370	7-60	3-73
Pyrene	336	7-54	378 interference or 0-52
ANS	355	7-60	3-73

a. 380nm used for dibrom and bromoform quenching studies.

ACKNOWLEDGEMENTS

We wish to express our appreciation to the Freshwater Biological Research Foundation for providing the facilities and instrumentation needed for this research. We are also deeply grateful to Mr. Richard Gray, whose dedication to and concern for the environment made this facility possible.

We also thank Professor John M. Wood for allowing us to use many of the facilities and supplies in his laboratory, Professor Samuel Kirkwood, University of Minnesota, for his helpful discussions, the University of Minnesota Graduate School for financial assistance, and to Mrs. Carrie Tietz for her assistance in proofreading and typing this manuscript.

We also thank the American Heart Association for their financial support.

LITERATURE

1 BARENBOIM, G.M. 1963. Interaction of Excited Biomolecules with Oxygen-1. Quenching of the Photoluminescence of Biomolecules by Oxygen and Nitric Oxide. Biofizika 8: 154-164.

2 BIERI, V.G., and HOELZYL WALLACH, D.F. 1975. A Study Using Paramagnetic Quenching of Fluorescence by Nitroxide Lipid Analogues. Biochim. Biophys. Acta 406: 415-423.

3 BOWEN, E.J., and METCALF, W.S. 1951. The Quenching of Anthracene Fluorescence. Proc. Roy. Soc.(London) A 206: 437-447.

4 CHEN, R.F. 1971. Fluorescence Quenching Due to Mercuric Ion. Interaction with Aromatic Amino Acids and Proteins. Arch. Biochem. Biophys. 142: 552-564.

5 COGAN, V., SHINITZKY, M., WEBER, G., and NISHIDA, T. 1973. Microviscosity and Order in the Hydrocarbon Region of Phospholipid and Phospholipid-cholesterol Dispersions Determined with Fluorescent Probes. Biochemistry 12: 521-529.

6 HORROCKS, A.R., KEARVELL, A., TICKLE, K., and WILKINSON, F. 1966. Mechanism of Fluorescence Quenching in Solution. Part 2-Quenching by Xenon and Intersystem Crossing Efficiencies. Trans. Faraday Soc. 62: 3393-3399.

7 KARIM AHMED, A. 1976. Environment 18: 6.

8 KATES, M. 1972. Techniques in Lipidology. Isolation, Analysis and Identification of Lipids. North-Holland Publishing Co., Amsterdam.

9 KAUTSKY, H. 1939. Quenching of Luminescence by Oxygen. Trans. Faraday Soc. 35: 216-226.

10 KIRKWOOD, S., and PHILLIPS, P.H. 1946. The Relationship Between the Lipoid Affinity and the Insecticidal Action of 1,1-bis (p-fluorophenyl) 2,2.2-trichloroethane and Related Substances. J. Pharmacology 87: 375-381.

11 KORNBERG, R.D., and McCONNELL, H.M. 1971. Inside-Outside Transitions of Phospholipids in Vesicle Membranes. Biochemistry 10: 1111-1120.

12 LAKOWICZ, J.R. Unpublished Observations.

13 LEE, A.G., BIRDSALL, J.M., METCALFE, J.C. 1973. Measurement of Fast Lateral Diffusion of Lipids in Vesicles and Biological Membranes by ^{1}H Nuclear Magnetic Resonance. Biochemistry 12: 1650-1658.

14 LENARD, J and ROTHMAN, J.L. 1976. Transbilayer Distribution and Movement of Cholesterol and Phospholipid in the Membrane of Influenze Virus. Proc. Natl. Acad. Sci. 73: 391-395.

15 LEONHARDT, H., and WELLER, A. 1961. Fluorescence Quenching Studied by Flash Spectroscopy. In: Luminescence of Organic and Inorganic Materials. Hartmeet, Kallman and Grace (Eds.) Wiley, N.Y.

16 MATAGA, N., OKADA, T., and EZUMI, K. 1966. Fluroescence of Pyrene-N,N-dimethylaniline Complex in Non-polar Solvent. Mol. Phys. 10: 203-204.

17 MEDINGER, T., and WILKINSON, F. 1965. Mechanism of Fluorescence Quenching in Solution Part I. Quenching by Bromobenzene. Trans. Faraday. Soc. 61: 620-630.

18 MELNIKOV, N.N. 1971 Chemistry of Pesticides. Springer Verlag, N.Y. Chapter 5: 75.

19 ROLLEFSON, G.K., and BOAZ, H. 1948. Quenching of Fluorescence in Solution. J. Phys. Colloid Chem. 52: 518-527.

20 SCANDELLA, C.J., DEVAUX, P. and McCONNELL, H.M. 1972. Rapid Lateral Diffusion of Phospholipids in Rabbit Sarcoplasmic Reticulum. Proc. Nat. Acad. Sci. U.S.A. 69: 2056-2060.

21 STICKEL, L.F. 1973. Pesticide Residues in Birds and Mammals, Chapter 7. Environmental Pollution by Pesticides (Edwards, C.A., Ed.). Plenum Press, NY.

22 SVESHNIKOFF, B. 1936. The Quenching of Fluorescence of Dye Solutions by Foreign Compounds. Acta Physiochimica U.R.S.S. IV: 453-470.

23 TSONG, T.Y. 1975. Effect of Phase Transition on the Kinetics of Dye Transport in Phospholipid Bilayer Structures. Biochemistry 14: 5409-5414.

24 TURI, J.S., HO, N.F.H., HIGUCHI, W.I., and SHIPMAN, C. Jr. 1975. Systems Approach to Study of Solute Transport Across Membranes Using Suspension Cultures of Mammalian Cells. III: Steady-state Diffusion Models. J. Pharm. Sci. 64: 622-626.

25 WAGGONER, A.S. and STRYER, L. 1970. Fluorescent Probes of Biological Membranes. Proc. Natl. Acad. Sci. 67: 579-589.

26 WARE, W.R. and NOVROS, J.S. 1966. Kinetics of Diffusion-Controlled Reactions. An Experimental Test of the Theory as Applied to Fluorescence Quenching. J. Phys. Chem. 70: 3246-3253.

27 WARE, W.R., WATT, D., and HOLMES, J.D. 1974. Exciplex Photophysics. I. The α-Cyano-Naphthalene-Olefin System. J. Amer. Chem. Soc. 96: 7853-7860.

28 WEISS, J. 1939. Photosensitized Reactions and the Quenching of Fluorescence in Solutions. Trans. Faraday Soc. 35: 48-64.

29 ZIMMERMAN, O.T. 1946. DDT, Killer of Killers. Industrial Research Service, Dover, NH.

30 ZINSSER, H. 1935. Rats, Lice, and History. Boston Little Brown, 88.

DISCUSSION

ALPER: In the experiments you described on the diffusion of oxygen in DNA, in what form was the DNA?

LAKOWICZ: The DNA was a completely free form, that is, not associated with any proteins.

ALPER: Yes, but was it dry?

LAKOWICZ: No, in solution as a double helical DNA. The binding mode was intercalation into the center of the helix.

ALPER: It was double-stranded DNA? In high or low concentration? Was the solution very viscous or not very viscous?

LAKOWICZ: Very low concentrations.

ALPER: There is a lot of water tightly bound to the double helix (Luzzati, V., Nicolaieff, A., and Masson, F., J. Mol. Biol. 3: 185, 1961; Hearst, J.E. and Vinograd, J., J. P.N.A.S. 47: 825 & 1005,1961), and I just don't understand where your result came from.

LAKOWICZ: I am not sure where the water is in the double helix.

ALPER: Presumably the ethidium bromide presumably intercalated?

LAKOWICZ: Yes. There are various modes of bindings and we were careful to choose conditions which insured intercalation.

ALPER: Well, one would have thought that the oxygen could have arrived at the ethidium bromide molecule straight through the bound water.

LAKOWICZ: We thought the same thing when we did the experiment. We thought this would be a case where there would be no inhibition, and that's why the experiment was done.

SZABO: I think there are some extensive measurements in the partition quotient, and the paper that I know in particular is the one by Katz and Diamond, where they measured partition quite well with isotopes (Katz, Y. and Diamond, J. M., Thermodynamic constants for nonelectrolyte partition between dimirystoyl lecithin and water. J. Membrane Biol. 17: 101-120, 1974). What does the partition quotient mean? Does it mean you are absorbing the solute at the membrane surface or in the membrane interior? And, this I think you can eventually figure out. But what is very important to distinguish is the meaning of the partition quotient.

What is the meaning of the diffusion quotient that you measure in the membrane? Why doesn't it depend on the partition quotient, for example?

LAKOWICZ: Permeability measurements never separate diffusion from partition, because the flux of a molecule across the membrane is proportional to both its membrane concentration and its diffusion coefficient within the membrane.

SZABO: How are you able to separate them; that's the question.

LAKOWICZ: We were able to separate them by knowing the amount of material in the membrane.

WEINER: When you are talking about diffusion of a molecule through an aqueous medium into a membrane, and we are dealing with very large partition coefficients, does the unstirred layer play an effect? Does that have to be taken into account in making the calculation for your diffusion coefficient?

LAKOWICZ: With any technique there are various assumptions involved. At this point in time we don't know what the diffusion barriers at each region of the bilayer are. So, in order to answer the question, whether the majority of the quenching is coming from diffusion of pre-existing molecules in the bilayer or from outside, we will have to wait until we do those experiments in more detail.

What is nice about the fluorescence technique is that we eliminate the problems that exist with unstirred layers because we are always creating upon excitation a new population of clocks that are waiting to collide. So, it is an equilibrium situation.

SZABO: The diffusion coefficient that you measure is not the diffusion coefficient within the membrane. It is a diffusion coefficient or some transfer rate of an overall process. It could be from the access space to the membrane interior or within the membrane interior. And it is very difficult to know where it is exactly.

LAKOWICZ: We are not sure at this point. We separate the lateral diffussion across the face of the membrane from the trans-membrane diffusion. This is a very important problem, and we feel that we are getting to that point simply because the transport rate of probes from one side to the other can be made much slower than the transport rate of the quencher, so that we can use them as two independent populations.

HOFFMAN: Is that 270 nanosecond lifetime a single exponential?

LAKOWICZ: In order to do these experiments, there are two philosophies on measuring lifetimes. There is a pulse technique where one requires large amounts of time to gather data for a single lifetime. With biological experiments there is an inherent advantage in being able to measure lifetime in one second rather than one hour. So, we have a phase fluorometer which utilizes a phase shift technique, and we can measure lifetimes in a second. So that eliminates some of the single exponential information. But, since we operate on multiple frequencies, we can sort that out. These are close to single exponentials but not absolutely.

AUTHOR INDEX

SUBJECT INDEX